# Below-ground Interactions in Tropical Agroecosystems
## Concepts and Models with Multiple Plant Components

# Below-ground Interactions in Tropical Agroecosystems
## Concepts and Models with Multiple Plant Components

*Edited by*

**M. van Noordwijk**

*World Agroforestry Centre (ICRAF) SE Asia,
Bogor, Indonesia*

**G. Cadisch**

*Department of Agricultural Sciences,
Imperial College London, Wye, UK*

and

**C.K. Ong**

*World Agroforestry Centre (ICRAF) East and Central Africa,
Nairobi, Kenya*

CABI Publishing

CABI Publishing is a division of CAB International

CABI Publishing
CAB International
Wallingford
Oxfordshire OX10 8DE
UK
Tel: +44 (0)1491 832111
Fax: +44 (0)1491 833508
E-mail: cabi@cabi.org
Website: www.cabi-publishing.org

CABI Publishing
875 Massachusetts Avenue
7th Floor
Cambridge, MA 02139
USA
Tel: +1 617 395 4056
Fax: +1 617 354 6875
E-mail: cabi-nao@cabi.org

© CAB International 2004. All rights reserved. No part of this publication may be reprodced in any form or by any means, electronically, mechanically, by photocopying, recording or otherwise, without the prior permission of the copyright owners.

A catalogue record for this book is available from the British Library, London, UK.

**Library of Congress Cataloging-in-Publication Data**
Below-ground interactions in tropical agroecosystems : concepts and models with multiple plant components / edited by M. van Noordwijk, G. Cadisch, and C. Ong.
     p. cm.
Includes bibliographical references (p.    ).
  ISBN 0-85199-673-6 (alk. paper)
  1. Plant-soil relationships--Tropics. 2. Agricultural ecology--Tropics. I. Noordwijk, Meine van. II. Cadisch, G. (Georg) III. Ong, C. K. IV. Title.
  S596.7.B46 2004
  631.4--dc22
                       2003017139

ISBN 0 85199 673 6

Typeset in 9pt Meridien by Columns Design Ltd, Reading
Printed and bound in the UK by Biddles Ltd, King's Lynn

# Contents

| | |
|---|---|
| **Contributors** | ix |
| **Foreword** D.P. Garrity | xiii |
| **Editors' Introduction** | xv |
| **Acknowledgements** | xxi |

**1 Ecological Interactions in Multispecies Agroecosystems: Concepts and Rules** — 1
*C.K. Ong, R.M. Kho and S. Radersma*
- 1.1 Introduction — 1
- 1.2 Separating positive and negative interactions — 4
- 1.3 Assessing plant–environment–plant interactions — 4
- 1.4 A framework for a predictive understanding of multispecies systems — 10
- 1.5 Conclusions — 14

**2 Locally Derived Knowledge of Soil Fertility and Its Emerging Role in Integrated Natural Resource Management** — 17
*L. Joshi, P.K. Shrestha, C. Moss and F.L. Sinclair*
- 2.1 Introduction — 17
- 2.2 Local knowledge — 18
- 2.3 Soil classification — 21
- 2.4 Soil fertility — 24
- 2.5 Below-ground interactions — 31
- 2.6 Implications — 34
- 2.7 Building on local practice — 35
- 2.8 Recognizing the sophistication of local knowledge — 36
- 2.9 Realizing the limits of local knowledge — 36
- 2.10 Communication and empowerment — 37

**3 Models of Below-ground Interactions: Their Validity, Applicability and Beneficiaries** — 41
*R. Matthews, M. van Noordwijk, A.J. Gijsman and G. Cadisch*
- 3.1 Introduction — 41
- 3.2 Models incorporating below-ground interactions — 42
- 3.3 Validity, reliability and applicability — 47

|  |  |  |
|---|---|---|
| 3.4 | Beneficiaries and target groups | 51 |
| 3.5 | Relevance to larger systems | 57 |

**4 Tree Root Architecture** — 61
*F.K. Akinnifesi, E.C. Rowe, S.J. Livesley, F.R. Kwesiga, B. Vanlauwe and J.C. Alegre*

|  |  |  |
|---|---|---|
| 4.1 | Introduction | 61 |
| 4.2 | Root distributions | 62 |
| 4.3 | Simple indicators of root distributions | 69 |
| 4.4 | Tree root ideotype and response to management practices | 73 |
| 4.5 | Conclusions | 80 |

**5 Crop and Tree Root-system Dynamics** — 83
*M. van Noordwijk, S. Rahayu, S.E. Williams, K. Hairiah, N. Khasanah and G. Schroth*

|  |  |  |
|---|---|---|
| 5.1 | Introduction | 83 |
| 5.2 | Root growth, functional shoot–root equilibrium and local response | 86 |
| 5.3 | Problems and opportunities for measuring root dynamics | 91 |
| 5.4 | Empirical data on root growth and decay | 97 |
| 5.5 | Model representations of root dynamics | 101 |
| 5.6 | Management implications | 105 |
| 5.7 | Research issues and priorities | 106 |

**6 Opportunities for Capture of Deep Soil Nutrients** — 109
*R.J. Buresh, E.C. Rowe, S.J. Livesley, G. Cadisch and P. Mafongoya*

|  |  |  |
|---|---|---|
| 6.1 | Introduction | 109 |
| 6.2 | Mechanisms for nutrient occurrence in deep soil | 110 |
| 6.3 | Utilization of deep soil nutrients by plants | 112 |
| 6.4 | Quantification of deep nutrient uptake | 115 |
| 6.5 | Achieving more efficient use of deep soil nutrients | 119 |
| 6.6 | Preventing the accumulation of mobile nutrients in deep soil | 123 |
| 6.7 | Conclusions | 123 |

**7 Phosphorus Dynamics and Mobilization by Plants** — 127
*P.F. Grierson, P. Smithson, G. Nziguheba, S. Radersma and N.B. Comerford*

|  |  |  |
|---|---|---|
| 7.1 | Introduction | 127 |
| 7.2 | Phosphorus forms and fluxes: understanding what we measure | 128 |
| 7.3 | P mobilization and acquisition by plants | 135 |
| 7.4 | Modelling P bioavailability and uptake in agroforestry systems | 140 |

**8 Managing Soil Acidity and Aluminium Toxicity in Tree-based Agroecosystems** — 143
*M.T.F. Wong, K. Hairiah and J. Alegre*

|  |  |  |
|---|---|---|
| 8.1 | Introduction | 143 |
| 8.2 | Identifying the causes of acidity in low-input tropical agroecosystems | 144 |
| 8.3 | Transfer of organic alkalinity in agroecosystems | 149 |
| 8.4 | How do we implement this knowledge to manage acidity? | 155 |

**9 Uptake, Partitioning and Redistribution of Water by Roots in Mixed-species Agroecosystems** — 157
*M. Smith, S.S.O. Burgess, D. Suprayogo, B. Lusiana and Widianto*

|  |  |  |
|---|---|---|
| 9.1 | Introduction | 157 |
| 9.2 | Competition and complementarity for water use in mixed-species systems | 158 |
| 9.3 | Partitioning of water by plant root systems: theory | 160 |
| 9.4 | Measurement of water uptake and bidirectional flow | 162 |
| 9.5 | Control of water partitioning | 164 |
| 9.6 | Modelling of water uptake in mixed agroecosystems | 167 |
| 9.7 | Summary and conclusions | 169 |

| | | |
|---|---|---|
| **10** | **Catching and Competing for Mobile Nutrients in Soils** | **171** |
| | *G. Cadisch, P. de Willigen, D. Suprayogo, D.C. Mobbs, M. van Noordwijk and E.C. Rowe* | |
| | 10.1 Introduction | 171 |
| | 10.2 Nutrient mobility in soil solutions | 172 |
| | 10.3 Catching nutrients in single-species stands | 179 |
| | 10.4 Competing for nutrients in soil solution | 181 |
| | 10.5 Conclusions | 189 |
| **11** | **Below-ground Inputs: Relationships with Soil Quality, Soil C Storage and Soil Structure** | **193** |
| | *A. Albrecht, G. Cadisch, E. Blanchart, S.M. Sitompul and B. Vanlauwe* | |
| | 11.1 Introduction | 193 |
| | 11.2 Magnitude of below- versus above-ground plant imputs | 193 |
| | 11.3 C storage dynamics and determinants | 195 |
| | 11.4 How can SOC stocks be increased by below-ground inputs and activities? | 198 |
| | 11.5 Effects of SOC increases on different soil properties | 200 |
| | 11.6 Impacts of SOC increases on plant productivity | 203 |
| **12** | **Soil–Atmosphere Gas Exchange in Tropical Agriculture: Contributions to Climate Change** | **209** |
| | *L.V. Verchot, A. Mosier, E.M. Baggs and C. Palm* | |
| | 12.1 Introduction | 209 |
| | 12.2 Greenhouse gases | 209 |
| | 12.3 Trace gases and land use | 213 |
| | 12.4 Conclusion | 224 |
| **13** | **Benefiting from $N_2$-Fixation and Managing Rhizobia** | **227** |
| | *P.L. Mafongoya, K.E. Giller, D. Odee, S. Gathumbi, S.K. Ndufa and S.M. Sitompul* | |
| | 13.1 Introduction | 227 |
| | 13.2 Nodulation and $N_2$-fixation in the *Leguminosae* family | 228 |
| | 13.3 Rhizobial classification | 229 |
| | 13.4 Quantification of $N_2$ fixed by different legumes | 230 |
| | 13.5 Managing environmental constraints to increase $N_2$-fixation | 233 |
| | 13.6 The need to inoculate with rhizobia | 235 |
| | 13.7 Fate of fixed nitrogen | 238 |
| | 13.8 Conclusions | 241 |
| **14** | **Managing Mycorrhiza in Tropical Multispecies Agroecosystems** | **243** |
| | *T.W. Kuyper, I.M. Cardoso, N.A. Onguene, Murniati and M. van Noordwijk* | |
| | 14.1 Introduction | 243 |
| | 14.2 Arbuscular mycorrhiza | 248 |
| | 14.3 Mycorrhizal functioning in (multispecies) agroecosystems | 251 |
| | 14.4 The importance of the mycorrhizal network | 252 |
| | 14.5 Benefits of a perennial mycorrhizal system in multispecies agroecosystems | 253 |
| | 14.6 Mycorrhizas in models of nutrient uptake | 255 |
| | 14.7 Managing arbuscular mycorrhizal associations | 258 |
| **15** | **Nematodes and Other Soilborne Pathogens in Agroforestry** | **263** |
| | *J. Desaeger, M.R. Rao and J. Bridge* | |
| | 15.1 Introduction | 263 |
| | 15.2 Factors contributing to soilborne pests and diseases | 266 |
| | 15.3 Strategies for the management of soil pests based on general sanitation | 274 |
| | 15.4 The avoidance approach to the management of soil pests | 276 |
| | 15.5 The confrontational approach to the management of soil pests | 279 |
| | 15.6 Conclusions | 282 |

**16 Soil Biodiversity and Food Webs**    285
F.X. Susilo, A.M. Neutel, M. van Noordwijk, K. Hairiah, G. Brown and M.J. Swift
16.1 Introduction    285
16.2 Effects of land-use (change) on soil biota    290
16.3 Functions of soil biota in ecosystems    293
16.4 Food-web theory and below-ground biodiversity    296
16.5 Farmers and below-ground biodiversity: many questions remain    302

**17 Managing Below-ground Interactions in Agroecosystems**    309
M.R. Rao, G. Schroth, S.E. Williams, S. Namirembe, M. Schaller and J. Wilson
17.1 Introduction    309
17.2 When and where are BGI important?    310
17.3 Scope and options for managing below-ground processes    312
17.4 Conclusions    327

**18 Managing Movements of Water, Solutes and Soil: from Plot to Landscape Scale**    329
S.B.L. Ranieri, R. Stirzaker, D. Suprayogo, E. Purwanto, P. de Willigen and M. van Noordwijk
18.1 Introduction    329
18.2 Understanding the water balance as the basis for lateral flows    331
18.3 Trees, groundwater and salt movement    333
18.4 Consequences of subsurface flows for nutrient transport    336
18.5 Soil cover, runoff and its consequences for sediment transport    338
18.6 Discussion: scaling-up the effects of land-use change on river flow    344

**19 Soil and Water Movement: Combining Local Ecological Knowledge with that of Modellers when Scaling up from Plot to Landscape Level**    349
L. Joshi, W. Schalenbourg, L. Johansson, N. Khasanah, E. Stefanus, M.H. Fagerström and M. van Noordwijk
19.1 Introduction    349
19.2 Myths, misunderstandings and analytical frameworks    350
19.3 Case study 1: Sumberjaya, West Lampung, Sumatra (Indonesia)    352
19.4 Case study 2: Dong Cao catchment, Vietnam    355
19.5 Science-based models of watershed functions    357
19.6 Soil erosion – farmer perception versus simulation modelling    358
19.7 The gap between knowledge and practice    361
19.8 Discussion    362

**20 Challenges for the Next Decade of Research on Below-ground Interactions in Tropical Agroecosystems: Client-driven Solutions at Landscape Scale**    365
M. van Noordwijk, G. Cadisch and C.K. Ong
20.1 Introduction    365
20.2 Example 1. Lake Victoria basin    366
20.3 Example 2. Sumberjaya benchmark for watershed function conflicts    370
20.4 Example 3. Alternatives to slash-and-burn in the western Amazon basin    372
20.5 In praise of complexity?    374
20.6. Challenges for the future    378

**References**    381

**Index**    429

# Contributors

**Albrecht, A.**, IRD c/o World Agroforestry Centre (ICRAF) East and Central Africa, PO Box 30677, Nairobi, Kenya.
**Alegre, J.**, World Agroforestry Centre (ICRAF) Latin America, Av. La Universidad 795, La Molina, Apartado 1558, Lima, Peru.
**Akinnifesi, F.K.**, World Agroforestry Centre (ICRAF) Southern Africa, SADC-ICRAF Agroforestry Project, Makoka Agricultural Research Station, Zomba.
**Baggs, E.M.**, Department of Agricultural Sciences, Imperial College London, Wye Campus, Wye, Kent TN25 5AH, UK.
**Blanchart, E.**, IRD, Montpellier, France.
**Bridge, J.**, CABI Bioscience UK Centre, Bakeham Lane, Egham, Surrey TW20 9TY, UK.
**Brown, G.**, Embrapa Soja, C.P. 231, Londrina-PR, 86001–970, Brazil.
**Buresh, R.J.**, IRRI, DAPO Box 7777, Metro Manila, The Philippines.
**Burgess, S.S.O.**, Department of Integrative Biology, University of California, Berkeley, CA 94720, USA.
**Cadisch, G.**, Department of Agricultural Sciences, Imperial College London, Wye Campus, Wye, Kent TN25 5AH, UK.
**Cardoso, I.M.**, Department of Soil Sciences and Plant Nutrition, Federal University of Vitosa, Vitosa, Minas Gerais, Brazil.
**Comerford, N.B.**, Department of Soil and Water Science, University of Florida, Gainsville, Florida, USA.
**Desaeger, J.**, Department of Plant Pathology, The University of Georgia Coastal Plant Experiment Station, Tifton, GA 31793-0748, USA.
**de Willigen, P.**, Alterra, Green World Research, Postbus 47, 6700 AA Wageningen, The Netherlands.
**Fagerström, M.H.**, World Agroforestry Centre (ICRAF) SE Asia, PO Box 161, Bogor 16001, Indonesia.
**Gathumbi, S.**, MacArthus Agro-Ecology Research Centre, 300 Buck Island Ranch Road, Lake Placid, FL 33852, USA.
**Gijsman, A.J.**, CIAT, Apartado A, reo 6713, Cali, Colombia.
**Giller, K.E.**, Department of Plant Sciences, Wageningen University, PO Box 430, 6700 AK, Wageningen, The Netherlands.
**Grierson, P.F.**, Ecosystems Research Group, School of Plant Biology (Botany), The University of Western Australia, 35 Stirling Highway, Crawley, WA 6009, Australia.

**Hairiah, K.**, Faculty of Agriculture, Brawijaya University, Jl. Veteran, Malang 65145, Indonesia.
**Johansson, L.**, Rackarbergsgatan 100/350 752 32, Uppsala, Sweden.
**Joshi, L.**, World Agroforestry Centre (ICRAF) SE Asia, PO Box 161, Bogor 16001, Indonesia.
**Khasanah, N.**, World Agroforestry Centre (ICRAF) SE Asia, PO Box 161, Bogor 16001, Indonesia.
**Kho, R.M.**, Kamerlingh Onnesstraat 15, 6533 HK, Nijmegen, The Netherlands.
**Kuyper, T.W.**, Subdepartment of Soil Quality, Wageningen University, PO Box 8005, 6700 EC Wageningen, The Netherlands.
**Kwesiga, F.R.**, World Agroforestry Centre (ICRAF) Southern Africa, PO Box 128, Mount Pleasant, Harare, Zimbabwe.
**Livesley, S.J.**, Ecosystems Research Group, Botany Department, The University of Western Australia, Nedlands, WA 6907, Australia.
**Lusiana B.**, World Agroforestry Centre (ICRAF) SE Asia, PO Box 161, Bogor 16001, Indonesia.
**Mafongoya, P.L.**, World Agroforestry Centre (ICRAF) Southern Africa, PO Box 510089, Chiputa, Zambia.
**Matthews, R.**, Institute of Water and Environment, Cranfield University, Silsoe, Bedfordshire MK45 4DT, UK.
**Mobbs, D.C.**, Centre for Ecology and Hydrology, Edinburgh EH26 0QB, UK.
**Mosier, A.**, USDA Agricultural Research Service, Soil Plant Nutrient Research, 301 S. Howes, Rm 420, PO Box E, Fort Collins, CO 80522, USA.
**Moss, C.**, School of Agriculture and Forest Sciences, University of Wales, Bangor, Gwynedd LL57 2UW, UK.
**Murniati**, Forest and Nature Conservation Research and Development Centre, Jalan Gunung Batu 5, PO Box 165, Bogor, Indonesia.
**Namirembe, S.**, Faculty of Forestry and Nature Conservation, PO Box 7062, Kampala, Uganda.
**Neutel, A.M.**, Utrecht University, Utrecht, The Netherlands.
**Ndufa, S.K.**, Kenyan Forestry Research Instiute (KEFRI), Regional Research Centre Maseno, PO Box 25199, Kisumu, Kenya.
**Nziguheba, G.**, Laboratory of Soil Biology and Fertility, Catholic University of Leuven, Kasteelpark, Arenberg 20, B-3001 Heverlee, Belgium.
**Odee, D.**, Kenyan Forestry Research Institute (KEFRI), Regional Research Centre Maseno, PO Box 25199, Kisumu, Kenya.
**Ong, C.K.**, IRD c/o World Agroforestry Centre (ICRAF) East and Central Africa, PO Box 30677, Nairobi, Kenya.
**Onguene, N.A.**, Institute for Agricultural Research for Development (IRAD), PO Box 2123, Yaound, Cameroon.
**Palm, C.**, Tropical Soil Biology and Fertility Programme, Unesco-Rosta, UN Complex-Gigir, PO Box 30592, Nairobi, Kenya.
**Purwanto, E.**, Forestry Education and Training Centre, Jl. Gunung Batu, Bogor, Indonesia.
**Radersma, S.**, International Centre for Agroforestry, PO Box 30677, Nairobi, Kenya.
**Rahayu, S.**, World Agroforestry Centre (ICRAF) SE Asia, PO Box 161, Bogor 16001, Indonesia.
**Ranieri, S.B.L.**, World Agroforestry Centre (ICRAF) SE Asia, PO Box 161, Bogor 16001, Indonesia.
**Rao, M.R.** 111 ICRISAT Colony Phase-I, Akbar Road, Secunderabad-500 009, AP., India.
**Rowe, E.C.**, Department of Plant Sciences, Wageningen University, PO Box 430, 6700 AK, Wageningen, The Netherlands.
**Schalenbourg, W.**, c/o World Agroforestry Centre (ICRAF) SE Asia, PO Box 161, Bogor, Indonesia.

**Schaller, M.**, University of Bayreuth, Institute of Soil Science and Soil Geography, P.B. 101251, D-95440 Bayreuth, Germany.

**Schroth, G.**, Biological Dynamics of Forest Fragments Project, National Institute for Research in the Amazon (INPA), C.P. 478, 69011-970, USA.

**Shrestha, P.K.**, Local Initiatives for Biodiversity, Research and Development (LI-BIRD), PO Box 324, Pokhara, Nepal.

**Sinclair, F.L.**, School of Agriculture and Forest Sciences, University of Wales, Bangor, Gwynedd LL57 2UW, UK.

**Sitompul, S.M.**, Faculty of Agriculture Brawijaya University, Jl. Veteran, Malang 65145, Indonesia.

**Smith, M.**, CSIRO Sustainable Ecosystems, Davies Laboratory, PMB Aitkenvale, Qld 4814, Australia.

**Smithson, P.**, International Centre for Research in Agroforestry, Nairobi, Kenya. Current address: Berea College, Berea, Kentucky, USA.

**Stefanus, E.**, c/o World Agroforestry Centre (ICRAF) SE Asia, PO Box 161, Bogor 16001, Indonesia.

**Stirzaker, R.**, CSIRO Land and Water, GPO Box 1666, Canberra, ACT 2601, Australia.

**Suprayogo, D.**, Faculty of Agriculture Brawijaya University, Jl. Veteran, Malang 65145, Indonesia.

**Susilo, F.X.**, Faculty of Agriculture, Lampung University, Jl. Prof. Soemantri Brojonegoro No. 1, Bandar Lampung, Indonesia.

**Swift, M.J.**, Tropical Soil Biology and Fertility Programme, Unesco-Rosta, UN Complex-Gigiri, PO Box 30592, Nairobi, Kenya.

**Vanlauwe, B.**, TSBF/CIAT, Unesco-Rosta, UN Complex-Gigiri, PO Box 30592, Nairobi, Kenya.

**van Noordwijk, M.**, World Agroforestry Centre (ICRAF) SE Asia, PO Box 161, Bogor 16001, Indonesia.

**Verchot, L.V.**, World Agroforestry Centre (ICRAF) East and Central Africa, PO Box 30677, Nairobi, Kenya.

**Widianto**, Faculty of Agriculture Brawijaya University, Jl. Veteran, Malang 65145, Indonesia.

**Williams, S.E.**, c/o World Agroforestry Centre (ICRAF) SE Asia, PO Box 161, Bogor, Indonesia.

**Wilson, J.**, Centre for Ecology and Hydrology, Edinburgh Research Station, Bush Estate, Penicuik, Midlothian EH26 0QB, UK.

**Wong, M.T.F.**, CSIRO Land and Water, Private Bag No. 5, Wembley, WA 6913, Australia, and Soil Science and Plant Nutrition, The University of Western Australia, Nedlands, WA 6907, USA.

# Foreword

Complex agroecosystems are probably as old as humanity – certainly the 'Garden of Eden' has remained a strong symbol of the good life that a mixed tree plus crop system can provide.

Efforts to increase the amount of plant products that can be harvested led to success and to the combination of preparing the land, planting, caring and harvesting that we call agriculture. The number of plants growing on the same field became reduced, and most plants with a low harvest value became relabelled as weeds. Initially tree products still came from spontaneously established trees retained in the farmed landscape, or in the natural forest surrounding it. But a shortage of wood in more densely populated areas that needed timber for ship-building led to monoculture tree plantations.

Gradual crop selection for higher yield induced the 'domestication' of most of the crops on which the world still depends. In the 20th century, science-led breeding techniques and larger-scale distribution channels were the basis for big steps forward in agricultural productivity, and engendered the 'Green Revolution'. A reduction of complexity was seen to increase yield and provide more abundant food for the rapidly growing human population.

Critiques of the 'Green Revolution' approach included concerns about the use of agrochemicals, over the social equity impacts of the new technology, but also over the loss of complexity in agriculture. Monocultures became a symbol of an approach to agriculture that only focused on producing staple foods, and missed out on the multifunctionality of local food production for a varied and healthy diet, and provision of environmental services.

In the midst of this debate, 'agroforestry' was recognized as a new term for age-old practices. Twenty-five years ago a global centre for research on agroforestry (ICRAF) was established that would evolve into the World Agroforestry Centre, as we know it today. Its charter, written in 1978, refers to solutions to rural poverty and environmental protection that have lost none of their relevance and urgency, even as we approach the Centre's 25th anniversary. Agroforestry's potential for 'transforming lives and landscapes' still remains underutilized.

Initial enthusiasm for combining any tree plus any crop under the heading of 'agroforestry' did not last long, and the scientific study of plant–plant interactions and tree–crop interfaces became important. The expectation was to greatly improve the productivity of traditional forms of agroforestry, by designing new systems that would meet multiple objectives, while acknowledging the interactions between components. Ten years ago 'below-ground interactions' were seen as a black box that needed to be made transparent, as a basis for technology development and dissemination.

The chapters of this book show that considerable progress has indeed been made in our understanding of below-ground interactions, and their consequences for mixed cropping and agroforestry systems. It is gratifying to see the long list of authors and co-authors, including many fresh PhDs, that have solved parts of the puzzle.

Ideas on technology development and dissemination have gradually evolved. Expectations of substantial improvement over what farmers do to manage their local natural resources have been tempered by increased understanding of the unique and complex conditions that they face in managing smallholder tropical farming systems. Current paradigms place high values on local ecological knowledge, farmer-led technology development and collective action at landscape scale (including 'Landcare') – with less expectations for drastic technical 'design' of improved systems. Basic understanding of the interactions between components of the system can, however, greatly inform this process if it is well articulated.

A complex agroecosystem is not a target in itself, but a means to an end. The list of functions we all expect from agricultural landscapes has grown over time. These now include mitigation of global warming and the need for terrestrial carbon storage, and the aspirations for biodiversity conservation embedded in global conventions. Landscape-level water and soil movement in the uplands is a major issue for people living downstream. In the later part of the book these issues are introduced, and are linked to the understanding of interactions at plot level.

This book is targeted at a new generation of students, who as professionals can play a crucial role in facilitating research and development for and by farmers to make a better world. As the 'research needs' of all 20 chapters show, there is an enormous amount yet to be learned and discovered. This book provides a solid foundation upon which to build.

Dennis P. Garrity<br>
Director General<br>
World Agroforestry Centre (ICRAF)

# Editors' Introduction

Below-ground interactions are often seen as the 'dark side' of agroecosystems, especially when more than one crop is grown on the same piece of land at the same time. This book aims to review the amount of light the past decade of research has shed on this topic. It also aims to review how far we have come in unravelling the positive and negative aspects of these interactions and how, in dialogue with farmers, we can use the generic principles that are now emerging to look for site-specific solutions.

The basic concepts of 'competition', 'trade-offs' and 'complementarity' in tree–crop combinations may date back at least 200 years. In his *History of Sumatra* (1783), William Marsden contrasts a live-pole pepper agroforestry system used on Sumatra with the dead-pole system used on Borneo (a contrast that persists to this day). Note particularly the underlined text, which illustrates his understanding of the above concepts:

> ... The next business is to plant the trees that are to become props to the pepper... These are cuttings of the *chingkariang* (*Erythrina corallodendron*), usually called *chinkareens*... Trial has frequently been made of other trees, and particularly of the *bangkudu* or *mangkudu* (*Morinda citrifolia*), but none have been found to answer so well for these vegetating props. It has been doubted indeed, whether the growth and produce of the pepper-vine are not considerably injured by the *chinkareen*, which <u>may rob it of its proper nourishment by exhausting the earth</u>; and on this principle, in other of the eastern islands (Borneo, for instance) the vine is supported by poles, in the manner of hops in England. Yet it is by no means clear to me, that the Sumatran method is so disadvantageous in the comparison as it may seem; for, as the pepper plant lasts many years, whilst the poles, exposed to sun and rain, and loaded with a heavy weight, cannot be supported to continue sound above two seasons, there must be a frequent renewal, which, notwithstanding the utmost care, must lacerate and often destroy vines. It is probable also that <u>the shelter from the violence of the sun's rays</u> afforded by the branches of the vegetating prop, and which during the dry monsoon, is of the utmost consequence, <u>may counterbalance the injury occasioned by their roots</u>; not to insist on the opinion of a celebrated writer, that <u>trees, acting as siphons, derive from the air and transmit to the earth as much of the principle of vegetation, as is expended in their nourishment.</u>

Around the time that Marsden was writing, the concept of plant nutrition did not distinguish the different chemical elements. Indeed, it was not until 1880 that biological nitrogen fixation was described in scientific terms, although precursory ideas certainly existed. In some places, the ecological knowledge of farmers may still be based on principles similar to those Marsden applies when describing the interactions he observed. Indeed, Marsden's own thought processes were probably based on principles told to him during his stay in Sumatra.

Though we now have access to more detailed and analytical ecological concepts, indigenous knowledge systems should not simply be ignored. These new concepts should be used to complement and enrich the more practice-oriented knowledge systems of farmers, rather than to replace them.

The material selected for this book is intended to complement existing textbooks on intercropping (Vandermeer, 1989, 2002), agroforestry (Ong and Huxley, 1996; Young, 1997; Huxley, 1999; Franzel and Scherr, 2002), and biological $N_2$-fixation (Giller, 2001). To this end, this volume provides a synthesis of plant–soil–plant interactions in agroforestry, mixed pastures and intercropping systems – with a focus on processes that are relevant to many types of multispecies agroecosystems. Schroth and Sinclair (2003) recently reviewed concepts and research methods in the domain of *Trees, Crops and Soil Fertility*. The current volume was planned to complement this work with a more in-depth coverage of recent research results, as well as of the models that have emerged to integrate current understanding. Although agroforestry examples may dominate, the intention is to contribute to a better understanding of any multispecies agroecosystem. Agroforestry, with its greater complexity, simply provides a sound basis for the subject – because it is probably easier to simplify than to extrapolate to more complex interactions.

Although the principle of interaction between plants is not unique to tropical systems, we have chosen to concentrate on tropical agroecosystems as they have remained more complex than their temperate counterparts, resisting the onslaught of 'modernization'. They are also better studied and understood, although in the mix of perceptions, myths, oversimplifications and hypotheses, the observable reality remains hard to distinguish.

Each chapter provides an overview of key results and progress made with regard to research methods. This leads to an operational description of specific concepts in the form of simulation models. Within each chapter the main challenges that remain are discussed. Hopefully, a further generation of researchers will thus be stimulated to take up the challenge of linking basic research to practical applications in a wide range of systems, both with and without trees.

The book begins with an overview of the simple methods used to diagnose the net effect of below-ground interactions on overall plant (tree, crop, grass) performance. This sets the stage for a more detailed analysis of the contributory processes. Chapter 2 introduces the methods researchers use to explore farmers' knowledge of soil fertility and below-ground effects. The chapter also explores the way farmers explain the basic observations they make with regard to agroecosystem response to management and externally imposed variation.

Chapters 4, 5 and 6 focus on the root systems of trees and crops as key to our understanding of below-ground interactions. Chapter 4 reviews the spatial distribution of roots and the architecture of root systems, whereas Chapter 5 deals with the time dimensions of both root turnover and root dynamic response to changes in internal supply of, and demand for, nutrients and water. Chapter 6 gives an account of our current understanding of the processes of nutrient uptake from deep soil layers, a subject that has engendered considerable debate, and in which some empirical progress has been made.

The next block of Chapters (7–10) looks at root functioning in more detail. Chapter 7 reviews the uptake of phosphorus, which is the least mobile soil nutrient. Chapter 8 discusses related issues of aluminium toxicity in acid soils and the potential contributions of litterfall and organic–mineral interactions in multispecies agroecosystems, which can modify root development via detoxification of aluminium. Chapter 9 focuses on water uptake by interacting plants, and Chapter 10 discusses competition and complementarity issues in nutrient uptake. Chapter 10 also explores the potential of deep-rooted plants to act as a 'safety-net', and so enhance the recycling of nutrients.

The next group of Chapters (11 and 13–16) goes beyond plant–mineral-soil–plant interactions, and considers interactions via soil organic matter (Chapter 11), $N_2$-fixing symbionts (Chapter 13), mycorrhizal partners (Chapter 14), nematodes and other 'plant disease' organ-

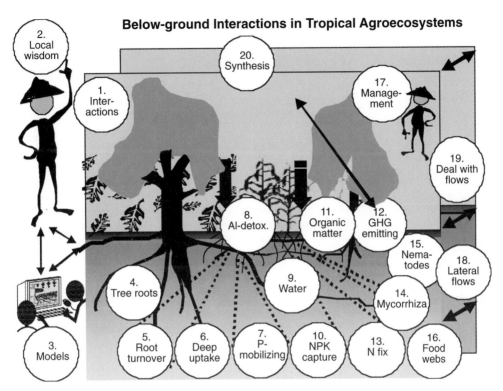

**Fig. 0.1.** Diagram of the structure of this book: its focus on below-ground interactions at the plot level (Chapter 1 and Chapters 4–16), and its exploration of linkages across plots to landscape scales (Chapters 12, 18 and 19). Specific attention is paid to farmers' knowledge and management options in Chapters 2, 17 and 19, and a critical discussion of modelling approaches is undertaken in Chapter 3.

isms (Chapter 15) and the below-ground foodweb as a whole (Chapter 16). In Chapter 17 we evaluate how the process-level understanding of below-ground interactions discussed in the previous chapters can contribute to farmers' management of real-world agroecosystems, by interventions at the plot (or farm) scale.

Considerable progress has been made with regard to understanding processes at the plot scale. However, this is not the only scale to consider. Erosion losses from an upper field may, in fact, constitute an input into lower fields. Chapters 18 and 12 explore interactions at the farm-to-landscape scale, via subsurface and surface flows of water (Chapter 18) and the emission and/or absorption of greenhouse gases (Chapter 12).

Chapter 3 and several of the subsequent chapters consider simulation models that can be used: (i) as tools, to synthesize existing concepts and explanations of below-ground interactions, and (ii) to provide a 'context' for research that focuses on specific processes. Chapter 3 also considers the issue of 'for whom models are built'. Chapter 20, the synthesis, returns to the issues of 'where' and 'how' our attempts to open the 'black box' of below-ground interactions can be of direct use in managing natural resources.

Because it exhibits so many interactions, the below-ground agroecosystem forms a fascinating backdrop to the detective work reported in this volume. Investigations have, in some cases, ripped the initial hypotheses apart and forwarded new explanations for the matters they considered – some of these explanations are surprisingly simple whereas others involve unexpected actors.

Usually a typology of agroecosystems that draws distinctions between agriculture, agroforestry, pastures, forests, fallows and crop fields takes up a substantial part of the introduction of any book on this subject. We have attempted to overcome this problem by adopting a rather radical perspective, which assumes the following:

- All agricultural systems are essentially the same, (see Box 0.1 for some of the nuances of this statement).
- All agricultural systems have essentially the same outcomes, (see Box 0.2 for some of the nuances of this statement).
- All agroecosystems (real or virtual) behave in the same way (as long as they are driven by the same set of inputs, see Box 0.3).

Awareness of ecological issues (including pollution, climate change and landscape-level issues) has recently increased, as has interest in organic agriculture. As a result, more and more scientists, policy makers and people in general are now looking at multispecies systems in terms of their potential to improve biodiversity and ecosystem functions, rather than only considering the economic benefits they provide. The broad theoretical and practical issues covered in this book provide realistic guidelines in terms of what can and cannot be expected of ecosystems composed of multiple species.

As editors, we would like to thank all those authors and co-authors who rapidly responded to our invitation to contribute to this book. We would also like to thank all the authors' employers and host organizations, who were kind enough to grant their staff sufficient time to transform into a readable book the notes and draft texts compiled at the initial workshop in Malang, Indonesia (hosted by Brawijaya University and ICRAF Southeast Asia). Finally, we are sincerely grateful to the Forestry Research Programme of the Department for

---

**Box 0.1.** All agricultural systems are essentially the same.

Except for their:

- Plant components: the types of crops, trees and weeds that share space and time

- Use of external inputs: the amounts and types of organic and inorganic inputs

- Spatial complexity: the differentiation in zones with different components

- Management interventions: rule-based interventions, triggered by (observable) conditions in the field

- Calendar of events: the timing of planting, pruning, weeding, harvesting, ploughing and slash-and-burn events (S&B)

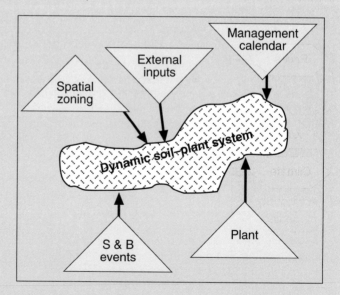

**Box 0.2.** All agricultural systems have the same outcomes.

Except for their:

- Sensitivity to: climatic and pest variability
- Sensitivity to: variability of prices
- Physical yields of: useful products that can be obtained
- Labour and cash requirements: for implementing the management interventions
- Environmental impacts: derived from (sub)surface flows of water, soil and nutrients, gaseous emissions and the C stocks on site

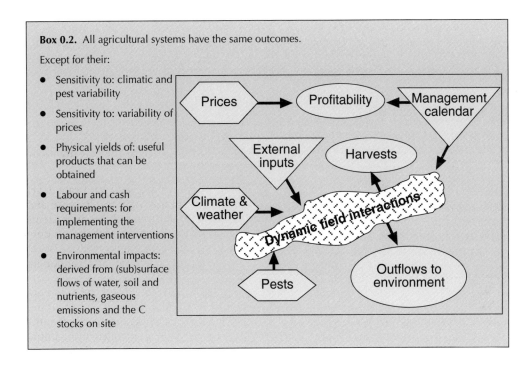

**Box 0.3.** Model hypothesis 1. All agroecosystems (real and virtual) essentially behave in the same way, if they are driven by the same set of inputs (parameters or real).

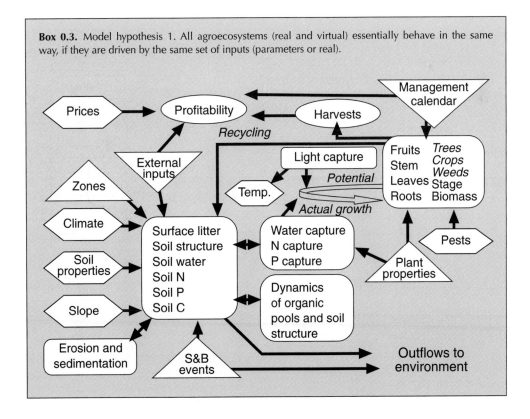

International Development (DFID-FRP), for supporting the project financially, and to CAB International, with whom it was a pleasure to work. Technical editorial support was very ably provided by Dr Sandy Williams and Dr Jim Weale (SCRIPTORIA Academic English Editing Services). Mrs Subekti Rahayu handled all secretarial support very efficiently. Finally, it only remains for us to say that, as its general editors, we sincerely hope that this work will stimulate research to such an extent that this book will be rendered completely out of date in 10 years' time ...

Meine van Noordwijk, Chin K. Ong and Georg Cadisch
Malang/Bogor, Nairobi, Wye

# Acknowledgements

This publication is an output from a research project funded by the UK Department for International Development (DFID) for the benefit of developing countries. The views are not necessarily those of DFID. The following Forestry Research Programme projects have supported research results that have substantially contributed to the various chapters:

| Author | FRP project number |
| --- | --- |
| David Odee | R7342 |
| Edwin Rowe | R6511, R6523 |
| Fergus Sinclair | R4181, R4594, R4731, R4850, R5470, R5651, R6322, R6523, R7188, R7227, R7264, R7635 (R7637 LPP) |
| George Cadisch | R6523, R7936 |
| Julia Wilson | R6321, R7342 |
| Ken Giller | R6523 |
| Laxman Joshi | R4731, R5470, R6322, R7264, R7635 |
| Meine van Noordwijk | R6348, R6523, R7188, R7315, R7936 |
| Mike Wong | R4754, R5651, R6071 |
| Peter Gregory | R5651 |
| Robin Matthews | R6348, R7342 |
| Roland Buresh | R6348, R7188 |

Apart from the team of authors and co-authors who have, during the writing workshop, been providing valuable comments and peer review on chapters in which they were not directly involved, we would like to acknowledge the following persons who have provided valuable comments on draft versions of the chapters: Alain Albrecht, Pia Barklund, Else Buenemann, Jan Goudriaan, Gerry Lawson, Claire de Mazancourt, Patrick Mutuo, Astrid Oberson, Cheryl Palm, Meka Rao, Shashi Sharma, Paul Smithson, Bruno Verbist and Julia Wilson.

Technical editorial support was very ably provided by Dr Sandy Williams and Dr Jim Weale (SCRIPTORIA Academic English Editing Services). Mrs Subekti Rahayu handled all secretarial support very efficiently.

# 1 Ecological Interactions in Multispecies Agroecosystems: Concepts and Rules

Chin K. Ong, Rhamun M. Kho, Simone Radersma

---

**Key questions**

1. Can agriculture mimic the beneficial functions associated with natural ecosystems?
2. What are some of the drawbacks of multispecies agroecosystems?
3. How can we know whether below-ground interactions are important?
4. How can we develop a predictive understanding of multispecies systems?
5. What are the basic rules we need to know?

---

## 1.1 Introduction

### 1.1.1 Agriculture as a mimic of nature?

The idea of designing a multispecies agroecosystem as a structural and functional mimic of natural ecosystems is appealing, because it offers a simple integrated principle for working towards sustainable agriculture. The hypothesis that diversity is associated with higher productivity is not new; as early as 1859 Charles Darwin asserted that

> it has been experimentally proved that if a plot of ground be sown with one species of grass, and a similar plot be sown with several distinct genera of grasses, a greater number of plants and a greater weight of dry herbage can be raised in the latter than in the former.

The search for new models for agriculture is particularly attractive where modern agriculture that is completely dependent on fossil energy and chemical inputs is unaffordable, unsustainable or no longer acceptable to our quality of life (Lefroy *et al.*, 1999). But natural systems are much more complex than any form of agriculture used today, where the trend is still one of reducing complexity. In a recent review, Vandermeer and his co-workers (1998) argue that multispecies agroecosystems are more dependable, in terms of production, and more sustainable, in terms of resource conservation, than simple ones. However, despite efforts made to prove this during the century and a half following Darwin's assertion, the evidence is not yet conclusive.

In his review of the issue in the context of the humid tropics, Ewel (1999) also concluded that it pays to imitate natural systems, especially with regard to the use of perennial plants to maintain soil fertility, protect against erosion and make full use of light, water and nutrients. However, there is a trade-off between maintaining full vegetative cover and promoting the growth of desirable plants. It may therefore be necessary to accept a lower 'crop' yield, because an intrinsic feature of natural systems is a high investment in structure. On the other hand, reducing the complexity of agricultural systems to promote 'short-term' productivity, by substituting external inputs for biological functions, tends to further increase dependence on such inputs. High levels of fertilizer tend to switch off specialized mechanisms of nutrient input and cycling, such as biological nitrogen fixation (Chapter 13, this volume) and proteoid roots (Chapter 7, this volume).

Ong and Leakey (1999) attempted to reconcile differences between recent agroforestry research and interactions between savannah trees and understorey vegetation in the semiarid tropics. Whilst the productivity of natural vegetation under savannah trees is generally higher than that of vegetation between the trees, the expectation derived from this observation – that growing trees among crops ('agroforestry') will boost crop yields – has generally not been confirmed in experiments. Van Noordwijk and Ong (1999) explained that because a high proportion of the above-ground part of mature savannah trees consists of a woody structure, rather than foliage, they provide shade and microclimatic improvement whilst incurring only a low water-use 'cost'. This means that the amount of water 'saved' by the resultant reduction in soil evaporation may more than offset the water 'lost' through tree transpiration. In recently planted trees ('new agroforestry') the amount of water used for transpiration exceeds the reduction in soil evaporation, so decreasing the amount of water available for crops. A basic problem is that savannah trees' investments in woody structure require time and energy, thereby reducing returns to farmers. Thus there is a penalty, which takes the form of a long wait before trees mature and before any benefit is seen in terms of understorey productivity.

Existing trees can be used for their beneficial effect, just as they have by millions of farmers in the parkland systems of West Africa. However, the idea that this beneficial effect can be emulated and further improved upon in 'designed' agroforestry is one that has led to much disillusionment.

How can we confidently relate the function of a mimic system to its structure when we know so little of the underlying processes that confer persistence and resilience on the natural ecosystems on which a mimic system is based? Perennial vegetation has considerable benefits in terms of maintaining such ecosystem functions as 'catchment hydrology' (Chapters 6 and 18, this volume), 'nutrient cycling' (Chapter 10, this volume), 'nitrogen fixation' (Chapter 13, this volume) and 'reduction of trace gases'. Such benefits encourage the use of agroforestry as a technique that allows sustainable land and water management in areas where high-energy-input or large-scale agriculture are impractical (Kidd and Pimental, 1992). Thus, information on the underlying processes in less complex systems (such as agroforestry) may provide important clues about the same processes in multispecies agroecosystems.

### 1.1.2 Multispecies agroecosystems

Multispecies agroecosystems (associations of two or more species growing together on the same piece of land in a certain temporal and spatial arrangement) are widespread throughout the tropics. In comparison with monocrops, such systems promise the following three benefits to farmers: increased productivity, increased stability and increased sustainability. The first potential benefit concerns total *productivity*, which can be higher, i.e. output of valuable products per unit of land and labour is increased, through reduced damage by pests and diseases (Chapter 15, this volume) and through better use of resources. A multispecies system often has a green canopy that is denser for longer than that of a monoculture, allowing it to capture light that would be lost in a monoculture. The mixed canopy may also reduce weed competition and reduce water loss by evaporation directly from the bare soil, leaving more water for productive transpiration. A deeper and denser rooting system in a multispecies system may exploit the soil more completely, increasing the potential for water and nutrient uptake (Chapters 4, 5, 6 and 10, this volume). Better soil physical properties and the reduction of runoff (Chapters 9 and 18, this volume) may conserve water, whereas enhanced soil biological activity and nutrient cycling may increase the availability of nutrients (Chapters 11 and 14, this volume).

The second potential benefit of multispecies systems is increased *stability*, i.e. sensitivity to short-term fluctuations is reduced by decreasing the risk of pests and diseases (Chapter 15, this volume) and by spreading those risks through species diversity (Chapter 16, this volume). If one plant component fails to produce, the production of the other plant components may compensate for it. In agroforestry systems, trees may increase the microscale variability in soil and in crop growth, which increases the probability that at least part of the crop will yield successfully.

The third potential benefit of multispecies systems is increased *sustainability*, i.e. long-term productivity is maintained by the protection of the resource base. This may be the result of, for example, reduced erosion, input of nitrogen through biological $N_2$-fixation (Chapter 13, this volume), retrieval of subsoil nutrients and/or reduction of nutrient losses through reduced leaching (Chapter 6, this volume). Productivity and sustainability are interrelated and have the potential to conflict, especially in nutrient-limiting environments. Increased productivity may imply an increased exploitation of the environment and a mining of nutrient resources. As a consequence, on infertile soils a productive system with high outputs will most probably not be sustainable without external inputs.

A potential drawback of multispecies systems is that plant components of a lower value to the farmer may compete too heavily with those of a high value. However, by species choice and arrangement, competition can be controlled/manipulated by appropriate management (Chapter 17, this volume). Van Noordwijk and Ong (1999) commented on the different ways ecologists and farmers perceive 'competition'. To the first, the term often refers to 'use of and dependence on common resources'; and, competition between individuals of the same species is generally stronger than that between individuals of different species. To the farmer, however, the yardstick against which competition is 'measured' is the 'value' derived from the growth of the different plants. If all plants have an equal value per unit biomass produced, as happens in a monoculture, there is no perception of the occurrence of competition. Competition becomes a problem if the difference in value increases. Companion plants of lower use/value can then become 'weeds' and competition is seen to increase, even though the resource base (in space and time) of these plants can partially differ from that of the crops.

In this chapter we introduce concepts and models that can be used to explore how and where multispecies agroecosystems may be able to improve the use of plant growth resources, using experience gained from recent agroforestry research. We will begin by unravelling the complex interactions that occur between trees and crops. We will then examine how resource availability influences competition between plants and, finally, we will define simple rules governing success or failure when mixing trees and crops. A list of the models and their application for above- and below-ground interactions are discussed elsewhere in this book (Chapter 3).

In the past, sweeping generalizations have been made, which suggested that any multispecies system was better than any monoculture. However, because we can now acknowledge that sufficient evidence exists to prove such supposition false, the current goal of interaction research is to determine which particular multispecies system will realize and maximize the potential benefits in any given environment. Before such systems can be promoted, however, we need to take into account the needs and constraints imposed by the socioeconomic and policy context. However, a number of recent books (Franzel *et al.*, 2001; Otsuka and Place, 2001) discuss these more applied issues, allowing this book to focus on the biophysical aspects of interactions in tropical agroecosystems, and especially their more elusive below-ground aspects.

### 1.1.3 Land Equivalence Ratio (*LER*)

Farmers' direct interests lie, particularly, in the 'production benefit'. The benefit of intercropping is most frequently (Vandermeer, 1989) quantified by the *LER*, which is defined as the relative land area in pure stands that is required to produce the yields of all products from the mixture. If the *LER* >1, then the mixture is more advantageous than separate monocultures. Of course, the amount of land used is only one of the production factors, which include labour, energy and total cost. However, multispecies systems that do not provide a gain in efficiency in terms of land use have little chance of being more efficient in these other aspects.

In tree–crop systems (agroforestry), one component is dominant and perennial. However, farmers are often not concerned with maximizing both tree and crop components; rather, they are concerned with maximizing the annual crop's production whilst maintaining an acceptable level of growth in the tree component. Therefore, the production benefit can be expressed by *I*, the yield advantage of the annual crop component only (Sanchez, 1995; Ong, 1996; Rao *et al.*, 1998). It is defined as the increase in crop yield relative to the yield in monoculture. If $I > 0$, then the agroforestry system is, in terms of crop production, more advantageous than the monoculture.

## 1.2 Separating Positive and Negative Interactions

'Interaction' refers to the influence that one or more components of a system has on the performance both of another component of the system and of the overall system itself (Nair, 1993). Several classifications of interactions can be made, depending on the context and upon one's view. Besides an ecological classification, based on the net effect (positive, nil or negative) of each component (Anderson and Sinclair, 1993) and the agronomic partition between above-ground and below-ground interactions, more mechanistic classifications can be made. In terms of time, a distinction can be made between direct (i.e. instantaneous) and indirect interactions.

In the case of indirect interactions, a time period exists between the cause (e.g. depletion of soil water whilst that water supply is ample) and the result. So, the effect only becomes apparent later in the season when there is a shortage of water. According to the number of components involved, a distinction can be made in terms of two-way, three-way, etc. interactions. A two-way interaction involves two components (e.g. in a specific ecozone, the presence and the characteristics of one plant component may influence the production of another plant component). A three-way interaction involves three components, in which a two-way interaction in its turn interacts with a third component (e.g. in a broader context, in several ecozones, the influence of one plant component on another plant component may in turn be influenced by the environment).

## 1.3 Assessing Plant–Environment–Plant Interactions

Insight into plant–environment–plant interactions can be obtained using different approaches. This section describes: (i) the separation of simple effects; (ii) the use of resource capture concepts; and (iii) the use of a resource balance concept.

### 1.3.1 Separating simple effects

The net effect of one plant component on another plant component must be the result of positive (i.e. 'fertility') effects and negative (i.e. 'competition') effects. So, one approach by which we may obtain insight into this area is to separate and quantify these effects. This idea was formalized in the following equation (Ong, 1995):

$$I = F + C \tag{1.1}$$

where $I$ is the 'overall interaction' (i.e. the percentage net increase in production of one component attributable to the presence of the other component); $F$ is the fertility effect (i.e. the percentage production increase attributable to favourable effects of the other component on soil fertility and microclimate); and $C$ is the competition effect (i.e. the percentage production decrease attributable to competition with the other component for light, water and nutrients).

However, positive and negative component effects are highly site specific, and change with the environment (see Sanchez, 1995). Therefore, the insight gained with this approach does not contribute greatly to a predictive understanding applicable to a broad context, and thus to other situations.

After a modification by Ong (1996), the equation evolved (Rao et al., 1998) to give the following:

$$I = F + C + M + P + L + A \tag{1.2}$$

where $F$ refers to effects on chemical, physical and biological soil fertility, $C$ to competition for light, water and nutrients, $M$ to effects on microclimate, $P$ to effects on pests, diseases and weeds, $L$ to soil conservation and $A$ to allelopathic effects.

The advantage of Equation 1.2 is that it provides a comprehensive overview of the possible effects involved. However, as emphasized by those authors, many of these effects are interdependent and cannot be experimentally estimated independently of one another. Such interdependence is a serious drawback because, as a result of the overlap, quantification of the individual terms will most probably give a sum that exceeds $I$. Therefore, the equation cannot help one determine the relative importance of each term for a given system. Another drawback is that the interaction with the environment is not explicitly stated in the equation, but is rather implicitly contained in each term. In other words, Equation 1.2 approaches $I$ as a two-way interaction between two plant components. For a predictive understanding applicable to other ecozones it would be preferable if $I$ were to be approached as a three-way interaction, making the influence of the environment explicit.

Cannell et al. (1996) attempted to clarify the resource base of this equation. They argued that part of the 'mulch' effect of a tree is derived from light, water and nutrient resources, which the tree acquired in competition with the crop ($F_{comp}$). Another part of the mulch effect may result from the fact that the tree can exploit resources that the crop cannot ($F_{noncomp}$). Similarly, a proportion of the resources acquired by the tree in competition with the crop is recycled within the system, and may thus be used by a future crop ($C_{recycl}$). If $F_{comp}$ were based on the same resources as $C_{recycl}$, then in the long run the two terms would cancel each other out. The question of whether or not a tree–crop combination gives yield benefits therefore depends on: (i) the complementarity of resource use; (ii) the value of direct tree products – specifically those obtained in competition ($C_{nonrecycl}$) relative to the value of crop products that could have been produced with these resources; and (iii) the efficiency with which tree resources are recycled into crop products, a point specifically true for those resources obtained in competition with the crop ($C_{recycl}$).

The main advantage of Ong's method is its simplicity with regard to quantifying system performance as the result of a few main effects, which can be directly measured with a relatively simple experimental setup. But, there are disadvantages. The first is the lack of a timeframe. The assumption of Cannell et al. (1996) that $F_{comp} = C_{recycl}$ may be true once the system has reached equilibrium in the long term. However, before that stage, the fertility effect is more prone to delays than the competition effect, because of slow or low liberation of available nutrients from recycled material. Thus in the first few years, which are important for the assessment of the technology by the farmer, $F_{comp} < C_{recycl}$, and there is a strong possibility that $I$ will be negative. The equation does not allow for delayed effects, although it can be modified to include a short-term and long-term fertility effect. Van Noordwijk et al. (1998a) estimated the terms of such a modified equation, by including a treatment based on the removal of the tree and quantification of 'residual fertility' effects. Under such conditions, these long-term fertility effects were substantial, but so was the competition term.

Another disadvantage of the direct empirical approach is that the agroforestry system performance results of this equation cannot be transferred from one environment to another. Kho (2000a) developed a method to overcome the latter disadvantage of Ong's equation (see Section 1.3.3 below, 'The resource balance concept'). His method allows for the transfer of the performance results of a specified system from one environment to another, and is based on the sum of positive and negative factors similar to that used by Ong.

However, Kho's method can be used quite easily in a qualitative way, and complements Ong's method in terms of the transfer of performance results from one environment to another. The use of both methods together may give a reasonable idea of why systems involving tree–crop interactions perform as they do. However, even if the two methods are used to complement each other, there are still limitations to their usefulness. Kho (2000a) noted that his method focuses on resource–use interactions and is not applicable if pests, diseases or allelopathy (caused by the tree component of the system) are important factors determining the system's performance.

Neither Ong's nor Kho's method has a timeframe, and a timeframe is necessary to take into account delayed effects (such as those mentioned above) and long-term trends. Thus, an important aim of agroforestry in general is not addressed: the methods may show that a certain agroforestry system works better in certain environments than does a sole crop, but this does not mean that such a system is really sustainable over a longer period.

Another feature not covered by these methods is the performance of a tree–crop system that is highly dependent on interactions between factors; thus, a simple sum of positive and negative factors is not going to give the right result. An example of this is water–P interaction in P-fixing soils, in which P-transport to roots is decreased by decreases in soil water content. Decreases in soil water content by trees affects the environmental factor p, thus necessitating a $W*p$ term in Kho's equation.

In cases where long-term performance and interaction of tree factors and environmental factors are important, mechanistic research may be necessary to explain the functioning of the system, in order to understand its performance and be able to extrapolate that performance to longer time periods or to other environments. However, mechanistic research suffers from its own pitfalls. An important one is a loss of overall understanding as a result of focusing on one or two factors. Another related pitfall is the risk of getting lost in a multitude of detailed processes, without realizing that only a few factors may really play an important role in determining 80–90% of a system's performance in a certain environment.

These pitfalls of mechanistic research may, in turn, be (partly) overcome by starting to look at the system from the perspective of Kho's method. The first question we should ask is 'what are the main limiting factors in a certain environment (including pests and diseases)?' The second question is 'how do trees and crops influence each other in general (including influences through mechanisms like allelopathy) and particularly via limitation of the main resource?' Using Kho's method to look at the main environmental resource limitations and the effects trees have on the different resources, it is possible to prioritize the mechanisms that need to be looked at in more detail. In this way, the use of system-performance analysis methods can be a first step in determining the priorities for mechanistic research.

### 1.3.2 Use of resource capture concepts

One plant component may influence another by changing its capture of the most limiting resource. Another approach by which we may gain insight is, therefore, by modelling the capture of the limiting resource in a multispecies system. Biomass production (W) is the product of such capture, and the efficiency with which the captured resource is converted into biomass (Monteith et al., 1994; Ong et al., 1996):

$$W = \varepsilon_{conversion} \times Capture \quad (1.3)$$

where $\varepsilon_{conversion}$ is the conversion efficiency and *Capture* the capture of the specific resource.

Two assumptions are usually made when using this approach. First, the conversion efficiency is usually considered to be species-specific and conservative, which justifies the use of empirically determined efficiencies. Secondly, because biomass production W is calculated from *Capture* it is regarded as the 'dependent' variable responding to *Capture*, which is the 'independent' variable. These assumptions rely on the premise that the resource under consideration is the only limiting resource and that all other growth resources are in ample supply. In this specific case, the conversion efficiency is constant at its maximum value (Kho, 2000a), and the response of the plant can be entirely attributed to the increased capture of the resource being considered.

If other resources are also limiting (i.e. if the response curve is a smooth curve reaching a plateau gradually), the assumptions are no longer valid. First, theoretical as well as empirical evidence shows that increased limitation of other resources will decrease the conversion efficiency of the resource in question (Kho, 2000a). The conversion efficiency of nutrients equals (in the absence of nutrient losses from the plant) the reciprocal of the nutrient concentration in the plant. For nitrogen, phosphorus and potassium, the maximum concentration can be two to three times the minimum concentration (Van Duivenbooden, 1995), which shows that conversion efficiencies are not at all conservative. In a dataset of radiation conversion efficiencies (Azam-Ali et al., 1994), 72% of the total variance was found to be attributed to the environment and only 10% to species, which indicates that conversion efficiencies are determined more by environment than by species. Secondly, increased capture of one resource will always be accompanied by increased capture of other resources. If several resources are limiting, increased capture of other resources must also contribute to production, and it is not clear whether, or to what extent, increased capture of the resource under consideration is the cause or the effect of increased biomass production. Therefore, Kho (2000a) argued that the relation between production and capture is a correlation, not a causal relation.

An approach that uses Equation 1.3 is thus methodologically sound only if a strict 'law of the minimum' is applicable. However, this is a theoretical idealization that is, in reality, seldom true. In most environments, a crop responds to the increased availability of several resources, having for each resource a smooth response curve gradually reaching a plateau (de Wit, 1992) (Box 1.1). Limitation is thus indicated more by a point on a gradual, continuous scale rather than by a discrete yes/no variable (see Kho, 2000a, and the next section). One reason for this is that plants have a certain plasticity, which allows them to adapt their architecture in order to acquire the most limiting resource (Chapter 4), thereby making that resource less limiting. Another reason why a strict 'law of the minimum' occurs so seldom is that, with an increasing timescale (from hours to seasons and

---

**Box 1.1.** General principles on limiting resources and capture.

**1.** Limitation does not involve only one resource that, if saturated, is replaced by another resource. Usually several resources are limiting, in which case the relationship between biomass production and the capture of a single resource should be viewed as a correlation, not a causal relation.
**2.** The variation in conversion efficiencies between environments is larger than that between species, which shows that conversion efficiencies are determined more by the environment than by species.

beyond) and/or space (from a single plant organ to a field and beyond), the response curve is the sum of several individual response curves. Even if the latter are Blackman-type curves, this composite curve is, because of temporal and spatial heterogeneity, smooth (Kho, 2000a).

For the study of plant–environment–plant interactions, de Wit's approach is quite relevant, because plants alter the availability of several resources simultaneously and will, therefore, alter conversion efficiencies by changing resource limitations. Moreover, the 'most limiting' resource for a plant in monoculture is not necessarily the 'most limiting' resource for a plant in a multispecies system.

The above methodological weakness for predicting biomass production in multispecies systems may be avoided in two different ways. First, empirically obtained efficiencies should not be used as parameters (constants) in process-based models. Instead, efficiencies should be studied and modelled in relation to availabilities of the other resources, and treated as a variable in process-based models. Secondly, dynamic simulation models for different resources can be linked. These models should operate at such a small (detailed) scale that, in the integration step, Equation 1.3 can be replaced by a more realistic model (involving the capture of several resources). This is an enormous challenge. Until it is accomplished, however, the present models can give a mechanistic insight into the resource flows in a system and may be helpful in terms of evaluating the relative influence of alternative multispecies designs (species combinations, temporal and spatial arrangements, etc.) on resource flows.

### 1.3.3 The resource balance concept

One plant influences another by changing the availability of *several* resources in the environment of the other. The effect on production depends on the degree to which the resources concerned are limiting (Fig. 1.1).

In an environment where the resource is not limiting (right in Fig. 1.1), a change in availability does not have much of an influence on production (all other factors being constant). The more a resource is limiting (to the left in Fig. 1.1), the greater its influence. Limitation of a resource can be defined as the ratio between the slope of the response curve and the use efficiency of the resource (Kho, 2000a):

$$L = \frac{\partial W / \partial A}{W / A} \qquad (1.4)$$

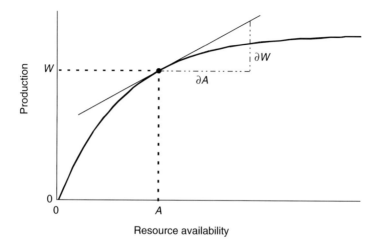

**Fig. 1.1.** Response curve of biomass production in relation to availability of a certain resource (all other factors equal) with slope ($\partial W/\partial A$) and use efficiency ($W/A$) for a specific environment (●).

where $L$ = limiting factor, $W$ = biomass, and $A$ = availability.

On the plateau, the slope equals zero; thus, the minimum value of limitation $L$ equals zero. Near the origin, the slope equals the use efficiency, so that the maximum value of limitation $L$ equals one. By rearranging Equation 1.4 it is clear that, for small changes, the *relative* change in production ($\Delta W/W$) equals the *relative* change in availability ($\Delta A/A$) multiplied by the limitation $L$ (all other factors being equal).

Each specific environment has, for a particular species, its own balance of available resources, and can be characterized by a set of response curves (Fig. 1.1; one curve for each resource) on which the environment concerned occupies a certain point. Therefore, for a particular species, each specific environment can be characterized by the set of limitations $L_i$ for each resource $i$. Kho (2000a) showed that the limitations of *all* resources ($CO_2$, radiation, water and all nutrients) most probably total one:

$$\sum_{i=1}^{n} L_i = 1 \qquad (1.5)$$

That is, if for a particular species in a specific environment the limitations of some resources are known and add up to one, it can be inferred that all other resources are not limiting. Alternatively, if the sum is less than one, it can be inferred that other resources are still limiting. A rough approximation of limitations can be obtained by analysing publications reporting use efficiencies and response to changed availability (see Kho, 2000a).

In general, a plant component does not change the availability of only one resource in the environment of another plant component; rather, it changes the availability of several resources (light, water, nitrogen, phosphorus, etc.). How does this influence production? If the availability of several resources changes simultaneously (e.g. leading to increases and/or decreases in availability), is biomass production then determined only by the (change in) availability of the most limiting resource? Or, by that of all limiting resources?

Within the temperature range at which a crop species can grow and reproduce (roughly from 0°C to 35°C for temperate species and from 10°C to 45°C for tropical species; Ong and Monteith, 1985), crop dry matter production ($W$) in a specific environment is a function of resource availability:

$$W = f(A_1, A_2, \dots, A_n) \qquad (1.6)$$

where $A_i$ is the availability of resource $i$ and $n$ is the number of all resources. Let $z$ denote an arbitrary management option that alters resource availabilities (some resources may be altered positively, others negatively, some by much, others less or not at all).

For the sake of argument, we will temporarily assume that the management option used can be applied gradually, using many small steps; but in practice this is not essential. The effect of the management option on production can be found by differentiating Equation 1.6 with respect to $z$. According to the chain rule:

$$\frac{dW}{dz} = \sum_{i=1}^{n} \frac{\partial W}{\partial A_i} \cdot \frac{dA_i}{dz} \qquad (1.7)$$

Multiplying both sides by $dz$, and expressing the differentials relative to their original value (i.e. dividing both sides by production $W$ and multiplying the right-hand side by $A_i/A_i$) gives:

$$\frac{dW}{W} = \sum_{i=1}^{n} \frac{\partial W}{\partial A_i} \frac{A_i}{W} \times \frac{dA_i}{A_i} \qquad (1.8)$$

Substitution of Equation 1.4 yields:

$$\frac{dW}{W} = \sum_{i=1}^{n} L_i \times \frac{dA_i}{A_i} \qquad (1.9)$$

which shows that, for small changes, the relative change in production ($\Delta W/W$) equals simply the *sum* of the relative changes in availability ($\Delta A_i/A_i$) multiplied by their limitation $L_i$.

Biomass production is not determined by the (change in) availability of only one resource (e.g. the most limiting) but by all the limiting resources. The contribution each resource makes to the relative change in production is proportional to both its degree of limitation and its relative change in availability. This result is the basis for the framework considered in the next section of this chapter.

## 1.4 A Framework for a Predictive Understanding of Multispecies Systems

With regard to growth resources, one plant component can have many effects on another plant component in a multispecies system (upper half of Fig. 1.2). A given multispecies system with a given management system has a particular canopy and root architecture in time and space. The mixed canopy architecture (leaf area index and extinction coefficients of each plant component in different layers) and root architecture (root length densities of each plant component in different soil layers) determine the *relative* ability of each plant component to acquire resources from their shared environment. For each plant component, this ultimately results in net effects on the availability of resources. For one plant component, the relative net change in the availability of resource $i$ equals:

$$T_i = \frac{\Delta A_i}{A_i} = \frac{A_{i;multi} - A_{i;mono}}{A_{i;mono}} \quad (1.10)$$

where $T_i$ is the relative net change in availability of resource $i$ because of the other plant component, $A_{i;multi}$ is the availability of resource $i$ to the plant component concerned in the multispecies system and $A_{i;mono}$ is that in the monoculture.

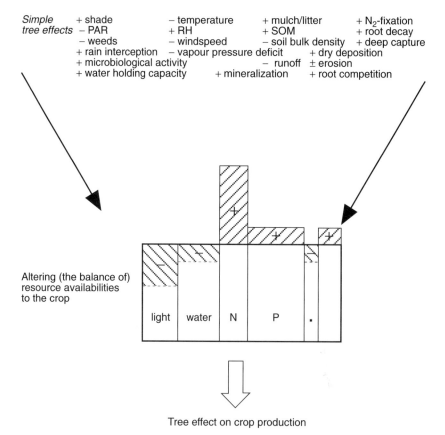

**Fig. 1.2.** Trees influence crop production by altering the balance of resources available to the crop (light, water, N and P). The height of each shaded area relative to the height of the rectangle represents the relative change in the availability of the resource ($T_i$). The width of each shaded area relative to the total width represents the limitation of the resource in the tree–crop interface ($L_i$). The sum of positive and negative shaded surfaces relative to the total surface of the rectangle represents the overall tree effect *I* expressed as a fraction of sole crop production. PAR, photosynthetically active radiation; RH, relative humidity; SOM soil organic matter. (Adapted from Kho *et al.*, 2001.)

Because these coefficients are determined by the mixed canopy and root architectures they are probably 'conservative' when applied to a particular multispecies system, soil depth and slope. They most probably serve as key characteristics of the particular multispecies system (with one for each plant component set), but this is a subject for further study.

When a particular multispecies technology is placed in a specific environment (in terms of soil, climate and topography), it interacts with that environment. If we approximate the differentials in Equation 1.9 by differences and substitute the definition of $I$ and Equation 1.11 into it we may expect for $I$ (Kho, 2000b; Kho *et al.*, 2001):

$$I = \sum_{i=1}^{n} L_i \times T_i \qquad (1.11)$$

Equation 1.11 is represented graphically in the lower half of Fig. 1.2. Note that Equation 1.11 only considers interactions related to growth resources. Allelopathy and effects caused by damage by pests and diseases fall outside its scope. Therefore, Equation 1.11 may be extended to:

$$I = \left( \sum_{i=1}^{n} L_i \times T_i \right) + P + A \qquad (1.12)$$

where $A$ is the relative change in production because of allelopathic effects ($A \leq 0$) and $P$ is the relative change in production (positively or negatively) because of the influence of the multispecies system on pests and diseases. $P$ is also a function of technology and of the environment. If there are no sources of pathogens in the environment, $P$ will be zero.

Equations 1.11 and 1.12 explain the production benefit associated with a multispecies system at a certain state (time and maturity). The relative net effects on resource availability ($T_i$) may change when the system grows to maturity. In a newly established simultaneous agroforestry system, competition for light may be relatively low, resulting in a negative effect close to zero. However, the rooting system of the trees is still superficial, resulting in low (negative) net effects on water and nutrient availability. As the system matures, the canopy and the rooting system of the trees develop. The amount of light available to the crop decreases (the net effect becomes more negative). The net effects on water availability may become either less negative (as a result of less competition) or positive (if the benefits offered by a more favourable microclimate and soil physical properties outweigh competition). The net effect on nutrient availability may also become less negative or positive (as a result of less competition and increased nutrient cycling). On sloping lands, the potential benefit of water and soil conservation is greater than on a flat surface. So, the increase with maturity of the net effects on water and nutrient availability may be stronger on sloping lands.

The more a resource becomes available in the environment of a multispecies system, the smaller its limitation (see Equation 1.4, Fig. 1.1 and Box 1.2). The more other limiting resources become available in the environment of that multispecies system, the smaller their limitation (see Equation 1.5). Combined with Equation 1.11 this leads to two rules (see Box 1.2) that can be viewed as counterparts to classic principles of crop production (Kho, 2000b).

These rules are helpful both for *predicting* the performance of a multispecies technology when it is extended to another environment and for *developing* a multispecies technology (Kho, 2002). For example, Kho (2000b) showed that, with regard to alley cropping technology, the net effect that trees have on the availability (to the crop) of light, water

---

**Box 1.2.** Simple rules for predicting the performance of multispecies systems.

**Rule 1.** The greater the availability of a resource in the environment of a multispecies system, the smaller its relative importance in the overall interaction.
**Rule 2.** The greater the availability of other limiting resources in the environment of the multispecies system, the greater the relative importance of a resource in the overall interaction.

and phosphorus is most probably negative, whereas for nitrogen it is most probably positive. Consequently, in a (sub-)humid climate on nitrogen-deficient soils, the overall effect of alley cropping will most probably be positive, because of the positive nitrogen effect and the negative net effects on other resources (Fig. 1.3a). In the same climate, but on acid soils, phosphorus is relatively less available; this will increase the negative phosphorus effect (Rule 1) and decrease the positive nitrogen effect (Rule 2), resulting in a negative overall effect (Fig. 1.3b).

When designing and developing a multispecies technology, we want to increase the share of positive net effects and decrease the negative net effects. This may be done in one of two ways. First, in the design phase, we can try to make the net effects on resource availability ($T_i$) both positive and as large as possible. This can be done by choosing the right species combinations, the right temporal and spatial arrangements and the right management techniques, which are a unique part of the technology (e.g. pruning). Herewith, process-based models may

(a) N-deficient soil

(b) Acid (P-deficient) soil

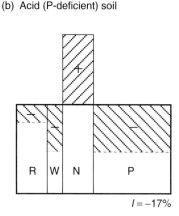

(c) N-deficient soil + N fertilizer

(d) Acid (P-deficient) soil + P fertilizer

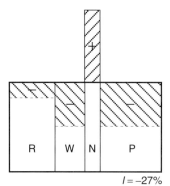 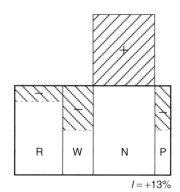

**Fig. 1.3.** Possible tree effect balances of alley cropping technology in a humid climate (a) on nitrogen-deficient soils; (b) on acid (phosphorus-deficient) soils; (c) on nitrogen-deficient soils with nitrogen fertilizer being provided to the alley crop and the sole crop; and (d) on acid soils with phosphorus fertilizer being provided to the alley crop and the sole crop. The relative net tree effects on availability of each resource ($T_i$) are equal in (a)–(d); only the environment (i.e. resource limitations, $L_i$) changes explain the different overall effects ($I$). R, radiation; W, water; N, nitrogen; P, phosphorus resources. (Source: Kho, 2002.)

be very helpful. These models should thus be built in such a way that the effect these factors have on (relative changes in) resource availability can be evaluated easily. Note that *relative* changes in availability are sufficient to explain a relative change in production (Equations 1.9 and 1.11). Therefore, absolute values of availability are not necessary, and the models are allowed to give predictions of availability that are based on one or more constant multiplication factors. With the calculation of the relative net change (Equation 1.10) the multiplication factors appear in the numerator and denominator and vanish from the equation. This property may thus reduce the effort necessary to produce such a model.

The second way to increase positive net effects and decrease negative effects involves the following. In the development phase, we can increase the limitations ($L_i$) of the positive net effects and decrease those of the negative net effects by choosing appropriate management options that are not unique to the technology, but that can be applied to both a multispecies system and a monoculture. Management options can be translated into effects on the availability of resources, which allows one to manipulate the share of the different resources in the overall interaction. For example, with regard to alley-cropping technology, options such as phosphorus fertilization, water-conserving tillage and/or weeding of superficially rooting weeds are probably appropriate. These management options will increase the availability of phosphorus and water, and thus will reduce the negative phosphorus and water effects (Rule 1) and increase the positive nitrogen effect (Rule 2). To illustrate this, compare Fig. 1.3b and 1.3d in terms of the effect that phosphorus fertilizer has when applied to both an alley-cropping system and a monoculture. On the other hand, external inputs of organic or inorganic nitrogen are probably inappropriate in such systems, because they will reduce the positive nitrogen effect (Rule 1) and increase the share of the negative effects (Rule 2; cf. Fig. 1.3a and 1.3c).

Simply being able to identify as positive or negative (+ or −) the net effects that plant components have (in terms of the availability of resources) on another, specific, plant component ($T_i$) can be very useful. In this sense, even very limited knowledge can be helpful for the prediction of overall interaction and the development of multispecies systems, in a qualitative sense. Fortunately, much of this information is already available, hidden in the literature on this subject. Kho (2000b) developed rules (see Box 1.2) to 'reveal' this information. By analysing the direction of the change of $I$, in response to a change in the availability of a resource (when all other factors remain constant), the sign of the net effect on this resource can be derived (Fig. 1.4).

For example, suppose that a particular agroforestry system has an $I$ equal to +5% in a season with good rains, but an $I$ equal to −25% in a season with poor rains. So, with decreased water availability, the overall interaction decreased. According to Fig. 1.4, the net effect that trees have on water availability is probably negative ($T_A < 0$), while the net effect trees have on another resource is probably positive ($T_B > 0$). Because $I$ was positive, the last statement (e.g. that there is a positive net effect on the availability of another resource) is most probably true. However, $I$ changed its sign, so we can be sure that the net tree effect on water availability in this particular system is negative.

The net effects that plant components have on resource availability ($T_i$) can easily be empirically quantified. The availability of a resource to a crop in a multispecies system and to a crop in monoculture are estimated and Equation 1.11 applied. Our interest lies not in absolute values of availability, but in relative changes in availability; hence, it is immaterial if the estimations of availability are biased with constant multiplication factors (Kho *et al.*, 2001).

When quantifying the limitations ($L_i$) we have to consider certain factors. Approximation of the differentials of Equation 1.9 resulted in Equation 1.11. This approximation is justified for small reductions in availability. However, especially in agroforestry systems, one plant component (the tree) may greatly influence resource availability to the other plant component (the crop). The limitations for the crop in the agro-

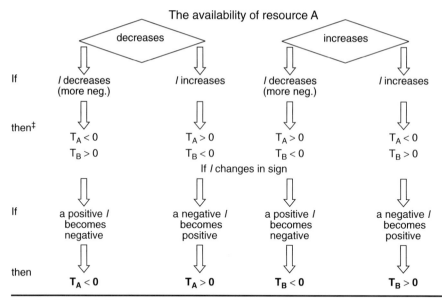

**Fig. 1.4.** Diagram used to derive the sign of the net effect on availability of a resource (other factors being equal). (Adapted from Kho, 2000b, with permission from Elsevier.)

$T_A$ refers to the net effect on changed resource; $T_B$ to that of another limiting resource.
‡ Both statements may be true. However, if $I$ is negative, the statement with the negative T value is most meaningful; if $I$ is positive, the one with the positive T value is most meaningful. If the overall interaction $I$ changes in sign, then certainty about the net effects is given.

forestry system may thus differ from the limitations for the monoculture, and the appropriate limitations for use in Equations 1.11 and 1.12 must be somewhere in between the values in the two systems. To be consistent, the partial derivatives in the limitation (Equation 1.4) should also be approximated with the right variations of the same order. For details and an example of the quantification of Equation 1.12, see Kho *et al.* (2001).

## 1.5 Conclusions

In this chapter, using experience gained in agroforestry research, we have used specific concepts and simple rules to explore how and where multispecies agroecosystems may be able to improve the use of growth resources. It is clear that simple rules (Box 1.2) are helpful both for predicting the general performance of a multispecies technology when it is extended to another environment and for developing a multispecies technology. For more subtle interactions, however, it will be necessary to use a combination of simple rules and a process-based model. For example, Radersma (2002) demonstrated, on a P-fixing Oxisol in western Kenya, that a 2–3% decrease in soil water content as a result of the influence of associated trees may cause a 30–40% decrease in maize production. In such situations, trees are likely to affect crop growth by inducing P deficiency through drying the soil. Further understanding of the processes underlying multispecies ecosystems are needed in order to evaluate their relative importance at the ecosystem and landscape levels (which are described in other chapters of this book).

It should also be pointed out that we have drawn our evidence from the limited number of species (usually two or three) currently used in agroforestry research. More progress is needed in the development of the theory and concepts dealing with how the number and composition of plant species influence ecosystem processes before we can extrapolate them to more complex systems. Recent advances in theoretical models based on simple but well-known mechanisms of interspecific competition may provide important new insights. For example, the models developed by Tilman and Lehman (2001) support the long-standing hypothesis that the number of species in a community may increase overall productivity, resource use and stability. However, they pointed out that traits that allow species to coexist when exploiting and competing for limited resources do not automatically give rise to greater productivity or stability (because these interactions are also highly dependent on species composition). See Chapter 4 for more discussions of the importance of species diversity and ecosystem performance. Finally, our current understanding of how multispecies agroecosystems function should be integrated into the enormous wealth of local ecological knowledge known to exist (see Chapter 2, this volume).

---

**Conclusions**

**1.** Well-chosen multispecies agroecosystems are probably more productive and sustainable than the best-performing monocultures. However, the evidence for this is not yet conclusive.
**2.** The drawbacks of multispecies systems include lower yields in annual crops (because of higher investments in vegetation structure) and a long wait for trees to mature and yield products.
**3.** Simple rules for predicting the general performance of multispecies systems exist, including rules for predicting the balance between above-ground and below-ground resource-use limitations.
**4.** For the prediction of a more specific level of performance, it is necessary to combine simple rules with process models.

---

**Future research needs**

**1.** How important are the number of species and species composition in determining overall productivity and stability?
**2.** How important is ecosystem structure in determining multispecies-agroecosystem function?
**3.** How can we incorporate local ecological knowledge into the scientific rules and concepts associated with modellers' ecological knowledge?

# 2 Locally Derived Knowledge of Soil Fertility and Its Emerging Role in Integrated Natural Resource Management

Laxman Joshi, Pratap K. Shrestha, Catharine Moss and Fergus L. Sinclair

---

**Key questions**

1. How can we explore local ecological knowledge beyond the 'classification' of soils, trees and crops?
2. How can local knowledge of soil and below-ground interactions be used to inspire and complement 'science'?
3. How does local ecological knowledge relate to actual farmer management decisions?

---

## 2.1 Introduction

The livelihoods of rural people often depend heavily both upon soil fertility and their ability to maintain and utilize it. It is reasonable to expect that such people will have observed soils and the processes surrounding their utilization very closely, so developing knowledge that they can use to predict the likely consequences of possible interventions. We assert here that it is useful to distinguish such locally derived knowledge from formal 'soil science' because, although local insights and scientific understanding may be comparable in some respects, the former may also differ in their scope and structure.

Most, if not all, of current 'soil science' was derived from the testing and refining of ideas and concepts that started as part of 'local knowledge'. However, during this process, the conditions of tropical smallholder farmers were unevenly represented. Local knowledge represents the current position of a local community in terms of its land use. Since local conditions vary and people have different objectives and levels of dependence on soil resources, local knowledge may vary from place to place. However, some commonality may exist when farmers have similar means of observation and farm in similar agroecological conditions. This makes documentation and analysis of local knowledge a key task in the development process. Appreciation of local knowledge is of fundamental importance to professionals seeking to assist the local development of sustainable land-use practices, both because it is necessary for effective communication with farmers and because it allows research and extension activities to be appropriately targeted at locally experienced constraints.

In this chapter we start by discussing the terminology and approaches that surround research on local knowledge. These reveal important differences in the emphasis of research following anthropological, as opposed to natural science, traditions. We advocate an interdisciplinary way forward that both distinguishes practical explanatory and predictive knowledge from cultural values and norms and seeks to use terminology that is as free as possible from associations with particular disciplinary traditions. Although this remains a controversial distinction, it has been incorporated within a knowledge-based-systems methodology that has been used to acquire local knowledge about soil fertility in several long-term participatory development initiatives. We use results from three such studies (undertaken in Nepal, Ghana and Indonesia) to illustrate key points about local knowledge concerned with soil classification, soil fertility and below-ground interactions, and discuss them in the light of more general literature concerned with local knowledge. We conclude by discussing the implications of these findings.

## 2.2 Local Knowledge

The terminology surrounding the study of local knowledge is rich, although people's choice of language often reflects the disciplinary context within which their work is grounded. For present purposes, we view knowledge as: an output of learning, reasoning and perception and a basis for predictions of future events; it is people's understanding and interpretation based on some explainable logic of supposedly general validity.

This does not necessarily imply any objective notion of absolute truth, but rather a particular interpretation of information and data. Although the semantics of the terms 'data', 'information' and 'knowledge' are debatable, we define 'data' here as a recorded set of either quantitative or qualitative observations; and 'knowledge' as a logical interpretation or explanation of data, acquired either directly or from other sources. We use the term 'understanding' to mean knowledge that is specific to the person who interprets it, regardless of whether they can articulate it or not, whilst 'knowledge' is used to mean understanding that can be articulated and so can be recorded independently of the interpreter, thus making its utility more general (Sinclair and Walker, 1998). 'Information' is a collective term that embraces 'data', 'understanding' and 'knowledge'. The knowledge a specified group of people has about a specified domain constitutes a 'knowledge system'.

The distinction between farmers' knowledge and practice has not always been recognized in the literature on this subject. This is most notable with respect to the body of work on ITK (Indigenous Technical Knowledge), which often describes people's actions rather than the underlying rationale driving them (IDS, 1979). Knowledge alone does not lead to action; conditions and constraints due to cultural norms, religious obligations, and economic and policy circumstances can all influence farmers' decisions, forcing them to act in an ecologically irrational manner. Moreover, agricultural practice generally unfolds over time (during a season, or over several years in the case of perennial crops), so that farmers may make many separate decisions about the cultivation, tending and harvesting of crops, each of which would be contingent upon the circumstances extant at the time that it is made. These build up a complex agricultural practice, in which it is difficult to disentangle ecological knowledge from other social and economic constraints by simply observing the result (Richards, 1989).

A generic conception of farmer knowledge systems concerned with natural resource management can also usefully distinguish pragmatic knowledge about how the natural world works (predicting outcomes of management interventions) from cultural values that modify the desirability of various outcomes (Fig. 2.1). The latter distinction is controversial, particularly when viewed from the anthropological tradition, which sees all knowledge as being culturally embedded (Ellen, 1998). However, it has

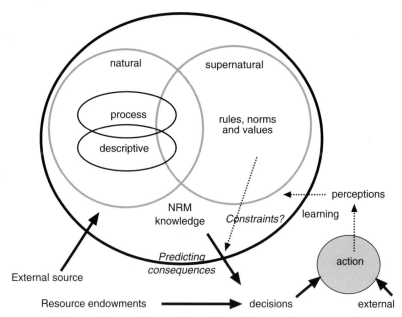

**Fig. 2.1.** Conceptual diagram of various forms of knowledge influencing farmers' natural resource management (NRM) decisions.

been found to be empirically useful in dialogue with farmers. Accepting these distinctions, knowledge of the natural world can be seen to comprise 'explanatory knowledge' (concerned with ecological processes) and 'descriptive knowledge' (concerned with the properties of the various components of agroecosystems, such as trees, crops and soils). This contrasts with 'supernatural knowledge', which consists of higher level, often spiritually based, explanations for the order of things. The latter may form the basis of the rules, norms and values assigned by culture, religion or other moral or social imperatives. This, in turn, often places constraints on people in terms of what they are prepared to do. Various examples of this may be offered. Muslims and Hindus do not eat pork or beef, respectively. Mayan farmers are reputed not to have sold maize, because they believed that maize was symbolically equivalent to human flesh (Asturias, 1949). The Hanunoo shifting cultivators in the Philippines use the interpretation of their dreams in their selection of cultivation sites (Conklin, 1957). In Zambia, in cases of a venomous snake bite, local people can readily articulate the mechanism by which a victim is affected, but, why that particular person met with the misfortune of being bitten requires a higher-level, supernatural explanation involving malice and witchcraft (Sinclair and Joshi, 2000). In practice, however, farmers tend to reply to pragmatic questions about the ecology of their farming systems with answers based on natural rather than supernatural explanations. Hence, most of the time, it is not difficult to separate the natural aspects of knowledge from the supernatural.

Local knowledge is also dynamic and continuously evolving, in that farmers learn both by evaluating the outcomes of their previous actions and by observing the environment. Farmers also augment their knowledge by interacting with other people and the media. This view contrasts with the ubiquitous use of words such as 'traditional' (Berkes *et al.*, 2000; Ford and Martinez, 2000) or 'indigenous' (Sillitoe, 1998) to describe rural people's knowledge, since they imply old, pristine knowledge systems that are culturally specific. In reality farmers' knowledge is likely to be hybrid in nature, with bits of knowledge being drawn from multiple sources. Indeed, many of the crops now cultivated by smallholder farmers are exotic, and have been introduced, together with some knowledge regarding their cultivation, from other parts of the world. For example, in the jungle rubber system in Indonesia (South-East Asia), smallholders now cultivate a South American tree introduced by colonial governments about a century ago (Gouyon et al., 1993). Local smallholders use technology that is, in part, derived from colonial plantation management, e.g. tapping techniques, but also from smallholder innovation, e.g. high-density planting and allowing secondary forest to regenerate around the rubber trees instead of clean weeding (Dove, 2000).

There is a long and still active tradition of defining local knowledge systems in opposition to scientific knowledge (Levi-Strauss, 1966; Sillitoe, 1998; Berkes et al., 2000). Various terms are encountered in the literature referring to this dichotomy – 'formal' vs. 'informal', 'western' vs. 'indigenous' and 'outsider' vs. 'insider'. However, the problem with this sort of frame of analysis is that, in most cases, the knowledge of local people is not some pristine indigenous perception of the world. It is more likely to have been interacting with external knowledge, at least to some extent, for the last 500 years or so (Agrawal, 1995).

It is very difficult, if not impossible in any meaningful way, to trace the origin of knowledge. Attempts to generalize about fundamental differences in local and scientific knowledge are fraught with difficulty. Assertions that a local people's knowledge is heuristic (based on rules of thumb that may have no explanatory basis) has not been borne out by research. It has been shown, in a range of cultural and agroecological contexts, that some of the understanding that farmers have involves mechanistic explanation of natural processes comparable with, and often complementing, scientific knowledge (Richards, 1994; Sinclair and Walker, 1999; Ford and Martinez, 2000). For these reasons we prefer to use the term 'local ecological knowledge' to refer to knowledge about agroecology held by people living in a particular locality. 'Locality', in this sense, may be defined socially as well as geographically. As shown in Fig. 2.1, such local ecological knowledge comprises both directly and indirectly acquired knowledge. Typically, it is the locally derived elements that differ from scientific knowledge in their level of aggregation (grouping according to perceived pertinence). Whereas science has emphasized reductive analysis, farmers tend to think more holistically, with limits imposed on their analysis by what they are able to observe and experience. This creates regularities in local knowledge of natural processes across cultures, as well as regularities in terms of how local knowledge contrasts with scientific understanding.

In summary, recent research into locally derived ecological knowledge across a range of agroecological and cultural contexts indicates that it often:

- has explanatory aspects, with a logical structure comparable with scientific understanding (Sinclair and Walker, 1999);
- has regularity regionally (Sinclair and Joshi, 2000), and across similar agroecosystems, in contrasting cultural contexts (Thorne et al., 1999; Roothaert and Franzel, 2002);
- has some complementarity with scientific knowledge (Thapa et al., 1995; Sinclair and Walker, 1999; Thorne et al., 1999);
- is holistic, but is also often agroecologically specific, being aggregated by the organisms and environmental context from which it was derived (Moss et al., 2001);
- can be readily articulated and recorded through structured discussions with local people (Sinclair and Walker, 1998).

Many proponents of the importance of local knowledge have promoted its use both in combination with scientific investigation and as a means of enhancing our overall ecological understanding. However, for some time, wide application of what local ecological knowledge had been acquired remained elusive, partly because of the difficulty associated with accessing much of the knowledge contained in reports, articles and theses. The development of formal methods for making explicit records of local ecological knowledge on computer (in a form that allows them to be flexibly accessed, evaluated and used) have made it easier to incorporate local knowledge in agricultural research and extension (Box 2.1; Walker *et al.*, 1997).

## 2.3 Soil Classification

The visually impenetrable nature of soil creates particular constraints, in terms of what farmers can observe below the ground, and makes classification of the medium itself a critical element of knowledge systems concerned with soils and their fertility. This is reflected in many past studies of local knowledge about soils, which have predominantly documented how farmers classify soils (Talawar and Rhoades, 1998). Soil colour and texture are the soil properties most commonly used in local classifications across different geographical regions, agroecological zones and cultures (Tamang, 1992; Joshi *et al.*, 1995b; Shah, 1995; Turton and Sherchan, 1996; Shrestha, 2000). As a result, local terms used to label soil types are often based on dominant soil colour (such as 'black', 'red', 'yellow' or 'white' soils), or on soil texture (such as 'sandy' or 'clay-rich' soils) or on combinations of the two. This is illustrated by a specific example from western Ghana, showing how collectors of *Thaumatococcus daniellii*, a valuable non-timber forest product, describe soil types in their vicinity (Table 2.1).

Ongoing research on local knowledge of soil both in western Nepal (Shrestha, 2000) and in Ghana (Sinclair, 2001) has found variation in the use of local terminology for soil types. In Nepal, this occurs when the soil colour is not distinct or where soils exist with a gradation of texture. Nepali farmers were generally able to differentiate more than ten soil samples into distinct classes, based on their colour and texture. However, groups of male and female farmers came up with different names for some of the intermediate soil types. As in Ghana, colour categories are often combined with categories of texture when describing specific soils (such as 'red clayey soil', 'red sandy soil', 'white

---

**Box 2.1.** Formal methods for knowledge acquisition.

Much of the understanding about local ecological knowledge presented in this chapter has been developed through the use of a knowledge-based-systems methodology for acquiring and evaluating local knowledge (Walker *et al.*, 1995). This comprises two major phases: the first involves gathering knowledge from people and recording it in an easily accessible form; the second investigates how widely this acquired knowledge is held in the community of interest (Walker and Sinclair, 1998).

In the first phase, ecological knowledge is collected from a small sample of deliberately chosen individuals thought to be knowledgeable about the domain of interest and willing to cooperate. The knowledge is collected through repeated, focused interviews with these key informants. Between successive interviews, knowledge is abstracted from records of the discussions with key informants and expressed as a series of unitary statements (written in simple, formal grammar) and terms. These are stored on computer in the form of a knowledge base, so that the knowledge is accessible and can be evaluated using tools for handling qualitative data, including automated reasoning procedures (Kendon *et al.*, 1995). Contextual information about who articulated the knowledge and the conditions under which each statement is valid are also stored (Sinclair and Walker, 1998). A customizable software package (AKT5 – the Agroecological Knowledge Toolkit, freely downloadable from www.bangor.ac.uk/afforum) provides the facilities necessary to explicitly record, access and evaluate local ecological knowledge. It has built-in features for representing hierarchical information, displaying synonyms and exploring cause–effect relationships.

In the latter phase, involving a test of generality (or distribution) of knowledge across multiple communities, a large randomized sample of people is drawn from the target community (as in Joshi and Sinclair, 1997) to explore how representative the knowledge base is. For details, including the rationale of the approach and a manual for the AKT5 software, see Dixon *et al.* (2001).

**Table 2.1.** Soil types as described by collectors of *Thaumatococcus daniellii* in the western region of Ghana. *Twi* words are given in italics. (Adapted from Waliszewski, 2002.)

| Broad categories | *Twi* name | | Texture | Comments |
| --- | --- | --- | --- | --- |
| | Specific soil types | Literal translation | | |
| *Asase denden* (hard soil) | *Ntetie* (clay) | *Asase fita* — White soil | Predominantly clay | Becomes waterlogged and sticky to the touch in the wet season, hard in the dry season. |
| | | *Asase koko* — Red soil | | Becomes waterlogged in the wet season, hard in the dry season. A red clay. |
| *Asase mremre* (loose soil) | *Afonwea* | Wet soil | Sand:clay (70:30) | Soil that is found near streams, wet but not waterlogged. |
| | *Anwea* | Sand | Pure sand | Pure sand, different from beach sand. |
| | *Asase tuntum* | Black soil | Organically rich | A soil that is found in the forest or near village rubbish tips after organic matter has decayed. Noted as being very fertile. |

sandy soil', and so on) and this may lead to some differences in the local terminology applied to different soil classes (Conklin, 1957; Carter, 1969; Kerven *et al.*, 1995; Joshi *et al.*, 1995b; Sandor and Furbee, 1996). The actual soil classification schemes are, therefore, generally determined by what soils occur locally, so that what is described as 'red soil' by farmers in one location in Ghana, for example, is often quite different from what those in another locality describe as 'red soil'.

People also often describe soils in terms of their suitability for certain crops. Such knowledge, and the resulting classification of soils, may vary according to the crops grown (Moss *et al.*, 2001). When one crop is dominant in an area (e.g. olives in Syria), farmers may evaluate land almost entirely on the basis of its suitability for that crop (Cools *et al.*, 2003). The Nepal study, however, revealed that, despite the use of different local terms to name the same soil type, there was remarkable consistency in farmers' knowledge about important soil properties (such as fertility, drainage, erosivity, manure requirement and moisture retention). This points to the importance of having a broader understanding of the context of local terminology concerning soils (beyond the identification of labels used for different soil types), in order to learn about local soil classification and discern both regularities across locations and local peculiarities.

Thus far, studies of local soil classification have shown that, to describe and classify soils, farmers use physical properties that are either visible to the naked eye or that can be sensed through touch (e.g. texture and structure), sometimes augmented by smell or even taste. In general, farmers make only rudimentary use of below-ground soil features and processes when classifying soil or explaining the underlying basis for their classification schemes. Farmers generally have difficulty explaining why red soil is red, what makes some soils sandy and others heavy or why sandy soil holds water for less time than clay-rich soil. Sandor and Furbee (1996), in a study made amongst the Lari people in the Colca Valley of Southern Peru, found that, although farmers were able with regard to the textural classification of their soil, there was no evidence that their soil knowledge went beyond practical considerations to, for example, explain soil genesis. Studies in both Nepal and Ghana show that farmers do possess good explanatory knowledge regarding the functional attributes of different types of soil that affect their agricultural capability. One obvious example is the link between farmers' perception of black soil as fertile and their knowledge of its high organic matter content. An example of such is offered by Nepal, where farmers apply animal manure to improve soil fertility. Farmers in a number of cases perceive that the darker the colour of black soil the greater its fertility (Tamang, 1992; Joshi et al., 1995b; Shrestha, 2000). Moreover, farmers in Landruk (west Nepal) differentiate soil that is black because of its high organic matter from soil that is black as a result of black mineral parent material in the subsoil layer.

Throughout the hills of Nepal, farmers use the terms *malilo* and *rukho* for fertile and unfertile soils, respectively (Thapa et al., 1995; Shrestha, 2000). These terms are used generically by farmers in a number of ways, for example to classify trees based on the effect they have on soil and crops. *Malilo* soils contain high levels of organic matter; are deep (with few or no stones); are soft and friable; retain moisture for a long time; absorb and hold nutrients added through manure; can be easily ploughed; and produce good, healthy crops with a high yield. *Rukho* soils, on the other hand, are sandy or stony; contain little or no organic matter; are shallow; retain moisture for only a short period; do not easily absorb and hold nutrients; are difficult to cultivate; and are associated with low crop yields. The Nepal study found that farmers perceive the high fertility of *malilo* soils to be an inherent property, related to texture. These soils are, therefore, potentially more productive than other soils. Farmers also perceive that some soils are inherently *rukho*, though they are not able to explain why this is so. *Kamere mato* (white calcareous soil with a large amount of mica), *Jogi mato* (reddish mixed coloured soil with mottling) and yellow clayey soil fall into this category. Farmers know that crop yield is low on these soils even if a large quantity of animal manure is applied to them.

Sandor and Furbee (1996) report that farmers in the Lari community, in the Colca Valley of southern Peru, use functional attributes to describe soil types, such as soils that are wet and 'rot roots' (clay-rich soils), that 'need much water' (excessively drained, coarse-textured soils), that are 'weak' or 'lazy', that 'need ash or fertilizer' or that can or cannot be used to 'grow maize'. This knowledge system is widespread in the eastern Andes, although even within the Colca Valley variation occurs in the names given to specific soil types and the knowledge associated with them. Similarly, Zuni Indians in New Mexico use terms for soils that emphasize surface condition or water infiltration and transport of parent material (Norton et al., 1998). As is also widely true in Ghana and Nepal, these terms may include reference to where soils are located in the landscape. Amongst the Zuni, *He'bik'yaye* refers to a sticky clay area with poor infiltration, *so:lana* to a coarse but fertile alluvial sediment that captures water, and *danaya so:we* to an organically enriched soil from below upland forest trees.

Similarly, in Ghana, soils in valley bottoms are often distinguished from upland soils even when similar in colour and texture, because they tend to be wetter. Various terms are also used in Ghana to characterize slope. In Nepal, *ghol* ('poorly-drained, wet land') is distinguished from *tar* or *tari* ('well-drained, dry land'). Aspect is also important in mountain farming systems. Therefore, farmers in Palpa in western Nepal distinguish two types of upper slope rainfed land (*bari*): *poshilo bari*, which faces the sun, and *ripyan bari* ('shadow land'). *Poshilo bari* receives sunshine for longer, and so is *ovano* ('warm') and healthy, and produces good crops if there is adequate rainfall. *Ripyan bari* receives less or no sunshine, is *chiso* ('damp and cold') and produces weak, low-yielding crops.

Temperature and moisture are often combined in farmers' descriptions of soils. In places as far apart as Ghana and Costa Rica, soil under trees is described as 'cool' – a perception connected with moisture retention (Moss et al., 2001; Stokes, 2001). In Sri Lanka, farmers with forest gardens describe plants as *seraiy* ('heating') or *sitelay* ('cooling') when referring to their competitiveness. They believe that cooling species conserve soil moisture and so compete less with neighbouring plants (Southern, 1994). Similarly, in Thailand certain ground flora species (such as *Eupatorium adenophorum*) are considered to be beneficial in jungle tea gardens because they keep soil 'cool' and 'moist' (*din yen*). Others (such as *Imperata cylindrica*) are thought to be disadvantageous because they promote 'hot soil' (*din ron*) (Preechapanya, 1996).

Jungle rubber farmers in Jambi province (Indonesia), classify soil as *tanah panas* ('hot soil') and *tanah dingin* ('cool soil'). Their assessment of a soil is based on how quickly it warms and the length of time for which it retains heat (Joshi *et al.*, 2003). A similar concept also exists among farmers in south Sumatra (Hairiah *et al.*, 2000b). Farmers associate the faster heating of *tanah panas* soils with their higher sand content. These soils are also known to be highly porous, and consequently to suffer more pronounced nutrient leaching. *Tanah dingin* soils, on the other hand, contain more organic matter and less sand: they remain relatively cool and are considered better soils for plant growth. Andean farmers also use *fria* ('cold') *and caliente* ('hot') to classify soils of different fertility; this classification corresponds (though not directly) to measured subsoil nutrient and soil humus content (van der Ploeg, 1989).

In summary, soil classification varies greatly with geographical location and, sometimes, among people with different priorities within a locality (such as men and women, or those who grow different crops). In some cases, details of soil description may reveal more generic concepts – such as that reported by Niemeijera and Mazzucato (2003) from eastern Burkina Faso – which underlie local systems of soil classification and which are common across locations and cultures. This makes it important to seek explanations for local soil classification, as only then can general patterns be discerned from local location-specific terms. Soils are described both in terms of easily appreciated physical attributes (such as colour and texture) and in relation to their agricultural function. Understanding soil classification and description and its local variations are fundamental to communicating effectively with farmers about soil fertility and belowground interactions, subjects explored in more detail in the following sections.

## 2.4 Soil Fertility

### 2.4.1 Concepts

Generally, farmers find soil fertility (primarily associated with the productive potential of land) to be a complex concept. Perception of soil fertility is often based on how well plants grow and yield in the medium. This localizes knowledge about soil fertility in terms of both the various soil types that occur (and so can be compared) in a particular locality, and the types of crops that a farmer has experience of growing, or considers to be a priority. It is common for the impacts of soil fertility on plant growth to be explained in terms of the provision of plant nutrients and water, with some understanding of how these are influenced by soil texture. But these are by no means the only locally expressed determinants of soil fertility – the role of solar radiation, weeds and soilborne pests and diseases, as well as how easy or difficult soils are to work, may also be involved.

The above can be illustrated using results obtained by ongoing research on local knowledge in bush-fallow farming systems across a range of locations in Ghana (Moss *et al.*, 2001). Here, fertile soil was described by farmers as 'land that crops do well on'. This was an aggregate concept relating to the ability of the soil to provide moisture and nutrients to plants. In literal terms, fertile soils were said to be 'strong' or 'fat' and there was an aggregated concept of soil nutrients – these were referred to as 'food' (Table 2.2). Farmers recognized changes in soil fertility over time (for example, when fertility is restored due to the accumulation of organic matter during the fallow phase), as well as spatial differences in soil fertility caused by differences in the inherent fertility of the soil due to the characteristics of the underlying parent material.

Ghanaian farmers also used the concept of soil fertility in relation to specific crops. So, for example, land that was good for cocoa was not good for rice, whereas land that was described as fertile in relation to maize cultivation was not considered adequate for cocoa. Specifically, farmers recognized the following general properties of fertile soil:

**Table 2.2.** Terminology used to describe soil fertility by farmers in five villages in the forest, transition and savannah zones of Ghana. *Twi* words are italicized. (Source: Moss *et al.*, 2001.)

| Local term | Literal meaning | Explanation |
|---|---|---|
| *Ahoōden* | 'Strength'/'power' | Fertility |
| *Seradeē* | 'Fat' | Fertility |
| *Aduane* | 'Food' | Soil nutrients |
| *Asaase ōkyene adeē* | 'Land that helps things' | Land that is good for crop growth/fertile (used only at Oda) |
| *Asaase a enyine* | 'Land that is mature or well grown' | Fertile soil found after a long fallow |
| *Asaase shesheshe* | 'Hot land' | Characteristic of infertile soil |
| *Enyunu* | 'Cool' | Characteristic of fertile soil |

- it has a high organic matter content, derived from the decomposition of a lot of vegetation;
- it is 'cool', even when the sun is shining on it;
- it is soft and friable, to the extent that a knife would easily penetrate it to some depth; and
- it harbours few weeds during the cropping phase, so that less weeding is required.

Conversely, infertile soil was described in terms of:

- having a low organic matter content;
- being hard or compact;
- being infested with invasive weeds that were difficult to control; and
- producing stunted plants, which are associated with low yields.

In the Ghanaian system, the key management options used to maintain soil fertility were the selection of crops appropriate to different types of land, and the use of fallows. It therefore follows that the qualities farmers associated with fertile soil were also those associated with soil on land that had been fallowed for a long time, and in which large amounts of organic matter had accumulated. Farmers reported that the decomposition of vegetation led to the formation of a black layer on top of the soil and that decomposed vegetation made the soil soft. They stated that this, in turn, increased water infiltration and, therefore, the subsequent availability of water to crops, and that it also reduced surface runoff and made land easier to cultivate, so increasing crop yield.

Similarly, the recognition of links between fertility, nutrient supply and soil texture is also common among farmers in Nepal (Shrestha, 2000). Soils of different textures are reported to interact differently with various factors of production. Fertile (*malilo*) soil is thought to have the capacity to absorb and retain large amounts of nutrients released from manure and make them readily available to crops, when there is adequate water. Clay-rich *garungo mato* ('heavy soil') exhibits this quality more than sandy *halka mato* ('light soil'). Farmers relate the differential water requirements of light and heavy soil to how they supply soil nutrients to crops. So, farmers rank the fertility of light and heavy soils differently, depending on rainfall, with the fertility of light soil being considered high when rainfall is moderate but low when rainfall is high. The opposite is considered to be true for heavy soil. Farmers explain this in terms of excessive rainfall washing away light soil, including the nutrients and manure it contains, which is detrimental to crop growth – typically inducing yellowing of the leaves. On the other hand, farmers perceive heavy soil as requiring a large amount of water to saturate and 'melt' it: only then will soil nutrients be available to crops. As a result of this knowledge, farmers apply more animal manure to heavy soil than to light soil, if they have sufficient manure to do so. Farmers describe light soils as 'coarse', 'granular' or 'loose', a quality that they believe both facilitates the movement of water through the soil and promotes good root growth, as root penetration and spread is easy. However, because the water retention capacity of such a soil is low, frequent rainfall or irrigation is required for good crop production.

Nepali farmers also possess knowledge about differential rates of crop germination and growth in light and heavy soils given different rainfall patterns. Light soils easily become moist, even with light rainfall: this results in early germination and the fast growth of crops. Heavy soil is hard when dry, and requires high and regular rainfall to moisten it; hence, crop germination and growth is slow in such soils if rainfall is low and irregular. This illustrates the conditional nature of much local knowledge about soils. Nepali farmers' perceptions of soil fertility are also influenced by their choice of crop. A survey in western Nepal (Joshi *et al.*, 1995a) found that farmers selected particular local varieties of rice to suit the fertility of a given piece of land: *Guruda* was planted in low-fertility soils, *Pakhe jarneli* in medium-fertility soils and *Battisara* in high-fertility soils.

Although it is perhaps not surprising that farmers involved in intensive crop cultivation have well-developed ideas about soil fertility, there is also mounting evidence for the existence of a sophisticated understanding of soil fertility among shifting cultivators and people operating fairly extensive forest farming systems. In her seminal anthropological treatise on the Bemba of northern Zambia, Audrey Richards (1939) describes how Chitemene shifting cultivators selected new fields using a combination of vegetation and soil indicators. A similar use of indicators is described in other detailed anthropological studies undertaken in Asia, such as Conklin's (1957) description of Hanunoo shifting cultivators in the Philippines and Dove's (1985) treatment of Kantu swidden agriculture in Kalimantan.

Conklin (1957) details eight soil criteria used for site selection by the Hanunoo: moisture content, sand content, rock content, general texture, firmness, structure in the dry season, stickiness in the wet season and colour. It is evident that these broadly correspond to the previous examples of soil fertility perception used in the more intensive, sedentary farming systems of contemporary Ghana and Nepal. In ongoing research in Sumatra, Joshi *et al.* (2003) present more explanatory knowledge held by farmers operating extensive jungle rubber agroforestry systems in Jambi (Fig. 2.2). As in Ghana and Nepal, the Indonesian farmers explain that various factors influence the fertility of soil by affecting nutrient content – referred to there, as in Ghana, as 'food' for plants. Interestingly, these factors include both the direct effects of weed competition and the positive contributions made by organic matter from decomposing weeds (post weeding), as well as the effects of leaching and fertilizer application.

### 2.4.2 Managing soil fertility

Farmers' manipulation of soil fertility varies in intensity (from the use of organic and inorganic fertilizer to the use of long fallows) as a result of differences in both their conceptualization of soil fertility and the sophistication of their understanding of associated ecological processes. In general terms, two concepts are widely used in the management of soil fertility. The first involves adding material thought to contain nutrients to the soil (a concept that includes fallows, where the material is grown *in situ*). The second involves selecting fertile sites in which to plant crops. Sometimes both approaches are combined, and it is common for local farming practices to involve the movement and concentration of nutrients on enriched crop fields.

Local knowledge of nutrient heterogeneity in soils and vegetation is exploited at a range of scales. At a fine scale, people recognize and exploit fertile microsites that occur in depressions or close to trees in the Sahel (Lamers and Feil, 1995). Conversely, the Akamba people of Kenya specifically fertilize areas of low fertility in their maize fields, to bring them more in line with crop performance in the rest of the field (Kiptot, 1996). At the other end of the scale, the traditional shifting cultivation used by the Bemba in Zambia involves pollarding trees in large 'outfield' areas of Miombo woodland, piling the wood and then burning it on a much smaller, ash-enriched 'infield' (Chidumayo, 1987). The ratio of outfield to infield area depends on the fertility of the site.

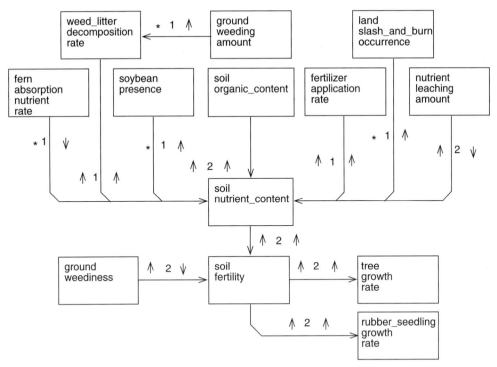

**Fig. 2.2.** Local perceptions of factors associated with soil fertility in jungle rubber agroforests in Jambi, Indonesia (output from the AKT5 software package). Nodes (boxes) represent named attributes of components of the agroecosystem. Arcs represent a causal influence by a node on another node (at the arrowhead of the arc), as specified by the arrows and numeral. Small arrows represent the direction of change of values of the independent attribute (left-hand side) and the impacted attribute (right-hand side): ↑ signifies an increase, ↓ signifies a decrease. Numerals specify 'symmetry': for example, '2' signifies that if ↑x causes ↑y, then ↓x causes ↓y; this does not apply for relationships marked '1'. The symbol '*' indicates a value other than an increase or decrease.

It is also common for farmers to use livestock to move nutrients across landscapes. This occurs in a wide range of marginal agricultural environments in Africa, where farmers improve the fertility of favoured fields by using livestock to collect and concentrate nutrients from extensive rangeland areas (Hilhorst and Muchena, 2000). Similarly, the fertility of privately owned farmland in the mid-hills of Nepal is maintained by nutrient transfers from common grazing and forest areas. It has been estimated that three to six times as much common land is required to support an equivalent area of cropland, because, whereas the cropland is enriched, the common land is depleted (Wyatt-Smith, 1982).

### 2.4.3 Application of nutrients

In the mid-hills of Nepal, the major influences farmers have on the fertility of rainfed cultivated land (*bari*) are the application of manure and their control of the influence trees have through both direct competition with crops and the fertilizing effect (*rukhopan*) of their leaf litter (Fig. 2.3). Here, leaf *rukhopan* refers both to speed of decomposition and the amount of nutrients released from leaf litter, with more *rukho* leaves being less useful as fertilizer. Similar rankings of tree species, in terms of the fertilizing power of their leaf litter, are found in other systems. For example, Akamba farmers in Katangi (Kenya) classify the litter of

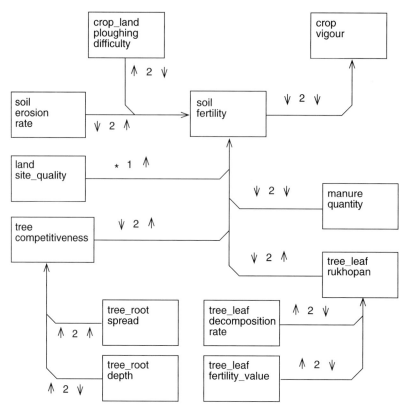

**Fig. 2.3.** Diagrammatic representation of Nepalese farmers' knowledge of factors influencing the fertility of *bari* (rainfed cultivated land) land soils (output from the AKT5 software package). Legend as for Fig. 2.2. *Rukhopan* is a concept involving an inverse contribution to soil fertility.

different tree species in terms of usefulness as a fertilizer. So, whilst at one end of the scale, mango (*Mangifera indica*) leaf litter is not considered to be a good green manure, the litter of *Balanites aegyptiaca* is highly valued as a fertilizer (Kiptot, 1996). Their knowledge of the comparative fertilizing power of leaf litters is derived from deliberate trials. Such trials generally entail the farmer applying litter to patches of crop fields that are lower yielding than the rest of the field. The effects had on fertility are then assessed in terms of the extent to which yield in those areas improve in subsequent years. By applying different leaf litters separately, these farmers are able to assess their relative performance.

In the mid-hills of Nepal, the most significant soil fertility management intervention undertaken by farmers is the use of animal manure, often mixed with crop residues and referred to as *Gobar mal*. Farmers are aware that animal manure is the main source of soil nutrients available to crops on the rainfed cultivated land (*bari*), whereas lower slope (*khet*) land benefits from the deposition of soil washed down from the upper slopes and is left fallow to regenerate its fertility. Farmers, therefore, consider quality and quantity of animal manure to be the key determinants of the fertility status of their *bari* soil. Such farmers know the value of well-decomposed animal manure, and regard such manure as being of a higher quality than partially decomposed or fresh manure. They explain that a well-decomposed manure is soft and friable, mixes easily with soil and provides nutrients immediately, whereas undecomposed manure does not mix well with soil and becomes caked when dry. This causes insect pest infestation of the soil to increase. They do not, however, appear to appreciate other ameliorating effects animal manure can have on soil (such as pH buffering and the raising of its cation exchange capacity). In one location (Palpa), farmers were also concerned about the speed at which crop residues left on their *bari* land decomposed. For example, wheat roots were considered to be very tough, taking up to a year to fully decompose and making soil *rukho* ('infertile') in the process. When left in the field after the harvest, these roots were thought to absorb water and make soil *ovano* ('dry'), which affects the growth of the subsequent crop. To cope with this, farmers have changed their harvesting practice. Now, rather than cutting stems at the base, they uproot the whole plant.

Nepali farmers who use chemical fertilizers, mainly those in the *Terai* (plains) and low hills, consider them to have both positive and negative effects (Tamang, 1992; Joshi *et al.*, 1995b). Continuous application of purely inorganic fertilizers was thought to harden soil and cause clod formation on ploughing. This, in turn, reduced crop germination and growth and increased the labour required to prepare the land. These farmers also perceive that the use of chemical fertilizers promotes heavy extraction of the residual nutrients from the soil, increasing the amount of fertilizer needed each year. In the long term, farmers think that such fertilizers gradually make soil *rukho*. For this reason, although they recognize that immediate increases in crop yields result from the use of inorganic fertilizer, the majority of hill farmers prefer not to apply inorganic fertilizers if they can afford to buy sufficient organic manure. Farmers are also aware that the negative effects they perceive can be reduced if chemical fertilizers are used in combination with animal manure (Tamang, 1992; Shrestha, 2000). Scientific investigations have confirmed that the use of organic manure and chemical fertilizers in combination better maintains the long-term chemical and physical properties of hill soils at optimal levels than the use of chemical fertilizers alone (Subedi and Gurung, 1995; Tripathi, 1996).

Farmers in many rural communities apply farmyard manure to improve soil fertility. However, careful management of manure and other fertilizers is essential as 'over-fertilization' can result in unwanted consequences. To the northeast of Nazret, in the Rift Valley of Ethiopia, farmers largely apply manure to maize fields but not to teff (*Eragrostis tef*), which has a lower yield potential and tends to lodge and suffer weed infestation if highly fertilized (Fujisaka, 1997). Ethiopian farmers also believe that rapid vegetative growth increases susceptibility to drought, and therefore limit their application of nitrogen fertilizer.

### 2.4.4 Soil 'coolness' and fertility

Farmers also relate soil temperature to fertility. As discussed above, 'coolness' and 'moisture content' are often conflated by farmers when describing soil. Collectors of non-timber forest products in Ghana identify, very straightforwardly, solar radiation and fire as the major forces that heat the soil, and soil moisture and shade as the major cooling influences (Fig. 2.4). They specifically distinguish the heating of soil gravel from that of bulk soil, and state that hot soil has a series of specific detrimental effects on the growth and yield of their plant of commercial interest, *Thaumatococcus daniellii*, by forcing water downwards in the soil profile, thus drying out the surface soil. In the midhills of Nepal, where a cool, dry winter is followed by a warm, wet growing season, the perception of soil temperature is more complicated. There, it is believed that soils can be either 'too cold' or 'too hot' for optimal crop growth. It is therefore thought, for example, that crop germination and growth will only be good when the soil has received adequate heat during the growing season. However, these farmers do not attempt to carry their explanation of this beyond their perception that seed germination and root and shoot growth are inhibited by a 'cold soil'.

Nepali farmers are aware that soil temperature is influenced by altitude, shade and aspect. These farmers also state the less obvious fact that soil temperature can be

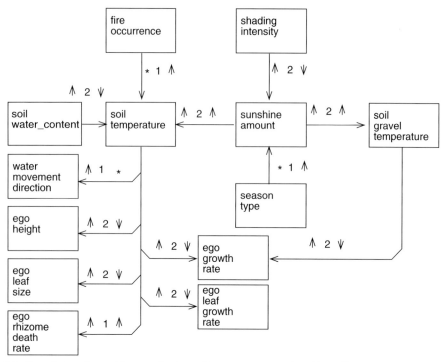

**Fig. 2.4.** Knowledge of causal factors affecting soil temperature articulated by collectors and cultivators of *ego* (*Thaumatococcus daniellii*) in the western region of Ghana (output from the AKT5 software package). Legend as for Fig. 2.2. Adapted from Waliszewski and Sinclair (2004).

increased or maintained in different ways, including the application of manure, ploughing or burning. High-altitude farmers in Landruk (western Nepal) explain that, in order to heat up the soil, they have to apply higher doses of animal manure than are required at lower altitudes. They state that animal manure emits heat and so maintains soil temperature. Another method used by local farmers to heat their soil is *mato pakaune* ('ripening of soil'), which is achieved by exposing subsurface soil to the sun by ploughing. Burning of trash, largely undertaken for hygienic reasons or to obtain ash, is also thought to heat the soil. Ploughing and burning are also perceived to kill harmful soil insect pests.

In summary, soil fertility as a concept held by farmers is generally both aggregated and pragmatic, and is conceived in terms of the extent to which soil supports the growth and yield of crops. This means that farmers' concepts of soil fertility are strongly conditioned by what crops are grown and may, in areas dominated by a single crop, only explain the suitability of soils for this crop. Most local knowledge systems disaggregate fertility into at least two components, related to the ability of soils to supply nutrients and moisture (as conditioned by soil texture and modified by organic matter content). In many systems, particularly those in which leaving land fallow is a strategy used to regenerate fertility, weed and soil pest burdens may be as important a determinant of fertility as the soil's nutrient and water contents. Farmers use organic material, including manure, compost and leaf litter, to improve soil fertility, and are generally aware of the different overall nutrient content and decomposition patterns associated with the different materials at their disposal. Farmers often have a good understanding of fine-scale variations in soil fertility, and either exploit fertile patches or practice a kind of precision farming, which targets fertilizer applications at less fertile patches. In many agricultural systems nutrients are concentrated on relatively small areas of enriched cropland, while other, often commonly owned, areas are depleted. Soil moisture and temperature are often confounded in farmers' explanations of the effects that sunlight, shading and fire have on soil fertility, and may, in some cases, also be related to nutrient release and cultivation.

## 2.5 Below-ground Interactions

It is evident from the preceding discussion of local knowledge about soil fertility that farmers possess explanatory knowledge about the processes by which plants affect soil and so compete with other plants growing close by and/or complement or facilitate their growth. It is also clear that farmers have knowledge of both individual interactive effects, such as uptake and release of nutrients and water (akin to ecophysiology), and the net outcome that results from the interactive effects on plant survival and growth when species are grown together (akin to classical plant ecology) (Anderson *et al.*, 1993).

A common problem in scientific discussions of local knowledge involves the need to come to terms with the tendency of farmers to think holistically. Such thinking often results in concepts that aggregate processes that straddle above- and below-ground interactions (rather than in concepts that view atmospheric and soil processes as discrete). The *rukhopan–malilopan* concepts used in Nepal (discussed above in relation to soil classification and fertility as applied to tree–crop–soil interactions) are a good illustration of this. As with many other concepts held by Nepali farmers, the local terminology used to describe this concept pivots on an axis, which, in this case, uses *malilo* to represent trees that enhance fertility and which do not compete much with crops, and *rukho* to refer to trees that reduce fertility and which compete with crops. The process of tree–crop competition is generally referred to as *rukhopan*, which emphasizes the competitive element. However, the same set of processes may be referred to as *malilopan* when the role trees play in soil improvement is being considered. The key processes Nepali farmers describe as being involved in tree–crop–soil interactions (Fig. 2.5) include the effects of: (i) water dripping from tree leaves (*tapkan*), which causes splash erosion of the soil, and so encompasses both above- and below-ground mechanisms; (ii) root competition for nutrients; and (iii) nutrient contributions from decomposing leaf litter. The net effect of these interactions, in terms of whether a particular tree has a positive or negative effect on crops and soil, depends on various attributes of the tree. In an extensive survey in eastern Nepal (Joshi and Sinclair, 1997), when asked which tree attributes affected how *rukho* a tree was, most farmers (> 70%) cited leaf decomposition rate and various root system attributes (density, depth and horizontal spread). However, almost half also cited leaf drip and/or shading effects as being above-ground influences by the tree canopy on soil and crop yield that they were unable to disaggregate (Fig. 2.6).

Data on the distribution of farmer's knowledge concerned with tree attributes that affect the degree to which they compete with crops across eastern Nepal not only illustrates the complexity and aggregation of farmers' knowledge about processes affecting interactions, but also the persistence of the *rukhopan* and *tapkan* concepts on a regional basis. In the same study (Joshi, 1998), a series of key concepts relating to tree–crop interactions and the nutritive value of tree fodder were found to be used widely throughout the Himalayan Hindu Kush region, although the tree attributes thought to affect these varied with site in accordance with how dependent people were on species exhibiting different attributes. Despite earlier emphasis placed on the differences that exist in the knowledge of people within sites (Rusten and Gold, 1991), few differences were found to occur as a result of gender or wealth. Similar knowledge in a community does not, however, imply similar priorities: the earlier studies that found differences among men and women used ranking procedures that may have conflated knowledge and priority.

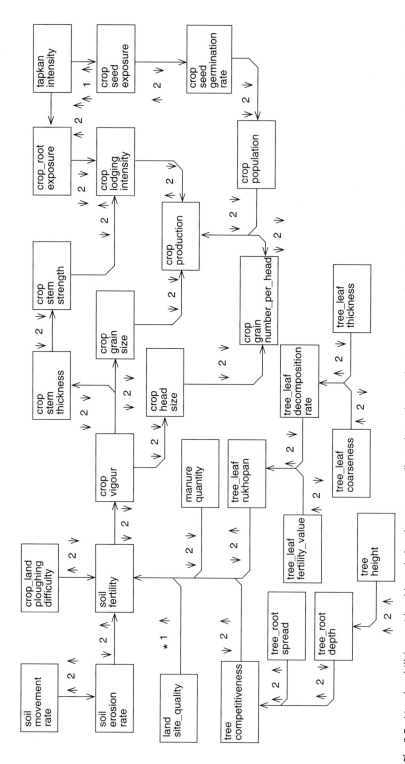

**Fig. 2.5.** Nepalese hill farmers' local knowledge about cause–effect relationships at the tree–crop interface (output from the AKT5 software package). Legend as for Fig. 2.2. *Rukhopan* is a concept involving an inverse contribution to soil fertility.

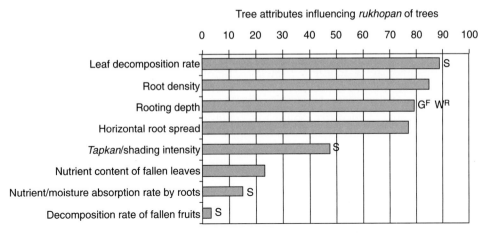

**Fig. 2.6.** Mean percentage of farmers mentioning tree attributes that affect the *rukhopan* of trees (a negative effect on soil fertility); data from interviews with a random sample of 221 farmers, stratified according to site, gender and wealth, in the eastern mid-hills of Nepal. The four sites differed in terms of forest access and remoteness. Significant differences due to site are marked 'S', due to gender are marked 'G' (superscript indicates which gender was higher) and due to wealth are marked 'W' (superscript 'R' denotes that wealthier farmers mentioned this attribute more often than poorer farmers). Adapted from Joshi (1998).

There are many other similar examples of local explanatory knowledge of plant interactions mediated via the soil. Rubber farmers in Jambi (Indonesia) use the concept of the relative competitiveness of rubber and non-rubber tree species, as well as other ground vegetation, to express the comparative advantage obtained when different species are grown together (Joshi *et al.*, 2003). Rubber is known to be a weak competitor: most other vegetation and tree species in the vicinity have a negative impact on it. Farmers recognized that 'competition' straddles both above-ground and below-ground resources, and were explicit about aerial competition being primarily for sunlight, whereas below-ground competition was perceived as occurring for both plant nutrients and moisture. As in Nepal, below-ground competitiveness was thought to be influenced by the morphology and activity of the root system. Tree and weed species exhibiting dense rooting in surface soil were considered to be more competitive, and hence were thought to affect rubber seedlings more severely than deep and sparse rooting species. These considerations were further modified by the health and state of the root systems of a given species.

So, plants under stress (such as transplanted rubber trees that have had their root system pruned drastically before transplanting) were known to be very poor competitors.

In northern Thailand, farmers engaging in the enrichment planting of tea in thinned forest deliberately plant close to particular forest trees, which they believe 'transfer' water and/or nutrients to the tea (Preechapanya, 1996). Similarly, in Ghana, farmers with multistrata cocoa agroforests believe that the soil around some forest trees retained in their fields remains moist, and are able to rank trees in terms of their water relations (Saunders, 2002). Both Thai and Ghanaian farmers cite water exudation from the cut stems of some tree species as evidence of their water-yielding properties. Until recently, such explanations were considered fanciful by scientists. However, mechanisms for the transfer of water and nutrients among plants are now the subject of active research. There is, for example, increasing evidence of hydraulic lift (see Chapter 9, this volume), in which water is taken up by trees from moist soil at depth and exuded into dry soil near the surface, where it may facilitate nutrient uptake both by the tree itself and by other plants in its vicinity (Caldwell and Richards, 1989). There is also evidence of mycorrhizal connections (see Chapter 14, this volume) between plants of different species, through which nutrients may be transferred (Finlay and Read, 1986).

Farmers' explanations of below-ground interactions are particularly intriguing because of the amount of interpretation involved, because such interactions can only be directly observed to a limited degree. It is noteworthy that, based on their observation of water exudation from cut stems, coupled with their observation of a tree's effect on soil moisture and the growth of other plants around it, the reasoning of farmers as far apart, both geographically and culturally, as Ghana and Thailand is similar as regards the impacts trees may have on soil moisture. Researchers have identified less detailed explanatory local knowledge concerning below-ground interactions than they have concerning above-ground interactions. This is partly due to the limited potential for observing below-ground interactions, and farmers themselves are often aware of these limits in terms of observation (Thapa et al., 1995). This may create important complementarity between scientific and local knowledge systems. For example, ongoing research in the western mid-hills of Nepal (Shrestha, 2000) has found that, although farmers were well aware of nutrient and soil loss through surface runoff, they were oblivious to the substantial nutrient losses caused by leaching (see Chapter 10), which have been measured by scientists (Gardner et al., 2000) but are not visible to farmers. This is particularly pertinent here, since technological interventions that reduce runoff are likely to increase leaching. It remains to be seen whether knowledge of leaching derived from scientific research will affect the farmers' approach to nutrient management. Farmers have already influenced experimental procedure in this research, however, by pointing out that lack of full cultivation of runoff measurement plots, which were too small for oxen to plough, was likely to make them unrepresentative of the local situation. Removing plot dividers to allow normal cultivation and then replacing them remedied this. Although the opacity of a medium often determines the limits of farmers' observations, other senses may also be involved. For example, farmers in western Nepal have a crude understanding of the process of nitrogen volatilization, in as much as they are aware that gaseous losses of nutrients occur from surface soil rich in manure or compost.

In summary, farmers often have mechanistic explanations for the way interactions affect plants and soil. This is particularly true in the case of mixed-cropping systems, where plants of very different stature, morphology and duration (such as trees and annual crops) are grown in close proximity. Farmers' concepts of plant interactions generally aggregate above- and below-ground processes, but are also often associated with underlying explanations of mechanisms that allow soil and atmospheric processes to be distinguished. The degree of detail apparent in farmers' knowledge of below-ground interactions is severely limited by their ability to observe soil processes. However, even though considerable interpretation is required to move from what can be observed to an explanation of how interactions happen, remarkable regularity exists with regard to local concepts of tree–crop interactions in different geographic and cultural contexts that share similar agroecologies. There is, generally, considerable complementarity between local and scientific knowledge of below-ground processes. Thus, there exists scope to build upon the rudimentary conceptual frameworks for understanding interactions farmers already possess, by introducing them to new knowledge obtained through research.

## 2.6 Implications

In the preceding discussion of local knowledge regarding soil classification and fertility and interactions mediated by the soil, we have established that farmers often have a sophisticated understanding of such issues, based largely on their own observations. Although notions of the description, classification and fertility of soils are heavily localized, underlying explanations of interactive processes can be generalized – although local knowledge tends to be both aggregated and limited by the methods of observation available to farmers. This means that knowledge may be specific to the soil and crop types within a locality, although some of the underlying explanatory knowledge about plant interactions may be open to wider extrapolation. In this final section, we show how both the considerable scope of local knowledge and its limitations are important to the participatory design of interventions that may both improve the productivity and sustainability of farming systems and facilitate interaction between researchers, extension staff and farmers.

Whilst studying local knowledge may be interesting in itself, the research reported here has been driven by development imperatives. A key criticism, from anthropological quarters, of the knowledge-based-systems approach advocated here is that it seeks local knowledge on a utilitarian basis (Sinclair and Walker, 1999). Indeed, a key criterion for the inclusion of items in a knowledge base is that they are useful, in as much as they could conceivably be used, in some form of reasoning process, to answer a question about the subject of the knowledge base. How then can local knowledge of soils be useful in agricultural development? The answer to this question can be broken down into three aspects: building on local practice, recognizing the sophistication of local knowledge and realizing its limitations. These aspects are cemented by the integrating principle of effective communication and empowerment. Each aspect and principle is discussed below.

## 2.7 Building on Local Practice

In some circumstances, interventions that build on local practice to improve soil management will stand a far higher chance of adoption by local farmers than entirely new technologies. It also makes sense to look for local solutions, which may be far less costly and risky than external introductions, before embarking on expensive research and extension. Indeed, it is often through understanding why farmers are not already employing locally known techniques for more sustainable soil use that we are able to identify key constraints within the system. Farmers often take actions that they know compromise sustainability, because they trade the negative impacts off against either the positive gains from the practice or the costs of taking alternative courses of action. For example, although hill farmers in Nepal know that large-leaved trees on crop-terrace risers promote splash erosion of soil, they still grow them, because the immediate benefits they provide in terms of fodder outweigh their negative effects in terms of erosion (Thapa et al., 1995). So, a comparison of local knowledge and practice is a powerful tool for identifying leverage points for research into areas in which farmers know they are making a trade-off. Farmers are, therefore, likely to be receptive to solutions such as, in this case, the use of smaller-leaved trees with similar fodder attributes to the currently grown large-leaved ones that promote soil erosion. Such analyses can also be important in distinguishing between requirements for extension and research. There is clearly no point in telling the farmers what they already know. To continue the present example, Nepali farmers already know that 'large-leaved trees cause soil erosion'. So, they require research, which they are unable to conduct themselves, that will address both their fodder needs and the need to reduce soil erosion.

It is also possible that local practices exist that can be built upon to address farmers' needs. For example, jungle rubber research in Indonesia identified the local practice known as *sisipan*, which involves rejuvenating rubber plots by gap replanting, instead of by slashing, burning and replanting at the whole-field level (Joshi et al., 2003). Encouraging this practice could have important impacts on sustainability, with respect to the maintenance of soil fertility, biodiversity and watershed functions over the extensive area (in Sumatra and northern Borneo) covered by jungle rubber – estimated to be around 3 million ha (Gouyon et al., 1993). It was evident that, as with many smallholder farming practices, use of *sisipan* was a response contingent on specific circumstances (such as a lack of capital and the risk posed by vertebrate pests in new plantations) rather than a one-off decision. So, a farmer might interplant new rubber seedlings in a mature jungle rubber stand for some years (*sisipan*) until he or she has sufficient resources to opt for the slash-and-burn technique. The key to making gap-replanting a more attractive option to farmers than slash and burn lies in improving the productivity of the gap-rejuvenated rubber. This requires a method for establishing high-yielding rubber clones in the shaded and competitive environment of a gap in a mature jungle rubber stand. Progress is being made in research that aims to incorporate higher yielding rubber clones into stands by grafting clonal buds on to seedlings planted in gaps. This is an example of how local knowledge systems can be explored and then combined with scientific research to generate a sustainable technology built on local practice.

## 2.8 Recognizing the Sophistication of Local Knowledge

The existence of reasonably sophisticated local explanations about soil processes also has profound implications for what research should be considered relevant to farmers. Where farmers have a detailed understanding of tree–crop interactions (such as in the Nepalese hills), fundamental research undertaken on mechanisms of interaction will clearly be perceived as relevant by farmers and thus will be easier to communicate to them. What is perhaps astounding is not that farmers have been found to hold sophisticated knowledge about below-ground interactions but that, until recently, researchers have not appreciated this. This has led to the erroneous assumption that adaptive research is more relevant to farmers than more fundamental research. That the opposite may be true is suggested by mounting evidence of both a high degree of sophistication in the local understanding of interactions and farmer experimentation.

Farmers are probably better able than researchers to conduct adaptive research. However, it is difficult for them to tackle more fundamental research issues, because of limits imposed both by the observational techniques available to them and the extent to which they can vary the environment – not least because they have to obtain a living from that environment while, at the same time, conducting their research. This realization affects both what type of research is considered useful in support of farmer innovation and the form in which research results are communicated to farmers. Adaptive research tends to lead to prescriptive technology packages, whereas farmers may actually want flexible new knowledge and components that they can adapt to their needs. This requires a shift away from 'extension of prescriptions' towards 'extension of principles'. Enhancing the local knowledge system, through new research identified via analyses of the local knowledge initially held, may build capacity more generally. A richer knowledge system may reduce vulnerability, by ensuring that local communities are better able to cope with any new stresses and problems, including ones that have not been specifically anticipated.

## 2.9 Realizing the Limits of Local Knowledge

Despite growing interest in, and recognition of, local knowledge in research and development initiatives, it is important not to romanticize it. This is particularly true with respect to soil processes, since it is evident that the observational limits imposed by the nature of the soil medium results in severe restrictions in terms of what farmers can see and hence understand from their own experience. This makes scientific knowledge and the research that generates it a potentially powerful tool for use in assisting farmers to manage soils more sustainably. It is clear that there is much that farmers still need to know to improve their livelihoods and that there are significant contributions that science can make. For example, as alluded to earlier, several years of research at the Agricultural Research Station at Lumle (western Nepal), undertaken in conjunction with the Queen Mary and Westfield College, University of London (UK), has revealed that nutrient losses through leaching are much greater than had previously been thought to be the case (Gardner et al., 2000). Ongoing participatory research on soil and water conservation in west Nepal indicates that, although they know a lot about surface processes (runoff), hill farmers have little or no understanding of leaching. Sharing of knowledge about leaching losses with farmers (see Chapters 6 and 10) has motivated them to experiment with hedgerow-planting deep-rooted crops, to trap and recycle the leached nutrients, a technique that had previously been very difficult to convince them to adopt. Similarly, recently conducted research at the Rubber Research Institute of Sri Lanka has revealed that not only can banana be intercropped with rubber to provide early returns before the rubber trees can be tapped, but that rubber yields are higher under intercropping than under monoculture (Rodrigo et al., 2001). Now, further research, involving both the evaluation of local knowledge and continued agronomic trials (on-station and on-farm trials), is looking at both a wider range of intercropping options and the requirements for effective extension of rubber intercropping technology to improve the livelihoods of smallholder farmers (Senevirathna, 2001).

## 2.10 Communication and Empowerment

Effective communication is a prerequisite for participatory research and the effective extension of new technical interventions. It is clear that farmers' knowledge about soils is often localized, in terms of being aggregated, with regard to the soil and crop types found in their vicinity or of particular importance to them. This makes effective communication a far from trivial need, since a one-to-one correspondence is unlikely to exist between scientific terms and the terms used by farmers. Conventionally, rather than learning and using local terminology when communicating with farmers, researchers and extension staff have expected farmers to learn the scientific nomenclature and concepts encapsulated in the recommendations and technology packages extended to them. Respecting local knowledge (see Box 2.2 for some notes on ethics and intellectual property rights issues related to local knowledge), by taking the trouble to learn about it, can be an important part of developing a productive participatory relationship with a local community, and may help to empower local articulation of research and extension needs, as well as providing the 'tools' for understanding what has been articulated. The recent identification of concepts that are common across large regional domains (such as *rukhopan–malilopan* in Nepal) and the existence of similar knowledge in culturally and geographically different places that share agroecological circumstances (such as the common understanding of water relations of trees in Thailand and Ghana) suggests that learning and using farmer concepts may not be as daunting as it might first appear. The existence of some degree of regularity in farmer knowledge across cultures allows the use of frameworks for knowledge acquisition, thus speeding up the process of gaining familiarity with the knowledge system in new localities.

Research on local knowledge is in an active phase: a key area for investigation is the need to explore how universal farmers' knowledge is. A number of studies point to regularities in knowledge across cultures, as mentioned earlier with respect to tree water relations in Thailand and Ghana and the knowledge of 'local theories of soils' among farmers in eastern Burkina Faso (Niemeijera and Mazzucato, 2003). Regularities are also evident with respect to tree fodder evaluation in Nepal and Kenya (Thorne et al., 1999; Roothaert and Franzel, 2001). A more extensive test of the hypothesis that farmers in similar agroecological circumstances develop similar knowledge is required and, if proven, should pave the way for the more general use of local knowledge in developing the research and extension agenda and in communicating with rural people.

> **Box 2.2.** Local knowledge: respect, ethics and intellectual property rights.
>
> Exploration and use of local knowledge has received much attention with regard to the debated subject of ethics and intellectual property rights (IPR). Although this applies mostly to unique knowledge held by indigenous communities and individuals (particularly knowledge about medicinal plants and local germplasm), the need to respect, value and acknowledge sources of local knowledge should not be underestimated in any local knowledge study. Here are some guidelines, adapted from norms developed for researchers and interviewers by the Alternatives to Slash-and-Burn (ASB) programme at the World Agroforestry Centre (ICRAF).
>
> **Principles:**
> 1. Respect local people's rights to privacy, dignity, safety and convenience.
> 2. Respect and follow local rules and culture.
> 3. Ensure no harm is caused to the interviewees and their community as a result of interviews and published material.
> 4. Be honest and accurate when representing your findings – this includes the use of quotations and photographs.
> 5. Do not make promises that cannot or will not be fulfilled.
> 6. Obtain verbal consent from subject regarding publication of photographs and/or quotations.
>
> **Interview guidelines:**
> Where a translator is required, the translator should be made aware of this procedure and should comply with ethical standards.
>
> *Pre-interview*
> 1. Make an appointment with the subject for a time and place that are suitable for him/her.
> 2. Be on time for the interview – but be prepared to wait for the interviewee.
> 3. Present the interview's objectives in a clear and transparent manner, and explain the likely output of the interview (such as publications).
> 4. Ensure the subject that any information they wish to keep confidential will be withheld from publication. The interviewer must take particular care when subjects are involved in activities that could be construed as illegal (e.g. timber harvesting on State land or cocoa growing).
>
> *Post-interview*
> 1. Formally acknowledge the subject, and all field staff assisting in the interview, in the list of contributors to the publication (except as per point 4 above).
> 2. Where possible, submit a draft copy of the publication, in an accessible form (i.e. translated), to the subject for his/her review and approval prior to final publication.
> 3. Provide a final copy of publication to the subject and to any institutions, communities, and/or individuals of the subject's choice.

Another area of active research surrounds local terminology. Although it is fairly easy to pick up qualitative terms for attributes of plants and soils, in some cases farmers evidently have knowledge of sufficient detail to be represented quantitatively. This has been shown, for example, with respect to Nepali farmers' knowledge about the nutritive value of tree fodder (Thorne *et al.*, 1999). Expanding rapid knowledge acquisition methods to incorporate the capture of any local understanding of continuous variables is required in order to explore this further. Perhaps most importantly, further research is urgently required on the impact of exchanges of knowledge between researchers and farmers, and also on the farmers' capacity to cope with changing environmental and market conditions. Greater understanding of these two issues will increase the benefits gained from working with local knowledge, in terms of sustaining rural livelihoods.

**Conclusions**

**1.** Based largely on their own observations, farmers generally have a sophisticated understanding of soil properties, extant local variations and a utilitarian basis for classifying their soils. Farmers also have an explanatory understanding of soil fertility and interactions mediated by plants.

**2.** While notions of the description, classification and fertility of soils are heavily localized, underlying explanations of interactive processes may be of a more general and cross-cultural nature.

**3.** A better understanding by researchers and extension agents of the scope of local knowledge is important to the participatory design of interventions that might improve productivity and sustainability of farming systems, as well as in facilitating interaction with farmers.

**Future research needs**

**1.** How universal is farmers' explanatory ecological knowledge across cultural backgrounds and ecological zones? A more extensive test of the hypothesis that farmers in similar agroecological circumstances develop similar knowledge can pave the way for more general use of local knowledge in developing the research and extension agenda and communicating with rural people.

**2.** Can the knowledge acquisition methods used be modified to incorporate continuous variables into their local scales of measurement, in addition to the qualitative statements that can already be captured?

**3.** How does the ecological knowledge currently held by farmers interact with their capacity to sustain rural livelihoods and cope with changing environmental and market conditions?

# 3 Models of Below-ground Interactions: Their Validity, Applicability and Beneficiaries

Robin Matthews, Meine van Noordwijk, Arjan J. Gijsman and Georg Cadisch

---

**Key questions**

1. What lessons can be learnt from modelling mixed-species and monocrop systems?
2. What models are currently available in the public domain?
3. How valid and reliable are existing models for complex agroecosystems?
4. Who is likely to use and benefit from explicit models and what contribution can they make?

## 3.1 Introduction

We all use models, to such an extent that we may not even think of them as models at all. Each one of us carries around in our mind a mental model of the way we perceive the world to work – this model is an abstraction of what we call 'reality', although philosophers would argue whether an objective reality even exists. We are also surrounded by visual models, such as maps or pictures, and are likely to use abstract arithmetical or algebraic models in our daily activities. As children, we probably played with models of cars or people. Computer simulation models are just another type of model, consisting of abstract mathematical representations of processes occurring in nature. The feature common to all these different types of models is their ability to provide a way of understanding the world around us, allowing us to interact with it.

Since the 1960s, many simulation models have been constructed that describe the way different crops grow and develop in relation to their physical environment. The ways in which these models have been applied in relation to tropical agriculture have recently been reviewed by Matthews and Stephens (2002). So far, greatest use has been made of such models by the research community as tools for organizing knowledge gained in experimentation. Simulation models are often also put forward as potential tools for decision support; certainly previous experience has shown that their use in this way has had major impacts in the areas of irrigation scheduling and pest management, in that they have changed the growers' way of thinking (see Cox, 1996). However, the original aim that underlies their use as operational decision support systems (DSSs) has not always been achieved – as soon as farmers learn the optimal management regime for their crops, they have no further use for an operational DSS. Models also have a useful role to play as tools in education, both as aids to learning the principles of crop and soil management, and also in helping students to develop a 'systems' way of thinking, to enable them to appreciate that their speciality is part of a larger system (Graves *et al.*, 2002).

The number of models that describe processes in mixed-species cropping systems is substantially lower than the number of models that describe monocrop systems, partly because of their increased complexity, and partly because modelling efforts have focused on developed agriculture where monocrops are the norm. The productivity of mixed-species systems is the result of the many interactions between the different plant species in the system and their environment. If our aim is to increase the productivity or stability of such systems, it is logical to, first, understand and quantify these interactions, particularly those for water and nutrients and, secondly, to use this understanding to make changes to a particular system to achieve our aims. For example, agroforestry is not beneficial in all situations, and competition between trees and crops can result in reduced crop yields. Moreover, the long growing period of trees means that it is time-consuming, and hence expensive, to find out whether a particular system is likely to be successful or not. Models provide a way of evaluating the likelihood of success of these changes, both before the real system is interfered with and within a reasonable timeframe.

In other chapters of this book, the current state of our knowledge of below-ground interactions in mixed-species production systems has been described and, in some cases (e.g. Chapter 10), this knowledge has been incorporated into simulation models. Drawing on lessons learnt from the modelling of both mixed-species and monocrop systems, in this chapter we discuss some of these models and consider how valid and reliable they are, who is likely to use and benefit from them, and what contribution such models can make.

## 3.2 Models Incorporating Below-ground Interactions

Various models that incorporate simultaneous below-ground interactions between different plant species, and whose descriptions have been published in the literature, are given in Table 3.1. We have purposely not included models in which the below-ground interactions are temporal in nature (e.g. crop rotations), as any two or more crop models with soils components could, in principle, be run sequentially in order to simulate these types of system. Instead, we have focused on models that have addressed direct below-ground interactions between two or more species. The list is not intended to be exhaustive, but is rather intended to show some of the types of models available. These are discussed below.

### 3.2.1 SCUAF

SCUAF (Soil Changes Under AgroForestry v4.0, Young *et al.*, 1998) is a nutrient-cycling model with an annual timestep, and is used to predict medium-term changes (10–20 years) in soil properties under specified agriculture, agroforestry and forestry systems within given environments. The model includes soil erosion, soil organic matter, and nitrogen and phosphorus cycling processes, as well as competition for nutrients between trees and crops. Although plant growth is included, this is specified by the user as an input in the form of annual increments in biomass. Water uptake and use is not simulated, due to the low resolution of the timestep. An example of this model's use is the analysis made of the economics of hedgerow intercropping in the Philippines (Nelson *et al.*, 1997).

### 3.2.2 WaNuLCAS

WaNuLCAS (Water, Nutrient, Light Capture in Agroforestry Systems; van Noordwijk and Lusiana, 2000) simulates tree–soil–crop interactions in a range of agroforestry systems in which trees and crops overlap in

Table 3.1. List of some models incorporating simultaneous interspecific below-ground interactions.

| Model name | Reference | Scale (subplot, plot, farm, region, country) | Time-step | Dimensionality (1, 2, 3) | Strengths | Known limitations | Target users | Examples of applications |
|---|---|---|---|---|---|---|---|---|
| SCUAF | (Young et al., 1998) | Plot | Yearly | 1 | Simplicity and ease of use; includes SOM, N and P dynamics | Simplistic simulation of plant growth | Researchers | Bioeconomic modelling (Nelson et al., 1997) |
| WANuLCAS | (van Noordwijk and Lusiana, 2000) | Plot, landscape | Daily | 1, 2 | Multispecies systems, interactive competition for $H_2O$, N and P | Complex; tree canopy development; no nitrification module | Researchers | Safety-net or filter efficiencies (Cadisch et al., 1997, and Chapter 10, this volume) |
| HyPAR | (Mobbs et al., 1998) | Plot | Daily | 1, 2, 3 | 3-D simulation of light interception, below-ground competition | | Researchers | Characteristics of successful agroforestry systems (Cannell et al., 1998) |
| HyCAS | (Matthews and Lawson, 1997) | Plot | Daily | 1 | Includes SOM, N and P dynamics; compatible with DSSAT databases | Homogenous canopy, not validated, not user friendly | Researchers | MSc projects (Cranfield University) |
| COMP8 | (Smethurst and Comerford, 1993b) | Root/plot | Variable | 2 | Includes P and K | | Researchers/ forest industry | Tree/weed competition (Smethurst et al., 1993) |
| WIMISA | (Mayus et al., 1998a) | Plot | Daily | 2 | 2-D compartments of soil water calculations | Does not include nutrient limitations or soil erosion; LAI development and microclimate needs improving | Researchers | Tree–crop interactions in windbreaks in Sahel (Mayus et al., 1998b) |
| APSIM | (McCown et al., 1996) | Plot | Daily | 2 | Modularity due to object-oriented techniques; components well tested | Mixed-species model not validated; source code not available | Researchers, extension agents, consultants, farmers | Cropping systems, agroforestry, windbreaks (Huth et al., 2002) |

Continued

Table 3.1. Continued.

| Model name | Reference | Scale (subplot, plot, farm, region, country) | Time-step | Dimensionality (1, 2, 3) | Strengths | Known limitations | Target users | Examples of applications |
|---|---|---|---|---|---|---|---|---|
| CROPSYS | (Caldwell and Hansen, 1993) | Plot, farm | Daily | 1 (canopies are 2-D) | Individual crop models well tested; compatible with DSSAT databases | Maximum of two species; not well validated | Researchers, educators | |
| ALMANAC | (Kiniry et al., 1992) | Plot | Daily | 1 | Process-based, but designed for practical applications | Two species are simulated sequentially in each time step | Researchers | Crop–weed interactions (Debaeke et al., 1997) |
| GAPS | (Rossiter and Riha, 1999) | Plot | Daily, some processes <daily | 1 | Modularity due to object-oriented techniques | Not widely tested | Researchers, educators | |

SOM, soil organic matter; LAI, leaf area Index; DSSAT, Decision Support System for Agrotechnology Transfer.

space and/or time. It runs on a daily timestep, and has a four-layer soil profile and four spatial zones. Within each of these layers and zones, the water, nitrogen and phosphorus balance is calculated, including uptake by a crop (or weeds) and up to three different types of trees. The model is written in the STELLA modelling environment, making it easier for users to modify parameters and add extra model components than would be the case for models developed in traditional programming languages. WaNuLCAS has been used both for teaching and research (e.g. Cadisch *et al.*, 1997; see also Chapters 9 and 10, this volume).

### 3.2.3 HyPAR

HyPAR (Mobbs *et al.*, 1998) is a combination of the Hybrid tree model (Friend *et al.*, 1997) and the PARCH crop model (Bradley and Crout, 1994), and can be used to simulate tree–crop interactions in agroforestry systems under a range of soil, climate and management conditions. The model operates on a daily timestep, and simulates the biomass production and partitioning of both the tree and the crop; water uptake and use (vertical redistribution of soil water, infiltration, drainage and soil water evaporation); and competition for nitrogen between the two species. The current version (v3.0) includes routines to represent disaggregated canopy light interception and 3-D competition for water and nutrients between the roots of trees and crops (Chapter 10, this volume). An example of its use is the analysis of the rainfall requirements for successful agroforestry systems made by Cannell *et al.* (1998).

### 3.2.4 HyCAS

The HyCAS model (Matthews and Lawson, 1997) simulates competition for resources (light, water, nitrogen and phosphorus) in tree–cassava agroforestry systems on a daily timestep. The model is based on two other models, the HYBRID tree model (Friend *et al.*, 1997) and the sole-crop GUMCAS cassava model (Matthews and Hunt, 1994), which are integrated in a way similar to that used to integrate the two different models used in the HyPAR model. The use of the GUMCAS model structure as a base provides compatibility with the DSSAT (Decision Support System for Agrotechnology Transfer) standard format for input and output files (Hunt *et al.*, 1994), and hence access to the large database of weather and soils data compiled by the IBSNAT project (International Benchmark Sites Network for Agrotechnology Transfer; Tsuji and Balas, 1993) and by other modelling groups. To our knowledge, apart from its use during initial testing and verification, the HyCAS model has not been used for any specific application.

### 3.2.5 COMP8

COMP8 (Smethurst and Comerford, 1993b) simulates, based on solute transport theory, nutrient uptake by both competing and single root systems. The model calculates the volume of soil allocated to each root system and the concentrations of solute at the root surfaces. It also allows each root system to have a different absorbing power, as experimental evidence has indicated that uptake per unit surface area of root varies between species (Smethurst and Comerford, 1993a). The model was used as a research tool to study competition for P and K between pine trees and various weeds (e.g. Smethurst *et al.*, 1993), but does not seem to have been used since then.

### 3.2.6 WIMISA

WIMISA (WIndbreak MIllet SAhel; Mayus *et al.*, 1998a) simulates crop growth as influenced by trees growing as windbreaks. The model is two-dimensional, and simulates the growth of a number of crop rows as a function of the local incident solar radiation and soil water. Soil water flow is simulated in two dimensions, to account for horizontal gradients due to different water extraction by trees and crop and horizontally varying evapotranspiration. Competition for water is expressed by distributing available soil water between trees and crop in proportion to their uptake rates in a non-competitive situation. Water uptake is calculated on the basis of root length density distribution. The model has been used to analyse experimental data from windbreak/millet experiments in Niger (Mayus et al., 1998b).

### 3.2.7 APSIM

The APSIM (Agricultural Production Systems Simulator) suite of crop simulation models began life as a collection of point-based models for sole crops and pastures (McCown et al., 1996). Some initial attempts were made to link these individual crop models together, in order to study competition for light and soil resources in intercropped plant species (Carberry et al., 1996). More recently, a tree stand module has been added (Huth et al., 2001) and, with developments in the intermodule communication procedures that allow the simulation of spatial entities, work has started on linking crop and tree modules together to simulate agroforestry systems (Huth et al., 2002). This has been done using object-oriented programming techniques, with different crop rows and soil compartments being represented by different instances of the same submodel classes. Simple rules for partitioning tree transpiration demand between soil compartments are based on the soil water supply of each compartment, and on the assumption that tree root density decreases proportionally to the square of the distance from the tree. Although the approach gives realistic predictions of the behaviour of each of the components, it has yet to be validated against observed data (Huth et al., 2002).

### 3.2.8 CROPSYS

CROPSYS (Caldwell and Hansen, 1993) is a process-level simulation model that is designed to predict the performance of multiple cropping systems across genotype, soil, weather and management combinations. The central core of the model is a soil module that simulates the basic processes of the water balance and nitrogen balance on a daily timestep continuously for long time periods. Crops come and go, and are represented by crop modules based on the CERES and SOYGRO families of crop models (maize, rice, wheat, barley, sorghum, millet and soybean). Crop processes include light interception and photosynthesis, dry matter partitioning, phenology, root system development, and growth in canopy dimensions and leaf area. When two species share the field at the same time, the model calculates competition for light and competition for water, nitrate and ammonium by soil layer. Input and output files follow the DSSAT standard format definitions. CROPSYS has been further developed into an object-oriented hierarchical framework called JanuSys (Caldwell and Fernandez, 1998).

### 3.2.9 ALMANAC

ALMANAC (Agricultural Land Management Alternatives with Numerical Assessment Criteria; Kiniry et al., 1992) is based on the EPIC model (Williams et al., 1984), and simulates plant growth, water balance and nutrient balances for two or more competing species. Competition for soil water and nutrients is based on the current rooting zone of each species and on the demand exerted by each species. If the available water in the rooting zone is less than the potential evapotranspiration, the species planted first has first access to, and will fulfil its needs from, what water is available; the second-planted species can use what remains. Among other things, ALMANAC has been used to investigate maize–soybean intercrops (Kiniry and Williams, 1993) and crop–weed interactions (Debaeke et al., 1997).

### 3.2.10 GAPS

The GAPS (General-purpose Atmosphere–Plant–Soil Simulator; Rossiter and Riha, 1999) model simulates intercrop competition for light and water between a number of species. It contains a number of different crop modules that are based on existing, published, models, and also a module for fast-growing trees. Communication between modules is handled by a simulation driver. The various crop models within GAPS have different ways of calculating the amount of water extracted from the soil. In the simplest calculation method, the amount of plant-available water in each soil layer is supplied in proportion to the crop's total root length density in that layer. In a second method, the amount of plant-available water is supplied in proportion to an exponential function that decreases with depth; in a third method, the crop equilibrates its leaf and water potentials with the soil water potential and takes up water according to the potential gradient. In each approach, the amount of available water is reduced as it is taken up by each crop in turn. GAPS is used as a research and teaching tool.

## 3.3 Validity, Reliability and Applicability

### 3.3.1 Model limitations

The models described above all have limitations that may restrict their usefulness for a particular application, partly because they are, by definition, a simplification of a complex reality, but also because our knowledge of all the processes involved is incomplete. Most models, for example, assume that roots are distributed regularly in some way – a common assumption is that root density declines exponentially with depth. In reality, this is often not the case, as roots preferentially follow cracks in the soil, and dry, infertile or compacted soil zones restrict root growth while compensatory root growth occurs elsewhere. Failure to take into account the clumping of roots will cause an overestimate of the ability of the root system to extract water (Passioura, 1983). Similarly, Comerford et al. (1994) found that K uptake was overestimated unless the root spatial pattern was included in the model. The plasticity of the root system shape and size, particularly as it is affected by the presence of roots of other species, is also not incorporated into most models. One root system may restrict the growth of another, through competition for water and nutrients, or through allelopathic influences.

Additionally, not all of the roots in a root system may be active at any one time in terms of water or nutrient uptake. Robertson et al. (1993), for example, estimated that only 11% of wheat roots were active in terms of nitrate uptake even when no nitrogen fertilizers were applied. Similarly, water-uptake models based on the consideration of plant and soil water potentials predict that water should be taken up at all depths in proportion to root length density. However, observations show that it is actually often taken up preferentially from the surface layers, even when the soil, at depth, is fully wetted and there are roots present there. It would seem that plants have some mechanism for controlling the activity of different parts of the root system over and above the physical processes that govern water and nutrient uptake.

However, it should be remembered that these limitations do not necessarily prevent a model from being used for a particular purpose, provided that the limitations are taken into account when interpreting its output.

### 3.3.2 Do models reflect conditions in farmers' fields?

Apart from the limitations of below-ground models that were discussed above, there exists the question of whether models adequately reflect conditions in farmers' fields in developing countries. Most models are developed using data from controlled experiments. Indeed, this is an essential part of the research process – in order to understand the influence of a particular factor, other factors must be held constant. However, the real situation in agroecosystems is more complex, and often involves a large number of factors all interacting together. The challenge for modellers, therefore, is to be able to capture this complexity in their models.

Reviewing the SARP (Systems Analysis for Rice Production) project at the International Rice Research Institute, Mutsaers and Wang (1999) noted that most of the crop simulation models used were originally developed for monocrops grown under highly uniform conditions (resulting from high levels of external inputs) and that the focus was on maximizing yield. As such, they were not well equipped to deal with cropping systems in lesser developed countries (LDCs), the characteristics of which include

- limited control by the farmer of factors determining production (including water);
- management practices that are often aimed at risk reduction rather than yield maximization;
- limited or no use of external inputs;
- high within-farm and within-field soil variability;
- high potential for weed and pest infestation;
- cultivation of mixtures of species.

The modelling of mixed-species systems, particularly in relation to the below-ground interactions described in this book (e.g. WaNuLCAS, Chapter 10, this volume), goes some way towards addressing the last two characteristics listed. However, the other points made by Mutsaers and Wang (1999) remain valid; most crop models, for example, have been developed for conditions of high-input agriculture. Thus, practices that are common in developed countries (DCs; e.g. fertilization and irrigation) are included in the models, but practices common in LDCs (e.g. green-manure systems) are not. The widely used DSSAT crop models, for instance, recognize a range of different fertilizer types, but lack detailed soil organic matter (SOM) routines, which are crucial when describing low-input systems, in which most nutrients taken up by plants are derived from the decomposition of crop residues or SOM, rather than from applied fertilizers. The soil N transformation submodel in all of the DSSAT models is based on that used in the PAPRAN pasture model (Seligman and van Keulen, 1981), from which it was adapted to work with crop models (Godwin and Jones, 1991). This SOM/residue module assumes only two OM pools, a fresh organic matter pool (e.g. crop residues, etc.) and an older humic pool, but does not differentiate between very old and more recent fractions of this humic pool. Similarly, it does not include the possibility of a litter layer on the soil surface, and cannot, therefore, describe systems in which large amounts of organic material accumulate without physical incorporation into the soil, such as, for example, green manuring or forest systems. For this reason, a new SOM/residue submodel, based on the well-tested CENTURY soil organic matter model (Parton et al., 1988; Kelly et al., 1997), was recently added to the DSSAT models (Gijsman et al., 2002a). Figure 3.1 shows the significant improvement in the ability of the new model to predict changes in soil organic carbon evident in a 40-year dataset from Rothamsted in the UK.

Thus, it is important that the development and use of models for predicting the functioning of agroecosystems in LDCs do not perpetuate a research tradition that may be more relevant in orientation and method to agriculture in DCs, or to richer farmers in LDCs who can afford to use intensive agricultural management practices. Scientists developing and using models for tropical agricultural systems need to make the conceptual links that will make their models relevant to the conditions that occur in the fields of resource-poor farmers.

### 3.3.3 Long-term processes

In order to assess the degree of sustainability possessed by a particular system, there is a need to quantitatively understand how the processes determining production interact with soil characteristics, environmental conditions and management practices. The main limitation attached to the use of process simulation models for the analysis of long-term trends is that they have not yet been thoroughly validated, particularly over a long enough time-span to judge their long-term behaviour. This is partly due to the shortage of good-quality long-term data, although such experiments do exist in the UK (e.g. Rothamsted, ~ 140 years), the USA and elsewhere, and can give valuable insights into soil fertility issues and the sustainability of crop yields. However, these experiments are the exception rather than

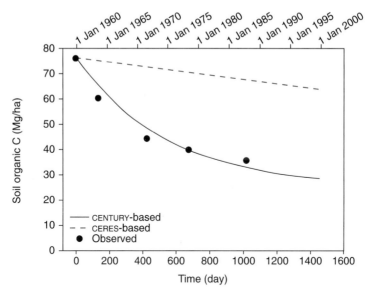

**Fig. 3.1.** Soil organic carbon content of the 0–23 cm layer of the soil in the Rothamsted Highfield bare-fallow experiment (Jenkinson et al., 1987), as simulated with the CERES-based (dashed line) and CENTURY-based (solid line) SOM/residue modules.

the rule, and, to our knowledge, none have been run in developing countries. They also have their limitations: they are laborious and time consuming, often the many variables needed to validate the models thoroughly have not been measured, and they generally take too long to give results (in relation to the timeframe available for making decisions). Moreover, variability in environmental conditions makes it difficult to extrapolate specific results from one time and place to other environments (van Keulen, 1995). Despite these problems, recent efforts have been made to compare a number of soil organic matter models against data from long-term field experiments (e.g. Smith et al., 1997a).

A second limitation to the use of long-term models for tropical agroecosystems is our incomplete knowledge of many of the biophysical processes underlying these systems, or our inability to incorporate these processes into the models. Most crop models were originally designed to describe crop growth and soil processes over one season, and the relatively simple relationships generally employed are usually adequate for this time period. However, we just do not know whether all of these relationships are sufficient to describe soil changes over much longer time periods. Error propagation within the models may be another potential problem – a small error may be relatively unimportant over a single season, but over several seasons it could accumulate and result in a substantial error at the end of the run. This is clearly demonstrated in the comparison of the CERES-based simulation with data measured over 40 years (Fig. 3.1). So far, little work has been done to investigate the magnitude of such errors.

### 3.3.4 Spatial variability

Spatial variability is an inherent characteristic of tropical agroecosystems. Specific agronomic practices further increase spatial variability (e.g. by nutrient transfer from grass strips to tree lines in orchards, tropical hedgerow intercropping, or by injection of slurry or fertilizer into the soil). Animals also increase spatial variability – often they have preferences for certain spots (e.g.

shade or near a water source). Many of the management decisions made by small farmers living in heterogeneous environments make use of spatial variability on their farms, such as growing different crops on different patches of land, abandoning part of their land, or focusing their efforts only on those patches with the highest returns to investment of labour or inputs (van Noordwijk et al., 1998e).

Most of our current models are one-dimensional (i.e. vertical) and do not handle spatial variability well, if at all. There is a clear need to develop existing models further, or to construct new ones, in order to address this limitation. Unfortunately, the structure of many existing models does not facilitate transformation to spatially explicit versions, as their linear nature restricts them to being run in sequence many times, in order to simulate each patch of land in turn (e.g. Basso et al., 2000). This makes it difficult to simulate simultaneous interactions between patches of land (e.g. soil, or water flow down a gradient). The easiest course of action in such cases is usually to build a new model from scratch. Conversion of a one-dimensional model to a two- or three-dimensional one would be made much easier by the use of a 'systems dynamics' structure (e.g. Jones et al., 2001), in which the calculation of rates of change of state variables in the model are separated from the integration (or updating) of these variables.

### 3.3.5 Data requirements

Simulation models describing crop growth generally need input data on weather and soil conditions, crop characteristics, management information, and possibly on the occurrence of pests and diseases. Complex ecosystem models may require even more input data. Some of these data will be the same over large areas, whereas some will vary at much smaller scales. Obtaining such data at a scale which is sufficiently detailed that it represents the actual variability existing in a field or a region can be difficult and time consuming, and may deter potential users of a model, particularly if they want quick answers. For new sites or plant species, such data may not even be available. Thus, there is a need on one hand for databases that can provide inputs for a diverse set of conditions and cropping systems, and on the other there is a need for approaches that facilitate an estimation of parameters from more easily measurable characteristics.

For weather variables, equations have been developed that give estimates of conditions at a given location, based on the location's coordinates and altitude, and interpolation of weather data from several nearby weather stations (Jones and Thornton, 2000). With such methods one can, in principle, obtain reasonable weather data for almost any site on the globe, although, of course, models can never account for all local variation.

For soil parameters, relationships have been developed to estimate parameters that are difficult or laborious to measure, using more easily measured characteristics (for example, soil water release parameters can be estimated using measurements of soil texture, bulk density and soil organic matter). These so-called 'pedotransfer functions' (see also Chapter 9, this volume) exist for many different soil types or regions (e.g. Canada – de Jong, 1982; Australia – Minasny et al., 1999; and the USA – Saxton et al., 1986), but to date have not been developed for many soils in LDCs. Though it is tempting just to use any pedotransfer function for any soil, the estimates given by each can vary widely (Gijsman et al., 2002b); one should thus carefully consider which method is most appropriate for which soil and, if possible, combine it with some actual measurements.

With the compilation of a large dataset, comprising over 2000 entries largely from tropical species, recent advances have been made regarding the quality characterization of organic resources, i.e. cover crops, agroforestry prunings, manure and crop residues (Palm et al., 2001). The compilation of root length densities and rooting depths for agroforestry trees in this book (Chapters 4 and 5) will help fill another knowledge gap.

There is also a need to make input data more 'user-friendly' for users who may not be familiar, in a scientific sense, with a model. For example, rather than requiring specific numerical values of parameters to be input, ranges such as 'high', 'medium' or 'low' could be requested, and the model itself could then convert these into numerical values that it could use. However, some thought needs to be given to reconciling the different perceptions different people have – what one user may call 'high' may be what another calls 'low'. This problem of perception means that some numerical definition may still be necessary. Modellers should also appreciate that the units required are often not intuitively obvious to the end-users, but rather follow the specific model design. Options allowing users to choose their own units could solve this problem.

## 3.4 Beneficiaries and Target Groups

In the context of research on mixed species production systems in the tropics, the ultimate beneficiaries will, in most cases, be subsistence farmers and/or smallholders. However, it is unlikely that this group will use models directly – it is more likely that such models will be used by, and therefore will be targeted at, researchers, consultants or educationalists in LDCs.

In Fig. 3.2, we have attempted to show, in a simplified form, how models might contribute to the flows of information between the main groups involved in the agricultural systems of LDCs. The uppermost level represents the people involved in developing the models, who are, currently, mainly scientists in DCs or scientists working at the International Agricultural Research Centres (IARCs), although there are an increasing number of models being built by national scientists in LDCs. The second level represents the direct users of the models, i.e. those who actually take a model, run it, and interpret the results. In our classification, these may be consultants (in either DCs or LDCs), and scientists and educationalists in LDCs. The third level represents groups of people who may, potentially, benefit from the models' output, but

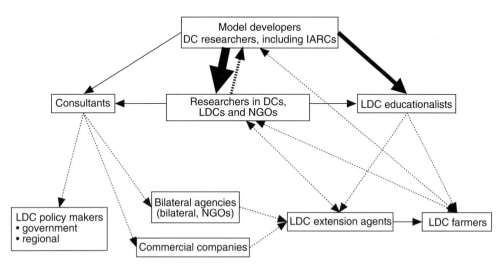

**Fig. 3.2.** Postulated relationship of simulation models to flows of information between various stakeholders in agricultural development. Solid arrows represent information encapsulated in simulation models themselves, dashed arrows represent flows of information by other means, but arising from the use of models. Thickness of arrow is an estimate of relative information flow rate. Overlap may occur in certain cases, e.g. researchers may also be consultants. IARC, International Agricultural Research Centre; DC, developed country; LDC, lesser developed country; NGO, non-governmental organization. Reproduced from Matthews and Stephens (2002) with permission.

who are unlikely to use the models directly. These include policy makers at various levels, staff in aid agencies or commercial companies, extension personnel and farmers. We recognize that the boundaries between the groups are not clear cut and that, in many cases, the same people may be fulfilling two different roles; model developers, for example, may use their models to fulfil an educational role in graduate classes, or as tools for consultancy work. Similarly, there may well be cases in which people within development agencies or non-governmental organizations (NGOs) use the models directly. Nevertheless, as a workable framework for considering how models may fit into the overall development process, we believe this is a useful starting point. In the following section, we discuss the relevance of simulation models to each group in turn.

### 3.4.1 Researchers

The ways in which crop–soil models have been applied in a research context have been recently reviewed by Matthews *et al.* (2002a). There is no doubt that, so far, the largest uptake and use of models has been by the research community, both in DCs and LDCs. This is because models are primarily research tools – for most scientists, the incorporation of knowledge into simulation tools is a process taken for granted. For them, models are a way of organizing and utilizing information, which, in turn, can help to identify gaps in their knowledge upon which they can focus further experimental research. There is, however, some concern that much use of models is really just confirming what is already known, rather than adding new knowledge (Matthews and Stephens, 2002). Sinclair and Seligman (2000) make the point that many papers on crop modelling merely calibrate models to local conditions, and have proposed three criteria that should be met if papers on crop modelling are to be published: (i) a clear statement of the scientific objective and a defined domain of relevance; (ii) a mechanistic framework; and (iii) an evaluation of the scientific innovation of the model.

### 3.4.2 Consultants

It is difficult to know the extent to which agricultural and policy consultants use simulation models as, due to client confidentiality, such work is not generally published. Examples of models in the DSSAT suite (Decision Support System for Agrotechnology Transfer; Tsuji *et al.*, 1994) that are being used in consultancy work in South Africa have been discussed by Stephens and Middleton (2002) and include the following:

- CERES-Maize was used to simulate the production potential and risk associated with maize on two farms for an organization wishing to buy a commercial farm for small-scale farming development. The modelling helped them to decide which farm to buy.
- The DSSAT package was used, on behalf of a fertilizer company, to gauge the optimum level of nitrogen application for a particular field.
- CERES-Maize is used in yield estimation for the Orange Free State Department of Agriculture. These data are then used by the National Crop Estimates Committee (NCEC) to provide information for FAO (Food and Agriculture Organization) and SADAC (Southern African Development Community) early-warning systems.
- A simulation study has been made of the impact of climate change on South African maize production, the results of which will be used by a mitigation team as the basis for plans to minimize the impact of climate change.
- CERES-Maize was used to determine the potential for, and risk of, growing maize on rehabilitated soils. This study was undertaken on behalf of a mining company, which had bought land from farmers on the understanding that, at the end of the opencast mine activities, the land would be returned to farming activities. There are also plans to use the model to help the mining company

monitor whether they are on target in terms of restoring the original potential of the site.

In each case, the modelling work was carried out on behalf of the clients by the modelling consultants – the clients did not use the models themselves.

### 3.4.3 Educators and trainers

Educators and trainers are also important end-users of simulation models (Graves *et al.*, 2002). They may either use models to help illustrate to their students particular processes (e.g. root growth, water uptake, etc.) and the effects these have at higher levels of analysis, or they may use modelling as part of the process by which students learn systems analysis techniques, in which case the students may build their own models or use existing models to provide information about a component of a larger system, such as an agroforestry system or a farm.

There are several advantages for students. For example, the speed and sensitivity of 'experiments' can be increased, and complex relationships and interactions can be more easily understood. 'What if?' scenarios can be developed to allow students to learn heuristically. Educational institutions can also benefit, through a reduction in the need for expensive laboratories and equipment, and through the more efficient use of instructors' contact time with students. However, there are also disadvantages to the use of simulation models. Students may end up believing that the model is some kind of 'reality', and may fail to learn essential field and laboratory skills. They may also waste time struggling to operate the models, rather than understanding the lessons they can convey. At the institutional level, disadvantages can include the time and costs involved in developing, selecting or adapting appropriate models, and in modifying courses to incorporate their use. Not all institutions may have adequate computer resources, particularly those in LDCs.

### 3.4.4 Policy shapers and makers

There may also be scope for the use of simulation models to support strategic decisions at a larger scale. However, the extent to which this would be successful will depend on institutional issues and the level of training available to the decision-making staff. In DCs, there tends to be a greater commitment to the acceptance of new knowledge and to the promotion of new practices. This allows technology to be advanced more rapidly (Tollefson, 1996). In the institutions of LDCs there may be resistance to a new technology, especially if it is seen to pose a threat to the existing system. Decision makers who are very busy and who are already dubious about the value of models may find the opportunity cost of learning unacceptable. Spedding (1990) makes the point that policy makers are generally sceptical of systematic methods: 'they are alarmed at the idea of it being publicly known where they are trying to get to, except in the most general terms, in case they never arrive!'.

It should be remembered that the interest of policy makers does not necessarily lie with the best state of an agroecological system, but more with the impact that will be made by the type of interventions they have in mind. Similarly, development and donor agencies are, increasingly, under pressure to demonstrate the value of their proposed intervention for the community and to seek supportive measures from policy makers. However, demand for the assessment of solutions has shifted away from purely biophysical effects and towards the impact such solutions have on people's livelihoods.

### 3.4.5 Extension staff and NGOs

Extension personnel and NGOs are an important link in the chain that links researchers and farmers. In DCs, the former often provide farmers with a human interface with computerized DSSs. In farming, much advice comes from trusted advisers: substituting such advisers for a computer may be off-putting to many farmers (Knight, 1997). For example, Blokker (1986) found that DSSs designed for direct use by farmers (as distinct from those interpreted by an extension officer) were generally not appreciated by farmers and only had a marginal influence on their decision making.

In LDCs, however, it is less likely that extension personnel will have access to a computer, in which case computerized DSSs would not be appropriate. The information encapsulated in the form of a computer model is not, therefore, likely to flow, in that form, further than research and education groups. However, this does not mean that such information flow need stop at these groups. Information gained from research is more likely to reach extension staff and NGOs in other forms – such as research reports, brochures, posters, training workshops, verbal communications on field days, radio broadcasts, or via informal contacts with research staff.

The influence extension staff and NGOs have on poverty alleviation can be very significant, as they are in direct contact with the farmers being targeted. It is the efficiency and speed with which they are able to transmit information to the farmers that will partly determine whether a particular technique is likely to be adopted or not. They also have another important role to play in the transfer of information in the opposite direction (from farmer to researcher) so that research activities are relevant to the real problems faced by farmers, not just to problems that researchers perceive farmers to have.

Unfortunately, extension services in many LDCs are badly under-equipped in terms of staff, transport and accommodation (Tollefson, 1996), not to mention access to computing facilities. This situation may well change in the future, however, as computer technology becomes cheaper and the skills to operate them become more widespread: computerized DSSs may well then become more relevant to extension staff.

### 3.4.6 Farmers

It seems unlikely, for three reasons, that, in the short to medium term, there is much potential for the on-farm use of computer-based DSSs by smallholder farmers in LDCs. First, the time when they will achieve the financial ability to purchase and run a computer is a long way off in the case of most subsistence farmers: many do not even have an electricity supply available with which to run a computer. Secondly, the level of education needed to successfully operate a computer is likely to limit uptake. Daniels and Chamala (1989) found in Australia that farmers' interest in computers was related to their level of education – those with higher levels of education were more interested, whereas those with less formal education preferred to go by experience. Even in DCs, poor computer literacy among farmers has hindered the uptake of IT systems (Hamilton et al., 1991). In LDCs, where rural education is often of a low standard and where even the educational level of extension workers is low, the constraints are even greater. LDC farmers would require a huge amount of training and support to begin to use the systems in a useful way. Thirdly, it is, anyway, not at all certain that answers to the sort of questions that farmers are most likely to ask could be provided by operational DSSs. Nevertheless, despite all of these constraints, opportunities for the rural poor to participate in the information revolution are being explored – for example, fishermen in southern India are obtaining weather forecasts and wave-height predictions from the internet via centrally located computers in their villages (Le Page, 2002). It may be only a matter of time before output from crop–soil models is made available in the same way.

If it is anything to go by, the experience in DCs of using simulation models as operational decision support tools has shown that, rather than being useful as operational DSSs in their own right, they are probably more useful as research tools that provide solutions to constraints: these solutions can then be developed into simple rules-of-thumb. In Australia, for example, the SIRATAC dial-up crop management system was developed in the 1970s to help farmers make better tactical decisions with regard to spraying for cotton pests (Macadam et al., 1990). The system's use by growers increased steadily in the early 1980s, but subsequently fell into decline as growers, having developed their own rules-of-thumb from it for the best times for spraying, found that they had no further use for it. There is no doubt that the models at the heart of the system had a major impact in terms of improving practices within the cotton-growing industry, and that without their use cotton-growing practices would not have been economically viable. However, the SIRATAC system's usefulness lay more in the provision of the underlying knowledge for optimal management rather than in its use as a tactical decision-support tool.

An emerging approach is the use of models in a participatory way, as pioneered in Australia and Zimbabwe by the APSRU (Agricultural Production Systems Research Unit) group. The APSIM (Agricultural Production Systems Simulator) package of models is currently being used to help farmers, policy makers, extension agents and researchers to improve their understanding of the trade-offs necessary between different crop and cropland management strategies under scenarios of climatic risk (Meinke et al., 2001). Rather than focusing on a particular optimal strategy, the model is used to explore the consequences of various cropping practices that are suggested by extension personnel and by the farmers themselves, who are aware of their own labour and capital resource constraints. The modelling aspect is important, as it would not be possible to undertake such an analysis, either on farm or at a research station, in a reasonable time frame. Researchers found the model useful, as it made them more aware of the constraints faced by smallholders, and suggested new lines for research. A similar approach was suggested by Beinroth et al. (1998). In this approach, a model could be used to explore the trade-offs necessary between domestic requirements, irrigation demand and downstream use of river water in Colombia. This involved regular discussions between stakeholders, allowing new scenarios to be formulated and simulated in an iterative manner until a consensus was reached.

Clearly, by interacting directly with farmers, the flows of information that occur between model developers, users and beneficiaries are likely to improve. All have much to learn from each other; the models may be able to suggest improvements to existing practices, but farmers will be able to temper these suggestions with their practical experience. Of all the target groups discussed, direct interaction with farmers probably has the greatest potential to improve rural livelihoods, although the numbers of people whose livelihoods are actually improved as a result will depend strongly on the dissemination of such improvements outwards to others not directly involved.

### 3.4.7 How do we ensure uptake and impact of simulation models?

Matthews et al. (2002b) considered the route by which models will have an impact on the process of improving the livelihoods of farmers to consist of three phases: (i) the applicability of the models to particular problems; (ii) the uptake of models by end-users; and (iii) the translation of this use of models into a measurable impact. Failure to reach any one stage will prevent models from having any final impact.

#### 3.4.7.1 Model applicability

Limitations inherent in the models themselves, some of which were discussed above, may prevent them from being applicable to certain problems. For example, a model assuming a uniform canopy could not be used to investigate the distribution of light in a spatially heterogeneous, mixed-species canopy. The inapplicability of models to real-life problems is one major factor that limits their wider use; most models have been developed as research tools, and several have been modified for use as decision support systems, but they still address problems perceived by researchers rather than farmers (Stephens and Middleton, 2002). A farmer is probably not that interested in knowing that he/she can obtain 2% more yield by applying fertilizer on March 25 rather than 1 week later if the main constraint is whether the fertilizer will be delivered at all!

### 3.4.7.2 Model uptake

Uptake of models very much depends on the needs of a particular end-user. Stephens and Hess (1996) classified a number of constraints to the uptake of the PARCH model by researchers in East Africa as 'intellectual', 'technical' or 'operational'. Often, the outputs of simulation models may be too complicated (both in terms of language and amount of information given by the model) for direct use by the beneficiaries (e.g. farmers, policy makers, etc.). In part, this problem is related to the mental models that each user has of the same system – a soil scientist's concept of soil fertility, for example, is usually very different from that of a farmer. The scientist may focus only on its nutrient status, without considering its physical characteristics (Corbeels et al., 2000), whereas the farmer's perception of soil fertility is not limited to its nutrient status, but is often related both to integrative characteristics (such as the soil's ability to produce good crops), and to soil characteristics that they can actually see or feel (see also Chapter 2, this volume). Thus, recommendations derived from models and other scientific assessments need to be translated into a language that is easily understood by the beneficiaries. Efforts in this direction have been made by Giller (2000) who translated scientific findings based on the Organic Resource Database (Palm et al., 2001) into parameters that could be understood by farmers (Fig. 3.3).

### 3.4.7.3 Model impact

The impact a model has is difficult to quantify, particularly as it may occur over different timescales (Collinson and Tollens, 1994). For this reason, factors that enhance the likelihood that a model will have an impact are difficult to identify. However, Matthews et al. (2002b) reviewed a number of examples, where, in their view, crop simulation models had had some impact. They listed the following characteristics as having some influence:

1. Involvement of competent modellers.
2. Working in multidisciplinary teams.
3. Participatory approach with practitioners.
4. Having a clearly defined problem.
5. Demand for solutions from a target group.
6. Long-term commitment by funding sources.
7. Quantification of risk in variable environments.
8. The need for quick answers.

They noted that, in nearly all of the examples they discuss, the only common factor was the involvement of competent modellers. This suggests both that modellers should be an integral part of a team involved in the overall

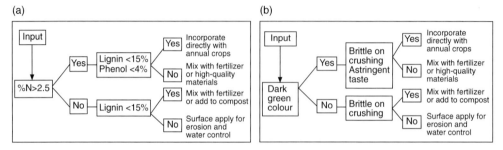

**Fig. 3.3.** Recommendations for use of organic resources in agriculture based on (a) scientific (chemical) plant quality attributes (Palm et al., 2001), and (b) translated into easily observable parameters defined by farmers (Giller, 2000).

process of livelihood improvement, and that it is important that they also have the opportunity to enter into dialogue with farmers and other target groups. In the participatory modelling approach being developed by the APSRU group (discussed above), clients are involved in projects – this includes not only farmers, but all decision makers involved in agricultural development. If the clients are farmers, collaborative experiments are conducted, and the results are extrapolated in time using models to show the long-term consequences of the farmers' actions. If the clients are researchers, models are used for extrapolation of the research results in space and time or across environmental conditions (e.g. different soil types or weather conditions), in order to add value to expensive research. The approach described by Robertson *et al.* (2000), which was used to develop new cropping strategies for mungbean in response to changed external factors in Australia, is a good example of how this added value can be achieved. Van Noordwijk *et al.* (2001) have coined the term 'negotiation-support tools' for models used in this way.

Development of user support groups to provide help and model updates to users of existing models is one way of trying to encourage the maintenance and development of these skills after the main project has finished. Reviewing the SARP (Systems Analysis for Rice Production) project at the International Rice Research Institute (IRRI), Mutsaers and Wang (1999) found, first, that, despite the scale of the project and the foresight that the project's designers appear to have shown, modelling skills among the collaborating national scientists were being lost and, secondly, that the use of models was not likely to continue unless there were continued interventions from 'advanced' organizations. Lessons can perhaps be learnt from the experience of the IBSNAT project (Tsuji and Balas, 1993), whose DSSAT family of models are probably the most widely used family of crop simulation models in the world today. Part of the success of this family of models was no doubt due to the size of the project, and to the participatory and interactive relationship that existed between model developers and model users during the course of the project; the continued uptake and use of the models must, however, be due to the technical support that is still available, even though the project ceased in 1994. Users and developers still keep in touch via a listserver, so there exists a broad base of support, which is not dependent on one or two people. Users with a problem can post a query on the listserver and, usually within a day or two, can receive help and advice from other users or from the developers of the model.

In such support groups, emphasis should be placed on the applications of the appropriate models to solving practical problems of importance in the research areas of the members. For example, agroforestry models could be used to optimize the spatial arrangements of particular agroforestry systems, taking into account both biophysical and socioeconomic aspects that are influential. However, it is important that models contribute to the solving of a clearly defined problem, rather than just to the confirmation of what is already known by farmers.

## 3.5 Relevance to Larger Systems

It is important that models of below-ground interactions in mixed-species systems are seen as part of larger systems. The reductionist approach to science has been very successful in adding to our store of knowledge about the way the world functions, but there is a growing awareness that systems are more than the sum of their parts, and that they can only be understood fully by taking into account the complexity of 'emergent' behaviour in addition to the behaviour of their individual components (e.g. Coveney and Highfield, 1995). Improvements in one component of a system do not necessarily have a desired result at a higher scale. A good example of this is the promotion of *Mucuna* as a cover crop in Honduras in the 1970s in order to help intensification of cultivation, thereby raising yields and reducing the need for farmers to clear more areas of forest (Buckles and Triomphe, 1999). Those farmers practising the technology were able to grow twice as much maize on less land; but the resulting improvement in the local economy attracted an influx of migrants into the area, so that overall deforestation rates continued to increase at the same pace (Humphries, 1996). The key in this case is to understand the interactions and linkages between individual components of a system and how these relate to behaviour at a higher level.

### 3.5.1 Enhancement of livelihoods

In development circles, there has been a growing realization that single-factor-based research has not been able to address many of the problems faced by poor people, and that a much more multidisciplinary approach is required. For this reason, several international development organizations are currently promoting the use of the Sustainable Livelihoods (SL) framework as a way of thinking about objectives, scope and priorities for development, in order to enhance progress being made in terms of the elimination of poverty (Ashley and Carney, 1999). Such thinking has grown from the recognition that it is fruitless to try and solve technical problems without, at the same time, addressing the socioeconomic pressures against which they are set.

The main feature of the SL approach is that, instead of focusing on natural resources or commodities (as has been the case in the past), it places people at 'centre stage' and considers people's assets (natural, human, financial, physical and social capital) and their external environment (trends, shocks, and transforming structures and processes; see Fig. 3.4). Households adopt various strategies in order to achieve certain outcomes, such as increased financial income, increased food security, and a better quality of life. A key concept is that of 'sustainability' – a livelihood is defined as sustainable when it can cope with, and recover from, stresses and shocks, and maintain or enhance its capabilities and assets both now and in the future, while not undermining the natural resource base (Carney, 1998). As such, the SL framework encourages researchers to think about the whole livelihood system, rather than just some part of it.

In relation to developing agriculture, therefore, there needs to be more emphasis placed by the modelling community on problem-solving approaches, and on making *people* more central to their way of thinking. On one level, this means thinking of the problems faced by ordinary people in LDCs, and constructing and applying their models to address, and to contribute to solving, these problems. For this to be effective, modellers need to both define clearly who the beneficiaries of their mod-

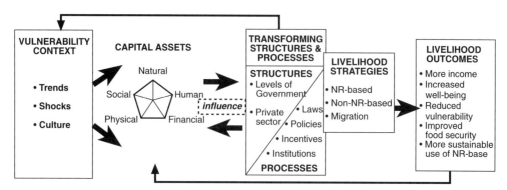

**Fig. 3.4.** The Sustainable Livelihoods framework (from Carney, 1998). Reproduced from Matthews and Stephens (2002) with permission.

els are and enter into dialogue with these people so that the final product is tailored to their needs. As part of this process, on one level, an increased effort needs to be made to disseminate the outputs of models, over and above the dissemination of the models themselves. On another level, there exists the need to consider people as integral components in the systems being modelled. The SL framework offers a good starting point from which to develop this methodology. It is hoped that this would eventually lead to the development of tools that practitioners could use to identify what are the real constraints to improved livelihoods in LDCs, so that future projects would be more realistically focused, thereby increasing their chances of having an impact.

Other chapters in this book are generally aimed at specific biophysical processes associated with below-ground interactions. However, although this approach is valuable from the point of view of scientific research, farmers do not necessarily think in the same terms as researchers. Rather, they are more concerned with how particular practices relate to their broader livelihoods. For example, in considering whether or not to adopt a particular research product, such as tree planting for deep nutrient capture, the kind of questions he/she is more likely to ask are 'How will my livelihood benefit from this?', 'Will I produce more food for my family if I do this?', 'Will I earn more cash if I take this up?' and 'Will my family's quality of life be enhanced?' For researchers, thinking about the products of the research process in these terms will mean that that their research will be more likely to result in improvements to the production system. Perhaps improved food security can be obtained through the greater use of agroforestry systems, so that the risk of crop failure is reduced. Increased cash generation may be obtained through planting fruit trees alongside crops and then selling the produce in the market. Quality of life could be enhanced by means of a more varied diet or through a reduction in labour requirements for different agricultural practices. Further questions may relate to specific practices, for example, 'Is it better to try growing apples or bananas in this particular environment?' or 'Is *Mucuna pruriens* or *Canavalia ensiformis* the better cover crop for weed control?' This approach is still reductionist, in that the overall system has been reduced to its components. The only difference between this and traditional approaches is that the definition of the problem and its solution has not been restricted to biophysical processes, but also includes the socioeconomic processes of the system. The key point is that the farm and its environment are seen as a complex adaptive system, rather than as independent components arising from single-discipline perspectives.

Many issues of global concern can also be addressed through a sustainable livelihoods perspective. For example, changes in the global climate and a reduction of biodiversity are concerns that are attributed, in part, to the loss of forested area as a result of clearing for agriculture. At this level, stabilization of the interface between forest and agriculture is generally seen as desirable in terms of preservation of the forested area. One line of thinking is that, by developing ways to improve the livelihoods of people at the forest margins, their need to move on and clear more forest will be reduced, which will contribute to solutions to the global problems (e.g. World Bank, 1992). In this way, improved productivity, through the adoption of resource management recommendations derived from, for example, a better understanding of below-ground interactions in mixed-species agroecosystems, could contribute to reducing the rate of deforestation. However, as has already been pointed out, it should also be recognized that the situation is not necessarily as simple as this. First, it is unlikely that the adoption of improved techniques alone can bring stabilization, without a concomitant improvement in infrastructure (i.e. in roads, hospitals, schools) and markets. Secondly, Johns (1996) noted that, where agriculture was successful in areas surrounding forest reserves, migration into the area was also increased and worked against biodiversity. The example of *Mucuna* in Honduras (discussed above) supports this observation. It is by developing models of these higher-order processes, in which models of below-ground interactions may well be a component, that we will understand these systems better.

---

**Conclusions**

**1.** A range of multispecies models exists with good below-ground process descriptions, but they are not widely validated.
**2.** To date, the users of simulation models are mainly within the research community, although they sometimes interact with indirect beneficiaries such as consultants, educationalists, policy makers, extensionists and farmers.
**3.** The impacts that crop–soil simulation models have had as decision support systems have, until now, mainly been the result of their contribution to the learning processes of practitioners and the subsequent development of rules of thumb.
**4.** Models should not be viewed in isolation, or thought of as being the sum of their individual components.
**5.** New approaches are being developed to include stakeholder participation and livelihood concepts.

---

**Future research needs**

**1.** Expanded databases for tropical systems (e.g. pedotransfer functions, root systems' characteristics and plasticity).
**2.** Models with an integrated livelihood perspective.
**3.** Integrative 'DSSs' (decision support systems) with the involvement of stakeholders.
**4.** More user-friendly input parameters.
**5.** Modularity and compatibility of components from different models.
**6.** Validation of models for a wide range of ecosystems and regions.

# 4 Tree Root Architecture

Festus K. Akinnifesi, Edwin C. Rowe, Steve J. Livesley, Freddie R. Kwesiga, Bernard Vanlauwe and Julio C. Alegre

---

**Key questions**

1. Where are the roots of all plant components that interact in complex agroecosystems? That is, in complex agroecosystems, what is the spatial distribution of tree and crop root systems?
2. To what extent can the genetically predetermined rooting pattern of a plant species be influenced by soil conditions, climate, companion species and management practices?
3. Are there simple rules that underlie root system architecture, and can these be used to reduce the time, labour and capital costs of the conventional sampling techniques used in root research?

---

## 4.1 Introduction

One reason for combining trees and crops is to capitalize on the greater capacity of trees to take up water and nutrients from deep soil layers (van Noordwijk *et al.*, 1996). Trees with dense mats of shallow roots are likely to compete more with crop plants for water and nutrients than are trees with deep roots. Thus key questions, with regard to tree root systems, are 'what depth do they reach to in the soil?' and 'what proportion of the tree root system lies within the shallow soil preferred by crop roots?' It has been shown that an abundance of tree roots in the rooting zone of an associated arable crop does not always imply strong competition that is detrimental to the crop (Chapter 15, this volume; Schroth, 1998). Indeed, de Montard *et al.* (1999) showed that tree root abundance was reduced by the presence of grasses. However, generally, competition is likely to be more severe when trees have shallow root systems that occupy the same rooting depths as the crops with which they are associated (Schroth, 1995, 1998; Akinnifesi *et al.*, 1999a; Chapter 10, this volume). As well as being of importance when considering competition, the amount and distribution of roots in the soil has great bearing on soil organic matter inputs and on soil structure.

Despite current interest in agroforestry, there is still a paucity of information about the form, function and dynamics of tree and crop root systems and about their interactions. In agroforestry research, most empirical assessments continue to be based on data gathered only from the above-ground components of the system. The body of information on root distribution and function is generally inadequate, although some major reviews of root research are emerging (Schroth, 1995, 1998; Jackson *et al.*, 1996; van Noordwijk *et al.*, 1996; Akinnifesi *et al.*, 1999a) and many of the knowledge gaps have been highlighted. Being unable to understand or predict the distribution, dynamics and complex interactions that occur between root systems, in tree–tree and tree–crop systems, remains a major obstacle to the better design and management of agroecosystems, and so to a subsequent improvement in farmers' livelihoods. Attempts to translate positive, but complex, tree–crop interactions into firm concepts and into recommendations relevant to farmers' needs have been constrained by our poor understanding of interacting root systems – especially in terms of the causes and effects of root plasticity, competition, and root responses to resource availability, resource surplus or conditions of stress (Akinnifesi *et al.*, 1999a).

In this chapter we aim to synthesize information on root distributions from several studies conducted throughout the tropics. We will also review methods used to assess root distributions using simple measurements of proximal root direction and branching characteristics.

## 4.2 Root Distributions

In multispecies agroecosystems (for example, multiple-cropping or agroforestry systems) the distribution of plant roots depends upon plant genotype, soil properties, nutrient status and plant vigour as well as on other factors. Early root studies focused on crops under monoculture (Pages *et al.*, 1989; Hairiah *et al.*, 1992), pastures and, in some cases, plantations or mixed forest ecosystems (Cuevas and Medina, 1988; Jonsson *et al.*, 1988). Jonsson *et al.* (1988) compared the rooting patterns of maize with five tree species, and concluded that these may compete with crops for soil resources because of a similarity in the distribution of their root systems.

### 4.2.1 Maximum rooting depth

Trees adapted to semiarid conditions are well known for having deep root systems; the roots of *Prosopis cineraria* and *Faidherbia albida*, for example, penetrate soils to great depths. However, very deep roots are also found in tropical rainforest ecosystems (Schroth, 1998). Lieffers and Rothwell (1987) reported a positive correlation between root penetration and depth of the water table. In many studies, root sampling has only taken place to a depth of 1–2 m and, therefore, no estimate of maximum rooting depth can be made. To date, the most extensive and comprehensive inventory of tree rooting depths was undertaken by Stone and Kalisz (1991), and involved 49 families, 96 genera and 211 tree species. This study's observations were made under diverse substrate conditions and used diverse methods, including caves, mines, wells, root excavation, etc. From the study, it was evident that tree roots could penetrate to great depths under favourable soil/site conditions; rooting depths found included 61 m in *Juniperus monosperma*, 60 m in *Eucalyptus* spp., 53 m in *Prosopis juliflora*, and 45 m and 35 m in *Acacia raddiana*. Trees with rooting depths of more than 15 m but less than 35 m were, in a decreasing order of magnitude, *Rhus viminalis*, *Pinus ponderosa*, *Pinus edulis*, *Andira humilis* and *Fraxinus* sp. These results indicate the inherent ability of many tree species to develop far-reaching roots in the absence of mechanical soil restrictions and fertility restrictions. Roots, particularly those of perennials, often extend to great depths in freely drained soils, and are only shallow where soil conditions restrict root growth (Savil, 1976). The downward penetration of tree roots is commonly limited by mechanical impedance, particularly where a hard iron pan has formed as a result of waterlogging and consequent anoxia, but also as a result of dry subsoil or chemical limitations, such as aluminium toxicity in acid soils (Stone and Kalisz, 1991).

Toky and Bisht (1992) found, in a dry site in India, that *Prosopis cineraria* and *Eucalyptus tereticornis* have deeper root systems than other species, an adaptation that enhanced their growth and their chances of survival. At Machakos, Kenya, Mekonnen et al. (1997) measured the maximum root depth of 15-month-old *Sesbania sesban* trees in a fallow and found it to be > 4 m. Trees with roots with the capacity to penetrate soil up to a depth of 5 m (Archer et al., 1988; Rowland, 1998) have been reported to have their roots limited to superficial zones in some sites (van Zyl, 1988; van Huysstein, 1988; Hairiah et al., 1992). Recently, Chaturvedi and Das (2003) examined the root distribution of eight 5-year-old multipurpose trees in north Bihar, India (*Acacia lenticularis, Acacia nilotica, Albizia procera, Dalbergia sissoo, Pithecellobium dulce, Sesbania grandiflora, Senna fistula* and *Syzygium cumini*). Among the eight species studied, the deepest rooting depth was obtained by *Acacia nilotica* (2.7 m) followed by *D. sissoo* (2.5 m); the shallowest rooting depth occurred in *Acacia lenticularis* (1.01 m). The shallow root systems at the site were similar to the rooting systems of 4-year-old *Dialium guineense* and *Pterocarpus* spp. in Ibadan, Nigeria (Akinnifesi et al., 1999a). Few agroforestry tree species with deep roots have been reported in the tropics (see Table 4.1). Most studies of agroforestry rarely sample rooting depths beyond 2 m, as they mainly target the rooting depths of crops. This fact makes it quite difficult to draw conclusions concerning rooting depth of trees. One study, however, found that roots of *Gliricidia sepium* attained a depth of 5.6 m in Makoka, Malawi (Table 4.1). In general, shallow-rooted trees are commonplace in strongly acidic soils; rooting depth is often dictated by mechanical soil impedance in drier soils.

### 4.2.2 Root distribution and depth

Schroth (1998) identified the need to distinguish between rooting depth and root distribution in the entire soil profile. Shallow-rooted plants that exhibit a rapid decline in root mass, length or density with increasing soil depth may be more competitive than those that have a lot of fine roots in the topsoil and yet maintain a substantial proportion of roots in the deeper soil layers (Schroth and Zech, 1995a). An abundance of tree roots at soil depths that are below the feeding zone of most annual crops may transfer more deep resources to the surface (see Chapter 6, this volume), meaning that the trees survive better during long periods of water stress. The proportion of the different fine root systems that lie in different soil layers is shown for several tree species and soil types in Table 4.1. These data clearly demonstrate differences between species and between soil types. For instance, over 80% of the fine roots of *Pterocarpus mildbraedii*, growing on an Alfisol in Ibadan, were located in the top 30 cm of soil, whereas, in the same soil, the roots of *Nauclea latifolia* were mostly found below a depth of 60 cm. The data collected also demonstrate that assumptions about the root depth of a species may not hold when it is planted on a different soil type. The roots of *Gliricidia sepium*, growing on a Ferric Lixisol (pH 5.4) in Makoka, Malawi, were concentrated at a depth of 30–60 cm; in an Ultisol (pH 4.8) in Lampung, Indonesia, the roots of this species were highly concentrated in the topsoil. Table 4.2 gives indicators of the rooting characteristics for 32 tree species from 51 profile root excavations. Surface root proliferation is high in 33% of all cases, whereas subsoil root proliferation was found to be high in 23.5% of cases. Only 25% of the cases were found to have more than 60% of their roots at the surface. Deep rooting was prominent in only 18% of all cases; most trees fell into either the 'shallow' or 'intermediate' categories (Table 4.2). However, most trees (86%) had their fine roots unimodally concentrated in the top soil layer (0–30 cm). Only 10% of the cases were shown to have exhibited bimodal root peaks. Tree root length density typically shows an exponential decline with soil depth (Ruhigwa et al., 1992; Schroth and Zech, 1995a; Akinnifesi et al., 1999b).

The root length density of *Senna siamea* at Glidji, southern Togo, was found to be 1.45 cm/cm$^3$ at a depth of 0–15 cm, but < 0.4 cm/cm$^3$ at a depth of 25–200 cm (Vanlauwe et al., 2002). Its roots were concentrated close to the tree trunk in the surface layer (0–10 cm soil depth), and its root

Table 4.1. Root distribution patterns of selected trees in the tropics.

| Species | Tree age (years) | Season[a] | Latitude | Altitude (m) | Soil type | Rainfall (mm) | Prod. syst./trt[c] | Tree management | Tree spacing (m × m) | Max. root depth (m) | Layer with most roots (cm) | % Fine root distribution | | | | | Source[d] |
|---|---|---|---|---|---|---|---|---|---|---|---|---|---|---|---|---|---|
| | | | | | | | | | | | | 0–30 | 30–60 | 60–100 | >100 | | |
| Acacia catechu | 6 | SR-U | 29°10′N | n/a[b] | n/a | 900 | T | Unpruned | 5 × 5 | 1.5 | 0–15 | 23.5 | 14.7 | 26.0 | 35.8 | 1 |
| Cajanus cajan | 4 | LR-U | 2°30′S | n/a | Podzol | 1800 | T+C | Pruned | 0.5 × 4 | 1.5 | 30–50 | 11.8 | 51.0 | 19.6 | 17.6 | 2 |
| Cajanus cajan* | 4 | LR-U | 2°30′S | n/a | Podzol | 1800 | T+C | Pruned | 0.5 × 3 | 1.4 | 0–20 | 85.9 | 6.7 | 6.0 | 0.4 | 2 |
| Clitoria fairchildiana | 4 | LR-U | 2°30′S | n/a | Podzol | 1800 | T+C | Pruned | 0.5 × 4 | 1.4 | 10–20 | 43.9 | 14.6 | 34.1 | 7.3 | 2 |
| Dactyladenia barteri | 6 | LR-U | 4°51′N | 150 | Ultisol | 2400 | T+C | Pruned | 1 × 4.5 | 1.5 | 0–40 | 59.1 | 29.5 | 11.4 | 0.01 | 3 |
| Eucalyptus tereticornis | 6 | LR-B | 29°10′N | n/a | n/a | 900* | T | Unpruned | 5 × 5 | 2.0 | 15–30 | 45.7 | 19.7 | 18.5 | 16.1 | 1 |
| Gliricidia sepium | 9 | LR-U | 15°30′S | 1030 | Ferric Lixisols | 850 | T+C | Pruned; no fert. | 0.9 × 1.5 | >3.0 | 40–50 | 15.6 | 33.9 | 21.6 | 28.8 | 4 |
| Gliricidia sepium | 9 | LR-U | 15°30′S | 1030 | Ferric Lixisols | 850 | T+C | Pruned/fert. | 0.9 × 1.5 | >3.0 | 40–50 | 20.8 | 52.5 | 13.0 | 13.7 | 4 |
| Gliricidia sepium | 6 | WS-U | 15°30′S | 1030 | Ferric Lixisols | 1045 | T+C | Pruned; no fert. | 0.9 × 1.5 | 5.6 | 0–30 | 32.5 | 16.0 | 10.0 | 42.2 | 5 |
| Gliricidia sepium | 5 | WS | 4°30′S | 100 | Ultisol | 2200 | T+C | Pruned | 0.5 × 4 | n/a | 0–10 | 85.9 | 9.9 | 4.2 | n/a | 10 |
| Leucaena leucocephala | 4 | LR-U | 2°30′S | n/a | Podzol | 1800 | T+C | Pruned | 0.5 × 4 | 1.4 | 80–90 | 25.2 | 17.8 | 31.4 | 25.2 | 2 |
| Leucaena leucocephala | 6 | LR-B | 29°10′N | n/a | n/a | 900 | T | Unpruned | 5 × 5 | 1.2 | 100–125 | 61.8 | 25.2 | 4.5 | 8.5 | 1 |
| Leucaena leucocephala | 2.4 | LR-B | 30°20′N | 517 | Udic Haplustalf | 1700 | T | Unpruned | 3 × 3 | >1.2 | 30–60 | 30.6 | 35.5 | 20.0 | 14.1 | 6 |
| Leucaena leucocephala | 10 | LR-B | 7°30′N | 1200 | Alfisol | 1300 | T+C | Pruned | 0.25 × 4 | >2.0 | 0–15 | 46.5 | 13.3 | 13.8 | 26.4 | 7 |
| Peltophorum dasyrrachis | 5 | WS | 4°30′S | 100 | Ultisol | 2200 | T+C | Pruned | 0.5 × 4 | n/a | 0–10 | 57.4 | 26.2 | 16.4 | n/a | 10 |
| Pterocarpus mildbraedii | 4 | LR-B | 7°30′N | 1200 | Alfisol | 1300 | T+C | Pruned | 0.25 × 2 | >2.0 | 0–20 | 80.5 | 12.5 | 6.0 | 0.9 | 8 |
| Nauclea latifolia | 4 | LR-B | 7°30′N | 1200 | Alfisol | 1300 | T+C | Pruned | 0.25 × 2 | >2.0 | 60–90 | 22.2 | 9.2 | 36.9 | 31.9 | 8 |
| Enterolobium cyclocarpum | 4 | LR-B | 7°30′N | 1200 | Alfisol | 1300 | T+C | Pruned | 0.25 × 2 | >2.0 | 10–30 | 53.2 | 11.3 | 17.4 | 18.2 | 8 |
| Prosopis cineraria | 6 | LR-B | 29°10′N | n/a | n/a | 900 | T | Unpruned | 5 × 5 | >2.0 | 15–30 | 28.2 | 15.3 | 24.7 | 31.8 | 1 |
| Senna siamea | 4 | LR-U | 9°37′N | n/a | Rhodic ferrasol | 1289 | T+C | Pruned | 0.5 × 4 | >2.0 | 10–20 | 69.7 | 9.0 | 11.3 | 10.0 | 9a |

| Senna siamea | 6 | LR-U | 4°51'N | 150 | Ultisol | 2400 | T+C | Pruned | 1 × 4.5 | 1.5 | 0–20 | 68.7 | 9.4 | 14.4 | 7.6 | 3 |
| Senna siamea | 5 | LR-B | 6°15'N | n/a | Rhodic ferrasol | 876 | T+C | Pruned | 0.5 × 4 | >2.0 | 10–20 | 48.3 | 37.5 | 7.1 | 7.1 | 9b |
| Senna siamea | 4 | LR-U | 7°22'N | n/a | Rhodic ferrasol | 1250 | T+C | Pruned | 0.5 × 4 | >2.0 | 0–10 | 53.8 | 12.8 | 20.5 | 12.8 | 9c |

[a] SR, Short rains; LR, long rains; U, unimodal; B, bimodal; WS, wet season; DS, dry season; [b] n/a, data not available; [c] production systems/treatment; T, tree; T+C, trees+crop; Fert, fertilizer added.
[d] Source: 1. Toky and Bisht (1992), Hisar, India; 2. F.K. Akinnifesi (unpublished data), Sao Luis, NE Brazil; 3. Hauser (1993), Onne, SE Nigeria; 4. W. Makumba and F.K. Akinnifesi (in preparation), Makoka, Malawi; 5. Rowland (1997), Makoka, Malawi; 6. Dhyani et al. (1990), Doon Valley, India; 7. Akinnifesi et al. (1995), Ibadan, SW Nigeria; 8. Akinnifesi et al. (1999a), Ibadan, SW Nigeria; 9. Vanlauwe et al. (2002), (a) Glidji, southern Togo; (b) Amoutchou, central Togo; (c) Sarakawa, northern Togo; 10. Rowe et al. (2001), Lampung, Indonesia (Soil pH for sites: 1 = n/a; 2 = 4.6; 3= 4.0; 4 = 5.4; 5 = 5.4; 6 = 5.5; 7 = 6.0; 8 = 6.2; 9a = 5.3; 9b = 5.3; 9c = 5.2; 10 = 4.8. Soil Org. C (%) at surface 0–20 cm: 1 = n/a; 2 = 1.42; 3 = 1.49; 4 = 0.76: 5 = 0.89; 6 = 0.58; 7 = 1.41; 8 = 1.24; 9a = 0.31; 9b = 0.29; 9c = 0.43).

**Table 4.2.** Indicative estimates[a] of rooting characteristics of 32 selected tree species and subspecies in 51 profile root distributions.

| Characteristics | Indicator | % of cases (n = 51) |
| --- | --- | --- |
| Surface root proliferation (0–30cm) | High | 33.33 |
| | Intermediate | 43.14 |
| | Low | 23.53 |
| Subsoil root proliferation (> 30cm) | High | 23.53 |
| | Intermediate | 43.14 |
| | Low | 33.33 |
| % Root in top 30 cm | High | 25.49 |
| | Intermediate | 47.06 |
| | Low | 27.85 |
| Rooting depth | Deep | 17.65 |
| | Intermediate | 33.33 |
| | Shallow | 49.02 |
| Root stratification (peaks) | <0–30 cm | 86.27 |
| | 30–60 cm | 5.88 |
| | >60 cm | 7.85 |
| Distribution pattern | Unimodal | 76.47 |
| | Bimodal | 9.80 |
| | Multimodal | 13.73 |

[a] Surface root proliferation (Lrv (root length density) < 0.5, low; 0.5–1.0, intermediate; > 1, high). Subsoil root proliferation > 30 cm depth (Lrv < 0.1, low; 0.1–0.5, intermediate; > 0.5, high). Rooting depth (< 1.5, shallow; 1.5–2.0, intermediate; > 2.0, deep). % Root concentration in top 30 cm (< 30%, low; 30–60% intermediate; > 60%, high). Root stratification (peak at < 0–30 cm; 30–60 cm; > 60 cm depths).
Source: Data re-analysed from Toky and Bisht (1992); Ruhigwa et al. (1992); Akinnifesi et al. (1999a,b); Akinnifesi et al. (1995); Odhiambo et al. (2001); Dhyani et al. (1990); Hauser (1993); Vanlauwe et al. (2002); Rowland (1998); Smith et al. (1999a); Makumba et al. (2001).

density at lower depths was generally low (<0.4 cm/cm$^3$). Mekonnen et al. (1999) reported that, at two sites (Ochinga and Muange) in the highlands of Kenya, the root length density of *Sesbania sesban* generally decreased with distance from the tree row and with depth. This general trend for a decline in tree root density as soil depth increases has been reported to occur in several multipurpose tree species, e.g. in *Prosopis chilensis* (Jonsson et al., 1988), in *Grevillea robusta* and *Gliricidia sepium* in Kenya (Odhiambo et al., 2001), in *Senna siamea* in three regions of Togo (Vanlauwe et al., 2002), in 13 woody species in Ibadan, Nigeria (Akinnifesi et al., 1999b), and in eight tree species in India (Chaturvedi and Das, 2003). However, it has been reported that maximum root concentrations can occur below a soil depth of 20 cm (Young, 1997; Schroth, 1998; Makumba et al., 2001). In a study reported by Chaturvedi and Das (2003), between 73% and 95% of root biomass was confined to the top 50 cm of the soil. This trend seems to contrast with the root distribution found in *Paulownia elongata*, which was found, in China, to have less than 2% of its roots distributed in the top 0–20 cm of soil (Zhu Zhaohua et al., 1986, cited by van Noordwijk et al., 1996). However, several investigators have observed root distribution patterns that are somewhat atypical for multipurpose trees, under diverse conditions. Such trends have been observed in *Leucaena leucocephala* in Ibadan, Nigeria, with maximum root densities occurring at a soil depth of 20–30 cm (Akinnifesi et al., 1995); *Clitoria fairchildiana* in northeastern Brazil (F.K. Akinnifesi, unpublished data) and *Acacia seyal* in Mali (Groot and Soumare, 1995).

As much as 50% of the total root length of some species has been reported to occur at a depth of 0–20 cm, e.g. *Calliandra calothyrsus* (Jama *et al.*, 1998a), whereas a proportion of total root biomass as small as 2% has been recorded at this soil depth in the case of *Paulownia elongata* (Zhu Zhaohua *et al.*, 1986, cited by van Noordwijk *et al.*, 1996). Root proportions of less than 10% have been reported in the 0–20 cm soil layer in other studies (Table 4.1). Root distribution among multipurpose trees grown on an Ultisol differed both laterally and vertically, with *Dactyladenia barteri* having fewer superficial roots and a deeper rooting system than the other species studied (Ruhigwa *et al.*, 1992).

By comparing fine root biomass in agroforestry and forestry ecosystems, Szott *et al.* (1991) found that, although some agroforestry systems may have twice the fine root biomass of annual cropping systems, they generally have less root biomass than natural forests. Nair *et al.* (1995) noted that, even when grown in pure stands, common multipurpose species (e.g. *Leucaena leucocephala*, *Senna siamea*, *Prosopis cineraria* and perhaps *Grevillea robusta* and *Paulownia elongata*) have less than 10% of the fine root biomass of corresponding natural forests in semiarid environments.

### 4.2.3 Lateral root spread

Depending on their environment, some astounding distances have been reported, in terms of lateral root spread, for some tree species. In semiarid Kenya, Rao *et al.* (1993) showed that the roots of 4.5-year-old *Senna siamea* trees extended laterally to about 9 m from the stem. Similarly, in a strongly acidic Ultisol in southeastern Nigeria, in which water was non-limiting, the lateral roots of this same species spread up to 15 m (Hauser, 1993). Stone and Kalisz (1991) showed the lateral spreads of the following species to be remarkably extensive, in the range of 30–50 m: *Quercus* spp., *Ulmus* spp., *Acacia koa*, *Juglans nigra*, *Adansonia digitata* and *Nuytsia floribunda*. The lateral spread of roots of other species, such as *Prosopis cineraria*, *Pinus ponderosa*, *Cariniana pyriformis*, *Hevea brasiliensis*, *Pinus sitchensis*, *Betula papyrifera*, *Pinus* spp., *Cupressus* sp. and *Eucalyptus camaldulensis*, ranged from 20 to 30 m.

Recent studies showed that of the eight species evaluated by Chaturvedi and Das (2003), three (i.e. *Acacia lenticularis*, *Dalbergia sissoo* and *Sesbania grandiflora*) had a symmetrical root distribution whereas the root distribution of each of the others was asymmetrical. All the species had roots close to the trunk, which branched away from its base in a fan-shaped pattern. Horizontal root spread varied between 8 m, in *Pithecellobium dulce*, and 1.7 m, in *Syzygium cumini* (at 5 years of age). The horizontal spread of the roots was between 1.1 and 1.6 times greater than the horizontal spread of the crown in *Acacia nilotica*, *Dalbergia sissoo*, *Pithecellobium dulce* and *Syzygium cumini*. A similar range of root : crown spread ratios was estimated by Prasad and Mishra (1984) for *Tectona grandis* (1.1) and *Terminalia tomentosa* (2.0) in 5-year-old trees. Toky and Bisht (1992) found root : crown ratios of between 1.23 and 1.26 in 6-year-old specimens of *Prosopis cineraria*, *Eucalyptus tereticornis* and *Populus deltoides*. Akinnifesi *et al.* (1999b) reported that the spread of the roots of 13 multipurpose trees in Ibadan ranged from less than 1 m to more than 2 m (at 4 years of age). Root excavations in Machakos, Kenya, found lateral tree roots that extended up to 5–15 m away from the tree trunk (Young, 1997). A lateral root spread of 5 m was reported for *Dactyladenia barteri*, and a lateral root spread of 15 m was reported for *Senna siamea* in a very humid site (an Ultisol) at Onne in southern Nigeria (Hauser, 1993). At the same site, Ruhigwa *et al.* (1992) reported that most of the roots of the four woody species grown in Onne (*Alchornea cordifolia*, *Dactyladenia barteri*, *Gmelina arborea* and *Senna siamea*) were concentrated in the top 0–20 cm of the soil. In the study, large-diameter roots were observed at the soil's surface in *Gmelina* and *Alchornea* hedgerows. Under semiarid conditions in Mali, the roots of *Acacia seyal* reached a distance of 25 m from the tree trunk, seven times the crown radius, and the roots of *Sclerocarya birrea* averaged 5.8 times the crown radius, and reached a maximum distance of 50 m from the tree trunk

(Groot and Soumare, 1995). However, the roots of some trees are confined to an area close to the tree trunk, e.g. *Eucalyptus camaldulensis*, having an almost uniform root biomass up to 80 cm depth, at about a 1 m radius, and peak root biomass at a depth of 80–100 cm (Jonsson *et al.*, 1988). Akinnifesi *et al.* (1999a) have shown that lateral root spread varied among 13 species studied, ranging from 0.6 m in *Albizia niopoides* to 3.5 m in *Nauclea latifolia*. *Alchornea cordifolia*, *Grewia pubescens* and *Triplochiton scleroxylon* had root spreads of more than 2 m from their trunks. In terms of spread, Chaturvedi and Das (2003) observed varied root : shoot ratios in multipurpose species. These ranged from 0.19–0.22 in *Sesbania grandiflora*, *Acacia lenticularis*, *Dalbergia sissoo* and *Acacia procera*, to 0.33–0.59 in other species (*Acacia nilotica*, *Albizia procera*, *Pithecellobium dulce*, *Senna fistula* and *Syzygium cumini*).

Similarly, in a study of the root activity pattern of wild jack tree (*Artocarpus hirsutus*), which used $^{32}$P, Jamaludheen *et al.* (1997) showed that most of the physiologically active roots of the tree were confined to the 30–70 cm soil layers, although the taproot might reach deeper layers. Horizontal root spread was rarely beyond 2.3 m during the first 10 years of tree growth. The study showed that, although lateral spread may reach 2.3 m, the effective or active roots (76% of the total root biomass), which are responsible for water and nutrient uptake, were confined to a 0.75 m radius, and only 6.2% of active roots occurred at a distance of 2.3 m from the tree.

### 4.2.4 Root size class distribution

Toky and Bisht (1992) showed that the number of primary, secondary and tertiary roots differed among the 12 species they studied. *Eucalyptus tereticornis*, followed by *Populus deltoides*, had the most abundant roots, mainly in terms of the highest number of primary and tertiary roots. *Acacia nilotica* had the highest number of primary roots. In an Ultisol in Sumatra, Indonesia, Hairiah *et al.* (1992) showed that *Gliricidia sepium* and *Senna siamea* had relatively few branch roots but these were, however, thick; *Calliandra calothyrsus*, in contrast, had numerous, thin roots.

Generally, most studies have dealt with fine root distribution in diverse ecosystems (Table 4.1). This is due to the early assumption that fine roots represent more than 95% of total root length. According to Bohm (1979) and Vogt and Persson (1991), the two main components of the tree root system are the main structural roots and the fine roots. The main structural roots are analogous to a skeleton in relation to the above-ground part of the tree: they form a base for the support and anchorage of the plant and may represent about 90% of its total biomass, whilst comprising less than 5% of its root length. In contrast, the fine root component (roots < 2 mm in diameter) consists of the long exploratory branched root systems, including the root hairs. The roles that the fine root network plays in nutrient and water uptake by the plant have been well reviewed (Schroth, 1995, 1998; Akinnifesi *et al.*, 1999a,b; Chapter 10, this volume). The root size distribution of 13 woody species in Ibadan showed that, with the exception of *Tetrapleura tetraptera* (45%), the fine root system (< 2 mm) makes up 86–99% of the total root abundance, whereas coarse roots constitute less than 10% of total root abundance in these species (Akinnifesi *et al.*, 1999a). Table 4.1 gives the distribution patterns of the fine root systems at different soil depths.

The root size distribution of *Leucaena leucocephala* was found to vary with fallow length, management and cropping intensity (Akinnifesi *et al.*, 1995). Generally, fast-growing woody species tend to have deep, extensive root systems. It has, therefore, been proposed that proper knowledge of the structural root systems and root architectures of trees is important and useful in selecting desirable species in agroforestry (Akinnifesi *et al.*, 1995, 1999a; Schroth, 1995).

Simultaneous agroforestry systems require rigorous research into tree species selections based on root architecture. According to van Noordwijk and Purnomosidhi (1995), the desirable root architecture is different for sequential and simultaneous agroforestry systems. Large tree roots, especially large storage roots in a specific region of the soil profile, are not necessarily very competitive; fine, rapidly absorbing roots, which 'turnover' at intervals of 14 days (or even more frequently), are the most competitive type of root (Eissenstat, 1992; Smucker *et al.*, 1995). The occurrence of large woody lateral roots will hinder seedbed preparation, cultivation and/or tillage, and may increase growth resource competition with food crops in simultaneous systems (Ruhigwa *et al.*, 1992; Akinnifesi *et al.*, 1999a). Trees with lateral roots that are confined to a distance of less than 1 m from the tree trunk are desirable for simultaneous agroforestry systems, especially those that require tillage of the inter-row spaces (Ruhigwa *et al.*, 1992; Akinnifesi *et al.*, 1999a,b).

The effects of large-diameter roots on crop nutrient uptake are not well known, but recent observations, made at a wetland tree–rice cropping system site in Bangladesh, found that crop yields declined as the number of large roots (> 5 mm in diameter) increased (Hocking, 1998). Pruning these larger roots had a significant, positive effect on the yield of the associated rice crops. However, the study was not able to demonstrate whether the reduction in rice yields resulted from the physical restrictions that the large tree roots imposed on the root growth of the rice or from the associated shading of the growing rice.

## 4.3 Simple Indicators of Root Distributions

### 4.3.1 Proximal root direction

The direction taken by the main roots that emerge from the stem base, known as the proximal roots, may provide insight into the location of the root system. These roots are easily revealed by excavating the soil from around a tree base: this can be done without causing substantial harm to the tree. This may be a quick way by which to assess both the rooting depth of trees and the effects that management techniques have on tree root systems. By applying an 'index of root shallowness' to 11 trees in Indonesia (calculated as: $Ds^2/\Sigma DH^2$, where Ds is tree stem diameter and DH the diameters of all roots descending at an angle of less than 45°), van Noordwijk *et al.* (1996) found that mango was the most deep-rooted species and *Pterocarpus integer* the most shallow rooted. However, this index depends on the existence of a reasonably constant relationship between stem diameter and total proximal root diameter. Using the index of root shallowness to compare *Gliricidia sepium*, *Grevillea robusta*, *Melia volkensii* and *Senna spectabilis* in Machakos Kenya, Ong *et al.* (1999) found that the index differed substantially according to tree age, and could not accurately predict tree–crop competition for trees that were more than 4 years old. However, a modified index, generated by multiplying $\Sigma D^2_{horizontal}$ by dbh (where dbh is defined as breast height diameter and D as the horizontal proximal root diameter) and taking tree size into account, gave a better prediction of competition in simultaneous systems.

### 4.3.2 Taprooting systems and vigour

Few studies have sufficiently quantified the rooting depth of the taproot system of trees in complex agroecosystems. Akinnifesi *et al.* (1999b) showed that, in Ibadan, the depths to which the taproot of 13 agroforestry tree and shrub species extended ranged from shallow (< 1.0 m) in *Dialium guineense*, *Pterocarpus erinaceus*, *Alchornea cordifolia*, *Grewia pubescens*, *Pterocarpus santalinoides*, *Millettia thonningii*, *Nauclea latifolia*, and *P. mildbraedii*, to medium (1.0–1.5 m) in *Triplochiton scleroxylon*, *Albizia niopoides* and *Tetrapleura tetraptera*, to deep (> 1.5 m) in *Enterolobium cyclocarpum* and *Lonchocarpus sericeus*. At all depths, *Enterolobium cyclocarpum* and *Lonchocarpus sericeus* had the most vigorous taproot system (as expressed by the taproot diameter and the taproot volume). These findings showed that taproots are generally situated at a depth that is shallower than the depth of the entire fine root system. Taproot volumes ranged from 0.68 $cm^3/m^3$ of soil in *Dialium* to 58.7 $cm^3/m^3$ of soil in *Enterolobium* (Table 4.3). Significant differences in root volume occurred among the 13 species in all soil layers. Form factor (defined as the factor that, when multiplied by stem basal area and root length, gives an

**Table 4.3.** Structural root volume and index of root taper (form factor) of the lateral and tap root systems of seven tree species grown on an Alfisol in Ibadan, 1994 (Akinnifesi et al., 1999b).

| Species | Height (m) | Root basal area (cm²) | Form factor | | Structural root volume (cm³/m³) | |
|---|---|---|---|---|---|---|
| | | | Lateral | Tap | Lateral | Tap |
| Albizia niopoides | 1.52 ± 0.09 | 84.3 ± 11.3 | 0.24 ± 0.09 | 0.15 ± 0.05 | 3.72 ± 0.9 | 13.1 ± 2.0 |
| Dialium guineense | 0.49 ± 0.06 | 4.5 ± 0.7 | 0.36 ± 0.10 | 0.01 ± 0 | 1.26 ± 0.1 | 0.68 ± 0.01 |
| Enterolobium cyclocarpum | 2.16 ± 0.17 | 98.6 ± 17.9 | 0.66 ± 0.11 | 0.42 ± 0.02 | 85.20 ± 12.2 | 58.71 ± 9.3 |
| Grewia pubescens | 1.36 ± 0.12 | 8.9 ± 2.5 | 0.33 ± 0.02 | 0.03 ± 0.01 | 17.16 ± 1.5 | 1.69 ± 0.6 |
| Lonchocarpus sericeus | 1.87 ± 0.09 | 69.4 ± 8.4 | 0.40 ± 0.02 | 0.23 ± 0.08 | 6.39 ± 1.0 | 32.57 ± 3.5 |
| Millettia thonningii | 0.97 ± 0.06 | 8.2 ± 2.2 | 0.30 ± 0.07 | 0.05 ± 0.01 | 5.83 ± 1.0 | 3.34 ± 0.7 |
| Nauclea latifolia | 1.29 ± 0.10 | 7.8 ± 0.9 | 0.75 ± 0.20 | 0.04 ± 0.01 | 140.72 ± 11.7 | 3.05 ± 0.7 |
| Pterocarpus mildbraedii | 0.87 ± 0.05 | 9.6 ± 2.0 | –[a] | 0.07 ± 0.01 | – | 4.73 ± 0.9 |
| Tetrapleura tetraptera | 0.69 ± 0.08 | 22.8 ± 4.8 | 0.75 ± 0.26 | 0.13 ± 0.09 | 9.55 ± 2.7 | 9.11 ± 1.1 |
| Triplochiton scleroxylon | 0.53 ± 0.04 | 34.6 ± 9.7 | 0.64 ± 0.17 | 0.15 ± 0.07 | 19.50 ± 3.0 | 12.08 ± 1.2 |

[a] No woody lateral roots >2 mm in diameter.

estimate of root volume adjusted for rate of taper or deviation from cylindricity) was low in *Dialium, Alchornea, Grewia, Pterocarpus erinaceus* and *Nauclea* and *Millettia*, indicating that the roots of these species taper rapidly away from the trunk. The $R^2$ value for the regression of taproot size against fine root abundance was generally more than 85%, except in the case of *Tetrapleura tetraptera* (Akinnifesi *et al.*, 1999a). The study showed that taproot size could be used to estimate the proliferation of fine root systems.

Vanlauwe *et al.* (2002) showed that the taprooting system of *Senna siamea* was different in three sites located in the savannah-derived ecosystems in the southern, central and northern regions of Togo. Taproot diameter was significantly larger in *Senna siamea* trees grown at Glidji (southern Togo) than in specimens grown at Amoutchou (central Togo) and Sarakawa (northern Togo), i.e. the taproot diameter was greater at the wetter sites than it was at the drier sites (see Table 4.1). Taproot penetration was depressed at Amoutchou due to the occurrence of a hardpan at a soil depth of 100 cm. In the other two sites, taproots extended beyond a depth of 2 m. Taproot vigour was greater at Glidji and the taproot tapered more rapidly than in Amoutchou. Shallower taprooting depths were exhibited by the seven tree species grown on an Ultisol in Indonesia (Hairiah *et al.*, 1992). Whilst *Senna siamea, Peltophorum pterocarpa* and *Albizia falcataria* tended to have a vertical taproot system, the taproot systems of *Gliricidia sepium* and *Erythrina orientalis* were slanting, while that of *Calliandra calothyrsus* was horizontal and superficial (0–30 cm). It is not known whether such patterns are due to the effect of damage caused to the taproots before they were transplanted or to a genetic feature of a species. Shallow rooting depths, generally of less than 1 m on the site, may be due to a strong, acid-soil-related effect.

### 4.3.3 Fractal branching models of root systems

The total length and biomass of tree root systems are important when considering their competitiveness and contribution to soil organic matter. However, these parameters are difficult to determine because root systems are, spatially, very variable. Several workers have developed methods, based on analyses of root branching characteristics, for assessing root length and biomass without undertaking intensive sampling (Fitter and Stickland, 1992; van Noordwijk *et al.*, 1994). These methods aim to predict the biomass and/or length of roots subtending from a root of a given diameter, thus enabling the prediction of whole-system properties from the diameters of the main or proximal roots. Root systems can be seen to develop according to simple rules that determine both the length to which a root grows before branching and the relative size of the branches. Underlying such an analysis is the 'pipe stem' theory propounded by Leonardo da Vinci and elaborated by Shinozaki *et al.* (1964). This rule states that the cross-sectional area of a fluid transport structure remains the same after branching, i.e. that the ratio of cross-sectional area before and after branching, p (previously denoted as $\alpha$), is 1 (van Noordwijk *et al.*, 1994; van Noordwijk and Mulia, 2002). Other descriptors essential for estimates of root system length and biomass are: (i) the average ratio of the cross-sectional area of the larger branch to the total cross-sectional area of both, or all, branches (denoted $q$); (ii) the average link length; and (iii) the degree to which links change diameter from one end to the other (i.e. their taper). To estimate root biomass the specific gravity ($g/cm^3$) of roots is also required.

The type of root architecture described by the above has become known as a 'fractal root branching model', since if branching parameters remained constant with root diameter the result would be a fractal structure (Fig. 4.1). In real root systems, variation occurs in $q$, $\alpha$ and link length. Much of this variation is random (Oppelt *et al.*, 2001), but there may also be trends with root diameter (Rowe, 1999). Branching rules are certain to break down for the finest roots, so the 'self-similarity' of the structure only applies over a limited range of root diameters. Ozier-Lafontaine *et al.* (1999) found that, for *Gliricidia sepium* root systems, root diameter

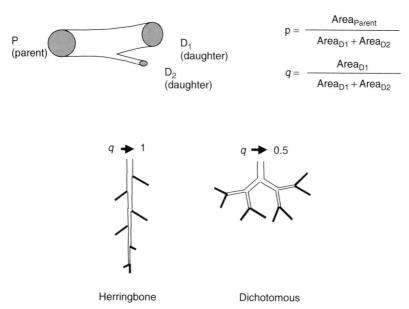

**Fig. 4.1.** Parameters needed for fractal root system modelling. For branches with two daughter links, $q$ varies between 1 and 0.5, representing 'herringbone' and 'dichotomous' root systems respectively.

had no consistent effect on p and $q$; the attenuation parameter p was increasingly variable at smaller diameters. Nevertheless, with some modifications, the fractal model can readily be applied to data from real root systems in order to estimate whole-system characteristics. Root biomass is comparatively easy to estimate, since larger-diameter roots make up most of the biomass and these obey fractal branching rules, which are more consistent and easily measured than those of fine roots.

The majority of total root length, by contrast, consists of fine roots, which are not easily characterized using the fractal modelling approach. Ong *et al.* (1999) found that fractal modelling worked well for predicting structural root system characteristics for *Grevillea robusta*, *Gliricidia sepium*, *Senna spectabilis* and *Melia volkensii* in semiarid Kenya, but seriously underestimated the length of fine roots (< 2 mm). Smith (2001) incorporated random variation in branching parameters into the fractal model, but the model still underpredicted total root length for *Grevillea robusta* by an order of magnitude. That author concluded that it is only possible to estimate root length using fractal branching rules alone if the algorithm is calibrated to adjust for errors in parameter estimation. In order to predict fine root system length, it may be necessary to combine the fractal method for estimating coarse root length with measurements of the average ratio of fine to coarse roots in soil cores. This approach was used by Rowe (1999), and provided an estimate of *Gliricidia sepium* root length per unit ground area that matched root length per area measured using soil cores. However, this model overestimated *Peltophorum dasyrrachis* root length per area. Further research is needed to determine whether the ratio of fine to coarse roots varies with season or soil type.

In order to fully characterize the topology of root systems, it is also necessary to record branching angles. Such work is labour intensive, but allows the important question of the vertical stratification of roots in the soil to be addressed. Ozier-Lafontaine *et al.* (1999) incorporated branching angles into a fractal model and produced estimates of *Gliricidia sepium* root length density distributions that were reasonably good matches to distributions measured by soil-based sampling. In a study by Chaturvedi and Das (2003), eight species had an angle of branching that was greater for primary roots (60–80°) than for secondary roots (47–64°) and which was greater in the top soil horizons than it was in the lower soil horizons, indicating that the spread of primary roots was greater than that of secondary roots. Relationships have been shown to exist between maximum rooting depth and shoot dry weight (van Noordwijk *et al.*, 1996) and between root biomass and breast-height diameter in tropical trees. Working with eight multipurpose tree species in India, Chaturvedi and Das (2003) found that dbh was significantly correlated at the 0.01 probability level with root biomass ($r = 0.58$), as well as with diameter × height ($r = 0.58$). Akinnifesi *et al.* (1999b) also reported that fine root density was strongly related to the distal soil depth and to the sum of squares of taproot diameter (> 5 cm) in 13 agroforestry tree species. The work of Eshel (1998) on dwarf tomato (*Lycopersicon esculentum*) showed that a whole (3-D) root system is amenable to fractal analysis. Testing fractal analysis of root architecture using *Gliricidia*, Ozier-Lafontaine *et al.* (1999) admitted that the fractal approach is amenable to modification, in comparison with explicit models, and is easy to apply. From a practical standpoint, coupling the fractal model with techniques that facilitate non-destructive calibrations of $\alpha$ and $q$ has proved useful (e.g. using electric capacitance for the *in situ* estimation of the size of root systems). Studies are needed that compare fractal analyses of various tree species grown under a range of external conditions. Such an analysis is cardinal to the determination of genetic and environmental (G × E) responses.

## 4.4 Tree Root Ideotype and Response to Management Practices

### 4.4.1 Response to soils and other site conditions

The distribution of the root system of a plant in space and time is a function of both genetic characters and site conditions. It is very difficult, if not impossible, to compare the root behaviour of trees and crops in varying environments, since virtually no systematic root experimentation has been undertaken to cover a wide geographical range. The ideal approach is to compare roots from isolated experiments from various sites. In general, tree root distribution is often limited by mechanical impedance, by anoxia, by moisture stress, and/or by soil temperature extremes (Stone and Kalisz, 1991).

The roots of 4- to 6-year-old specimens of *Grevillea robusta* (*Proteaceae*), grown on a relatively fertile soil (in comparison with most sites cited in this chapter) in Machakos, Kenya (1°33'S; Alfisol; unimodal rainfall (RF) 782 mm pa; 1560 m a.s.l.), reached a maximum depth of 1.6 m (Smith *et al.*, 1999a). This should be compared with a growth of 2.7 m in Nyabeda, Western Kenya (0°06'N; Alfisol; bimodal RF, 1800 mm; 1330 m a.s.l.; Smith *et al.*, 2001). Vanlauwe *et al.* (2002) observed a similar effect, also caused by site, on the differential distribution of a tree root system. *Senna siamea* roots in Glidji, southern Togo (6°15'N; RF 2400 mm pa), were more prolific at the soil's surface, and extended to a greater depth than was the case in Amoutchou, central Togo (1250 mm; 7°22'N), or Sarakawa, northern Togo (9°37'N), which had less rainfall, similar soil types and varying latitudes (Table 4.1). The large difference that existed in surface root proliferation between sites indicates that site conditions may play a significant role in root distribution patterns.

A high proportion of superficial root systems are associated with soils with low soil moisture levels (Coutts, 1983); deep root systems are likely to occur in areas subjected to drought. However, root : shoot ratio (on a biomass basis) tends to increase in soils with low levels of nutrients or high water stress. These findings contrast with the observations made by van Zyl (1988) of the root systems of grapevines in the dry region of South Africa. Plants growing in sandy and hilly areas tend to have shallow roots, especially under conditions of low rainfall and soil fertility and when the temperature of the upper soil layer is high.

It has generally been observed that tree roots are confined to the topsoil in strongly acidic soils (Hutton, 1990; Hairiah et al., 1992). According to Hutton and de Sousa (1987, cited by van Noordwijk et al., 1991b), *Leucaena leucocephala* (cv. cunningham) roots responded negatively (with reduced growth) both to strong soil acidity (pH 4.5–4.7) and to Al toxicity in Brazil, even after liming (2 t/ha). At Sao Luis Maranhao, *Leucaena leucocephala* (K8) grew well on a site with a pH of 5.0 (F.K. Akinnifesi, unpublished data, 1999). Hauser (1993) reported that *Senna siamea* grown on a strongly acidic soil (Ultisol) at Onne, near Portharcourt in Nigeria, exhibited a root growth that was apparently highly prolific. At that site, *S. siamea* had a rooting density of 16–35 roots per 100 $cm^2$ soil, a rooting depth that exceeded 1.6 m and a lateral root spread of 15 m from the tree trunk (Hauser, 1993). Hairiah et al. (1992) observed that, in an Ultisol in Sumatra, Indonesia, *S. siamea* had a taproot that extended to a depth of up to 70 cm, but had lateral roots that were confined to the top 20 cm of the soil. M.P. Gichuru (1990, unpublished results) excavated the root systems of *Senna siamea* growing in an Ultisol at Onne, Nigeria, and found that considerable vertical root restriction existed, which favoured extensive lateral root spread and horizontal growth of the taproot. These examples of the confinement of roots to the surface layers contrast sharply with the remarkably deep root distribution systems exhibited by *Gliricidia sepium* in Makoka, Malawi, where the fine roots penetrated to a soil depth of up to 5.6 m (Rowland, 1998). In soil zones characterized by a deep rooting depth, recycling of exchangeable bases accumulated by trees, especially those from newly weathered rock and leachates, might help to reduce soil acidity. The main limitation to root development in such soils is the high availability (toxicity) of Al and Mn ions in the subsoil. Considerable variation in tolerance to Al and Mn exists between crops and trees. The tolerance of different crops is based on physiological processes, e.g. fixation of Al ions in root cell walls, interactions with calcium and phosphate uptake and a local change of pH around the roots (Hairiah et al., 1992).

### 4.4.2 Response to moisture stress

The net result of genetic and environmental interactions contributes to the heterogeneous distribution of functional roots within the soil (Smucker and Aiken, 1992). According to Smucker and Aiken, the growth, respiration, exudation and turnover rates of roots associated with multiple morphologies at different phenological stages and soil environments affect the development of many symbiotic associations within the rhizosphere, which in turn modify the efficiency of root absorption of water. Drought stress may exert many regulatory effects on the biochemistry, physiology, and rates of growth of roots. These tree-root signals can cause changes in anion–cation concentration, hormonal contents (e.g. a reduction in cytokines and an increase in abscisic acid), pH and proteins and carbohydrates (Davies and Zhang, 1991). Water absorption efficiencies are a function of root distribution, rooting depth and of the spaces between absorbing root surfaces (Smucker, 1993). A relatively slow water flow, in unsaturated soil conditions, requires a more uniform distribution of roots within the soil matrix. Greater root uniformity minimizes the path lengths travelled by water to the surfaces of actively absorbing roots.

Scientists have expressed concerns that maximum, sustainable yields are at risk when extraordinarily large quantities of photoassimilates are allocated to root systems by plants subjected to short- and long-term soil water deficits (van Zyl, 1988; Smucker and Aiken, 1992). In pot experiments, the root biomass of grapevine was shown to have increased with moderate water stress (van Zyl, 1988). Accelerated branching in fine roots of less than 0.12 mm diameter seems to be induced by localized water deficits (Smucker and Aiken, 1992). Although the mechanisms of stress-induced branching are unknown in perennial woody species, an advancement of knowledge in these aspects may provide explanations for the major causes of tree–crop competition in drier regions.

In an experiment conducted at a bimodal rainfall (moisture limited) site at ICRAF's Machakos station in Kenya (ICRAF, 1994), it was shown that maize yield reductions due to intercropping ranged from 0 to 53% over eight seasons. However, the root systems of maize, grown in both monocropping and alley-cropping systems, showed different patterns. In *Leucaena*–maize alley cropping, the combined root density was two to six times greater than it was in a sole maize crop, which may increase overall nutrient use efficiency when intercropping with trees. It was demonstrated, however, by this study, that the presence of tree roots may not always reduce crop root growth in tree–crop spatial associations.

Tree rows can act as both a 'water pump' and/or a 'biological drain'; during early rains, taproots may redistribute topsoil water to the drier subsoil (Burgess *et al.*, 1998; Chapter 9, this volume). Under water-stress conditions, trees can absorb water from deeper soil horizons and release it into a drier, shallow soil layer during the night. Moreover, the concepts of 'root signal' and 'root communication' are beginning to gain attention in agroforestry, whereby roots can respond to water stress by sending a signal that can close the stomata in the leaves, thus minimizing water loss through evapotranspiration (Davies and Zhang, 1991). This concept needs to be verified and tested in other tree-based production systems and environmental gradients, in order to better understand the below-ground water dynamics involved.

### 4.4.3 Soil compaction

It is known that, under favourable environmental conditions, most plant species may develop a characteristic root system (van Huysstein, 1988). However, unfavourable conditions in the rhizosphere can bring about marked alterations to both trees and/or crop-root system(s), and can, thereby, exert certain limitations on the root–shoot balance in agroforestry associations (Archer *et al.*, 1988; van Zyl, 1988). Whether in a sole stand or in associations, optimum yields are difficult to obtain in shallow, compacted, dry and infertile soils, even with the best management. In soils where conditions in the deeper layers are unfavourable for root development, soil preparation may induce a larger root volume, due to increased growth and proliferation (branching) (Unger and Kaspar, 1994). Certain restrictions have been reported to impede root penetration in many South African grapevines, bringing production to a level that is unacceptable for economic production (Conradie, 1988; van Huysstein, 1988). Root restrictions, imposed by soil compaction, crusting and/or suboptimal microsite conditions in the rhizosphere, often result in a reduction in both shoot and root size (Bravdo *et al.*, 1992).

The adverse effects of compact soil horizons on root growth and yields have been recognized for many years (van Zyl, 1988; Unger and Kaspar, 1994; Meroto and Mundstock, 1999). A compact zone at a shallow depth, which prevents root penetration, is highly detrimental to plant growth and yield, and may provoke greater competition between the root systems of woody and non-woody components in simultaneous agroforestry. These effects may be more dramatic for soil conditions that depend solely on precipitation for the supply of water, or in rainfed systems in which erratic precipitation occurs, as in semiarid sites for example. In extreme situations, a compact zone that has developed in a dry soil may cause root girdling as a result of constrained radial root growth (Unger and Kaspar, 1994). There remains a gap in our knowledge of the effect that soil impedance (compaction) has on root growth in agroforestry systems.

Schroth (1995) reported a case, in central Togo, in which the roots of 4-year-old *Senna siamea* and *Acacia auriculiformis* trees penetrated a compacted hardpan layer, whereas the roots of *Cajanus cajan* could not penetrate the same layer, because it had a less vigorous rooting system. A similar trend was observed with *Senna siamea* roots grown in sites with hardpan or iron concretions in Amoutchou in central Togo (Vanlauwe *et al.*, 2002). Similarly, observations made in semiarid Kenya (rainfall 740 mm pa) showed that *Leucaena leucocephala* roots, unlike those of maize, penetrated beyond soft rock layers at a depth of 1.7 m (Govindarajan *et al.*, 1996). Smucker *et al.* (1995) reported opportunistic growth accumulations of roots by *Leucaena leucocephala* at profile depths that contained finer soil pores, slower water movement, nutrient accumulations and other, favourable edaphic factors that stimulated root growth. In a study aimed at simulating the effects of root impedance on root growth and nodulation in woody leguminous species, Akinnifesi *et al.* (1999c) showed that *Enterolobium cyclocarpum* was better able to penetrate soils with a greater root impedance (with a high bulk density and an artificial wax layer deliberately placed as an impediment to root growth) than *Leucaena* grown under the same conditions. This was because *Enterolobium cyclocarpum* has the greater root vigour, as indicated by its lower specific root length (SRL). Limited evidence suggests that species with high SRL will possess a greater plasticity in root growth and a better physiological capacity for water nutrient uptake. However, such roots will have a lower longevity, and they will exhibit high mycorrhizal dependence and a low capacity to withstand soil compaction and/or water stress (Eissenstat, 1992). The presence of a wax barrier, especially at a greater bulk density, reduced the root length and biomass in *Leucaena leucocephala*, but increased SRL in both species (Akinnifesi *et al.*, 1999c). *L. leucocephala* may invest its root biomass more efficiently than *Enterolobium cyclocarpum*, while it seems likely that the latter is better able to withstand soil compaction impedance and drought conditions (Akinnifesi *et al.*, 1999c). Using wax substrates, Taylor and Gardner (1960) showed that the penetrative abilities of the roots of legumes and non-legumes were not significantly different. The ability of roots to penetrate wax depends on wax rigidity, plant type, species, and soil density above the wax (Taylor and Gardner, 1960; Akinnifesi *et al.*, 1999c). These studies indicate that perennial root systems may be the less affected, as long as their first few roots can penetrate below the indurated layers to access water and/or nutrients needed to sustain tree growth. Since tree roots are often established in the profile before crops are seasonally introduced, crops are less able to compete with most trees for the limited resources available in compacted or dry soil environments.

### 4.4.4 Soil temperature

Root system expansion is a temperature-dependent process. Soil temperature controls the growth and initiation of new roots (Kaspar and Bland, 1992). Below-ground temperature affects root growth, branching, orientation and turnover (Kaspar and Bland, 1992). In dry regions, as soil warming spreads downward during the growing season, deeper soil layers become progressively more suitable for root growth. This fact, rather than plagiotropic or geotropic root growth, may explain the deeper root depth of trees in moisture-stressed areas. However, rate and capacity of downward root penetration may differ among tree species in different soil types. Maize root growth showed a sigmoidal, negative response to temperatures below 20°C or above 30°C. The optimum temperature for growth was 25°C to 30°C (Kaspar and Bland, 1992). There are some indications that low and high temperature ranges and tolerance vary in crop roots. Kaspar *et al.* (1981) reported the occurrence of genotypic variation, for both plagiotropic angle and temperature sensitivity, among soybean lines. The effects of high temperatures may be important in semiarid agroforestry, because of soil or microclimate temperature changes associated with the presence of trees.

### 4.4.5 Genotype-by-environment (G × E) interactions

Spatial and temporal distribution of tree root systems is usually influenced both by plant genetic characters and local site conditions (Akinnifesi *et al.*, 1999a). G × E interactions are very important in understanding how tree root configuration and distribution patterns may be affected by varying site conditions. There exists no standard or overriding analytical procedure for assessing G × E interactions that can be applied to all the trees that have so far received below-ground scrutiny. An understanding of G × E interactions gives a practical advantage insofar as it increases efficiency (in terms of time, labour and cost expended in the screening of tree roots) and reduces the number of representative sites that are required. However, it is not certain that global root research has enough data to support G × E strategies that could guide the process of selecting species based on their below-ground adaptation, competitiveness and nutrient recycling efficiency. That is, the selection of suitable tree species for diverse agroforestry uses could be limited by the lack of data from systematically conducted global trials (information from separate, individual trials is largely incomplete and non-uniform, as their objectives were often completely different). Quim (1996) proposed a stratified approach for assessing G × E interactions, which consisted of: (i) routinely estimating the stability of a trait across environments; (ii) measuring the stability of leading genotypes, in order to identify stable parents across environments; and (iii) conducting site-specific selection to identify material suitable for specific local environments. In crop breeding, this approach is a realizable goal for annual crops because such crops can be easily and rapidly tested across a wide range of environments. However, the challenge is much greater for agroforestry tree species because they require so much more time to grow and be screened than short-duration arable crops.

This chapter's review of the many published studies on root distributions reveals that management practices can be more important than G × E interactions; greater control of such management practices should therefore improve the efficiency of the process used to select species that are adapted to broader or more localized sets of environmental conditions. No study has, as yet, reported on the wide, inherent variability of root systems at species, provenance or cultivar levels, which may occur in different soil types subjected to a host of management systems during the different seasons and climates (Akinnifesi et al., 1999a). Until clear boundaries are identified between production systems, management practices, genotypes, and sites within regions, any effort to extrapolate to other locations the root data obtained at one particular site will remain a difficult task.

It is assumed that, when trees with a different genetic composition perform differently, despite the fact that they grew in the same environment, and were subjected to the same conditions, any differences may be attributed to their differing genetic codes. Several studies that have compared different species on the same site have reported varied performances that could only be attributed to the subjects' inherent genetic characters.

### 4.4.6 Production systems and tree husbandry

Tree rooting characteristics may change with crop husbandry, tree management and site conditions, all of which can be exploited in tree selection (van Noordwijk et al., 1991b; Schroth, 1995, 1998; Akinnifesi et al., 1995, 1999a,b,c,d). However, it is not always true that less competitive tree root systems are desirable in all agroforestry systems (de Montard et al., 1999). For example, in the *taungya* system (Nair, 1993) the contrary may even be desirable, i.e. crops that are less competitive than the tree are often preferred, as timber production is the major interest in this system (Akinnifesi et al., 1999a). In this case, vigorous tree growth is not considered to be an anathema; rather, crops with compatible root systems are often selected as intercrops with valuable timber species. In degraded lands that need to be rehabilitated by reforestation, being well-rooted, even insofar as having both superficial and deep roots, may be an advantage for quick establishment and erosion control.

### 4.4.7 Effects of pruning and lopping

In an acid soil (Ultisol) in Lampung, Sumatra, Indonesia, Hairiah et al. (1992) showed that the ratio of stem diameter to the biomass of the whole root system was relatively high for trees pruned at 50 cm above the ground. In simultaneous agroforestry systems, it may be possible to manipulate the tree root system, and thus competition for below-ground nutrients, early in the season and at determined periods thereafter, by removing or reducing the tree's shoots (see Chapter 17, this volume). Shoot pruning will reduce the root size and abundance and perhaps the tree's root activity in the soil (Akinnifesi *et al.*, 1995; Nygren and Campos, 1995).

Pruning the tree stem at a greater height has been found to decrease the number of proximal (i.e. large primary) lateral roots and increase root diameter (van Noordwijk *et al.*, 1991b). This may be due to the associated increase in the nutrient reserves available for root development. The allocation of a large biomass fraction to below-ground organs, and especially to structural storage roots, will increase the tolerance of trees to shoot pruning. A low pruning height and more intensive pruning has generally been suggested as a way of reducing tree–crop competition in simultaneous agroforestry systems. More field data are needed to determine the trends in root development in trees following pruning (of both shoot and roots) to orchestrate planting and management schedules in ways that minimize competition. The application of techniques that will optimize root functions and root interactions is needed, through selecting compatible tree–crop combinations, through shoot and/or root pruning, and through management techniques that ensure a higher shoot : root ratio of the arable component (Akinnifesi *et al.*, 1995, 1996, 1999a,b; Schroth, 1995, 1998).

In a root study in a *Leucaena*–maize cropping system in Ibadan, it was shown that regular shoot pruning of the hedgerows significantly reduced both the fine root density (61%) and the root diameter of *Leucaena leucocephala*, in comparison with an adjacent plot that had not been pruned for 4 years (Akinnifesi *et al.*, 1995). Further reduction in the trees' fine root proliferation in the top 100 cm soil layers was achieved by maize cropping for eight seasons, when compared with fine root proliferation in a short fallow or unpruned fallow. In addition, the intrusion of maize roots into soil directly under *Grevillea robusta* hedgerows has been reported (Huxley *et al.*, 1994), thus suggesting that a major volume of the crop roots may explore the soil beneath the hedges. The role of tree roots as a source of nutrients for the crop (resulting from root decay and the release of decomposition products), and the reduction in leaching due to the presence of tree roots cannot be overlooked or ruled out in agroforestry associations. This may have led to a large error in the interpretation of results from earlier root competition studies, as the number of tree roots found in the crop area was often used as a measure of competitiveness, and the synergistic effects (e.g. provision of nutrients to the crop) were ignored.

### 4.4.8 Effect of spacing

The effects of tree–crop competition in agroforestry can be minimized by increasing tree spacing or density (Akinnifesi *et al.*, 1999d). In an experiment on an Acrisol in northeast Brazil, it was found that a narrower spacing between tree rows of 2 m (compared with a wider inter-row spacing of 3 m) led to a greater maximum rooting depth and a greater fraction of superficial roots in pigeon pea (*Cajanus cajan*). The narrow tree spacing also reduced the root abundance, above-ground growth and yield of the intercropped maize (Akinnifesi *et al.*, 1999d). The availability of an unrestricted soil volume is probably an important factor dictating the size and distribution of the root systems. The size of a root system is largely diminished by increased interplant competition at higher densities. Generally, wide spacing is characterized by smaller root systems and horizontally spreading roots. Closely planted trees will produce a reduced amount of biomass and/or yield per tree than more widely planted trees, because of their smaller root systems. However, this is compensated for by the more intensive exploitation of the soil and nutrient resources and the available space, thus increasing the productivity per hectare (Akinnifesi *et al.*, 1999c). Managing spacing is also an effective way to reduce below-ground resource competition between crops and trees (Chapter 17, this volume). Growing trees with wider intra- and inter-hedgerow spacings can be effective for simultaneous systems in the semiarid tropics where soil moisture is a major limitation. A root experiment at Samford pasture station (27°32′S), northwest of Brisbane, Australia, has shown that, as tree spacing decreased, competition caused the trees to have deeper and denser rooting systems (Eastham and Rose, 1990). These observations suggest that intraspecific competition may alter the distribution of roots and their growth and hence their potential to exploit soil water and nutrients. The increased root proliferation often observed in closely spaced trees will also impact negatively on the growth and yield of companion crops in agroforestry systems.

### 4.4.9 Effect of intercropping, tree fallows and fertilization

Closely associated tree and crop roots often compete when the nutrient depletion zones around the roots overlap in the soil mass (Akinnifesi et al., 1996; Schroth, 1998). The root systems of rice crops were significantly deeper in sole rice plots than in mixtures of rice with *Leucaena leucocephala*, *Clitoria fairchildiana*, *Cajanus cajan* and *Inga edulis*, in an acid soil in Sao Luis, Brazil (F.K. Akinnifesi, unpublished data, Sao Luis, Brazil, 1999). However, the total abundance of rice roots was significantly lower in the sole rice plot than in the tree/crop mixture. Recent studies by Smith et al. (1999a) confirmed that, even with *Grevillea robusta*, which had been earlier acclaimed as 'non-competitive', root competition with maize was unavoidable, although pruning reduced its impact. Maximum root abundance of both crop and tree coincided in the upper profile layers. Significant suppression of both root length and downward root growth of maize occurred in the tree–crop mixtures compared with maize-only plots, and there was no spatial separation of the root zones of both the tree and the crop. The authors concluded that the previously held assumption, that *Grevillea robusta* is deep rooted with few superficial lateral roots, was incorrect for well-established trees.

Recent reviews of tree–crop root systems have indicated that competition may be two-way (i.e. tree on crop and vice versa), and in some cases crops may have a greater competitive advantage (Schroth, 1998; Akinnifesi et al., 1999a; Smith et al., 1999a). In hedgerow intercropping with *Senna siamea*, maize roots were reported to have outnumbered the tree roots by two to three times in the top 30 cm, even in plots where crop root density was low (ICRAF, 1994). Recently, root-space partitioning was observed between *Eucalyptus deglupta* and coffee in Costa Rica, showing no evidence of negative effects caused by *E. deglupta*, despite the fast growth of the tree (Schaller et al., 2003). The root system of coffee was reported to have restricted the rooting space of trees. These results invalidate the generally held belief that trees always have a greater competitive advantage than the crop component when grown in associations.

Only a very few studies have reported the effects of fallows and fertilization on root growth and distribution. Tree root growth in fallows can be assumed to be similar to shoot growth in fallows or plantations, i.e. generally increasing with the fallow length or tree age. Akinnifesi et al. (1995) have shown that the root abundance of *Leucaena leucocephala* increased with the length of fallow and decreased with an increased intensity of cropping and the removal of pruning residues.

The effects of increasing soil fertility, especially N, can be an increased fine root proliferation in the surface root mat layer of a forest, as indicated by the studies by Cuevas and Medina (1988). A similar effect, as a result of N accumulation from the addition of *L. leucocephala* prunings to the soil, was shown to have increased maize root growth to a level greater than that in the control plot (no added prunings) in Ibadan (Akinnifesi et al., 1995). Similarly, P application enhanced root proliferation in maize (J. Alegre, Peru, unpublished, 2001). On an acid soil, N application reduced surface root abundance in *Senna siamea*, whereas the effect on *Dactyladenia barteri* was negligible in the surface soil (0–30 cm depth) but increases in root abundance were found in the lower soil layers (35–95 cm) (Hauser, 1993). Figure 4.2 shows the effects of N application rates on *Gliricidia sepium* and

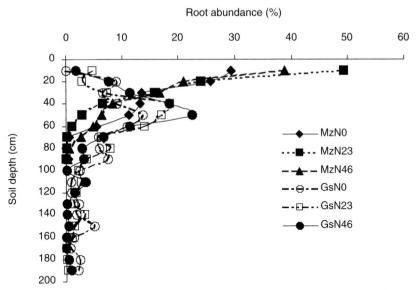

**Fig. 4.2.** Effect of N application on root abundance of *Gliricidia sepium* and maize in a simultaneous fallow intercropping system in Makoka, Malawi (Gs, *Gliricidia*; Mz, maize; N, nitrogen fertilizer at 0, 23 and 46 kg N/ha) (Makumba et al., 2001).

maize root distributions at Makoka, southern Malawi. Differential root stratification was observed between maize and trees at the site, and may explain the non-competitive association of *Gliricidia sepium* with maize in Malawi. Abundance of roots of both the trees and the maize were enhanced by N application.

## 4.5 Conclusions

Substantial amounts of information have been generated over the years on tree root architecture in diverse ecosystems. Root distribution patterns of trees were shown to be different among species and soil types. For example, *Senna siamea* exhibited differential root distribution patterns in three ecosystems in Togo. On a low altitude and high rainfall site *Gliricidia sepium* had 85% of its roots in the top 0–30 cm soil layer, but had less than 33% of its roots at the same depth in a drier environment in southern Malawi. Lateral and taproot volume vigour of 13 trees planted at the same site in Ibadan showed the effect of genotype on tree rooting systems, with lateral root volumes per m³ of soil ranging from 1.3 cm³ in *Dialium guineense* to 140 cm³ in *Nauclea latifolia*.

Associative or desired ideotype tree species can be selected based on root architecture. In Ibadan, *Enterolobium cyclocarpum* has superior lateral and taprooting systems. Therefore, that species would be suitable for reforestation or as a fallow species, where fast growth and the ability of trees to withstand erosion may be important. *Gliricidia sepium* is a suitable ideotype for simultaneous tree–crop intercropping systems. *G. sepium* is thought to have deeper roots (which are thus in a different soil layer to the crop roots) in low rainfall and high altitude environments. *Nauclea latifolia* is an example of an ideotype unsuited to association with food crops, due to its great lateral root proliferation and large root diameters. *Lonchocarpus sericeus* might be more suitable in simultaneous intercropping systems, because of its deep taproots and few lateral roots (which are maintained close to its trunk). It is suggested that, by exposing tree roots, it is possible to infer the root distributions of different species.

## Conclusions

**1.** Root distribution patterns of trees differ among climatic zones and soil types in predictable ways, with deeper root systems occurring in more seasonably dry environments.
**2.** Associative or desired ideotype tree species can be selected based on root architecture.
**3.** By exposing tree roots close to the stem, it is possible to infer the root distributions of different species.

## Future research needs

**1.** Selection of trees based on rooting architecture can only be meaningful at the species level at present, because data are not available for other levels. Thus, systematic investigations of tree root systems at provenance, clonal and cultivar levels are warranted.
**2.** More comparisons of root distributions with depth and lateral distance for the same species under different soil, intercrop and management conditions are needed to determine whether effects can be generalized.
**3.** There is a need to validate fractal models, and predictions based on proximal root direction, against measured soil-based root distribution measurements.

## Acknowledgement

The authors wish to thank DFID for financing their visit to Malang (Indonesia) in order to compile this chapter. The chapter benefited from useful inputs from M. van Noordwijk, D.M. Smith, D. Suprayogo, G. Schroth, K. Hairiah and W. Makumba during the planning stage. The first author wishes to thank the Canadian International Development Agency (CIDA), Canada, for providing the enabling environment under the Zambezi Basin Agroforestry Project.

# 5 Crop and Tree Root-system Dynamics

Meine van Noordwijk, Subekti Rahayu, Sandy E. Williams, Kurniatun Hairiah, Ni'matul Khasanah and Goetz Schroth

---

**Key questions**

1. When do roots grow? Is internal resource allocation to roots predictable? How do interactions in more complex systems affect this?
2. Where will root expansion take place? How quickly can root systems adapt their distribution to changing soil conditions (water, nutrient patches)?
3. How long do (fine) roots live? Is root turnover a major source of C input to the below-ground food web and to $C_{org}$ in the soil?
4. How can (or should) root system dynamics be represented in simulation models?

---

## 5.1 Introduction

Whereas the preceding chapter focused on the spatial aspects of tree roots, we will now consider the dynamics of root growth and decay in those crops, weeds, grasses, shrubs or trees that form a part of tropical agroecosystems, elaborating on van Noordwijk et al. (1996). Root turnover is important in the functioning of plants and agroecosystems for a number of reasons:

- Below-ground allocation of C (energy) may be around one-third of the C (energy) in the plant as a whole (Jackson et al., 1997; Wu et al., 2001).
- C (energy) provided by roots is a major source of C for the food web of soil biota.
- As roots are lost to 'rhizovory' (consumption), plants need to invest continuously in roots to maintain root length density.
- Through various channels, below-ground plant C allocation contributes to soil organic matter ($C_{org}$).
- Uptake by individual roots leads to 'depletion zones' being formed around them, and a situation whereby uptake is limited by the supply of nutrients via diffusion and mass flow (see Chapter 10, this volume); new roots can, therefore, start with a higher initial uptake rate.
- Through continuous root mortality and new root growth, a plant can adjust both the total size ('functional equilibrium') and the location ('local response') of its root system. This allows

it to increase its access to the (currently) most limiting resource (van Noordwijk et al., 1996), whilst maintaining its maintenance respiration cost at a level that can be sustained.
- As all plants have the ability to adjust their root systems (though the 'functional' and 'local' responses may differ in velocity and intensity), competitive interactions for below-ground resource capture should be seen as confrontations between organisms with different (long-term) strategies rather than just (short-term) tactics (which are determined by current root distribution, above-ground demand and supply of the below-ground resources).

The last decade has seen major progress in the quantification of root turnover at the plant and ecosystem levels (partially as a consequence of fears concerning the future impacts of elevated atmospheric $CO_2$, which have provided an impetus for the funding of such research). However, most, if not all, of the methods used to measure such dynamics are problematic (as we will discuss below); and we do not yet have comprehensive, well-tested models that relate all aspects of root turnover (as mentioned above) to the genotype and environment of plants. Of particular note is the fact that the rate at which root systems can adjust both their total size and spatial distribution (in order to meet the demands of the shoot and exploit supply in the soil) is still more a topic of speculation than of hard facts.

The standing biomass (or length) of the roots of any tree (perennial) or crop (annual) is the difference between cumulative root growth and cumulative root decay, from the time the plant begins to grow to the time of observation (Fig. 5.1). In annuals we can follow the process of root growth and decay from beginning to end because, almost by definition, all roots will die at the final harvest of the above-ground parts, which means that cumulative decay is equal to cumulative growth. In perennials, observation usually starts with an existing root system, and it may be found that their standing root biomass changes little, despite substantial growth and simultaneous decay.

The term '*root (biomass) turnover*' has been defined in various ways (Gill and Jackson, 2000; Schroth, 2003). However, different definitions usually all refer to 'cumulative root decay' (or 'root growth', assuming a steady-state size of the root system) divided by the average, maximum or minimum root

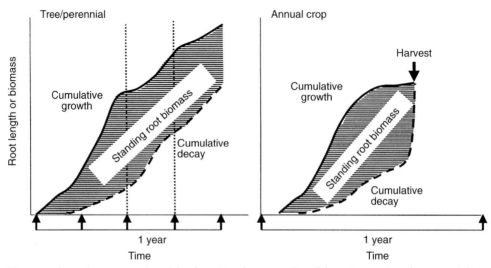

**Fig. 5.1.** Schematic representation of the dynamics of root growth and decay in trees (or other perennials) and in annual crops.

mass during the period. As Schroth (2003) discussed, use of the average will lead to more stable estimates. The same term ('turnover') is used to indicate different operational definitions in different publications, and care is needed when comparing the data obtained.

De Willigen and van Noordwijk (1987) defined *'root length turnover'* as the cumulative amount of roots that decayed during an observation period (1 year for perennials, one growing season, up to harvest, for annuals) divided by all roots that could potentially have decayed. The latter equals the standing root length at the start of the year *plus* the cumulative amount of roots that grew during the observation period. This definition results in a value for root turnover of 1 only where the standing root length becomes 0 at the end of the observation period. This definition can be used directly with measurement techniques that follow the fate of individual roots (such as minirhizotrons, see below), but not with methods based on sequential destructive sampling (see below).

Resource availability in the soil is dynamic in nature, with: (i) water entering from above as well as through subsurface lateral flows; (ii) nutrients either being released in the litter layer on top of the mineral soil or gradually weathering in the subsoil; and (iii) relative resource availability in the topsoil and subsoil changing with weather and seasons. The dynamic nature of soil resource availability therefore calls for a high degree of flexibility in root systems (higher than that normally found in the above-ground parts of plants). Such flexibility can be observed both when comparing plants of the same genotype growing on different sites, and when studying a single plant over its lifetime. Van Noordwijk *et al.* (1996) have already discussed the basic concepts of 'functional shoot/root equilibrium' and 'local response' in the context of agroforestry. In this chapter, we will explore how such concepts of the dynamics of tree–soil–crop interactions can be represented in simulation models.

Dead roots may be as important as live ones to the functioning of complex agroecosystems. Although the quantity of structural organic matter contributed to the soil by dead roots is generally less than the amount that arrives at the soil surface via litterfall, the specific location of decaying fine and coarse roots means that they contribute more to aggregate stabilization and the creation of soil structure (via biogenic macropores). The voids left by the partially decomposed remains of root systems can facilitate the growth of subsequent plant roots and their symbionts. An example of this is given in Fig. 5.2, which shows how the decaying roots of previous forest vegetation can provide a microenvironment that facilitates nodule development in subsequent tree plantations. In acid soils in the humid tropics, old tree root channels can play an important role in crop root penetration, water infiltration, the protection of

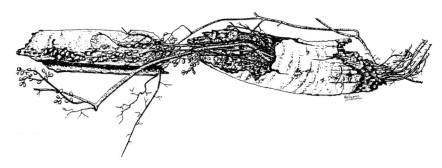

**Fig. 5.2.** *Acacia mangium* root growing inside a decaying tree root – a remnant of previous forest vegetation at a site in southern Sumatra (Indonesia). Inside this decayed tree root, the *A. mangium* roots had many root hairs and were profusely nodulated, whereas there was far less nodulation in mineral soil (drawing by Wiyono).

roots from Al toxicity and nitrogen management (van Noordwijk *et al.*, 1991a). Lucerne (alfalfa) has long been known to be a beneficial crop when planted in rotations with crops that have difficulties penetrating the subsoil. In a recent study, Rasse *et al.* (2000) found that lucerne root systems increased saturated hydraulic conductivity ($K_{sat}$) by 57%, total porosity by 1.7%, macroporosity by 1.8% (v/v) and the water recharge rate of the soil profile by as much as 5.4% per day. The enhanced soil structure resulted from the more severe drying/wetting cycles induced by lucerne, as well as from root turnover. The large increase in $K_{sat}$ relative to the increase in porosity suggests that root decay specifically increases the connectivity of macropores rather than their volume as such (see also Chapter 10, this volume).

The rate of root decay is also important with regard to the timing of landslide risks following forest conversion on sloping land. In the temperate zone, structural tree roots (which help to anchor the upper soil layer and the root mat to the subsoil) may take 5–10 years to decay. According to Sidle (1992), landslide risks following forest conversion peak in this period. In the tropics, with higher soil temperatures and more rapid decomposition, structural tree root decay may occur twice as quickly (e.g. 2–5 years), but no solid data exist yet to confirm such timing.

The rest of this chapter will describe patterns of root growth and root decay, before discussing: (i) empirical methods that may be used to quantify root dynamics; and (ii) the representation of these dynamics in simulation models. We will also discuss differences between different land-use types in root production and turnover – as the changes in root production and turnover that potentially result from a change in land use can trigger further changes in the below-ground ecosystem. This chapter is therefore meant to form the basis for the following discussions in this volume: root function (Chapters 6 to 10), below-ground carbon (Chapter 11), biological $N_2$-fixation (Chapter 13), mycorrhizas (Chapter 14), nematodes (Chapter 15) and below-ground food webs (Chapter 16).

## 5.2 Root Growth, Functional Shoot–Root Equilibrium and Local Response

### 5.2.1 Shoot : root ratios

Brouwer (1963, 1983) formulated the hypothesis that plants maintain a 'functional equilibrium' between shoot and root growth. He connected patterns in shoot : root ratios expressed by plants of the same genotype growing in different environments to a simple, hypothetical physiological mechanism. The growth of both shoot and root meristems requires resources acquired both above ground (carbohydrate) and below ground (water and nutrients), but the priority of access to these above- and below-ground resources differs between root and shoot meristems. When water and nutrients are in short supply, root growth can thus be favoured; where products of photosynthesis are in short supply, leaf growth can thus be favoured. Brouwer's hypothesis of shifts in the relative allocation of growth resources to root and shoot growth emphasized the ecophysiological functionality of these shifts, hence the name 'functional equilibrium'. Although the physiological mechanisms used by plants are certainly more complex than Brouwer formulated, this hypothesis is in line with broad patterns in relative root allocation across ecological zones (Chapin, 1980; Sanford and Cuevas, 1996). Shoot : root ratios in tropical forests range from 0.7 (on poor spodosols) to 2 (in tropical deciduous forests where seasonal water shortages occur), 4 (in montane forests) and 8 (in lowland humid forests). In wet ecosystems, the extremes are shoot : root ratios of 1 (for mangroves with well-adapted below-ground systems) and 100 (in riparian forests without a substantial below-ground compartment).

Broadly speaking, the main 'choice' for below-ground resource capture by plants is between exploring large areas of surface soil with lateral roots, and focusing on the capture of deep resources through a predominantly vertical orientation of roots. Mixed strategies are also possible, of course. Knapp (1973), Schulze (1983) and Breman and Kessler (1995) reviewed data on life histories and shoot–root allocation across ecological zones. Along a gradient from arid to humid climates, a number of shifts occur. The most arid environments are dominated by opportunistic, short-lived herbs with high shoot : root ratios that avoid drought conditions by setting seed early. Perennial strategies in extreme desert conditions are possible only for plants that hardly protrude above the ground (e.g. 'living stones'). Extreme differences in shoot–root allocation strategies can thus occur close together geographically, depending on the intervals between rainfall events (which are critical to perennials) and intensity of rainfall (which are especially relevant to the annuals that need to complete their life cycle on the water provided by a single rainfall event).

With slightly higher rainfall, dominance of the opportunistic herbs can be replaced by dominance of persistent perennial species that have drought-tolerance mechanisms and generally low shoot : root ratios. In many (semi)arid environments, the perennial strategy is only successful when deep roots can access below-ground water stores harvested from a substantial area or derived from long-distance subsurface flows. In such cases, plant establishment may depend on relatively wet episodes in the climatic cycle, which allow some of the roots to reach these deep resources. With increasing rainfall, the relevance of the 'lateral' strategy increases in the 'parklands' of the savannah zone. In this zone, a low tree density above ground is supported by a near-complete exploration of the topsoil by tree roots, which may extend to a distance of 50 m around a tree's stem. With a further increase in rainfall, an overall increase in shoot : root ratio is possible, and this is reflected both in an obvious increase in tree density and in increases in tree size, which occur to cope with the increased competition for light.

Shoot : root ratios are successively higher in savannah systems, tropical deciduous forests and lowland rainforests, increasing in each as the associated water supply increases. However, exceptionally deep-rooted trees in the humid tropics do occur and may be able to benefit from the light intensities of the 'dry' or 'bright' season for new leaf expansion (van Schaik et al., 1993; Nepstad et al., 1994; Wright, 1996). Where humidity is sufficiently high and the forest vegetation sufficiently dense, the herb layer is essentially replaced by an epiphyte layer inside the tree canopy (Holbrook and Putz, 1996). As these epiphytes have no access to the soil (which could be used as a buffer against temporary droughts), they may either 'avoid droughts' by physiologically shutting down between rainfall events, or by investing in large root systems, thus replicating the 'choice' in strategies obvious in the desert margin. In the rainforest zone, water availability may allow roots to focus on the surface layers, where most nutrients are available. In fact, on nutrient-poor soils (where plants resorb most nutrients from their leaves before litterfall, and thus where 'litter quality' is low and decomposition slow) a substantial part of the root system may be found in the 'root mats' within the surface layer, on top of the mineral soil. Buttress and stilt roots emerge above the soil surface to provide stability to the tall trees, as opportunities for the growth of taproots, and thus for below-ground anchorage, are limited. Under such circumstances, nutrient cycling can occur without the involvement of the mineral soil with its strong chemical (Al and Fe) sinks for P (Tiessen et al., 1993). Where litterfall is the major nutrient resource, above-ground deposits of litter in stem forks become an asset (Nadkarni, 1981). De Foresta and Kahn (1984) and Sanford (1987) described tree roots creeping up the trunks of their neighbours to benefit from above-ground litter deposits in Amazonian forests.

A recurrent theme in this tour of the world's biomes is the importance of seasonality, in the relative advantage of perennial versus short-lived strategies (with ample opportunity for coexistence in transition zones) and also in the shifts between lateral and vertical emphasis in root exploration. Within the life cycle of plants, a 'choice' must be made between the opportunistic strategies of fine roots (diebacks in unfavourable periods followed by regrowth when growth-limiting resources reappear) and root maintenance during periods of low activity.

In the past, agriculture was guided by the principle 'the more roots the better the crop growth'. However, evidence provided by simple agricultural systems points towards this being an overstatement (van Noordwijk and de Willigen, 1987; van Noordwijk et al., 1996), indicating that the highest level of crop production may be obtained in systems with relatively small root systems, which (even in terms of absolute size) are smaller than the root systems produced with a suboptimal water and nutrient supply. Selection pressure for high yields under monoculture may thus result in lower allocation to roots, although there is little evidence to date that such an effect is associated with high yielding cultivars of any of the major food crops. In intercropping situations, however, having a more extensive root system is valuable to the component species, as such extensive root systems increase competitive strength, even if they have limited benefits for system-level productivity. Models such as HyPAR (Mobbs et al., 2001) and WaNuLCAS (van Noordwijk and Lusiana, 1999) can evaluate both total water and nitrogen capture by combined root systems and the way these resources will be shared between the different plants (see Chapters 9 and 10). The 'functionality' of increased allocation of energy (or carbohydrate) resources to roots can thus be evaluated more quantitatively. It may well be that the 'functional equilibrium' response of plants has stronger 'functionality' for a plant when it grows in a multispecies situation, rather than in a monoculture.

When applying the 'functional equilibrium' concept in models (see below) it becomes evident that, in most situations (except in the case of small plants), increasing root length will be too slow a response to allow a meaningful immediate reduction of current stress conditions. This implies one of two things: (i) either the plant must rely on 'early-warning' indicators, rather than on the onset of stress, to signal that allocation should be modified (with all the usual uncertainty of what 'early-warning' signals actually mean in a fluctuating environment); or (ii) that the plant's response is only 'adaptive' when multiple stress/recovery cycles are involved.

Experiments with trees beyond the seedling stage have provided mixed (neutral or positive) results for the 'functional equilibrium' hypothesis. Joslin et al. (2000) compared 'normal' plots in a mixed deciduous forest at ambient rainfall with treatments where throughfall was diverted, so creating treatments with zero and two times the ambient throughfall. They found little change in net fine root production or standing root biomass, but there was some indication that both production and decay of fine roots in the wet treatment were higher than in the other treatments.

Seasonal drought in tropical moist forest may be the cue for fine root death and turnover. It may also trigger root growth in deeper layers, to access subsurface water and/or nutrients. Yavitt and Wright (2001) examined these possibilities by measuring fine root (<2 mm diameter) biomass and the timing of root growth and decay in an old-growth tropical moist forest on Barro Colorado Island (Republic of Panama) in the fifth year of a dry-season-irrigation experiment. Mean fine root biomass (at a soil depth of between 0 cm and 30 cm) was 3.7 Mg/ha within the control plot, versus 2.9 Mg/ha within irrigated plots. The direction of this change is in line with the functional equilibrium theory, but may be related to a faster rate of decay rather than to decreased allocation to new root growth. Average root longevity was estimated to be 1.14 years in the control and 0.82 years under irrigation.

The functional equilibrium hypothesis implies that (relative and potentially absolute) allocation to roots under non-stressed conditions is less than maximum. This implies that the energy/carbohydrate costs of root growth and maintenance constitute a substantial part of the overall carbon economy of the plant.

### 5.2.2 The carbon economy of the plant

The carbohydrate costs of developing and maintaining root systems are certainly not negligible (Buwalda, 1993). Therefore, an adequate representation of the amounts of photosynthate used in the growth, maintenance and uptake activity of roots is an essential part of accounting for the difference between 'gross' and 'net' photosynthesis, the former representing $CO_2$ entering stomata, the second the increase in above-ground biomass. The way below-ground respiration ($CO_2$ release) is partitioned over growth, maintenance, uptake and rhizosphere respiration is still subject of debate. It is difficult to draw a distinct line between roots and their co-habitants in the soil (Fig. 5.3), because sloughed-off root cap cells maintain metabolic activity while detached from the plant, mycorrhizas provide a continuum from plant to fungal tissue, and cell walls leak soluble carbohydrates. Therefore, no simple operational definition of the root–rhizosphere boundary exists that allows unequivocal measurements of the respective respirational activities.

New methods for separating the root from the rhizosphere component of total respiration make use of 'reporter genes', which indicate the specific activity of rhizosphere bacteria. So, further progress can be expected in this research area (Killham and Yeomans, 2001).

Jackson et al. (1997) estimated that as much as 33% of the global annual net primary production (NPP) is used for fine root production. Changes in the production and turnover of roots in forests and grasslands, in response to rising $CO_2$ concentrations, elevated temperatures, or altered precipitation or nitrogen deposition, could be key links between plant responses and longer-term changes in soil organic matter and ecosystem carbon balance (Norby and Jackson, 2000). The potential relevance this has for the global change debate has triggered substantial new research efforts to quantify root turnover (Vogt et al., 1998; Gill and Jackson, 2000). The results of experiments in which the $CO_2$ and/or the temperature around the shoot were increased have not been easy to interpret, as the short-term physiological response of plants tends to be overtaken by longer-term plant coordination effects and limitations of resource supply. A clear difference can be expected (van Noordwijk et al., 1998b) between situations in which water is the limiting below-ground resource and situations in which nutrients are the limiting factor. Increased $CO_2$ concentrations allow for a change in the physiological water use efficiency (amount of water lost in transpiration

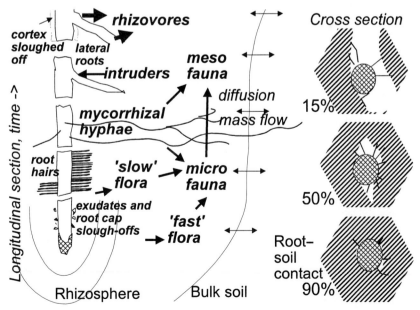

**Fig. 5.3.** Schematic view of events during the life of a single root axis (based on Clarholm, 1985, and Dhillion and Zak, 1993) (left-hand side of the figure), and a cross section (right-hand side) highlighting the range of root–soil contact situations that is likely to exist in structured soils (van Noordwijk et al., 1993).

per unit $CO_2$ absorbed), and thus allow higher plant growth rates to occur without a change in shoot/root allocation in instances where water is the limiting resource. Where nutrient supply is limiting, higher growth rates under elevated $CO_2$ will lead to (or increase) nutrient stress, and we can expect the allocation of carbohydrates to roots to increase (Pritchard and Rogers, 2000; Woodward and Osborne, 2000). In many tropical soils, P supply will probably remain the dominant limitation, in which case increased root allocation may be expected.

### 5.2.3 Optimum root longevity?

Costs and benefits of root turnover can be viewed from the perspective of a plant's carbon or energy balance, i.e. how much energy is needed to maintain roots in periods of low root activity (e.g. when the soil layer around the roots has dried out), compared with the energy costs entailed in making new fine roots as soon as conditions favour uptake again (Pritchard and Rogers, 2000)? Van Noordwijk et al. (1998b) calculated that for a growth respiration rate of 2 g $CH_2O$/g root tissue and a maintenance respiration rate of 0.03 g $CH_2O$/g root tissue per day, the break-even point (where maintaining roots in dry soil is as costly as letting them die and replacing them with new ones when conditions become favourable) would be about 60 days. Maintaining and rapidly revitalizing existing roots may give a competitive edge over plants that have to re-establish their fine root systems from main axes; so, we may assume that this 60-day estimate is low. Eissenstat et al. (2000) reviewed data on maintenance respiration in tree roots and found a value of 0.03 g $CH_2O$/g root tissue per day at 23.5°C for sugar-maple roots with a root N concentration of 1.5%. Roots with a 0.6% N concentration were found to respire at only 20% of this rate, whereas roots with up to 4% N were found to respire at four times the rate of roots with a 1.5% N concentration (probably due to a higher protein content). If the growth respiration rate were to be the same for these different types of roots, it would suggest that the break-even points occur at 300 days for roots containing low N concentrations (0.6%) and 15 days for roots containing 4% N. Maintenance respiration data given in the literature on this subject are, however, variable. Recently, Rasse et al. (2001) quoted values for fine and coarse tree roots of only 0.0006 and 0.0002 g $CH_2O$/g root tissue per day for, respectively, beech (*Fagus sylvatica*) and Scots pine (*Pinus sylvestris*) trees. These values would suggest much longer root lifespans at the break-even point than the examples discussed earlier.

### 5.2.4 Local response

Van Noordwijk et al. (1996) discussed the way plant roots respond to nutrient-enriched zones in the soil by enhancing local branch root development. This 'local response' is intricately linked with the nutrient and carbohydrate supply in the plant as a whole, and disappears if the plant as a whole is already well stocked with the nutrient locally available. Such branch root development is thus not just a response to local conditions, and can be understood in terms of competition between root meristems for carbohydrates and nutrients (either from external or internal sources).

The local response thus reflects coordination at the level of the root systems as a whole, rather than simply a mechanistic response to local conditions. Direct influences of local soil conditions on root growth do occur, however, as too much moisture may cause aeration problems, whereas too little may cause increased mechanical impedance and thus difficulties in terms of roots penetrating soil layers. Plants may differ in the degree to which these 'local responses' are expressed, even if we could make a comparison at equal internal nutrient supply and plant water status. In a study involving both grass and trees, nutrient enrichment of cores with NPK increased grass fine root production but, surprisingly, decreased oak root densities (Cheng and Bledsoe, 2002). A full mechanistic understanding of this differential response is lacking thus far, and simulation models have to rely on (over)simplified representations of these responses.

## 5.3 Problems and Opportunities for Measuring Root Dynamics

**Summary of this section: new perspectives on methods for studying root system dynamics**

**1.** Sequential destructive sampling methods: a lot of research time is still wasted on insufficiently replicated sampling methods that give inconclusive results.
**2.** Minirhizotron methods: use of improved schemes for analysing the data lead to higher turnover estimates than a previously used method, which used cumulative growth and decay data.
**3.** Experimental manipulations: root exclosures and modified soil patches (ingrowth cores) can be used to test hypotheses on root–root interactions.

Despite all the efforts made, commonly used methods for estimating root turnover can lead to biased results (systematic errors), uncertainty and wide confidence intervals. This is true both for those methods based on sequential sampling and for those based on repeated observations of individual roots using minirhizotrons (Vogt et al., 1998). Good introductions to the various root research methods discussed here can be found in Bengough et al. (2000), for sampling strategies; Oliveira et al. (2000), for auger sampling and ingrowth core methods; Smit et al. (2000a), for root observations at transparent interfaces with soil; and Hooker et al. (2000) for the measurement and analysis of fine root longevity. Schroth (2003) summarized how these methods can be applied in agroforestry research.

### 5.3.1 Sequential sampling

The idea underlying sequential sampling is a simple one: if one repeatedly samples root density in the same field or (agro)ecosystem, an increase in the value between two sampling dates indicates root growth whereas a decrease indicates root decay. As it is certainly possible that both an increase and a decrease occur in a given interval, one may expect such a method to give a conservative (under)estimate of root turnover. The method can, however, also overestimate root turnover, as the measurement of root density at any point in time contains 'measurement error' or uncertainty, due to the spatial variability of the roots in soil. In response to this uncertainty (for point estimates), the method has been modified to include a statistical test (normally a $t$-test), which can be used to assess whether a later data set differs 'significantly' from the one before. If the null hypothesis that two samples in a chronological sequence come from the same population is rejected, one records an increase as indicating net growth and a decrease as indicating root decay. However, a study by Singh et al. (1984) showed that serious (positive or negative) bias can occur in this method depending on the pattern of root growth and decay, the sampling interval and the number of replicates used. The method is, however, still commonly used without the potential for bias being properly acknowledged.

A simple way to understand the problem associated with this method is to realize that all estimates of root density (except the first and last of the series) are used twice in the estimation of differences and, thus, that the subsequent $t$-tests are not independent of each other. A root density estimate that, by chance, is rather high (i.e. one based on samples taken in areas of high root density) is likely to lead to the possibly erroneous conclusion that root growth occurred in the preceding interval, and root decay occurred in the subsequent interval. Although a restriction of the estimates to the 'statistically significant' differences takes out many of the relatively small differences from the summary of root turnover, it will include most of the large ones, even in the absence of any real change, simply due to the sampling errors. Figure 5.4 gives an example of the problem, which is derived from a spreadsheet calculation (available from the authors on request). The calculation includes sampling errors based on the 'coef-

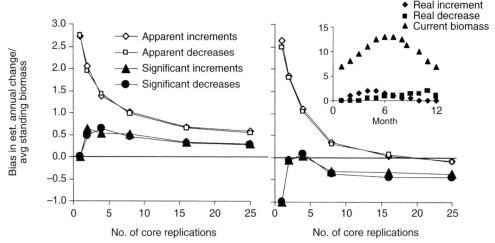

**Fig. 5.4.** Bias in estimates of annual root turnover based on a simulation of sequential core sampling, as a function of the number of replicate samples at each time interval. Different variants of the method were simulated, including either all apparent increments and decreases, or only those that indicate 'statistically significant' changes (with a $t$-test and 95% confidence limits). Results are shown relative to the average standing biomass, as averages for 25 years of observation using a monthly sampling scheme, and a conservative estimate of the coefficient of variation for individual cores (0.4) was used. On the left-hand side no change in roots was supposed to occur; on the right-hand side an annual turnover of 1 was assumed, along with one period where root growth dominated and one period where decay dominated (see insert for assumption about the root dynamics).

ficients of variation' that normally apply to root-core samples, even in 'homogeneous' vegetation (0.4–1.0; Bengough et al., 2000). For any number of replicates we can, therefore, estimate the sampling error of the mean, and simulate the data collection process using random numbers (assuming a *normal* distribution of the results per sample). Without restricting the increments to those that are 'statistically significant', the bias will be large in studies with a small number of samples, but still substantial in studies where 25 replicates would be used. Restriction to the 'statistically significant' changes still leads to a considerable positive bias (around 40% of average standing biomass) in the absence of change, and a negative bias of similar size in a simulation including a seasonal growth and decay pattern, even when 25 replicates are used. We can conclude that the bias obtained will be substantial when using this method, and that its sign (positive or negative) depends on the actual pattern of growth and decay being estimated. This means that we cannot easily apply a bias correction. Despite all the hard work that would go into such a sampling exercise, the results are likely to be disappointing or, worse, misleading.

Better results can usually be obtained (Schroth, 2003) with a compartment-flow model (Sanantonio and Grace, 1987), which applies estimates of the decay rate of dead roots in order to estimate outflow from the 'dead root' pool, derives the root decay from the difference between expected and measured pool size of dead roots, and then derives root growth from the difference between expected and measured live root pools. The method is obviously sensitive to the way the distinction between 'live' and 'dead' roots is drawn in data collection, the accuracy with which this distinction is made and the appropriateness of the assumption that the root decay rate is constant. Uncertainty in estimates of standing biomass and necromass pools, due to spatial variability and inadequate replication, still influence the turnover estimates derived in this way. Despite this generally acknowledged weakness in the methodology, sequential destructive sampling remains in use and is referred to by many published results. In a recent study in coffee gardens in Costa Rica, Chesney and Nygren (2002) encountered difficulties when applying the compartment-flow model devised by Sanantonio and Grace (1987) to fine root data, because necromass measurements did not match predictions made on the basis of a litterbag root decay constant of 0.00826/day. The main problems affecting their data are, probably, the erratic root biomass values they obtained, which do not fit any model and suggest undersampling of the spatial variation.

### 5.3.2 Minirhizotrons

Methods based on the repeated observation of individual roots growing in places where they can be observed (using rhizotrons, root observation boxes and minirhizotrons in the field) have become the main point of reference in studies of root dynamics (Hooker et al., 2000; Smit et al., 2000a). Observation techniques vary from tracings to photographic or video imagery. In subsequent analyses of change (using a number of visual criteria for root 'decay') a number of methods have been used, although all are related to the daily probability of survival (Box 5.1). Hooker et al. (2000) described methods

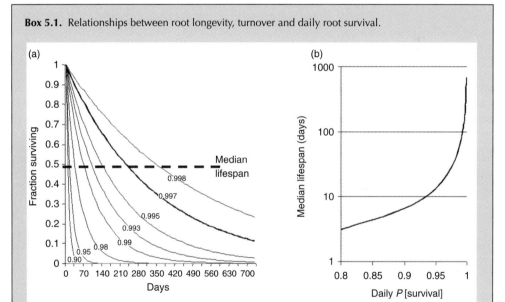

**Box 5.1.** Relationships between root longevity, turnover and daily root survival.

**Fig. B5.1.** (a) Expected survivorship curves for cohorts of roots with different values for daily probability of survival (P[survival]). (b) Relationship between daily probability of root survival and expected median root lifespan.

If $p$ = daily probability of survival for a unit root,
$dG/dt$ = daily growth rate,
$G$ = cumulative amount of newly formed roots,
$dD/dt$ = daily death (decay) rate,
$D$ = cumulative amount of decayed roots,
$L$ = median life span roots (number of days that 50% of the roots survive to) and
$S$ = standing biomass,
then we can derive the following relations for a steady-state population of roots, where $dG/dt$, $dD/dt$ and $p$ can be approximated as constants.

*Continued*

**Box 5.1.** *Continued.*

The relation between half-life time and daily survival probability is:

$p^L = 0.5$, or $L = \log(0.5)/\log(p)$ \hfill (1)

On any day the following relations must hold for a steady-state system:

$dG/dt = dD/dt = (1 - p) S$, or $S = (dG/dt)/(1 - p) = (dD/dt)/(1 - p)$ \hfill (2)

With these definitions, we can derive the basic turnover rate $r$ as:

$r = (dD/dt)/S = 1 - p$ [per day] \hfill (3)

Where turnover is expressed on a different time scale (e.g. per year), the $p$ value has to be adjusted (e.g. $p$[survival for 1 year] = $p$[survival for 1 day]$^{365}$ if $p$ can be assumed to be time-independent). The root length turnover (RLT) of de Willigen and van Noordwijk (1987) after $N$ days of observation, and for a steady-state population, can be calculated as:

$RLT = D_{fin}/(S_{start} + G_{fin}) = N p /(S + N p) = 1/(1 + S/(N p))$

and thus increases with the length of the observation period $N$.

The 'root length replacement ratio' (RLRR) of de Willigen and van Noordwijk (1987) is:

$RLRR = G_{fin}/S_{fin} = G_{fin}/(S_{start} + G_{fin} - D_{fin}) = 1/(1 + (S_{start} - D_{fin})/G_{fin}) = N p / S$

and also increases with the length of the observation period. So a value of 1 is obtained if all roots present at the start have decayed at the end of the observation period, while all of the new growth remains present.

The daily probability of survival, $p$, may indeed be the most efficient indicator, as it relates directly to other measures.

---

based on following the development and fate of individual roots or cohorts of roots first seen at the same observation time. In the practical applications of this method there tends to be a problem, in so far as most cohorts are small. This is especially problematic if one wishes to analyse individual replicate samples to test the significance of certain experimental factors.

Van Noordwijk et al. (1993) used a simpler approach based on cumulative root growth and decay patterns, with logistical distributions fitted through both. The time between the date by which 50% of seasonal root growth had occurred (by interpolation) and the date by which 50% of seasonal roots had decayed is used as an indication of 'median root longevity'. Though simpler in its resulting data structure, the latter method may, however, lead to bias and uncertainty in the resulting estimates of root longevity, as shown in a recent analysis (M. van Noordwijk et al., unpublished) and summarized in Fig. 5.5. For the study considered in Fig. 5.5, a number of possible growth curves for roots were used, together with stochastic predictions ($p$) of root decay, which were based either on a homogeneous probability of decay for any root during a standard interval or on a decay rate that depends on root age.

In the absence of random variation or age effects on root survival, root decay following a single pulse of root growth is correctly described by the method tested. Median lifespan and probability of survival are returned by the procedure without appreciable bias (Fig. 5.5). For exponentially increasing root growth patterns the bias is also very small, but for linear or logistical root growth the median longevity is overestimated by up to 3 weeks, whereas the saturation pattern (initial rapid root growth, at a declining rate) is overestimated by up to 8 weeks. The lower the weekly survival probability, the larger the bias in estimates of $p$. Whilst the method handled a single-pulse situation adequately, a double-pulse situation is particularly prone to bias, depending on the interval between the two pulses.

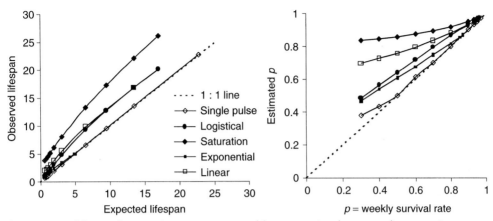

**Fig. 5.5.** Bias (difference between input parameters used for constructing data sets and output estimates derived from those data sets) in estimates of median lifespan and weekly survival when the data processing method of van Noordwijk et al. (1993) is simulated for a range of time patterns of root growth.

A new method (M. van Noordwijk et al., unpublished) was designed for estimating a daily survival rate $p$ that is valid regardless of the root growth pattern. In the analysis involved in this new method, the observed cumulative root growth pattern was used as an input. The shape of the cumulative root decay curve was predicted for a range of values of $p$. The method (Fig. 5.6) then includes the selection of the $p$ value with the least 'lack of fit', and a test for indications of the time-dependence of $p$ (a spreadsheet with the whole procedure preprogrammed is available from the authors). The method yields an estimate of $p$ that can be converted to give the time by which 50% of any cohort of roots can be expected to have died, indicated here as 'median longevity' of roots. Of course these estimates primarily refer to the longevity of roots at the observation surface: their relevance for roots growing in undisturbed soil remains a subject of debate.

Tierney and Fahey (2001) compared minirhizotron estimates of fine root longevity and production in the forest floor of a temperate broadleaf forest with observations made using surface windows (without any access tubes or modification of the root environment). Their conclusion, that the two methods gave the same survival rate for fine roots (< 0.1 cm in diameter), is comforting for all minirhizotron studies. Annual fine root production in the northern hardwood forest studied was approximately equal to standing biomass (for a median root longevity of 314 days), and had previously been underestimated using root ingrowth cores.

### 5.3.3 Other methods and comparisons between methods

#### 5.3.3.1 Ingrowth cores

Qualitative data on the relative patterns of root growth during a year can be obtained by repeatedly inserting fresh 'ingrowth cores' into the soil and measuring the amount of roots that colonize each core. However, the absolute growth rates obtained cannot be directly compared, as it is impossible to avoid disturbance: existing roots will be cut when the core is inserted, and the repacked soil in the ingrowth core will differ in structure from the surrounding soil. Furthermore, ensuring contact between the surrounding soil and an ingrowth core is problematic, however much care one takes.

However, the ingrowth core technique may be most suitable for comparisons between cores that have been deliberately modified. Hairiah et al. (1993) exchanged subsoil and topsoil in a study of aluminium tolerance and avoidance by the velvet bean (*Mucuna pruriens*), with and without the

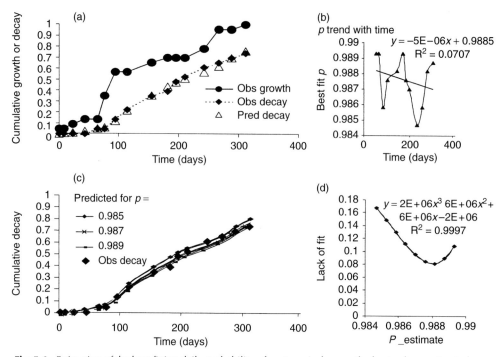

**Fig. 5.6.** Estimation of the best fitting daily probability of root survival, $p$, on the basis of reconstructed decay curves, given the observed root growth pattern at individual minirhizotrons (a); the procedure includes a check for trends of $p$ with time (b), and calculations of the expected decay curve for a range of $p$ values, to minimize the lack of fit (c, d).

addition of P to the soil. Williams (2000) compared a number of nutrient additions to ingrowth cores in young rubber plantations in the expectation that the relative 'local response' obtained would help to identify nutrient stress in the tree in a sensitive way. However, due to an exceptionally long dry season, little actual root growth occurred and the results did not confirm that this method can be used.

### 5.3.3.2 Pulse labelling

Root allocation of carbohydrates can be traced by the use of 'pulse-labelling' techniques, utilizing either the radioactive isotope $^{14}C$ or the stable isotope $^{13}C$ (or both). For example, on a highly productive temperate pasture in New Zealand, Saggar and Hedley (2001) used a $^{14}C$ pulse-labelling technique to measure seasonal changes in the assimilation and partitioning of photoassimilated C in plant, root and soil components of the agroecosystem. Of the net primary production rate of 32.8 Mg of C/ha/year of C, at the end of the year 18.2 Mg were found to have been respired, whereas 6.5 Mg remained in shoot biomass, 6.8 Mg in root biomass and 1.3 Mg in the soil. The half-life of C in the below-ground system (no distinction could be made between C in live roots, the rhizosphere and dead roots) differed between seasons, and was found to be 111 days for autumn roots, 64 days for spring roots and intermediate for the other seasons.

From a pulse-labelling experiment conducted on basket willow (*Salix viminalis*), de Neergaard et al. (2002) concluded that 41% of the $^{14}C$ recovered had been allocated to below-ground pools. Roughly 20% had been allocated to root biomass after 5 months (although this had peaked at 25% after 3 months). A further 9% was accounted for by root + soil respiration (mostly in the first month), 2% by microbial biomass and 10% (gradually increasing) by soil organic matter pools. Up to 13% of the carbon in the microbial biomass pool had apparently been assimilated by the willows in the preceding 4 weeks.

*5.3.3.3 Root litterbag incubation studies to look at root decay*

Root decay can be measured using the general approach of the 'litterbag' method for studying the decay of above-ground litter (van Noordwijk, 1993). However, the method is not as straightforward for root decay as it is for above-ground litter, as:

- roots decay when in close contact with the soil. So, separating them from soil to obtain their initial weight and then repacking them in pre-sieved (rootless) soil may lead to biased results, even if care is taken to let the soil's temperature and soil water content fluctuate in line with the surrounding soil;
- unlike above-ground litter, decaying roots cannot be intercepted at the time of 'litterfall', so the initial stages of decay are more difficult to capture. If one begins with samples from the pool of dead roots in the soil, results may be biased towards the more recalcitrant fractions.

The practical aspects of this technique and the results obtained have been discussed by Henrot *et al.* (1996). Van Noordwijk *et al.* (1998a) compared the decay of above- and below-ground inputs for four hedgerow species and found root decay to be slower than that of above-ground litter in all species. However, the relative ranking of these species in terms of root decay did not match that found for above-ground litter decay.

*5.3.3.4 Comparisons of methods*

Hertel and Leuschner (2002) compared four methods for estimating fine root production in a *Fagus–Quercus* mixed forest. They found differences of more than an order of magnitude between the four methods. Fine root production estimates derived by sequential coring in conjunction with compartmental-flow calculations were larger than those derived from sequential coring with maximum–minimum calculation of root production. Estimates obtained using either method were larger than those derived using the ingrowth core method and a recently developed root-chamber method for individual fine roots. A C-budget model for the site implied that 27% of net carbon gain was allocated below-ground, with a fine root production that was closest to (though 20% lower than) results for the sequential coring technique with the maximum–minimum method of calculation.

## 5.4 Empirical Data on Root Growth and Decay

Gill and Jackson (2000) compiled and analysed a global data set of 190 studies on root turnover estimates across climatic gradients and vegetation types, based on sequential root biomass estimates. The data refer to various methods and conditions, and a substantial scatter is thus to be expected. Root turnover was, for this overview, defined as annual root production divided by the maximum standing root biomass and thus was expressed on a per-year basis. Root turnover estimates increased exponentially with mean annual temperature, in the cases of: (i) fine roots in grasslands and forests; and (ii) total root biomass in shrublands – though for each vegetation type a large share of the variation observed remains unexplained. The $Q_{10}$ value (the increase in process rate for a 10°C increase in temperature) for root turnover was 1.4 for fine roots (< 5 mm diameter) in forests, 1.6 for fine roots in grasslands and 1.9 for shrublands. After adjusting for the temperatures of the sites, there was no statistically significant relationship between turnover and precipitation. The slowest average turnover rates in the Gill and Jackson (2000) data set were found to occur in the whole tree root systems (0.10/year), followed by shrubland total root systems (0.34/year), and fine root systems in grasslands, wetlands or forest (all within the narrow range of 0.53–0.56/year). For tropical versions of these vegetation types, fine root turnover was typically 0.6–0.9/year. Data for forests show a strong relationship between root diameter and turnover, with average turnover being 1.2/year in roots of diameter 0–1 mm, 0.52/year in roots of diameter 0–5 mm, and turnover decreasing to 0.1/year for roots in the 0–10 mm diameter class (note that all of these classes may be indicated as 'fine tree roots' in various studies).

Root longevity can be positively correlated with mycorrhizal colonization and tissue density, and negatively related to nitrogen concentration, root maintenance respiration and specific root length (Eissenstat et al., 2000). Hooker et al. (1995), however, reported that mycorrhizal roots live less long than non-mycorrhizal roots. Branched root axes, even if they have a small diameter, tend to live longer than unbranched axes of the same diameter (literature reviewed in Eissenstat et al., 2000). Pritchard and Rogers (2000) reviewed published data on root longevity observed using the minirhizotron technique. For annual crops in the temperate zone, the values published range from 130 days (or virtually the whole lifespan of the crop) for leeks, winter wheat and sugar beet, to 24 days for groundnut and grain sorghum. For herbaceous perennials (i.e. lucerne and sugarcane), published values of mean root lifespan range from 14 to 131 days. Among fruit trees, apple roots (which have a relatively small diameter, a low tissue density and exhibit little lignification of the exodermis) have much shorter lifespans (50% of new fine roots may die within 2 weeks of being formed) than the roots of citrus (50% survive for about 300 days), the latter exhibiting completely opposite physical traits (Eissenstat et al., 2000).

A comparative study has also been made of pine (*Pinus resinosa*) and poplar (*Populus tristis* × *P. balsamifera*). The median longevity of fine pine roots (with a specific root length of 16 m/g, and a standing fine root biomass of 0.62 Mg/ha) was found to be 291 days, as compared with 149 days for poplar roots (with a specific root length of 57 m/g, and a standing fine root biomass of 0.36 Mg/ha; Coleman et al., 2000). Where pine had half the average total root length per unit area, it maintained more root biomass; associated net $CO_2$ efflux into the soil was also slightly higher. Munoz and Beer (2001) measured fine root biomass (less than or equal to 2 mm) and productivity over 1 year in 16-year-old plantations of cacao (*Theobroma cacao*), shaded by 15-year-old *Erythrina poeppigiana* or *Cordia alliodora* and planted on a deep alluvial soil in Turrialba, Costa Rica. A fine root biomass of approximately 1.0 Mg/ha varied little during the year, giving (at the beginning of the rainy season) maximum values of 1.85 Mg/ha in the cacao–*C. alliodora* system and 1.20 Mg/ha in the cacao–*E. poeppigiana* system. Annual fine root turnover was close to 1.0 in both systems. Fine root production by both *C. alliodora* and *E. poeppigiana* (maximum of 205 and 120 kg/ha per 4-week period, respectively) was greatest at the end of the rainy season, whereas that of cacao was greatest at the beginning of the rainy season (34–68 kg/ha per 4-week period).

Some results obtained from minirhizotron studies in agroforestry experiments in Indonesia are summarized in Tables 5.1–5.3. The first data set was obtained in three long-term cropping system trials in Lampung (Hairiah et al., 2000c; Table 5.1). The shortest fine root lifespans and highest daily turnover rates (1 − daily survival probability) were found to occur in the leguminous cover crop *M. pruriens*, followed by groundnut and maize. The median longevities of all three species were between 18 and 25 days, whilst all three had a turnover rate of about 5%/day. The various trees studied had median fine root longevities of about 100 days and turnover rates of about 1%/day. The longevity and turnover rate of the perennial grass *Imperata cylindrica* fell between those of the annual crops and trees in these experiments.

Using the same methodology, two experiments (in Jambi province, Sumatra) on the early phases of rubber agroforestry systems were also compared (Tables 5.2 and 5.3). The first experiment yielded a daily turnover of around 0.5% and a median lifespan of 290 days. Neither weeding intensity nor position with respect to the tree had any statistically significant influence on these esti-

**Table 5.1.** Root dynamics as observed using minirhizotrons at the Biological Management of Soil Fertility (BMSF) site in North Lampung, Sumatra, Indonesia, in cropping system trials during the period 1996–1999. Data from six minirhizotrons per species, with daily survival rate and median lifespan (age expected to be reached by 50% of each cohort of roots) evaluated for individual replicates.

| Plant species | Remarks | Daily survival probability[†] | Median lifespan (days) | Daily turnover (%) |
|---|---|---|---|---|
| *Gliricidia sepium* | Regularly pruned in alley cropping | 0.9904[a] | 116.8[a] | 0.96 |
| *Peltophorum dasyrrachis* | Regularly pruned in alley cropping | 0.9877[a] | 121.1[a] | 1.23 |
| *Flemingia congesta* | Regularly pruned in alley cropping | 0.9916[a] | 96.1[a,b] | 0.84 |
| *Zea mays* (maize)[*] | In alleys or as monocrop | 0.9435[c] | 25.2[c] | 5.65 |
| *Arachis hypogaea* (groundnut) | In alleys or as monocrop | 0.9547[b,c] | 20.4[c] | 4.53 |
| *Mucuna pruriens* var. *utilis* | As cover crop in rotational system | 0.9534[b,c] | 17.7[c] | 4.66 |
| *Oryza sativa* (upland rice) | In alleys or as monocrop | 0.9614[b,c] | 33.8[c] | 3.86 |
| *Imperata cylindrica* | As weed on fallow land | 0.9844[a,b] | 55.2[b,c] | 1.56 |
| Grand mean | | 0.9679 | 59.8 | |
| Probability | | $P = 0.005$ | $P < 0.001$ | |
| Standard error of difference between means | | 0.0188 | 30.1 | |

[†]Means in one column labelled with the same letter are not significantly different at the 5% level.
[*]A separate test confirmed that maize root dynamics were not significantly different when maize was alley cropped (between three different species of hedgerow trees), intercropped with cassava or planted as a monocrop.

**Table 5.2.** Root dynamics as observed using minirhizotrons in a rubber (*Hevea brasiliensis*) agroforestry experiment in Rantaupandan, Jambi, Indonesia (Williams, 2000), during 1997 and 1998, with weeding intensity as the main experimental factor and the distance from the tree as the sampling position. Data from four minirhizotrons per sampling position and weeding intensity combination, with daily survival rate and median lifespan evaluated for individual replicates.

| Experimental factor/sampling position | Daily survival probability | Median lifespan (days) | Daily turnover (%) |
|---|---|---|---|
| 'High' weeding intensity | 0.9960 | 241 | 0.40 |
| 'Low' weeding intensity | 0.9955 | 344 | 0.45 |
| Within tree row (0.25 m from tree) | 0.9953 | 328 | 0.47 |
| Between tree rows (1.5 m from tree) | 0.9962 | 257 | 0.38 |
| Grand mean | 0.9958 | 293 | |
| Probability | NS | NS | |
| Standard error of difference between means | 0.00248 | 95.2 | |

NS, not significant.

**Table 5.3.** Root dynamics as observed using minirhizotron in a rubber (*Hevea brasiliensis*) agroforestry experiment in Sepunggur, Jambi, Indonesia, during 1997 and 1998, with two types of planting material (grafted clones of PB260 and GT1-derived seedlings) and fertilizer level (none, and recommended levels of N + P) as the main experimental factors and the distance from the tree as the sampling position. Data from two minirhizotrons per sampling position, planting material and fertilizer level combination, with daily survival rate and median lifespan evaluated for individual replicates.

| Experimental factor/sampling position | Daily survival probability | Median lifespan (days) | Daily turnover (%) |
|---|---|---|---|
| PB260 | 0.9889 | 106 | 1.11 |
| GT1 | 0.9861 | 71 | 1.39 |
| Within tree row (0.25 m from tree) | 0.9884 | 110 | 1.16 |
| Between tree rows (1.5 m from tree) | 0.9866 | 68 | 1.34 |
| Grand mean | 0.9875 | 89 | |
| Probability | NS | NS | |
| Standard error of difference between means | 0.00335 | 21.3 | |

NS, not significant.

mates. However, with regard to the median longevity, the standard error of the difference between means was about one-third of the mean, so the discriminatory power of the experiment was not very high with regard to such treatment effects.

In a second rubber experiment undertaken in the same province during the same time period, however, daily turnover was substantially higher at 1.2%, while the median lifespan was only 90 days. Again, no effects could be identified as having been caused by experimental factors (planting material or sample position). The difference between the two experiments, however, is remarkable. No solid explanation for this difference has yet been identified.

Overall, the relatively high variability between replicate minirhizotrons and the resultant low discriminating power of tests on treatment effects may provide a lesson for future research. That is, only large differences (such as those occurring between annual crops and trees) can be identified. For most situations, an 'order of magnitude' estimate, rather than a site- and management-specific value, is all that can be made using the methods currently available. So far, the simple model discussed above, containing a constant daily probability of survival for roots, appears to be consistent with the data, at least in the humid tropical environment in which the data for Tables 5.1–5.3 were collected.

### 5.4.1 Root turnover and consequences for uptake

In theory, rapid root turnover is beneficial because new roots produced in unexploited soil replace roots previously surrounded by depletion zones. By investigating the influence of P supply on total root production and root mortality during the barley growing season, Steingrobe *et al.* (2001) assessed the benefits of a more rapid root turnover on P acquisition. They found that shoot development and grain yield were reduced in a '−P' treatment. However, the standing root system in that treatment was nearly the same size as that in a '+P' treatment, and root production and turnover were greater. Through model calculations, the authors have shown that root renewal by continuous growth and mortality can contribute to P uptake efficiency. The physiological mechanism behind the higher root turnover is not yet clear, however. For a discussion of the long-term benefits of root foraging in heterogeneous environments, where patch depletion and root turnover may limit the long-term rewards of root foraging to perennial plants, see Fransen and de Kroon (2001).

## 5.5 Model Representations of Root Dynamics

Simulation models can treat below-ground resource capture at different levels of sophistication (Fig. 5.7; van Noordwijk and de Willigen, 1987):

**0.** Models that use empirical resource capture efficiency coefficients to represent the relationship between water and nutrient supply in the soil and the dynamics of plant growth (sometimes known as models 'without roots').

**1.** Models that differentiate between soil layers and use empirical data on relative root distribution to predict resource capture potential in each zone. Root distribution can be schematized via an exponential decrease with depth (Jackson et al., 1996) or its two-dimensional elliptical variant (van Noordwijk et al., 1995). Root length density can also be given as an 'independent' parameter for each layer or zone, and change of root length densities with time can be imposed on the basis of crop age.

**2.** Models that consider plants as organisms that have the capacity to adjust both the total amount of their roots (to complement the internal balance between above- and below-ground resource capture), and the location of new root growth (to ensure the growth of those parts of their root system with the best opportunities to take up the resource most limiting overall plant growth).

Models at level 0 have been successfully applied to many crop monocultures, and are the basis of Kho's approach to tree–crop interactions (Chapter 1, this volume). However, such models cannot give an account of below-ground interactions between plants at the process level. In the same way that *'pedotransfer'* functions (Chapter 9, this volume) allow estimates of quantitative soil parameters to be made on the basis of simple indicators (such as soil texture), we may need *'rhizotransfer'* functions that allow reasonable estimates to be made of the main root parameters without too much new data collection. The global data sets on root distribution (Jackson et al., 1996; Chapter 4) and root turnover (see

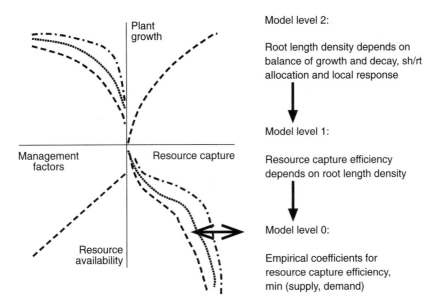

**Fig. 5.7.** Schematic representation of the relations between management factors, resource availability, resource capture (showing three different situations with different lines) and plant growth; three 'levels' for modelling resource capture are indicated; sh/rt = shoot/root (after van Noordwijk and de Willigen, 1987).

quotes above) form a starting point for this, but further work is needed.

Temporal development of the distribution of roots in the soil is important for the functioning of a root system, especially in models that use root length density distribution, rather than presence/absence of roots, as the basis for the predictions they make (de Willigen *et al.*, 2000). One way to describe root proliferation is to consider it as a process of diffusion with a first-order sink term (equivalent to a time-independent, daily probability of root survival) accounting for decay. De Willigen *et al.* (2002) derived analytical solutions for the two-dimensional diffusion of roots in both a rectangular area and a cylindrical volume. Root dry matter enters the soil domain at the plant base at a particular location on the soil's surface. The distribution patterns obtained strongly depend on the ratio of the diffusion coefficients in both horizontal and vertical directions. When dry matter permeates evenly across the complete surface (an approximation of what occurs in a relatively closely spaced crop), a steady state eventually results in which root length density decreases exponentially with depth, as is often found in experiments and natural vegetation (Jackson *et al.*, 1997).

The TRAP model (Rasse *et al.*, 2001) for 'Tree Root Allocation of Photosynthates' was developed to predict the partitioning of photosynthates between the fine and coarse root systems of trees in a series of soil layers. TRAP simulates root system responses to soil stress factors affecting root growth, such as temperature, soil penetrability, aeration and soil acidity. Validation data have been obtained from two Belgian experimental forests, one mostly composed of beech (*Fagus sylvatica*) and the other of Scots pine (*Pinus sylvestris*). TRAP accurately predicted ($R = 0.88$) nighttime $CO_2$ fluxes from the beech forest for a 3-year period. It also predicted total fine root biomass to within 6% of the measured values. Fine root turnover was predicted to be 2.1 Mg C/ha/year, with an annual root turnover of 1.0 for fine roots and 0.02 for coarse roots. The TRAP model focuses on the C balance of the tree, however, and does not include the effects of root growth and decay on the uptake of water and nutrients.

Both the HYPAR model (Mobbs *et al.*, 2001) and WANULCAS (van Noordwijk and Lusiana, 1999) can predict competition for water and nutrients between trees and crops (or other plants) at level 1 (see above). WANULCAS can also be used at level 2, although, when so doing, spatial root distribution is restricted to the exponential-decrease-with-depth or elliptical distributions (the parameters of which are treated as dynamic).

If nutrient (N or P) or water stress occurs, the relative allocation of growth reserves to roots can increase quickly, from say 10% to 90% of the daily used reserves. Allocation of growth reserves to roots can increase under mild stress, but usually the 'functional response' comes too late and is too slow to 'head off' the stress. Under nutrient or water stress, the acquisition of new reserves by the plant will be limited. So, under such stresses, absolute allocation to roots may only temporarily be higher than it is under a no-stress scenario. For a plant, the key strategy is to use 'early warning' signals (such as 'drought-signal hormones') and respond before the stress becomes serious. Quantitative indicators of stress that are not yet affecting current plant production take the place of these hormones in the models.

'Local response' is simulated in WANULCAS by a gradual change in the parameters of the elliptical root distribution, and is constrained by the total length of new roots that can be produced with the carbohydrates allocated. The intensity of change depends on both a 'responsiveness' parameter and the degree to which effective uptake per unit root length of the currently limiting resource differs between soil layers and zones. If roots in deeper layers are more effective (e.g. in the case of water stress), root distribution can shift to a configuration that involves a more gradual decrease of root length density with depth, or perhaps even to a configuration where root length density increases with depth.

If roots in the topsoil are more effective (e.g. when P uptake is the factor most limiting to plant growth and the topsoil has a sufficiently high water content to keep the P mobile), roots will expand (mainly) in the topsoil. The example given in Box 5.2 shows that rainfall distribution (when the cumulative rainfall is constant) can lead to substantial shifts in predicted root distribution, depending on the predicted re-wetting pattern of the soil. Under frequent but small amounts of rainfall, the model predicts that roots will accumulate in the topsoil, whereas less frequent, heavier rainfall events (which wet the entire soil profile) are predicted to induce deeper rooting. It also shows that, though the 'functional response' of increased root allocation may be limited in a monoculture (where it does not increase total resource capture), it can reduce the negative effects of competition and as such be 'functional' in a competitive situation. Most 'functional equilibrium' studies have, thus far,

---

**Box 5.2.** WaNuLCAS exploration of tree functional and local response.

A series of simulations was made for a moderately deep soil (1 m) with an annual rainfall of 1000 mm. Rainfall patterns ranged from '1 = every day 3 mm of rain' and '2 = every second day 6 mm', to '6 = every 32 days 96 mm'. As the potential evapotranspiration was assumed to be 4 mm/day, this environment would not provide enough water to avoid water stress, even if all rainfall were to be fully used. The rainfall patterns lead to situations of permanent moderate stress (rainfall pattern 1), alternations of sufficient water and severe water shortage (rainfall patterns 5 and 6) or intermediate patterns. In the overall water balance, with a decrease in the number of rainy days (through patterns 1 to 6), a decrease in the values for the interception and soil evaporation terms can be noted, while the contribution to ground water (deep infiltration) and runoff increases but remains small in absolute value. Cumulative tree water use tends to increase through rainfall patterns 1 to 6. If a grass sward is added to the simulations, canopy interception increases and thus the amount of soil water available to either tree or grass is reduced. The grass water use is predicted to benefit more from rainfall patterns 5 and 6 than the tree, causing a bell-shaped response curve for the tree.

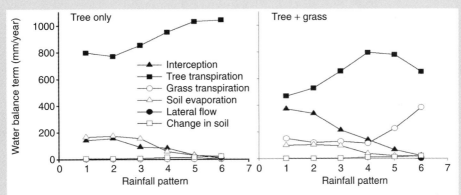

**Fig. B5.2.** Water balance for a range of WaNuLCAS simulations, in the absence of functional or local response of the tree, with and without a grass sward.

A sensitivity analysis was carried out on the two key parameters for the functional shoot/root balance and root distribution: 'Root_Allocation_Responsiveness' and 'local response'. Higher values of 'Root_Allocation_Responsiveness' lead to a more rapid shift of current growth resources to roots, at the expense of shoot growth, when the total uptake of water and/or nutrients falls short of current 'demand'. With increasing 'local response', root distribution shifts towards the soil layer and spatial zone in which roots are most successful (per unit root length) in taking up the most limiting resource. For both parameters, values of zero indicate no response, and values above 1 indicate a response that is more than proportional to the strength of the 'signal' (relative degree of below-ground stress for the Root_Allocation_Reponsiveness, and difference in actual uptake per unit root length for the 'local response', respectively).

*Continued*

**Box 5.2.** *Continued.*

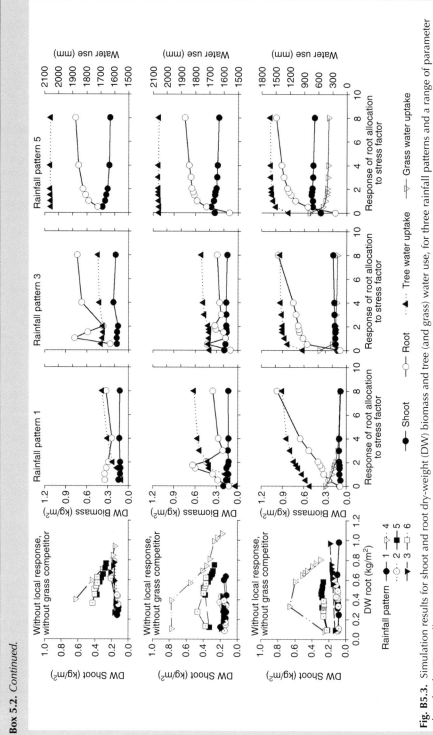

**Fig. B5.3.** Simulation results for shoot and root dry-weight (DW) biomass and tree (and grass) water use, for three rainfall patterns and a range of parameter values for the 'Root_Allocation_Responsiveness' parameter (see explanations Box 5.2). Simulations include situations with and without 'local response' (see text), and with and without competition from a grass sward.

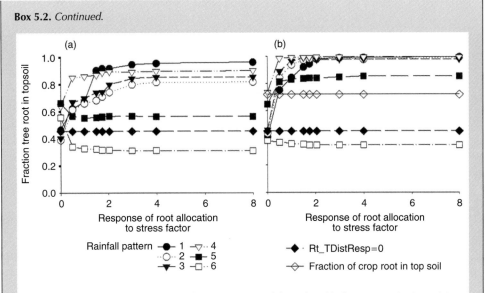

**Box 5.2.** Continued.

**Fig. B5.4.** Relative tree root biomass in the upper 25 cm of the soil profile for a range of values of the factor that governs the response to stress of the biomass allocation to roots, with (b) and without (a) a competing grass; the grass is assumed not to show a functional or local response, so it has a constant fraction of its roots in the topsoil; the line Rt_TdistResp = 0 indicates a situation without 'local response', so the 'response to stress' can modify total root biomass, but not root distribution for this setting.

been performed in monocultures, and they may thus have missed important aspects.

Dunbabin et al. (2002a,b) developed a 3-D model (WaNuLCAS is essentially 2-D), which includes plastic response to external nutrient supply, and which they parameterized for lupin and a variable nitrate supply. Uptake per unit root can double with a decreasing fraction of the root system inside a nutrient-rich patch (indicating considerable 'down regulation' of uptake in the normal situation), with preferential root development in enriched patches being responsible for further compensation and maintenance of the plant's total uptake capacity.

## 5.6 Management Implications

### 5.6.1 Changes in root production after land-use change

As standing root biomass and root turnover are substantial components of the below-ground ecosystem, changes in roots as a result of land-use change can be important. Idol et al. (2000) studied changes in the seasonal and spatial dynamics of root growth, mortality and decomposition that occur following the removal of standing forest vegetation. Four upland, temperate, deciduous forest stands in southern Indiana, USA, were compared (during the forest recovery phase) 4, 10, and 29 years after the forest overstorey trees were cut down. A mature stand (80–100 years since last harvest) was chosen to represent the preharvest conditions. A combination of soil cores and ingrowth cores were used to assess stand-level rates of root growth, mortality and decomposition. Root growth increased significantly after harvesting, but declined as the stand matured (if we may indeed interpret these data as a 'chronosequence'). In all stands, fine root mortality and decomposition were nearly equal to, or greater than, fine root growth.

Castellanos et al. (2001) examined the effects of slash-and-burn land-clearing of tropical dry forest and the establishment of pasture on fine root biomass and productivity in a site in Mexico. In the pastures (composed of *Cenchrus ciliaris, Panicum maximum* and *Andropogon gayanus*), both the productivity and mortality of fine roots (1 mm in diameter) were 30% lower than in the tropical dry forest in the top 5 cm of soil. However, this difference was partially compensated for by the pasture having more roots below that depth. In forest and pasture, fine root productivity in the top 5 cm of soil accounted for 86% and 76%, respectively, of the total fine root productivity throughout the sampled soil profile.

Rao et al. (2001) studied root turnover and nutrient cycling in native and introduced pastures in tropical savannah in the eastern plains (Llanos) of Colombia. Measurements of root production and turnover were made on two introduced (9-year-old) pastures (grass only, *Brachiaria dictyoneura* CIAT 6133; and a grass + legume mixture, *B. dictyoneura* + *Arachis pintoi* CIAT 17434), and compared with measurements made in a native pasture. Annual root production (biomass and length) was significantly greater in the introduced pastures than in the native pasture. Although root biomass turnover (2.2/year) and root length turnover rates (1.8–2.4/year) were similar among native and introduced pastures, the greater total annual root production (6 Mg/ha/year versus 2.4 Mg/ha/year) in the introduced pastures contributed to their strongly superior root turnover (and N and P cycling).

### 5.6.2 Management implications for multispecies agroecosystems

Knowledge of root distribution and dynamics can be used to increase the probability that applied nutrients (fertilizer) are preferentially used by the most economically important component(s) of the agroecosystem. Fertilizer should be placed closer to the tree trunk, rather than at the canopy edge (as is the current recommendation) in order to maximize P uptake by clove roots (Purbopospito and van Rees, 2002).

Seasonal differentiation exists, in terms of root activity, between trees and grass (Cheng and Bledsoe, 2002) and between trees and crops (Odhiambo et al., 2001). Munoz and Beer (2001) discussed the opportunities that exist for reducing nutrient competition between shade trees and cacao based on the different times at which their root growth flushes occur. In their view, competition could be minimized by early fertilization at the beginning of the rains, immediately after the shade trees were pruned. Schroth and Zech (1995b) showed that maximum tree root growth can even be pushed into the dry season through pruning.

Changes in above-ground phenology, including tree pruning, can have substantial impacts on root survival and on subsequent root patterns (van Noordwijk et al., 1996). The 'lung branch' technique, which involved retaining a single branch on pruned *Erythrina poeppigiana* trees, was observed to allow better fine root and nodule survival in a study by Chesney and Nygren (2002).

A major opportunity for 'managing' root turnover and thus for affecting the way turnover contributes to the overall functioning of an agroecosystem is the choice of species (and genotypes within that species) that will be planted. Literature on the genetics of root exudation (quality and quantity) has been assessed by Rengel (2002), who concluded that the best-studied phenomenon, thus far, may be the genetically controlled variation in citrate production in roots, which is linked to partial alleviation of Al-toxicity stress. P solubilization through organic chelating agents (such as citrate) and pH changes is relatively well understood, and there is good agreement between models and measurements of such (Kirk, 2002a,b; Chapter 7, this volume).

## 5.7 Research Issues and Priorities

Although the methods currently available still have major limitations and weaknesses, they can be used for further comparative studies. Actual data collection in tropical agroecosystems has been limited, and most of what we think we know is derived from temperate zones. Although the processes may be essentially the same, the quantities involved cannot be directly transferred to other conditions. Further studies on how root dynamics are affected by shoot growth (seasonality and phenology) and management (e.g. tillage) in monocultures and intercropping systems remain an important need. Thus far, we have little understanding of how heterogeneous root turnover rates are *within* a root system (e.g. is this affected by the position of the fine roots within the system or by the distribution of rhizovores?). Observed differences in root turnover between crop and tree roots may be related to differences in maintenance costs (linked to protein content).

Model representations can certainly be refined if further data become available that can be used to test predictions made concerning the 'plastic' response to stress factors and opportunities for local root activity.

---

**Conclusions**

1. When do roots grow? Generally 'ahead of' the above-ground parts of the plants; tree root growth can be asynchronous with crop root growth in seasonal environments. The functional shoot/root equilibrium needs time to adjust to new situations.
2. Where do roots grow? Definitely not just anywhere. Constraints are placed on this by the root system's branching pattern, and the need for connectedness between fine and coarse transport roots. Competition for resources within the plant can explain local response in (temporarily) favourable locations.
3. How long do fine roots live? From 2 weeks up to 1 year.
4. What can simulation models do? They can incorporate all these ideas into the prediction of below-ground interactions between plants; but parameterization for any specific situation is no simple task in the absence of effective 'rhizotransfer' functions.
5. Roots respond to nutrient-enriched zones in the soil, and such responses can be species-specific. This could be exploited by localized applications of fertilizer.
6. Root and shoot pruning have definite effects on root turnover and root distribution. This could be exploited by management operations.
7. The fine roots of trees live longer than those of crops (3–8 months vs. 2–4 weeks). So crop roots may respond more quickly to management interventions (although available data are very limited!).

---

**Future research needs**

1. How are root dynamics affected by shoot growth (seasonality and phenology) and management (e.g. tillage)?
2. How heterogeneous are root turnover rates *within* a root system (e.g. is this affected by the position of the fine roots within the system or by the distribution of rhizovores)?
3. How do environmental conditions and/or the ecological life history of a species or crop determine the plasticity of its root system, and how can we improve our methods to measure this in a more reliable way?

# 6 Opportunities for Capture of Deep Soil Nutrients

Roland J. Buresh, Edwin C. Rowe, Steve J. Livesley, Georg Cadisch and Paramu Mafongoya

---

**Key questions**

1. How and why do nutrients occur in deep soil layers?
2. Under what conditions can plants best utilize deep soil nutrients?
3. What realistic opportunities are there for extraction of deep soil nutrients?

---

## 6.1 Introduction

Soil is the reservoir from which plants obtain much of the nutrient elements essential for their growth. The capacity of soils to supply essential nutrients and the ability of plants to access and extract these nutrients are critical determinants of the productivity of tropical agroecosystems.

Nitrogen (N), phosphorus (P) and potassium (K) are required in relatively large quantities by plants. Plants typically take up N as either nitrate or ammonium, which form in soil as a result of the breakdown of soil organic matter, biological $N_2$-fixation (see Chapter 13, this volume), and fertilizer inputs. Plants take up P as phosphate ions, which form in soil as a result of their release from soil mineral reserves, as well as through the breakdown of soil organic matter and as a result of fertilizer inputs. Plants take up K as an ion ($K^+$), which forms in soil through its release from soil mineral reserves. Calcium (Ca), magnesium (Mg), and sulphur (S) are required by plants in relatively smaller quantities than N and K.

Tropical soils often do not supply sufficient plant-available N and P to meet the requirements of high crop production; strongly acid soils, which are common in the tropics, have a limited capacity to supply P, K, Ca and Mg (see Chapter 8, this volume). In addition, the capacity of a soil to provide nutrients can decline when land is cultivated without using nutrient inputs in a quantity sufficient to match the amount of nutrients removed in harvested products (Smaling *et al.*, 1999). This can increase the severity of nutrient limitations to plant growth.

The integration of perennials with annual agricultural crops can lead to greater spatial and temporal extraction of

© CAB International 2004. *Below-ground Interactions in Tropical Agroecosystems*
(M. van Noordwijk, G. Cadisch and C.K. Ong)

soil nutrients. The deep- and lateral-rooting systems of perennials can exploit a larger soil volume than the roots of annuals. In this fashion, perennials have the potential: (i) to retrieve nutrients from soil that is outside the effective root zone of annual crops; (ii) to intercept nutrients moving down soil profiles and moving laterally within farms and landscapes; and (iii) to access forms of nutrients not accessible to crops (Cannell et al., 1996). The uptake, by perennials, of soil nutrients that are either not extracted by annual crops or that are lost from the rooting zone of annuals represents the capture of nutrient resources that would not be used in a cropping system that consists solely of annual plants. Nutrients retrieved by perennials can be cycled within the agroecosystem and can, potentially, become available to annuals through the turnover of roots and the decomposition of leaf litter and pruned biomass. Such nutrients can also be exported from the agroecosystem in the form of the harvested products of the perennials (for example, fruit and timber).

The ability of perennials in agroecosystems to acquire nutrients that annuals alone cannot access could, potentially, increase the use and cycling of soil nutrients, increase total biological productivity, and help to protect water quality by preventing the flow of mobile nutrients (such as nitrate) into ground water and water bodies. In this chapter, we shall explain the mechanisms by which nutrients occur in deep soil layers and review the ability of plants to utilize deep soil nutrients. We will also outline the opportunities that exist for the extraction of deep soil nutrients, in order to increase the efficiency of nutrient use by plants while protecting the environment.

## 6.2 Mechanisms for Nutrient Occurrence in Deep Soil

Soils have reserves of nutrients stored in organic matter and minerals, but these nutrients typically occur in forms unavailable to plants. Only a small portion of each of these reserves is released into plant-available forms each year, through biological activity and chemical processes. Common plant-available forms include nitrate ($NO_3^-$), ammonium ($NH_4^+$), phosphate ($H_2PO_4^-$, $HPO_4^{2-}$), $K^+$, $Ca^{2+}$, $Mg^{2+}$ and sulphate ($SO_4^{2-}$).

The quantities of nutrients available for use by plants are determined by the supply of plant-available nutrients (soil reserves and fertilizer inputs), the uptake of nutrients by plants, and the loss of nutrients from the soil–plant system. The soil boundary of the soil–plant system is defined by the extent to which plant roots extract nutrients. Plant-available nutrients can occur in deep soil layers – including those below the nutrient extraction zone of annuals – either through movement from upper soil layers or through *in situ* formation.

The downward movement (i.e. leaching) of nutrients can occur when the formation of mobile, plant-available nutrients in upper soil layers is in excess of crop nutrient demand and rainfall is sufficiently high for downward water movement. A surge in the release of nutrients can occur after the addition of fertilizers, and a flush in the release of nutrients can occur through biological activity following the wetting of dry soil (Birch, 1958) and the decomposition of organic materials. In agricultural systems with annual crops, the supply of plant-available nutrients often exceeds plant demand at the onset of the cropping season, when the juvenile crop is developing, and between

---

**Perennials in agroecosystems have the potential to:**

1. Retrieve nutrients from soil outside the effective root zone of annual crops.
2. Intercept nutrients moving down soil profiles.
3. Intercept nutrients moving laterally within farms and landscapes.
4. Access forms of nutrients not accessible to crops.

cropping seasons, when the ground cover of plants is low or non-existent (Fig. 6.1). In systems with perennials, the supply of nutrients can exceed demand after perennials are cut or pruned.

Mobile nutrients with positive charge (cations) and negative charge (anions) are susceptible to movement below the rooting depth of annuals (Cahn *et al.*, 1993). Cations such as $Ca^{2+}$, $Mg^{2+}$ and $K^+$ are retained on the surfaces of negatively charged clay minerals and organic matter, thereby retarding their downward movement in the soil. Cations none the less accompany the movement of anions such as $NO_3^-$ and $SO_4^{2-}$ in the soil. The addition of N fertilizers and the consequent formation and leaching of nitrate can accelerate the leaching of $Ca^{2+}$, $Mg^{2+}$ and $K^+$, which move with nitrate to maintain charge neutrality in the soil (see Chapter 10, this volume, for information on the relative mobility of ions in soil).

Anions such as $NO_3^-$ and $SO_4^{2-}$ can move rapidly in soils. Some tropical soils, however, have positively charged surfaces – particularly in the subsoil – on which anions can be retained (Wong *et al.*, 1990a). This anion sorption can retard the downward movement of anions and result in their accumulation in lower soil layers, including layers below the rooting depth of annuals. The sorption of anions typically increases with soil depth, decreased pH, decreased soil organic matter, increased 1:1 clay minerals (i.e. kaolinite), and increased iron and aluminium oxides (Black and Waring, 1979). The accumulation of subsoil nitrate is favoured by large applications of N fertilizer in irrigated or high-rainfall areas with high anion retention in the subsoil. Michori (1993), for example, reported a huge accumulation of N (2200 kg N/ha as $NO_3^-$) at a depth of 1–5 m beneath irrigated and fertilized coffee in Kenya.

Plant-available nutrients can form in deep soil layers through the biological breakdown of soil organic matter and through chemical release from soil minerals. The biological breakdown of soil organic matter – referred to as mineralization – leads to the formation of nitrate, phosphate and sulphate. The rate of mineralization, expressed per unit of soil weight, is lower in deep soil layers than in topsoil, because soil organic matter content and biological activity are lower in deep soil layers. Deep soil layers can, however, account for a much larger weight and volume of soil than topsoil. Total mineralization below the topsoil in deep soils can consequently be substantial.

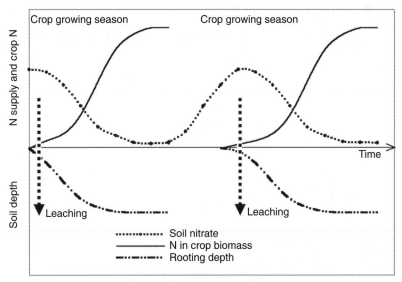

**Fig. 6.1.** Schematic diagram of the lack of synchrony between the supply of soil nitrate and crop demand for nitrate, which can lead to the leaching of nitrate to soil layers below the rooting zone of crops.

Ohlsson (1999), for example, estimated net N mineralization at a depth of 0.5–2 m to average 1.1 kg N/ha/day for four sites on deep acid soils in western Kenya.

The chemical breakdown of soil minerals can be an important source of phosphate, $Ca^{2+}$, $Mg^{2+}$ and $K^+$. The quantity of nutrients released depends on the weatherable minerals, and is relatively lower in soils with low weatherable minerals and a low base (Ca and Mg) status. In humid tropical systems, even where the parent material has a high content of bases, bases may not become accessible to plant roots. Burnham (1989) showed that, following the weathering of base-rich granite, a saprolite ('rotten rock') layer, which is not easily penetrated by roots, develops between the granite and the soil layer. As the weathering front develops, the released nutrients are leached and the pH drops before plants' roots can access them (Table 6.1).

The leaching and accumulation of nitrate in subsoil has been well documented in tropical soils. In the humid tropics, with >2000 mm annual rainfall, an estimated 25–33% of the nitrate derived from soil and fertilizer can be leached from annual crop systems (see Chapter 8, this volume). Nitrate can also be leached and can accumulate in the subsoil of weathered tropical soils receiving 1800–2300 mm annual rainfall (see Box 6.1) and 970–1200 mm annual rainfall (Wild, 1972). Phosphate, on the other hand, moves much more slowly than $NO_3^-$, $K^+$ and $Ca^{2+}$ in soil. The formation of phosphate from the weathering of minerals is also typically low in deep soil layers.

## 6.3 Utilization of Deep Soil Nutrients by Plants

Soil nutrients not accessible to annual crops can be extracted by perennials through two processes: (i) the retrieval of nutrients already present in layers below the effective extent of rooting of annual crops; and (ii) the interception of nutrients moving below or outside the rooting zone of annuals. In both cases, the distribution and density of roots, the demand

---

**Nutrients can occur in deep soil layers through:**

1. Downward leaching of mobile nutrients ($NO_3^-$, $SO_4^{2-}$, $K^+$, $Ca^{2+}$ and $Mg^{2+}$) when nutrient supply exceeds plant demand for the nutrient. This is favoured by high biological breakdown of soil organic matter and organic materials in topsoil, excessive inputs of fertilizer, high rainfall or excessive irrigation, and little or no plant growth.
2. Biological breakdown of soil organic matter in deep soil layers, releasing $NO_3^-$, $SO_4^{2-}$ and phosphate.
3. Chemical breakdown of soil minerals in deep soil layers, releasing phosphate, $Ca^{2+}$, $Mg^{2+}$ and $K^+$.

---

**Table 6.1.** Nutrient availability following weathering of granite near Kuala Lumpur, Malaysia. (Adapted from Burnham, 1989, and personal communication.)

| Component | Depth (m) | Clay (%) | Exchangeable bases (cmol$_c$/kg) | | | | pH |
| --- | --- | --- | --- | --- | --- | --- | --- |
| | | | Ca | K | Mg | Total | |
| Soil | 0.04 | 31 | 0.49 | 0.24 | 0.06 | 1.00 | 4.7 |
| | 3.00 | 40 | 0.27 | 0.23 | 0.04 | 0.68 | 5.1 |
| Saprolite | 8.00 | 30 | 0.11 | 0.15 | 0.06 | 0.44 | 4.5 |
| | 9.95 | 4 | 0.13 | 0.24 | 0.06 | 0.54 | 4.8 |
| | | | Total bases on abrasion (cmol$_c$/kg) | | | | |
| Granite | 10.00 | 0 | 42.5 | 104 | 39.5 | 266 | 8.0 |

> **Box 6.1.** Nitrate accumulation in deep, high-base-status soils in Kenya.
>
> Scientists initiating agroforestry experiments in farmers' fields in western Kenya between 1993 and 1995 observed considerable amounts of nitrate in deep layers of acid soils (Kandiudalfic Eutrudox and Kandiudalf). Nitrate at a depth of 0.5–2.0 m ranged from 70 to 315 kg N/ha in six farmers' fields grown with unfertilized maize (Buresh and Tian, 1997). Annual rainfall ranged from 1800 to 2300 mm. Aluminium saturation to a 2 m depth was < 10%, and was not a constraint to plant growth. However, growth of maize was severely limited by low plant-available P in soil. A subsequent survey of 96 unfertilized maize (*Zea mays*) fields across a range of soils and landscape positions revealed that 20% of the fields had > 70 kg N/kg as nitrate at a depth of 0.5–2 m (Shepherd et al., 2001). A survey of soil nitrate under perennial and annual cropping systems indicated markedly higher levels of subsoil nitrate under maize than under perennials such as trees, hedgerows and bananas (Shepherd et al., 2000).
>
> Six years of research in farmers' fields in western Kenya confirmed the occurrence of substantial nitrate in the subsoil of deep, high-base-status soils. Greater formation of nitrate from mineralization of soil organic matter than uptake of N by maize (Mekonnen et al., 1997) and high rainfall contributed to the leaching of nitrate. Once in the subsoil, the nitrate was sorbed on positively charged soil surfaces (Hartemink et al., 1996), which retarded further downward movement of nitrate. Subsoil nitrate tended to be associated with low-cation-exchange capacity per unit of clay (Shepherd et al., 2001). Unfertilized maize rooted to a maximum depth of 1.2 m and could not effectively extract the deep soil nitrate.
>
> Low maize production and a subsequent low uptake of N were at least partly associated with severe P deficiency. The build-up of nitrate below the active root zone of maize reflected a net loss of N from the soil–crop system, because the deep soil nitrate was not accessible to maize crops. Management practices such as P fertilization, which alleviated constraints to crop production, decreased the accumulation of nitrate in subsoil.

of plants for nutrients, and the distribution and concentration of plant-extractable nutrients and water will influence the extraction of nutrients by perennials.

Roots of perennials can undoubtedly extend beyond the rooting depth of annual crops. Many trees in the tropics can develop roots to > 25 m (Chapter 4, this volume). Roots of perennials can also exploit soil nutrients for a larger portion of the year than annuals, because perennials are present throughout the year whereas annual crops only grow for relatively short cropping seasons and are then absent for the remainder of the year. The roots of perennials might also have the ability to solubilize and extract nutrients in recalcitrant forms, which are not as readily available to annual crops (Chapter 7, this volume).

The retrieval and cycling of nutrients from soil below the zone exploited by crop roots has been referred to as 'nutrient pumping' (van Noordwijk et al., 1996). It is favoured when perennials have a deep-rooting system and a high demand for nutrients, when water or nutrient stress occurs in the surface soil, and when considerable reserves of plant-extractable nutrients or weatherable minerals occur in the subsoil (Buresh and Tian, 1997). Conditions conducive to 'nutrient pumping' include soils without physical or chemical barriers to deep rooting, the presence of perennials with rapid growth and a high nutrient demand, and soils with high levels of plant-extractable nutrients and water below the rooting zone of annual crops. These conditions were observed in deep soils in western Kenya, where nitrate accumulated in subsoil during periods of maize growth whilst perennials, grown in rotation with the maize, could effectively retrieve the subsoil nitrate 'lost' to maize (see Box 6.2).

Mobile nutrients can readily leach beyond the reach of crop roots both during early crop-growth stages and when soil chemical or physical barriers restrict root growth at depth. The intercropping of perennials with annual crops can reduce this leaching loss, provided that the roots of the perennials actively take up nutrients at the time of leaching events.

**Box 6.2.** Retrieval of subsoil nitrate in improved fallow systems.

The rotation of short-duration, fast-growing woody perennials with crops – referred to as 'improved fallows' – is a promising agroforestry system for the replenishment of soil fertility. Improved fallows have been extensively researched in western Kenya on acid soil, with no physical and chemical barriers to rooting and with 1800–2300 mm of rainfall per year. This research demonstrates that the rotation of a fast-growing and deep-rooting perennial (such as *Sesbania sesban*) with maize could, effectively, allow the nitrate that accumulated during previous maize crops to be taken up (Fig. B6.1). Maize – the staple food crop – is typically not fertilized by the farmers, and the roots of such unfertilized maize extended only to a depth of 1.2 m. It was found that soil nitrate to a depth of 4 m was 199 kg N/ha beneath continuous cropping of unfertilized maize. *Sesbania*, planted after maize, dramatically reduced the level of soil nitrate to 51 kg N/ha after 15 months of growth. *Sesbania* roots extended to a depth of > 4 m (Mekonnen *et al*., 1997).

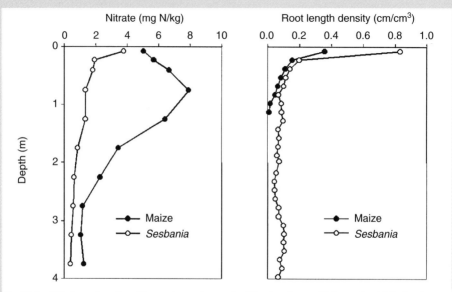

**Fig. B6.1.** Nitrate and root profiles under mature unfertilized maize and 15-month-old *Sesbania sesban* in western Kenya. Adapted from Mekonnen *et al*. (1997).

The retrieval and cycling of subsoil nitrate is strongly related to the N demand and rooting of perennials. In another study run on similar soil in western Kenya, perennials were grown on soil with a high level of subsoil nitrate that had accumulated during a previous cropping period. Fast-growing trees with high-root-length densities (such as *Sesbania* and *Calliandra calothyrsus*) rooted to a depth of > 4 m in 11 months (Jama *et al*., 1998a). Root length densities of ≥ 0.1 cm/cm³ extended to a depth of 2.2 m in the case of *Calliandra* and to a depth of 1.8 m in the case of *Sesbania*. During the 11 months of plant growth, the decrease in soil nitrate that occurred in the top 2 m of the soil (150–200 kg N/ha) corresponded to a large accumulation of N in above-ground biomass of *Calliandra* and *Sesbania* (312–336 kg N/ha). Slow-growing trees (such as *Grevillea robusta*) with a low demand for N only had root length densities of ≥ 0.1 cm/cm³ extending to 0.45 m depth, and soil nitrate increased rather than decreased during the 11 months after establishment of the trees.

The concept of tree roots capturing and recycling mobile nutrients leaching beyond the reach of crop roots is known as the 'safety net' (van Noordwijk *et al*., 1996).

The roots of perennials can act as a 'safety net' in agroecosystems that involve the intercropping of perennials and annuals. When perennials and annuals are grown together, their roots are likely to overlap in the upper soil layers. The roots of an actively growing perennial could be more effective than those of a young annual in capturing mobile nutrients in upper soil layers. As growth and nutrient demand by an annual crop accelerates, the possibility of competition for nutrients between the annual and perennial increases. The occurrence of a dense, active perennial root system beneath the zone of nutrient uptake by the annual crop would minimize competition while further reducing the likelihood of leaching loss. Favourable conditions for the safety-net role of perennials include soils with physical or chemical barriers to deep rooting and the accumulation of mobile soil nutrients (see Chapter 10, this volume). These conditions were observed in the humid tropics, where trees in a hedgerow agroforestry system demonstrated the safety-net function (see Box 6.3).

The roots of perennials and annual crops often occupy the same soil volume, and hence can compete for nutrients and water. The uptake and cycling by perennials of nutrients that would have been used by the crop in the absence of the perennial does not represent a net input of nutrients into the soil–crop system. It constitutes, instead, a redistribution of nutrients within the soil–crop system. This can occur when perennials have a considerable amount of lateral roots that extend through the root zone of adjacent crops. It might occur, for example, in environments where rainfall is not sufficient to recharge soil water below the crop rooting zone and a water table is not accessible to tree roots (see Chapter 9, this volume).

The opportunity for plants to take up leaching nutrients and nutrients already leached into deep soil layers is greater in the case of mobile nutrients (such as nitrate) than it is in the case of less mobile nutrients (such as phosphate). In addition, the critical root length density essential for taking up the relatively immobile phosphate ion is greater than the critical root length density essential for taking up more mobile ions such as nitrate, ammonium, $K^+$ and $Ca^{2+}$ (see Chapter 10, this volume). However, in the case of P, symbiotic mycorrhizal networks could make up part of the required root length density (see Chapter 7, this volume).

The uptake of nutrients from deep soil layers by perennials is typically lower in low-base-status soils (exchangeable Ca <1 $cmol_c$/kg) than in high-base-status soils (Szott et al., 1999). Root penetration into deep soil layers is often less in low- rather than high-base-status soils because of chemical and physical barriers to rooting. The concentration of weatherable minerals and plant-extractable nutrients is also typically less in the subsoil of low-base-status soils.

## 6.4 Quantification of Deep Nutrient Uptake

The retrieval, by perennials, of nutrients from below the root zone of annual crops is often one of the least-quantified nutrient fluxes in tropical agroecosystems. Nutrient retrieval from deep soil layers depends on the interacting factors of root distribution, plant demand for nutrients, soil nutrient and water concentrations, and nutrient transformations and movement in soil. Consequently, the presence of roots in deep

---

**The retrieval, by perennials, of nutrients from below the soil zone exploited by crop roots is favoured when:**

1. Perennials have a deep-rooting system.
2. Perennials have a high demand for nutrients through most of the year.
3. Water or nutrient stress occurs in the surface soil.
4. Considerable reserves of plant-extractable or weatherable minerals occur in the subsoil.

**Box 6.3.** The 'safety-net' function of hedgerow intercropping in the humid tropics.

Hedgerow intercropping was widely recommended in the 1980s as a way of incorporating the soil fertility benefits of perennials into annual cropping systems. Annual crops are grown between lines of trees, which are pruned regularly to prevent excessive competition. Despite this, the intimacy of the mixture means that competition for water, light, or nutrients can easily reduce crop growth, and many fast-growing tree species have proven to be unsuitable. Some hedgerow intercropping systems, particularly those in more humid areas, have, however, proved to be effective at maintaining soil fertility and crop yields. A study run in Lampung, Indonesia, suggests how competition may be minimized. In this field study, maize roots were mainly found in the upper soil layers and thus overlapped with the root system of the hedgerow tree *Gliricidia sepium* (Fig. B6.2). Both species mainly took up N from the upper soil layer, and were thus competing for this resource. Another hedgerow tree, *Peltophorum dasyrrachis*, in contrast, showed a more gradual decline in root length density with depth, and this was reflected in its N uptake distribution. *Peltophorum* is said to play a less competitive safety-net role in this system, since a large proportion of its roots were found in soil layers beneath those that contained most of the crop roots. The N in these deeper soil layers would be lost from a sole annual crop system, but *Peltophorum* is able to extract it and, eventually, return it to the topsoil via litter or prunings for possible use by a later crop.

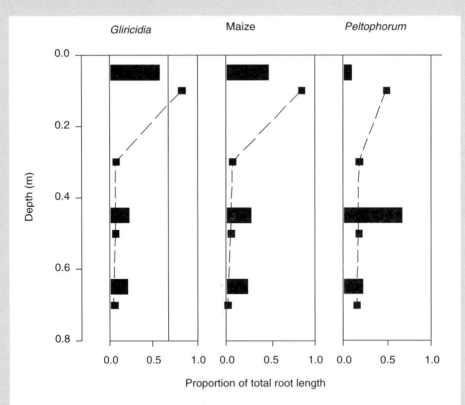

**Fig. B6.2.** Comparison of trees and maize in hedgerow intercropping systems in Indonesia on the proportions of total root length (square symbols) in four 0.2 m deep soil layers and the relative proportions of nitrogen taken up from $^{15}$N applied at depths of 0.05, 0.45 and 0.65 m (bars). Adapted from Rowe *et al.* (2001).

soil *per se* is not proof of nutrient uptake from deep soil layers, because the root length density, the concentration of plant-extractable nutrient, or the soil water content might be inadequate for nutrient uptake. A decrease in plant-available nutrients in deep soil is, by the same token, not direct proof of uptake from deep soil layers, because the decrease could result from the 'tying up' of nutrients into non-extractable forms or from loss by downward or lateral movement rather than by plant uptake. The measurement of nutrient retrieval from deep soil in agroecosystems can fortunately be facilitated using tracer techniques.

Three types of tracers are available: rare elements (such as $Sr^{2+}$ and $Br^-$); radioisotopes (such as $^{32}P$ and $^{35}S$); and stable isotopes (such as $^{15}N$ and $^{18}O$). Such tracers can be applied at different soil depths or at different distances from plants in order to determine the relative uptake by plant species in agroecosystems. Rare elements are easy to use and cost relatively little, but no major nutrient is directly simulated by rare elements. Radioisotopes are expensive and hazardous, but they are the only tracer option available for P. Stable isotopes provide an attractive method, particularly for N (Lehmann and Muraoka, 2001).

In a comprehensive study involving major tropical tree crops (IAEA, 1975), radioactive P was injected at different soil depths and distances from the stems of banana (*Musa* spp.), cocoa (*Theobroma cacao*), coconut (*Cocos nucifera*), coffee (*Coffea arabica*) and oil palm (*Elaeis guineensis*). The plant recovery of added P indicated that highest root activity for all the perennials occurred near the soil surface and close to the plants, even when soil conditions were ideal for extensive root development. The uptake of P from below a depth of 1 m represented only a small portion of total P uptake by the perennials. Another study, involving fruit trees on a deep acid soil in the Brazilian Amazon, provided little evidence that the trees retrieved large amounts of P from below a depth of 1 m, despite root activity at a depth of 1.5 m (Lehmann *et al.*, 2001a). There is, typically, little opportunity in tropical soils for trees to take up and recycle P from below the nutrient extraction zone of annuals, because of the relatively low concentrations of available phosphate and low root length density at depth, coupled with the low mobility of phosphate. Szott *et al.* (1999), in a review of the literature, estimated an annual retrieval of ≤ 1 kg P/ha from deep soil by fast-growing leguminous shrubs. The corresponding annual estimate for Ca was 10 kg Ca/ha in high-base-status soils (exchangeable Ca > 1 $cmol_c$/kg) and substantially less in low-base-status soils.

The stable isotope of N ($^{15}N$) has increasingly been used to assess root activity and uptake of subsoil N in agroecosystems consisting of pure and mixed plant stands. The plant recovery of $^{15}N$ injected at different depths and distances from perennials can be used to effectively assess the relative importance of different soil depths for plant nutrition and the relative differences that exist between plants in terms of the retrieval of N from deep soil. Use of the $^{15}N$ tracer method, for example, demonstrated that *Peltophorum* derived a greater proportion of its N from deep soil than *Gliricidia* (Fig. B6.2). The total uptake of native soil N from different soil volumes during extended time periods is, however, not effectively quantified with the tracer.

The net difference in subsoil inorganic N between agroecosystems that do and do not contain perennials can be used to approximate the retrieval of N by perennials from below the effective root zone of crops. This approach has been used with rotations of perennials and crops and with intercrops of trees and crops. In a rotational system at two locations with contrasting soils and rainfall in Kenya, Hartemink *et al.* (2000) found relatively little change in the amount of subsoil inorganic N during three cropping seasons of unfertilized maize, which produced limited total biomass because of P deficiency. *Sesbania* grown during the same time interval (15–16 months) depleted the inorganic N, which had previously accumulated at a depth of 0.5–2 m by 75–125 kg N/ha/year. The depletion of inorganic N at a depth of 1–2 m occurred at a rate of 50–75 kg N/ha/year. Despite its rapid growth and high N demand, *Sesbania* did not extract all inorganic N in the subsoil. The equivalent of 30–40 kg inorganic N/ha remained at a depth of 1–2 m when the *Sesbania* was harvested and followed by a maize crop.

Livesley *et al.* (2002) used changes in soil inorganic N to approximate N uptake by trees intercropped with maize on an acid Kenyan soil without chemical or physical barriers to deep rooting. Soil nitrate and ammonium were measured to a depth of 3 m and to a distance of 5.2 m from the tree rows, before and after maize cropping. At the start of the cropping season, a build-up of nitrate was present in both the topsoil and at a depth of 0.6–2 m (Fig. 6.2a). Maize utilized the topsoil nitrate, but had no net effect on nitrate below 0.6 m. *Senna spectabilis* utilized most of the subsoil inorganic N within 2 m of the tree row (Fig. 6.2b). Intercropping *Senna* with maize decreased inorganic N below 0.6 m by 53 kg N/ha during one cropping season.

Perennials can retrieve appreciable amounts of the N that accumulates in deep soils below the root zone of crops. Szott *et al.* (1999) estimated the annual retrieval of deep soil N by fast-growing leguminous shrubs to be 30–100 kg N/ha in a high-base-status soil.

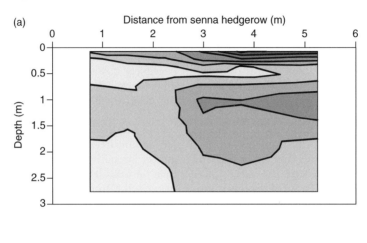

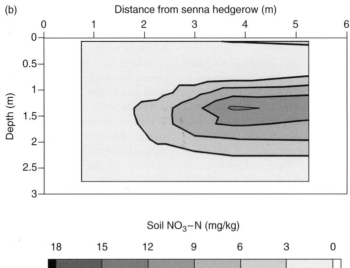

**Fig. 6.2.** Soil nitrate distribution with distance and depth from a row of *Senna spectabilis* before (a) and after (b) an adjacent maize crop on a deep acid soil in western Kenya. Adapted from Livesley *et al.* (2002).

After perennials take up accumulated subsoil N, the retrieval of additional N from deep soil will strongly depend on the rate at which plant-available N builds up in subsoil.

Many of those trees that have a demonstrated ability to retrieve subsoil nitrate in the tropics are $N_2$-fixing legumes. Both $N_2$-fixation and the retrieval of nitrate inaccessible to crops represent net inputs of N by trees to crop–soil systems. The uptake of soil nitrate can, however, conceivably reduce the input of N via $N_2$-fixation. Alternatively, $N_2$-fixation might reduce the uptake of subsoil nitrate. Research with lucerne (*Medicago sativa*) – a deep-rooted $N_2$-fixing perennial – in a temperate environment suggested that symbiotic $N_2$-fixation reduced the capacity of an $N_2$-fixing plant to utilize soil nitrate (Blumenthal and Russelle, 1996). Although this suggests that non-$N_2$-fixing perennials might be more effective in the utilization of subsoil nitrate, considerable evidence from the tropics indicates an appreciable retrieval of nitrate from outside the root zone of annual crops by fast-growing, $N_2$-fixing perennials (Boxes 6.2 and 6.3). Additional research is merited, in order to better understand the interactions and trade-offs that occur between uptake of deep N and biological $N_2$-fixation (Box 6.4).

## 6.5 Achieving More Efficient Use of Deep Soil Nutrients

Nutrients in subsoil can represent an unutilized or underutilized resource for plants. Their use is desirable for increasing the productivity of agroecosystems and for protecting the quality of ground water and water bodies. In this section, we shall illustrate several agroecosystems with the potential for more efficient use of nutrient resources in deep soil.

### 6.5.1 Intercropping perennials and annuals for complementarity in resource use

Intercropping, rather than rotating, soil-fertility-improving perennials with crops might increase the long-term efficiency of nutrient use from deep soil. Intercropped perennials are continually present in the agroecosystem, and their near-continual demand for nutrients can complement the episodic demand of annual crops for growth resources. The perennials can effectively use soil nutrients and water resources both between cropping seasons, and during cropping periods when nutrient supply exceeds crop demand. Perennials in rotational systems, on the other hand, can effectively utilize deep soil nutrients when they are actively growing (see

---

**Box 6.4.** A research opportunity.

'*Do $N_2$-fixing trees preferentially take up inorganic N inaccessible to crops rather than fix atmospheric $N_2$?*'

We put forward the following hypotheses for testing.

**1.** $N_2$-fixing trees primarily extend roots outside the rooting zone of crops not to capture N, but rather to capture other essential plant growth resources such as water and P. But, once the tree roots are outside the root zone of crops, they can preferentially take up accumulated plant-available N rather than fix $N_2$.
**2.** After extracting the available inorganic N from their root zone, $N_2$-fixing trees can revert back to $N_2$-fixation as a major source of N (see Chapter 13, this volume).
**3.** Retrieval of soil nitrate in preference to $N_2$-fixation can, beneficially, minimize nitrate transport into ground water and water bodies.

Research by Gathumbi *et al.* (2002a) on deep acid soils in western Kenya generally supports the second hypothesis. In a comparison of leguminous fallow species, they found similar uptake of soil N among the species, whereas $N_2$-fixation varied among species in order to meet the remaining plant demand for N.

Box 6.2). Nutrients can accumulate in subsoil and be lost during periods when perennials are absent from the agroecosystem.

An example of the potential for perennials in intercrops, as compared with rotations, arises from the unimodal rainfall region of southern Africa, where maize-based agricultural systems predominate. The rotation of maize with a fast-growing woody perennial, such as *Sesbania*, has been identified as an economic alternative to N fertilizers, particularly in remote areas such as eastern Zambia (Kwesiga *et al.*, 1999). *Sesbania*, grown as an improved fallow for 2–3 years between periods of maize cropping, can utilize subsoil nitrate and fix $N_2$. At the end of the fallow, *Sesbania* does not sprout after cutting. Consequently, it does not grow during subsequent maize crops and does not play a safety-net role in capturing and recycling mobile nutrients during the period of maize cropping (see Box 6.2 for related findings from western Kenya).

*Gliricidia* is a promising alternative to *Sesbania* for improved fallows in eastern Zambia. *Gliricidia*, unlike *Sesbania*, is a coppicing species that sprouts after cutting at the end of the fallow. It grows during the long dry season from June to November, when fields are typically bare, as well as during the maize-cropping season from December to May. The *Gliricidia* is cut at the start of the maize-cropping season and during the maize-cropping season, to prevent competition with the crop. The cut biomass of *Gliricidia* is placed on the soil surface, thereby cycling nutrients to maize as the biomass decomposes. *Gliricidia*–maize is consequently an intercropping system, whereas *Sesbania*–maize is a rotational system.

The comparative effects of a *Sesbania*–maize rotation and a *Gliricidia*–maize intercrop on soil nitrate are illustrated in Fig. 6.3. *Sesbania* and *Gliricidia* were grown for 3 years without maize, and then maize was grown for five seasons after cutting the fallows. *Sesbania* died after cutting, but *Gliricidia* sprouted after cutting and grew as an intercrop with maize during the five seasons. Nitrate accumulated below 1 m soil depth after five seasons of maize following the cutting of the *Sesbania*, but no such accumulation of subsoil nitrate occurred when *Gliricidia* was grown as an intercrop with maize. *Gliricidia* effectively provided a safety-net function with regard to nitrate. In the *Sesbania*–maize system, because of the absence of an active perennial, nitrate leached into deep soil below the effective rooting depth of the maize.

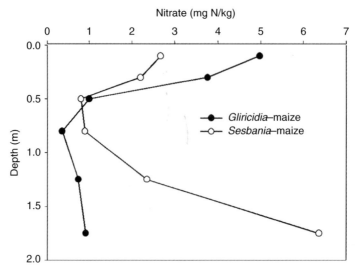

**Fig. 6.3.** Soil nitrate profiles following five consecutive maize crops either grown after *Sesbania sesban* or as an intercrop with *Gliricidia sepium* in Chipata, eastern Zambia (P. Mafongoya, unpublished data).

Furthermore, intercropping with *Gliricidia* increased topsoil nitrate. This reflects the recycling of soil N to upper soil layers through the decomposition of the biomass that was periodically cut from *Gliricidia* and placed on the soil surface.

The intercropping of perennials with crops might be more effective than the rotation of perennials and crops for long-term efficient cycling and use of soil nutrients, but additional factors must be considered in an overall assessment of the contrasting systems. The production of crops in intercropping systems can, for example, be reduced by competition, and the economic feasibility of intercrops can be strongly influenced by the potentially greater labour requirements for periodic cutting and management of the perennial (see also Chapter 17, this volume).

### 6.5.2 Incorporating annuals into underutilized zones of tree plantations

Young trees in plantations exploit only a fraction of the total soil volume. Cover crops or intercrops growing in the spaces between trees can increase the volume of soil occupied by roots, and thereby increase plant uptake of soil nutrients and reduce the leaching loss of nutrients. The cover crops and intercrops used typically decrease in importance, and even disappear from the system, as trees become larger. Opportunities could none the less exist for even greater exploitation of soil nutrients in the intertree spaces of plantations.

One example of an opportunity for the greater use of soil nutrients in tree plantations comes from the lowland humid tropics of Central Amazonia (Schroth *et al.*, 2000a). Well-drained, acid soils with a low cation exchange capacity – Oxisols (US Soil Taxonomy) or Ferralsols (FAO-UNESCO) – are common in the lowland humid tropics of South America and Central Africa. These soils generally have a low nutrient-supplying capacity as a result of high P fixation, low concentrations of weatherable minerals, and low base status. They can, however, have a relatively large N-supplying capacity. Heavy rainfall, common in the tropics, can leach nitrate, formed by mineralization and arising from applied fertilizer, into subsoil, where it can accumulate even under tree crops (Schroth *et al.*, 1999).

Schroth *et al.* (2000a) found considerable nitrate remaining in the soil between 15-year-old oil palms that received no N fertilizer and that had grown without a cover crop or intercrop for the previous 10 years (Fig. 6.4). High nitrate concentrations, par-

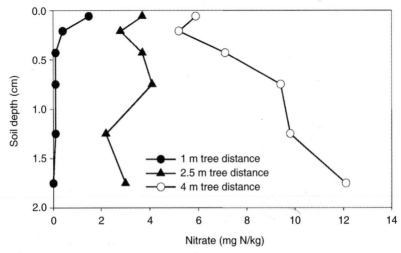

**Fig. 6.4.** Effect of distance from a tree on soil nitrate in a 15-year-old oil palm (*Elaeis guineensis*) plantation without N fertilization in Central Amazonia. Adapted from Schroth *et al.* (2000).

ticularly in deep soil, at a distance of 4 m from the trees indicated leaching loss of nitrate in the zones between trees. The findings suggest that greater inclusion of annual or semiperennial crops in the spaces between the longer living trees could exploit such underutilized N resources and reduce nitrate leaching.

Tree plantation systems should be managed to maximize use of plant-available soil N, particularly in the intertree spaces, while avoiding levels of inorganic soil N in excess of the plant demand for soil N. The inclusion of leguminous cover crops can contribute to increased N mineralization, which can lead to more nitrate than is required by plants in the systems. Deficiencies of nutrients such as P, which limit plant growth, should be eliminated. Broadcasting of fertilizer to overcome nutrient limitations to plant growth in the intertree spaces can be desirable both to ensure effective exploitation of soil N by intercrops and to promote lateral root development of the trees. Management of the agroecosystem to reduce nitrate leaching can have the associated benefit of avoiding cation leaching, soil acidification and ground-water contamination.

### 6.5.3 Mixing plant species for spatial complementarity

Improved fallows are a sequential agro-ecosystem in which perennials are deliberately planted in rotation with annual crops. Improved fallows, as researched and promoted in recent years, are typically monocultures of fast-growing legumes that are designed for the rapid replenishment of soil fertility (Sanchez, 1999a). The diversification of species used in fallows can, potentially, alleviate the build-up of pests and diseases associated with the extensive use of monocultures (see Chapter 15, this volume). The mixing of species in fallows might additionally offer opportunities for increased exploitation of above-ground and below-ground crop-growth resources.

Gathumbi *et al.* (2002b, 2003) examined whether the mixing of woody and herbaceous legumes, which had different growth and rooting patterns, could increase the utilization of soil nutrients by improved fallows on deep soils in Kenya. The mixing of *Crotalaria grahamiana* and *Sesbania* increased competition in the topsoil, leading to increased rooting of the mixture in the subsoil (Fig. 6.5). These findings suggest that

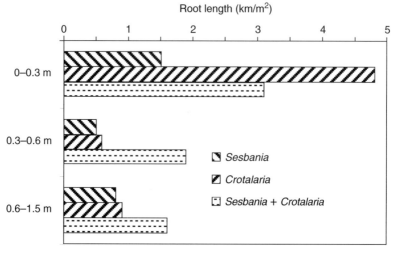

**Fig. 6.5.** Effect of mixing perennials on the distribution of root length after 6 months on a deep acid soil in western Kenya. Adapted from Gathumbi *et al.* (2003).

plasticity of rooting patterns, which enabled the species to modify their root distribution in response to competition when placed in mixtures, could enhance the exploration of soil and the exploitation of nutrients and water from the soil.

Gathumbi *et al.* (2002b) found that undersowing a creeping legume in fallows with an open-canopy, woody legume provided the greatest opportunity for using mixtures of species to increase utilization of deep soil nutrients. The undersowing of siratro (*Macroptilium atropurpureum*), for example, between rows of an open-canopy, woody species (such as *Sesbania*), increased total leaf area index and N accumulation in the fallow biomass.

## 6.6 Preventing the Accumulation of Mobile Nutrients in Deep Soil

It has now been well established that improved fallows, planted in rotation with annual crops, can effectively capture deep soil nitrate that accumulates during the cropping phase (see Box 6.2). Nitrate can accumulate in subsoil during the cropping phase as a result of leaching, particularly when the supply of topsoil nitrate exceeds the demand of the crop for N (Fig. 6.1). The rate of nitrate recharge in subsoil during the cropping phase can be a critical determinant of the magnitude of nutrient resource 'lost' to the crop and potentially retrievable by the perennial. It is also important in determining the optimal frequency of growing perennials in rotational systems for efficient use of soil N.

Factors other than N often limit the growth of crops that follow the perennials in rotational improved fallows. For example, many deep soils of the subhumid and humid tropics, in which nitrate can accumulate, can be deficient in P. Typically, perennials grown in rotation with crops on these soils cannot eliminate P deficiencies in the crops. The P fertilization of the crop in the rotation can be an economic means of increasing crop production, as demonstrated by Jama *et al.* (1998b) on a P-deficient soil in western Kenya.

Research in western Kenya has also demonstrated that P fertilization, used to eliminate P deficiency, can reduce nitrate accumulation in deep soil during continuous cropping of maize (Fig. 6.6). Increased growth of maize, as a result of P fertilization, led to greater removal of soil N by maize and a corresponding reduction in the build-up of soil nitrate. Similar observations have been made in P-fertilized pastures in Brazil (Cadisch *et al.*, 1994a). Management practices that alleviate constraints to crop production, such as P in the above example, can correspondingly increase crop demand for nutrients and thereby decrease the accumulation of mobile nutrients in subsoil. These findings suggest that at least some of the accumulation of subsoil nitrate observed in those farmers' fields that had a history of continuous crop production without fertilizer (see Box 6.1) was likely to have arisen from poor crop management and unbalanced plant nutrition.

## 6.7 Conclusions

The accumulation of mobile nutrients, particularly nitrate, in deep soil layers has been well documented in annual- and perennial-based systems. Much of this accumulation originates from nutrients leaching out of the effective root zone of crops, although mobile nutrients can also form in deep soil layers through mineralization of soil organic matter and as a result of their release from soil mineral reserves.

---

Opportunities for achieving more efficient extraction and use of deep soil nutrients include:

1. Intercropping of coppicing perennials with crops for soil fertility improvement.
2. Including more annual or semiperennial crops in the intertree spaces of tree plantations.
3. Mixing woody and/or herbaceous species in improved fallows.

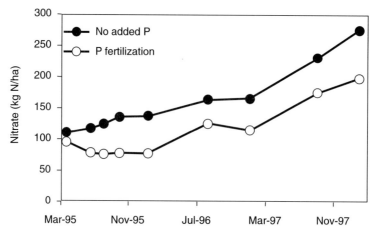

**Fig. 6.6.** Effect of P fertilization to maize in March 1995 on soil nitrate at 0.5–4 m depth during 3 years with four maize crops and two sorghum crops on a deep soil in western Kenya (R.J. Buresh, unpublished data).

The leaching of nutrients arises when the supply of nutrients is greater than plant demand for those nutrients. Consequently, some accumulation of mobile nutrients in deep soil reflects poor crop management, such as imbalanced crop nutrition in which the supply of one nutrient limits crop development and the supply of another nutrient becomes excessive, making that nutrient susceptible to leaching. Improved management of crops, which increases crop growth and demand for the nutrient, can reduce leaching and subsequent accumulation in deep soil. Greater synchronization of the release of nutrients from decomposing biomass and crop demand for nutrients can also minimize episodes of nutrient supply exceeding demand, resulting in leaching loss. A reduction in the leaching of anions (such as nitrate) has the associated benefit of reducing the leaching of accompanying cations.

The use of improved crop management practices, and balanced plant nutrition to alleviate nutrient deficiencies, can minimize the build-up of subsoil nutrients while increasing crop production. In such cases, less subsoil nutrients will, consequently, be present for retrieval and cycling by perennials grown simultaneously or sequentially with crops. Sustained nutrition of the perennial–crop system would, therefore, rely less on recycling of subsoil nutrients and more on biological $N_2$-fixation and fertilization.

Some plant-available, mobile and immobile nutrients will undoubtedly form outside the effective rooting zone of crops. Alternative management of annual crops will not reduce the accumulation of such nutrients or result in their utilization. The integration of deeper-rooting perennials could utilize these reserves of deep soil nutrients and could, potentially, increase the overall cycling and efficiency of nutrient use. Opportunities will increase with increased amounts of weatherable minerals, increased base status, and increased root length density in deep soil layers.

Much progress has been made in understanding and quantifying the processes involved in the retrieval of nitrate by perennials in agroecosystems. Less is known about the processes involved, and the magnitude of their effects, in the retrieval of cations and phosphate from deep soil. One research need that exists is the need to quantify these processes and assess their importance to nutrient balances and efficient nutrient cycling in agroecosystems. Such research will require innovative approaches, because the dynamics of formation and plant uptake are more challenging to measure in the cases of cations and phosphate than they are for nitrate.

Most measurements of nutrient capture from deep soil have been conducted in relatively small experimental plots. Such studies have helped improve our understanding of plant rooting, vertical movement of nutrients, and transformation of mobile nutrients in soil profiles. Lateral movements of mobile nutrients have, however, not been considered in these studies. The lateral movement of mobile nutrients through soil at farm and landscape levels could be important to nutrient cycling. The alleviation of such lateral movement through the strategic management of perennials within the mosaic of land use at the farm and landscape level could have the benefits of preventing pollution of ground water and water bodies. Measurements of the lateral movement of nutrients through soil at farm and landscape levels and the identification of opportunities for increased capture of these nutrients merit research (see Chapter 18, this volume).

In this chapter we have identified several potential opportunities for the use of nutrient resources in deep soil (Section 6.5) and for the prevention of nutrient loss to deep soil (Section 6.6). The modification of agroecosystems for increased capture and recycling of nutrients from deep soil and increased nutrient use efficiency does not, however, ensure greater economic returns and adoption of new practices by land users. A holistic approach is required, in which overall costs and benefits, system productivity and management, labour, and social issues are all considered. Appropriate practices to more effectively capture and use deep soil nutrients are likely to be specific to certain locations, but relatively universal principles, as outlined in this chapter, can help in the identification of appropriate practices.

**Conclusions**

1. Mobile nutrients, particularly nitrate, can accumulate and be retained in deep soil layers, usually after leaching out of the effective root zone of crops at times when the supply of nutrients is greater than current plant demand for those nutrients.
2. The use of improved crop management practices, and balanced plant nutrition to alleviate nutrient deficiencies, can minimize the build-up of subsoil nutrients whilst increasing crop production.
3. Deeper-rooting perennials can utilize such reserves of deep soil nutrients and thus increase the overall cycling and efficiency of nutrient use. Opportunities for deep capture increase with increasing amounts of weatherable minerals, and with increases in the soil base status, and root length density in deep soil layers.
4. Although the processes involved are now well understood, quantification (for a wide range of situations) of the origin and use of deep soil nutrient reserves is still needed, especially for less mobile nutrients such as cations and phosphate.
5. Lateral movement of mobile nutrients through soil at farm and landscape levels can be important to nutrient cycling and in preventing the pollution of ground water and water bodies.

**Future research needs**

1. Quantification of the processes associated with the retrieval of cations (Ca, Mg and K) from deep soil and an assessment of the importance of the retrieval of these nutrients to nutrient balances and productivity in agroecosystems.
2. A better understanding of the processes involved in, and the importance of, lateral movement of nutrients through soil at farm and landscape levels, as well as the identification of opportunities for increased capture of these nutrients.
3. The identification and promotion of economic and socially viable crop management practices that more effectively use nutrients and prevent their loss from the rooting zone of crops.

# 7 Phosphorus Dynamics and Mobilization by Plants

Pauline F. Grierson, Paul Smithson, Generose Nziguheba, Simone Radersma and Nick B. Comerford

---

**Key questions**

1. How do conceptual pools/fractions of P translate into operationally defined fractions of P availability?
2. Can root exudates mobilize significant amounts of P, and does this translate to plant uptake?
3. Once we can measure pools and fluxes in a practical way, will models be sufficient to predict P supply and uptake?

---

## 7.1 Introduction

Phosphorus deficiency is a major constraint to agricultural production in many regions of the world, including tropical Africa, Latin America, Australia and Asia. Phosphorus deficiency occurs in soils low in native P and/or with high fixation capacities (Mokwunye et al., 1986; Warren, 1992) or in soils that have been depleted of resources by intensive and repeated cultivation (Baanante, 1997; Fairhurst et al., 1999; Smithson and Sanchez, 2001). Considerable efforts have been made to assess the extent of P deficiency in soils throughout the tropics and to evaluate the potential of various P fertilizers, including phosphate rocks and organic mulches, to replenish soil P capital. In many tropical regions, the cost and availability of P fertilizer prohibits its use and, consequently, limits productivity. Cheaper sources of P, such as locally mined rock phosphates, are available, but these can prove to be relatively poor sources of P for annual crops, because frequently they are not very soluble. Although crops often perform poorly on tropical soils that have low levels of soluble P, many such soils contain considerable reserves of P that are fixed in unavailable or less labile forms. Certain plants, particularly wild plants and weedy species, perform well on such soils.

Available phosphorus might best be considered to exist in two forms: (i) the amount of P (inorganic P) that is desorbable in response to reducing the P concentration in solution via uptake at the bulk pH of the soil; and (ii) the amount of P (inorganic plus organic) that is available as a result of the biochemical action of plants, mycorrhizas or other soil microbes.

There is now mounting evidence that many plants are able to draw on different fractions of P, both inorganic and organic, by modifying their rhizosphere through the excretion of compounds such as protons, organic acids and enzymes. Such mechanisms have been demonstrated in some crop plants, such as white lupin (*Lupinus albus*) and pigeonpea (*Cajanus cajan*). Many indigenous species that grow on soils low in available nutrients also have highly developed mechanisms for both mobilizing and conservatively using P. Consequently, in addition to improving soil fertility for increased crop yields, a basic understanding of how some plants can grow without the addition of fertilizers is essential in order to utilize these species most effectively in agroforestry systems (Owino, 1996).

Many of the agricultural systems currently being converted to agroforestry are on highly weathered tropical soils that are severely depleted in mineral P and which have high P-fixation capacities. Detecting changes in pools of labile inorganic P on these soils is, therefore, difficult. Moreover, decomposition of organic matter and turnover of the microbial biomass are likely to be of significance in determining the bioavailability of P (Ewel *et al.*, 1991; Oberson *et al.*, 2001). This problem is compounded by the fact that, if they are not validated, conventional methods of estimating 'available inorganic' P, which rely on extraction with weak or strong acids or alkalis, can misrepresent the fraction of bioavailable P. Chemical indices of P bioavailability were developed to estimate the nutrient requirements of annual cropping systems, and often assume a static state (from single measurements of labile P) that has little relevance to perennial ecosystems growing on soils that at any one time contain small amounts of labile inorganic P but are replenished through rapid turnover. Few of these methods account for all the factors that determine bioavailability – for example, biogeochemical reactions, temporal and spatial variability in those reactions, and seasonality of plant growth and uptake.

## 7.2 Phosphorus Forms and Fluxes: Understanding What We Measure

Plants take up P from soil solution as orthophosphate ions: $H_2PO_4^-$ at slightly acid to neutral pH and $HPO_4^{2-}$ at pH > 7.2. The ultimate source of soil solution P is primary P minerals that, through weathering and biochemical processes, allow P to enter into, and be exchanged among, a number of fractions that can be broadly separated into the categories 'inorganic' and 'organic' (Fig. 7.1). Through dissolution, desorption and mineralization reactions, soil P enters the soil solution and becomes available for uptake by plants. Historically, most studies of P availability in relation to crop production have focused on the measurement of inorganic fractions. However, in agroecosystems that include perennials, P is primarily cycled through the organic components shown on the right-hand side of Fig. 7.1. Consequently, it is worth briefly revisiting here what is thought to constitute each P fraction, before we consider the rate-determining steps for transfers among fractions and how transfers affect the amount of P available to plants.

### 7.2.1 Phosphorus forms in soil

#### 7.2.1.1 Inorganic P ($P_i$)

In highly weathered tropical soils there are essentially no P-bearing primary minerals. The exceptions are those soils derived from surface phosphate deposits, or those soils to which finely ground phosphate rocks are added as an aspect of agricultural management. Phosphate rocks are primarily Ca-phosphates known as apatites, with the general formula $M_{10}(PO_4)_6X_2$ (where M is usually Ca and X is either $F^-$, $Cl^-$, $OH^-$ or $CO_3^{2-}$), and are the source of most phosphate fertilizers. The solubility of different apatites varies according to the fraction of the various anions contained in the mineral, with the ratio of carbonates to phosphates being used as an index of apatite solubility (McClellan and Gremillion, 1980). Though primary minerals are important P forms in

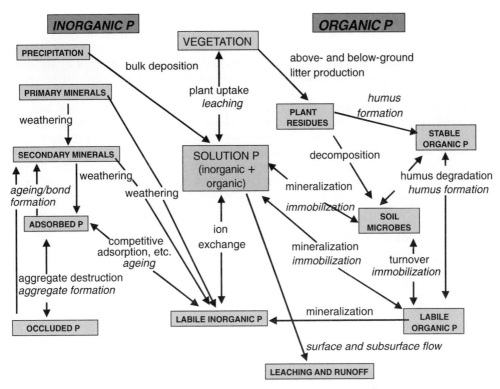

**Fig. 7.1.** Schematic representation of P fractions (boxes) and flows (arrows) in soil. Major processes controlling exchange among fractions are shown. Modified from Walbridge (1990).

phosphate-rock-treated soils, primary P minerals are of little interest in present-day tropical soils under low input or conventional fertilizer management.

Secondary phosphate minerals can be formed by *in situ* weathering of primary apatites in surface phosphate deposits, generally losing carbonate and forming less soluble secondary apatites. The general weathering sequence, which results in mineral assemblages of decreasing solubility, is from Ca-phosphates to Ca-Al-Fe-phosphates, and finally to Al-Fe phosphates (McClellan and Gremillion, 1980). In soils, dissolution of primary apatites or soluble fertilizers may be followed by re-precipitation of secondary phosphate minerals. The minerals formed depend largely on soil pH and mineralogy: in the acid soils common in the tropics, including Oxisols and Ultisols, various Al or Fe minerals may form, often represented generically as variscite (Al-P) or strengite (Fe-P). In calcareous soils, di- or tri-calcium phosphate may form. Ca-Al and Ca-Fe-Al phosphates may also be formed, for example crandallite (a Ca-Al-phosphate) as well as many others. All of these minerals are sparingly soluble, particularly the Al and Fe phosphates. Secondary phosphate minerals can provide a long-term and 'slowly available' reservoir of P.

While secondary phosphate minerals form in soils, soil P retention is much greater than can be accounted for by precipitation reactions, given the known solubility products of the various minerals. This type of P retention (known as adsorption or sorption) is thought to be the main short- and medium-term reservoir for the replenishment of soil solution P from the inorganic fractions (Fig. 7.1). Sorbed P is considered to be orthophosphate ions electrostatically and covalently bound to soil mineral surfaces. The process of sorption occurs very rapidly when P as $H_2PO_4^-$ enters the soil solution from any source, be it soluble fertilizer, phosphate rocks or mineralization of organic material. The $H_2PO_4^-$ ions replace the OH ions exposed at the edges or other surfaces of silicate clay minerals or of oxides or carbonate minerals. In acid soils, Al and Fe oxides are particularly active in the sorption process, since their crystal structures are always unbalanced with respect to charge; this is then balanced by $H^+$, $OH^-$ and other cations or anions. At the acid pH dominant in soils with substantial Al and Fe oxide content, such as Oxisols, the oxides carry a net positive charge and generally have a high specific surface area; they can, therefore, sorb substantial quantities of P (Sample et al., 1980).

Sorbed phosphate is only partially available to growing crops, becoming less available with the passage of time. It is for this reason that P fertilizer must usually be added at a rate several times higher than the crop removal rate, in order to provide adequate P nutrition. Consequently, large one-time additions of P may not provide equivalent cumulative crop responses over time as the same amount of P added in smaller annual increments (Linquist et al., 1996). Though large P additions may be useful and necessary to satisfy initial large P sorption capacities, both theoretical considerations and observed data show that frequent small additions eventually result in higher soil P levels than large one-time additions (Barrow, 1980; Cox et al., 1981; Otto and Kilian, 2001).

### 7.2.1.2 Organic P ($P_o$)

Until relatively recently, most research on soil P has tended to emphasize $P_i$ fractions. $P_o$ compounds, however, comprise a considerable fraction of total soil P – ranging from 20% to 90% (Dalal, 1977; Tate, 1985). The proportion of soil P in organic forms tends to increase with increased weathering, for example, from 7% of bicarbonate-extractable P in calcareous soils to 26% in highly weathered soils (Sharpley et al., 1987). The prevalence of highly weathered soils in the tropics makes organic P of greater potential importance in plant P nutrition than in younger temperate zone soils, and the processes that mediate the mineralization and turnover of $P_o$ are critical in determining P availability (Oberson et al., 2001). For example, in East Africa, crop growth was correlated with total soil $P_o$, and was not strongly related to typical soil-test $P_i$ (Friend and Birch, 1960). However, total $P_o$ in tropical soils is rarely measured, and is instead estimated indirectly by the summing of sequential fractions (e.g. Oberson et al., 2001).

$P_o$ includes P contained in living microbial cells, in microbial and plant debris, and in various products of organic matter decomposition. $P_o$ is not well described in terms of detailed structures, but general classes of compounds can be identified in soils. Inositol phosphates (sugar-like molecules with one or more phosphate groups replacing H) may comprise up to 50% of $P_o$ in soil. Smaller percentages of $P_o$ occur as nucleic acids (DNA, RNA and their derivatives), comprising less than 5% on average. Similarly, phospholipids, which are major components of cell membranes, account for around 1–10% of the total $P_o$ in soil. Other P monoesters and diesters, and polymeric forms of inositol phosphate, such as teichoic acid (a component of bacterial cell walls), have been isolated. However, about half of soil $P_o$ remains unidentified by classical chemical methods (Anderson, 1980; Paul and Clark, 1996).

Recent advances in $^{31}P$ nuclear magnetic resonance (NMR) techniques have improved identification of soil $P_o$ components. In some cases, specific organic P compounds can be identified (Pant et al., 1999). These include glucose-6-phosphate, glycerophosphate and phosphoenolpyruvate (PEP). In general, however, and particularly where P contents are low, P compounds tend to be lumped together in broad categories, primarily monoester and diester P. Recent studies dwell more on changes in these classes of compounds under different land-use or fertilization practices (e.g. Sumann et al., 1998; Zhang et al., 1999; Taranto et al., 2000). $^{31}P$-NMR lends itself to use in measuring the relative rates of depletion of various classes of compounds. For example, $P_o$ was depleted rapidly during the first 3 years of cultivation in northern Tanzania (Solomon and Lehmann, 2000), with diester P being relatively more depleted than monoester P. That diester P is mineralized with relatively greater ease than monoester P is a consistent result found in the majority of cases (Mahieu et al., 2000; Solomon and Lehmann, 2000).

Most NMR characterizations of soil $P_o$ have investigated P extractable by bicarbonate and/or NaOH solutions, which are various permutations of the sequential extraction procedures of Hedley et al. (1982), Tiessen and Moir (1993) and others. One may question whether the compounds are representative of the total $P_o$ pool; for example, Rubaek et al. (1999) found that NaOH extracted less than 10% of the total $P_o$ in some soil clay fractions they studied. The possibility that chemical changes occur in extracted compounds, due to hydrolysis or other reactions, should not be ruled out. In fact, most studies of soil $P_o$ fractions and transformations depend, at least in part, on some variant of sequential extraction procedures; in particular, these procedures almost invariably include a NaOH extraction step to remove $P_o$ from soil. Nevertheless, NaOH-extractable $P_o$ has been identified as an important source of available P in tropical soils that have not been supplied with mineral P fertilizers (Tiessen et al., 1992; Beck and Sanchez, 1994; Bentley et al., 1999).

### 7.2.2 Conceptual versus operationally defined fractions

Although much useful information has been gained over the years, the operationally defined pools extracted by any of the various schemes (e.g. Hedley et al., 1982) remain poorly characterized; their relevance to actual fractions or questions of bioavailability are questionable. The dichotomy that exists between our concepts of P pools and the techniques we use to estimate those pools remains large. This constitutes a stumbling block to any real understanding of soil P transformations (Tiessen and Moir, 1993; Gijsman et al., 1996).

Both $P_o$ and $P_i$ vary in their potential availability for plant uptake, and various pools or fractions (with half-lives in the soil that range from days to centuries) can be envisioned, and range from those immediately available to those that are essentially inert. For example, labile $P_o$ (see Fig. 7.1) is considered to be relatively available; however, the time frame of this availability is dependent on the lifetime of the plant and may equate to different $P_o$ compounds (e.g. inositol phosphate, phosphodiesterases, phospholipids) for different plants, and may be extracted according to different procedures. Similarly, various sequential extraction procedures remove increasingly recalcitrant forms of P.

Sequential extraction procedures have been attempted since the turn of the 20th century (Olsen and Khasawneh, 1980). The classic method used by Chang and Jackson (1957) epitomizes early efforts to characterize soil P, which dwelt primarily on inorganic Ca-, Al- and Fe-P fractions. More recent attempts, such as the Hedley fractionation (Hedley et al., 1982) and its variants, include, and even emphasize, $P_o$ as well as $P_i$ fractions. Many attempts have been made, with mixed results, to relate the operationally defined pools of the sequential extraction procedures to the agreed conceptual pools. For example, in unfertilized systems, bicarbonate- or NaOH-extractable $P_o$ have been found to be correlated with yields. Maroko et al. (1999) detected small increases in NaOH-extractable $P_o$ under leguminous fallows in western Kenya, and stated that this finding suggests a potential source of P for a subsequent crop. Linquist et al. (1997) found that soybean (Glycine max) yields were correlated with both NaOH- and bicarbonate-extractable $P_o$ in an unfertilized Hawaiian Ultisol, whereas NaOH-$P_o$ was the most important P fraction in explaining crop yields under the long-term unfertilized cultivation of a Peruvian Ultisol (Beck and Sanchez, 1994).

Field and pot studies, incubations and work with pure compounds, followed by extraction and quantification of changes in extractable P pools have been the means by which the pools are defined, with terms such as 'labile', 'moderately labile', 'slowly available', 'recalcitrant' and so forth being commonly used. In many cases, changes in $P_o$ pools under different management practices are quite small (Maroko et al., 1999), and are often of the same order of magnitude as the variability resulting from experimental/analytical error. Under short-term (< 1 year) leguminous fallows, changes in labile fractions may be so small that they are undetectable or inconclusive (Jama et al., 2000a; Nziguheba et al., 2000; Smestad, 2000). In many cases, the inability to measure changes in P pools may not be due to a lack of change, but simply to the large size of some pools that make small changes difficult to detect above analytical noise.

What has become increasingly clear is that, for many systems, simply following a standard chemical extraction procedure to estimate P availability can be a very limited approach. In recent years, increasing emphasis has been placed on measuring P transformations and fluxes rather than simply estimating pool sizes (e.g. Gijsman et al., 1996). In effect, we attempt to measure the 'arrows' representing P transformations rather than the 'boxes' representing pools in a schematic diagram outlining P pools and interchanges among the pools (Fig. 7.1). Measuring processes and their rates, rather than poorly defined pools that are sometimes large and thus insensitive to short-term changes, provides a far better understanding of the limits of P supply to plants and, therefore, of the potential for manipulating the system to maximize availability. For example, in one ongoing experiment in western Kenya, maize yields have been increased by the application of non-reactive Busumbu (Ugandan) phosphate rock, especially when combined with shrubby leguminous fallows of 9 months' duration (P. Smithson and Kimiti, unpublished data). Using standard extraction techniques (i.e. measuring the 'boxes' or pools), there was no evidence of increased dissolution of the phosphate rock, and no detectable changes in $P_o$ pools. In contrast, acid phosphatase activity in soil (an enzyme catalysing the process of mineralization) was significantly ($P < 0.001$) increased under fallow treatments in which inputs of organic matter were greater. So, for example, measuring the enzyme activity of soil, rather than the disappearance of reactants and appearance of products, may improve estimations of (potential) rates of reactions (Oberson et al., 2001). However, considerable caution should be applied, as the primary controls on P mineralization are not necessarily related only to plant P demand but also to a large number of other factors including the C : P or C : N : P ratio of the organic matter and abiotic effects, which include temperature and water potential (Umrit and Friesen, 1994; Gressel et al., 1996; Grierson et al., 1999; Grierson and Adams, 2000). Nevertheless, the measurement of rates of transformation (the arrows in Fig. 7.1) promises to offer more sensitive measures of P flux.

### 7.2.3 P cycling in agroforestry systems

The term 'P cycling' refers to the transfer of P from one component of an ecosystem to another (Fig. 7.1). Soil may contain a large amount of P, but the key parameter for fertility is the soil's capacity to supply sufficient solution P for plant growth. For short-lived plants, such as maize or wheat, the rate of replenishment of the soil solution pool that is required for growth is necessarily greater than that for longer lived plants, such as trees. Transfer of P into the soil solution is governed by a combination of biological and geochemical processes (the arrows in Fig. 7.1). In young soils, geochemical processes are dominant and solution P is derived mainly from soil inorganic P fractions. Input is very small in highly weathered soils, in which the predominant source of solution P is the mineralization of $P_o$. However, the interaction between P released from an organic form and the high adsorption capacity of many such soils is not well known. P cycling sustains P availability in natural ecosystems, in which decomposition of organic matter and mineralization of $P_o$ are the main sources of P for uptake by vegetation, and the P released by mineralization is largely controlled by the adsorption/desorption complex of the soil mineral phase. Transfer of P from vegetation to soil is ensured via litterfall and root decay. In agricultural systems without significant crop-residue or root inputs, particularly those on highly weathered soils, P cycling from organic matter may contribute little to the availability of P to crops, owing to the dominance of soil components such as sesquioxides and the large export of P with harvest. Nutrient cycling, however, plays an essential role in the P nutrition of tropical perennial crops, in which the turnover rates of litter are high.

The accumulation of P in plant biomass is a potential source of soil P. Whereas the entire plant biomass is returned to the soil in many natural systems, in agricultural or managed forest systems, a significant fraction of P uptake by crops is lost because of the removal of harvest products for human or livestock consumption. Under small-scale farming systems in Africa, harvesting removes almost all the P accumulated in cereal crops (Sanchez *et al.*, 1997). In contrast, loss of P as a result of the harvest can be relatively small in the case of peach palm plantations (between 6.4 and 31 kg P/ha/year; Deenik *et al.*, 2000; Lehmann *et al.*, 2001b and references therein). In agroforestry systems, roots might account for as much as 80% of the net primary production, where root turnover rates in a tropical dry climate could contribute up to 1.5–2.0 kg P/ha/year (Manlay *et al.*, 2002). Addition of plant biomass as green mulch can contribute to soil P availability, either directly, by releasing tissue P during decomposition and mineralization (biological processes), or indirectly, by acting on chemical processes regulating P adsorption–desorption reactions (Fig. 7.1). However, such an addition of plant biomass is unlikely to be sufficient to replace P removed from a site by the harvesting of whole crops (Palm, 1995).

Owing to the high chemical reactivity of $P_i$, there is as yet no accurate method for quantifying P mineralization in soils such as Oxisols (FAO Ferralsols), as mineralized P can either be rapidly sorbed on to the soil solid phase where it cannot be distinguished from initial inorganic P (Frossard *et al.*, 1996; Oberson *et al.*, 2001) or it can be assimilated by microbes (immobilization) (McLaughlin *et al.*, 1988; Oberson *et al.*, 2001). In addition to being a substrate for mineralization, soil organic matter contributes indirectly to solution P by complexing some ions, such as Al and Fe, which would otherwise constrain P availability. Decomposing organic matter may also release organic anions, which can compete with P for fixation sites and thus reduce its adsorption. The transfer of plant P to soil can be a very important step towards the improvement of P availability in agricultural production, as some species, with no specific agricultural use, can convert less-labile P into forms readily available to crops.

Agroforestry systems have focused on the enhancement of the use efficiency of soil P (i.e. increasing the amount of biomass or 'production' produced for a set amount of soil P) as a more cost-effective means of improving P availability to crops. The more extensive roots of trees and shrubs, in comparison with annual crops, increases exploration of a large soil volume, resulting in enhanced P uptake. A larger biomass (leaves and roots) is produced by tree and shrub species than by annual crops, resulting in an increased amount of P being recycled back into the soil (see Box 7.1). Practices used in agroforestry for recycling P, from plant uptake back to the soil, include the incorporation of plant biomass either on-site or elsewhere (biomass transfer) with resultant increases in crop yields and P availability (Niang *et al.*, 1996a; Jama *et al.*, 1997; Nziguheba *et al.*, 2000). The release of P from decomposing plant materials is regulated not only by the quality of the litter, which is generally defined by the C : N or C : N : P ratio, but also by physicochemical reactions; the modification of these reactions by incorporation of plant materials can enhance or depress P availability (Palm and Rowland, 1997). For an example of P availability in soil amended with green manures of different quality, see Fig. 7.2 (biomass transfer). In addition to the P content of plant tissue, water-soluble C also affects P release from green manure (Nziguheba *et al.*, 2000).

**Box 7.1.** Case study: the *Tithonia* story – early expectations and current realities.

*Tithonia diversifolia* (Mexican sunflower; Fam. *Asteraceae*) is widely distributed throughout the humid and subhumid tropics. *Tithonia* produces large quantities of leaf biomass with a high nutrient concentration, and has been considered as a potential source of plant nutrients (Buresh and Niang, 1997; Jama *et al.*, 2000b). Particular emphasis has been placed on the role of this shrub in P cycling and soil improvement, and there is now a considerable body of literature that focuses on this species. The higher P content in *Tithonia* leaves, in comparison with surrounding plant species, has raised speculation about the mechanisms involved in P acquisition by *Tithonia* and about the role that its green manure might play in improving P availability to crops.

Early results from western Kenya showed *Tithonia* to be superior to other common organic materials as a source of N for maize. In addition, there appeared to be synergies between *Tithonia* and added inorganic fertilizers (Gachengo *et al.*, 1999). A large-scale experiment followed, using *Tithonia* in combination with the locally available Minjingu (Tanzanian) phosphate rock. Early results demonstrated a dramatic maize yield increase with only modest additions of *Tithonia* (Sanchez *et al.*, 1997), and created interest both from a research perspective and from the perspective of its use as an immediate means to improve grain yields in western Kenya and elsewhere in the tropics. Among the proposed properties of *Tithonia* were the following:

- *Tithonia* is superior to inorganic fertilizers, and exhibits synergies with added inorganic fertilizers.
- *Tithonia* increases the solubility and effectiveness of phosphate rocks.
- *Tithonia* is a P accumulator species.

**The realities: useful green manure but no unique physiology**

After several years of study and testing, the reality seems to be that *Tithonia* is neither the unique plant some claimed it to be, nor the useless 'myth' based only on 'hype' claimed by others. As is often the case, the reality of the situation lies somewhere between these two perspectives. In the experiment detailed above, *Tithonia* was superior to urea as a source of N. However, the study site was also K deficient and, in later seasons, a K treatment was added to the original design. Most of the '*Tithonia* effect' could, in fact, be attributed to the addition of K in the *Tithonia* biomass. Interest in *Tithonia*–P interactions remained keen, however, owing to the fact that yields were generally higher when *Tithonia* was applied as a green manure than when equivalent inorganic NPK fertilizers were used in P-limited soils (B. Jama *et al.*, unpublished; Nziguheba *et al.*, 2002). When *Tithonia* manure was applied in conjunction with rock phosphate, Savini (1999) showed that *Tithonia* seemed to depress rather than enhance phosphate rock solubility, although the effect was temporary (Savini, 1999; Smithson, 1999).

*Tithonia* is similar to many other pioneer species, in that it produces abundant fine roots to scavenge available soil nutrients and accumulates those nutrients to relatively high levels in its leaves (around 0.4% P). However, P uptake rates by *Tithonia* are no greater than those of other species, and although P-starved plants take up P at initially high rates, the uptake rate falls off after a short time. *Tithonia* accumulates P at a rate that is proportional to the amount available in the soil, with stands on poor sites accumulating significantly less P and K than those on more fertile sites (George *et al.*, 2001). Overall, 'daisy fallows' seem to invade slightly better sites, rather than causing an improvement in those sites. Nevertheless, *Tithonia* and other composites often have large quantities of nutrients in their tissues. In South-East Asia, for example, farmers practice short-rotation fallows using another composite, *Austroeupatorium inulifolium*. In these fallows, dry matter, N and P accumulations were all about three to four times greater than those produced by fallows of native ferns or the invasive grass *Imperata cylindrica* (ICRAF, 1996; Smithson and Sanchez, 2001).

In summary, the incorporation of *Tithonia* leaf biomass into P-deficient soils can produce beneficial changes in soil P relations and in crop yields. In some cases, these beneficial effects are greater than those given by other, seemingly high-quality, organic materials (Fig. 7.2). However, major increases in yields, relative to inorganic fertilizers, have not materialized. As for soil improvement, results have been variable: on extremely poor sites *Tithonia* performs poorly and accumulates little P or other nutrients, demonstrating that there exists a direct link between *Tithonia*'s potential to acquire nutrients and the initial fertility of the site. Nevertheless, identification of plant species with the ability to recycle P into more available forms for crops should be considered a research priority.

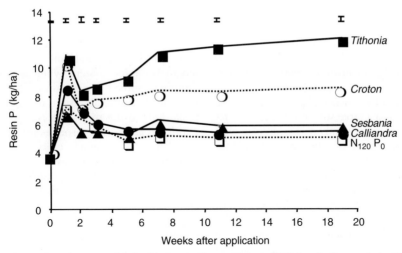

**Fig. 7.2.** Change in resin-extractable P with time in Kenyan Oxisols (FAO Ferralsol) amended with different green manures (5 t/ha dry biomass) or NPK fertilizer. Equivalent P loadings were 8 kg P/ha with *Sesbania sesban* (L. Merr.) and *Calliandra calothyrsus* (Meissner), 12 kg P/ha with *Croton megalocarpus* (Hutch) and 13 kg P/ha with *Tithonia diversifolia* (Hemsley A. Gray). Bars are standard errors of difference among means of treatments. Data from Nziguheba *et al.* (2000).

## 7.3 P Mobilization and Acquisition by Plants

Although crops often perform poorly on soils with low levels of soluble P, many such soils contain considerable reserves of P that are fixed in unavailable forms, both organic and inorganic. However, it is now generally accepted that many species, especially perennials and legumes, can utilize relatively non-labile fractions of P by modifying their rhizosphere through the excretion of organic acids and enzymes. These exudates convert the fixed P (as inorganic or organic forms) into soluble forms that can be used by the plant and, possibly, by other plants nearby or in subsequent rotations. The use of legumes to supply N has been promoted to overcome soil fertility constraints, but without adequate P this strategy can have only limited success (McLaughlin *et al.*, 1990; Vance, 2001). Improving the supply of P to legumes would therefore lead to better $N_2$-fixation, and hence to improved N nutrition in cropping systems, to increased yields and to reduced erosion. The use of fertilizers to supply P to African farms has been explored (Buresh *et al.*, 1997), but there has been relatively little attention given to the use of P-efficient plants, either in combination with poorly soluble P fertilizers or on soils with large P reserves but little 'available' P. Species or genotypes that are productive under these conditions may be able to access normally unavailable $P_o$, fixed soil P and sparingly soluble P. It is known that some plants are more efficient at acquiring P, owing to their root system morphology or their ability to form mycorrhizal associations. Others are able to modify the chemistry of their rhizosphere by, for example, excreting protons or organic acids, which solubilize fixed P, making it available for uptake (Randall *et al.*, 2001). In addition, there are wild species that are adapted to low-P soils that may be suitable for incorporation into multispecies systems where the opportunity exists for increasing the availability of the P source. Although many of these species exhibit root adaptations that produce a range of exudates that may increase the availability of P, these species also tend to exhibit greater use efficiency and conservation of P once it is obtained. It is in this context that we see an opportunity to advance our knowledge of the ability of organic-acid-secreting plants to obtain P from sparingly soluble soil and fertilizer sources, and to assess the value of such plants for P nutrition in intercropping or rotational systems, or in agroforestry.

The amount of P taken up by plants is influenced by an array of plant parameters and soil properties. Strategies for enhancing P acquisition by plants include: (i) increasing their access to a large soil volume, for species with an extensive root system and/or with mycorrhizal roots; (ii) the ability to reduce the soil solution's nutrient concentration to very low levels (i.e. a low $C_{min}$, where $C_{min}$ is the ion concentration in solution at the root surface where influx = efflux; Barber, 1995); and (iii) enhancing the mobilization of P by means of root-mediated modifications of the rhizosphere. Root-mediated changes that can influence P availability include pH changes, the production of chelating ions, and the exudation of phosphatase enzymes. Between 30% and 60% of net photosynthetic C is allocated to roots, while ~ 70% or more of root C is released as organic C into the rhizosphere in tree species (Lynch and Whipps, 1991). Exudation is influenced by plant nutritional status, with higher rates occurring under conditions of stress, and increases occurring with mechanical impedance of the soil. Root exudates also have an indirect effect on P availability, through their role as carbon substrates for the microbial biomass. Bacteria and fungi are known to solubilize $P_i$ (Richardson, 1994; Leggett et al., 2001), and it has also been suggested that the organic anions and phosphatase enzymes and phenolic acids they produce act as chemo-attractants for rhizosphere organisms. However, although the potential for root exudates to modify nutrient bioavailability is often discussed, more quantitative evidence is required to evaluate the significance of the effect that exudates have on nutrient uptake. The role of root exudates in P acquisition by plants is discussed in greater detail in the review by Randall et al. (2001).

### 7.3.1 Root morphology

Many species possess hairy roots or have the ability to rapidly establish a large root system, so maximizing uptake of P (and water) from the soil (e.g. *Tithonia*, Section 7.2.3). Root hairs and mycorrhizas increase the volume of soil explored, improve contact between the root and soil and effectively lengthen the root system, whereas closer and more widespread exploration of the soil system shortens the diffusion path for phosphate ions to the root surface. For example, *Eucalyptus gummifera*, a species indigenous to low-P soils, is extremely efficient in rapidly establishing an extensive root system (Mulligan, 1988). Other species may store P in their roots, an attribute that appears to be greatest in tree provenances from low-P soils (Mulligan and Sands, 1988). This is an important point, because it indicates that provenance selection of certain species can impact on the efficacy of root systems to acquire P, not just in crop species (e.g. Araujo et al., 1998; Nielsen et al., 1998) but also in agroforestry species such as *Grevillea robusta* and a range of *Eucalyptus* species. Slow early shoot growth relative to total root length, and high specific root length, have been proposed as one set of criteria for the selection of species and provenances of perennial species well-adapted to P-deficient soils (Ndufa et al., 1999). The majority of plants adapted to low-P soils form associations with mycorrhizal fungi, generally ectomycorrhizal (ECM) or vesicular–arbuscular (VAM) associations. Mycorrhizas are thought to benefit the host plant through the increased uptake of P; this is largely achieved by an extensive proliferation of hyphae to sites well beyond those that would come into contact with non-mycorrhizal roots.

The formation of proteoid or cluster roots and infection by mycorrhizal fungi appear to be alternative strategies for the enhanced uptake of water and nutrients (Lamont, 1986). The important agroforestry species *Grevillea robusta* and *Macadamia* spp., for example, have cluster roots, as does the leguminous crop white lupin. Cluster roots are non-mycorrhizal, but are usually very hairy with a large surface area. They are known to produce large quantities of organic acids (Section 7.3.3) and phosphatase enzymes (Section 7.3.4), which may enhance labile P forms in the soil solution (Grierson, 1992; Dinkelaker et al., 1995).

### 7.3.2 pH change

In solution, the form of phosphate ions is largely determined by pH (Lindsay, 1979; Section 7.2). Phosphate ions readily precipitate with metal cations (Fig. 7.1), and the type of mineral formed will depend on soil pH, as it governs the abundance of ions such as Ca, Fe and Al. Iron and aluminium phosphates tend to form under acidic conditions and increase in solubility as pH increases, whereas calcium phosphates form under alkaline conditions and increase in solubility as pH decreases (Lindsay, 1979). Consequently, any pH shift in the rhizosphere can have significant consequences for P bioavailability, the extent of which is dependent on the mineralogy of the soil. However, because of complex interactions among precipitation, sorption and ligand reactions in soil, it is still difficult to determine to what extent and even in what direction the bioavailability of soil P will respond to pH change (Geelhoed et al., 1999; Hinsinger, 2001).

Rhizosphere pH may differ from that of bulk soil by as much as three units (Grierson and Attiwill, 1989; Hinsinger, 2001). This difference in the pH of the rhizosphere and the bulk soil is primarily a consequence of the net balance between uptake of cations and anions (Marschner, 1995). The form of the N supply is the primary influence on the cation/anion uptake ratio, and thus on rhizosphere pH. Nitrate supply increases $HCO_3^-$ release, and therefore pH increases, whereas $NH_4^+$ tends to acidify the rhizosphere through greater release of $H^+$ (Chapter 8, this volume). In the case of tropical soils, which are predominantly acidic, it is unlikely that a further decrease in rhizosphere pH would be very effective in enhancing release of P. However, if a plant on acid soils is predominantly supplied with $NO_3^-$, the release of $HCO_3^-$ can enhance P uptake through increased solubilization (Gahoonia et al., 1992). Studies made on crop and perennial species in western Kenya have also demonstrated consistent and highly localized alkalinization of the rhizosphere of plants grown on acid Oxisols (Fig. 7.3). Since adsorp-

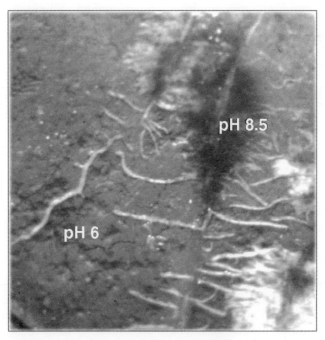

**Fig. 7.3.** Change in pH around new cluster (proteoid) roots of *Grevillea robusta* (Cunn.) growing in an acid Oxisol (FAO Ferralsol) from western Kenya (S. Radersma, unpublished).

tion is decreased with increasing pH, when Al is low desorption should be increased. If a plant is fixing $N_2$, the cation/anion uptake ratio will be large, meaning a net excretion of $H^+$. Consequently, the capacity of legumes to utilize rock phosphate may be greater than in plants fed $NO_3^-$ (de Swart and van Diest, 1987).

In some instances, rhizosphere acidification under low-P conditions may be at least partly attributable to exudation of organic acids (Section 7.3.3), although most of these organic acids are released as anions, and where concurrent acidification occurs it is probably the result of the release of $H^+$ to balance the net efflux of anions (Jones, 1998). The initial bulk-soil pH and the soil's pH buffering capacity are the main factors that determine the extent to which a plant can modify pH. Buffering capacity is primarily related to the organic matter content of the soil and less so to its clay content (Nye, 1986). Indirectly, the pH of the rhizosphere also changes in response to plant nutrient status, particularly under conditions of iron and phosphorus stress (e.g. Hoffland, 1989). Under conditions of low P, decreased uptake of nitrate, in comparison with a higher uptake of cations like Mg and Ca, means there is an increase in the cation/anion ratio and a subsequent net excretion of $H^+$.

The extent to which pH modification of the rhizosphere may enhance P uptake by plants is still difficult to assess. However, early work by Riley and Barber (1971) on soybean demonstrated a linear increase of P in shoots with decreasing pH. There have been few studies of this kind on perennial species important in agroforestry systems.

### 7.3.3 Organic anions: ligand exchange and dissolution of soil P

Over the last decade, a strong emphasis has been placed on identifying organic anions in the rhizosphere of many plants (see Dinkelaker et al., 1995; Randall et al., 2001). Organic anions released by plant roots, including amino acids, organic acids and phenolics, can increase P release from the soil complex. In particular, low molecular weight (LMW) organic anions, such as citrate, malate and oxalate, are known to increase P release through ligand exchange (Gerke, 1992). Of the carboxylic acids, citrate has a higher log formation constant for Al, Fe(III) and Ca complexes than other acids, and is likely to be most effective at releasing P (Jones and Darrah, 1994). Exudation of piscidic and malonic acids by pigeonpea roots (in addition to a range of carboxylic acids) facilitates the release of P from Fe-bound soil P, which is normally considered unavailable to plants (Ae et al., 1993; Ae and Otani, 1997). In a pot experiment, pigeonpea increased availability of P to maize plants that were subsequently sown in the pots (Arihara et al., 1991). Similarly, the exudation of citrate by white lupin, described in the landmark papers of Gardner et al. (1982, 1983), has been shown to benefit the P nutrition of the white lupin crop and other crops interplanted with, or grown after, the lupin (Gardner and Boundy, 1983; Horst and Waschkies, 1987; Hocking, 2001). Plants like white lupin and pigeonpea are able to draw on pools of soil P that are not immediately available to other species (Hocking et al., 1997). However, whereas piscidic acid is a very effective chelator of Fe(III), it has only a limited effect in terms of increasing calcium phosphate solubility. Consequently, the P-efficiency of this particular plant is dependent on the pH of the soil and on soil mineralogy (Ae et al., 1993).

Perennial species also produce organic anions. However, organic anion production by perennial species has not been as well studied as organic anion production by annual crops. Tree species, such as *Banksia integrifolia* (Grierson, 1992) and *Melaleuca cajeputi* (Watanabe et al., 1998), exude large amounts of citrate into their rhizosphere, particularly under low-P or high-Al conditions. The ectomycorrhizas of trees have also been reported to produce large amounts of organic acids that can increase the amount of labile P in the soil solution via a number of mechanisms, including solubilization, chelation or complexation (Malajczuk and Cromack, 1982). With regard to modifications to root morphology, quantification of the extent of P mobilization by organic anions and translation into P uptake by plants is still a major limitation to our understanding of the real effectiveness of root exudates (Kirk, 1999). Organic anions may mobilize P in the rhizosphere by a change in pH, a desorption of P, the chelation of metal ions (particularly Al and Fe (III)) or the formation of metal-chelate complexes (Gardner et al., 1982). Of these mechanisms, reactions involving metal chelates are the most important in acid tropical soils. However, much more needs to be known about the factors that control organic anion release from plant roots and their longevity in the rhizosphere, and their differing effects on P mobilization in different soils (Kirk, 1999).

The mechanisms described above, which might enhance P bioavailability, have the potential to be applied to tropical agroecosystems, in the same way that $N_2$-fixing plants have been utilized to increase N supply. For example, the inclusion of a green fallow consisting of legumes that are efficient at mobilizing P (e.g. pigeonpea, chick pea, cow pea, white lupin) might improve P supply to the food crop that follows in rotation, such as maize or wheat (Gardner and Boundy, 1983). Alternatively, tree species that can utilize forms of P that are unavailable to crops can provide alternative incomes through timber or fruit production (e.g. *Grevillea robusta*). The extent to which maize might or might not benefit from adjacent perennials is discussed further in Section 7.4.

### 7.3.4 P mineralization: phosphatase enzymes

Phosphatases catalyse the hydrolysis of phosphate esters derived from a broad spectrum of substrates. The most extensively studied phosphatase in soils is acid phosphomonoesterase (which is both plant and microbe-derived) because this enzyme catalyses the hydrolysis of organic phosphomonoesters (mineralization) to $P_i$, which can then be taken up by plants or microorganisms. The activity of extracellular phosphatase enzymes is pH dependent. Extracellular enzymes contribute to the decomposition of complex polymeric molecules into more assimilable monomers, and their synthesis by plants and microbes is inducible under conditions of low P. They are influenced by environmental conditions, by the physiological state of the plant, by root age and root position (i.e. activity is always greater at growing root tips), and are derived from both plants and microorganisms.

As mentioned in Section 7.2, phosphatase activity in the soil has been equated to demand for P, as an index of potential rates of P mineralization and an indicator of plasticity in plant/root response to P supply. However, the relationship between phosphatase activity and P uptake by plants remains unclear. Phosphatase activity in the rhizosphere increases under conditions of low $P_i$ (e.g. Firsching and Claassen, 1996) but may also occur with increasing $P_o$ (e.g. Tarafdar and Claassen, 1988; Grierson and Adams, 2000). In some instances, increased phosphatase activity in the soil also corresponds to increased $P_i$ (Grierson and Adams, 2000). Intracellular phosphatase activity of roots increases with decreasing P concentration in the tissue of higher plants (Helal, 1990), and there is increased root phytase activity (a specific type of phosphatase that hydrolyses phytate) under P-deficient conditions (Li et al., 1997). Part of the difficulty of interpreting phosphatase measurements of soil and the rhizosphere arises from the need to determine: (i) what percentage of the activity is plant-derived; and (ii) if what we measure is indicative of plant demand for P. In many instances, it is not. There is some evidence that intracellular mechanisms exist for the release of endogenous bound phosphates (*de novo* phosphatase synthesis in cell walls and vacuoles) under conditions of $P_i$ starvation, i.e. where there is an absence of an exogenous source. It is possible that this is controlled by multigene pathways (e.g. Glund and Goldstein, 1993). However, if there is a low demand for P, as there is in many plant species adapted to low-P soils, there may be no activation of the cellular mechanisms for phosphatase synthesis. Phosphatase enzymes exuded by plants and microbes can also be adsorbed by clays and occluded in humic compounds (Nannipieri et al., 1996) so becoming stabilized while retaining their activities. Consequently, the activity of phosphatases in the soil (including the rhizosphere) is not always controlled by the respective protein enzyme synthesized within the plant or microbe in response to P demand. However, enzyme activities can reflect the 'history' of the soil.

Although phosphatase synthesis and secretion are now well described with regard to crops such as white lupin (see Randall et al., 2001 and references therein), very little information is available on phosphatase production for most agroforestry perennials.

## 7.4 Modelling P Bioavailability and Uptake in Agroforestry Systems

Phosphorus bioavailability is a function of a finite number of processes. When these processes are defined for the plant species and soil of interest, it is possible to simulate or predict P uptake for a certain system. The most common soil-based, process-oriented approach to P uptake is explained in detail by the most recent editions of books by Barber (1995) and Tinker and Nye (2000). These ideas have been incorporated into a number of mathematical models, such as UPTAKE (Barber, 1995), COMP8 (Smethurst and Comerford, 1993a; Ibrikci et al., 1994) and SSAND (Adegbidi et al., 2001) and into the WaNuLCAS model (van Noordwijk and Lusiana, 2000). Although these models work differently, they are all fundamentally similar. They attempt to dynamically simulate the following three groups of processes, which result in plant nutrient uptake: (i) release of nutrients from the solid phase to the solution phase; (ii) movement by mass flow and diffusion through the soil solution phase; and (iii) uptake at the surface of a growing root system (see also Chapter 10, this volume).

Nutrient release to the solution phase is described by desorption and mineralization. Generally a desorption isotherm (such as a Freundlich isotherm) is used. Mineralization is an input that is based on one's knowledge of the bioavailability of $P_o$ under specific soil conditions. Mass flow moves the solution-borne nutrients to the root surface. In order to model this we require knowledge of the transpiration stream entering the average root and the concentration of the nutrient in solution. The amount of P taken up by the plant via mass flow is estimated to be around 5–10% of its needs, at least in younger soils (Marschner, 1995). In weathered and high-P-fixing soils, the contribution of mass flow is probably much less. If mass flow does not supply the demand of the plant root, a diffusion gradient is developed when the nutrient concentration at the surface of the root drops below that in the bulk solution. In that case, soil water content, diffusion path tortuosity, and the interaction of the nutrient with the soil surface determine the rate of diffusive flux of P to the root surface. Once the nutrient arrives at the root surface, it is taken up in accordance with a Michaelis–Menten-type relationship between the surface of the root and the external solution P concentration. Generally, the size of the root system changes with time, by either increasing the soil volume into which it grows (UPTAKE), or changing the root density within a soil volume (COMP8, SSAND and WaNuLCAS). Simulated results for P have been good, particularly for high levels of soil P (Smethurst and Comerford, 1993a). In soils with low levels of P, uptake by mycorrhizal fungi needs to be considered and is included in SSAND and WaNuLCAS.

The complexity of P pools, fluxes and processes (Fig. 7.1) and the lack of clearly defined operational pools make prediction of P availability and uptake extremely difficult, to such an extent that mechanistic modelling remains largely conceptual (see Darrah, 1993). However, the art of modelling is to know what to leave out so that one can concentrate on the few processes that can explain 80–90% of the outcome, e.g. root uptake rates, size of pool of solution P, mineralization rates, desorption/adsorption reactions (Box 7.2). As described in the previous sections, all of these processes are connected and are related to plant and microbial attributes, as well as to soil mineralogy and organic matter content and quality.

**Box 7.2.** Case study: application of the WaNuLCAS model to a Kenyan maize–tree system.

We describe here the application of the tree–crop interaction model WaNuLCAS (see also Chapter 10, this volume, for further details) in order to examine different mechanisms affecting maize performance in relation to distance to tree lines of *Grevillea robusta* and *Cassia spectabilis* in an agroforestry system in Kenya. The study site was on a deep, well-structured Oxisol (FAO Ferralsol), in which P was the nutrient most limiting to crop production. Soil water was not limiting and inputs of organic matter were low. This site was chosen specifically in order to test the model, which was based on the assumption that trees would affect crop growth by changing P availability, independent of water supply. The model included details of: (i) overall growth rate of the trees and P demand of the tree and the crop; (ii) root length densities and distribution and, in the case of the crop, their development as a function of above-ground growth; (iii) some measure of P availability and desorption constants; (iv) soil water contents, as determined by tree and crop uptake, which determine P diffusion; (v) the extent of mycorrhizal infection; and (vi) possible tree-rhizosphere effects on P availability/mobility and an assessment of its transfer to maize roots.

The WaNuLCAS model considers a mobile P pool derived from Bray (acid-fluoride) or Olsen (bicarbonate) extractions, with P availability modified according to the adsorption characteristics of a particular soil. A second pool is the 'immobile P', which only feeds into the mobile P pool by a weathering parameter or by rhizosphere-P mobilization. Rhizosphere effects are modelled by transferring P from the 'immobile pool' to the 'mobile pool' and/or changing the $K_a$ (the adsorption constant) of the mobile pool. If P fractions in the rhizosphere soil of the tree did not differ from the bulk soil and the tree did not respond to the addition of P fertilizers, then the P demand of the tree is set at zero. To what extent a maize crop might profit from having parts of its root system in the rhizosphere of the tree at the time of transfer of P from the immobile to the mobile P pool is described by a tree-P-mobilization parameter. This parameter calculates the increase in available P in the rhizosphere due to: (i) organic anion concentrations as measured in the rhizosphere; (ii) the extra available P this anion loading provides; and (iii) pH increases and the extra available P resulting from the pH increase. The tree-P-mobilization parameter is then multiplied by: (i) the fraction of the total soil volume that is the rhizosphere of the tree roots for each zone and depth (derived from root length density, root diameter, the distance of the rhizosphere and total soil volume); (ii) the fraction of total soil volume that was maize root (derived from root length density, average root diameter and total soil volume); and (iii) a 'syn-location' parameter (derived from root-wall images), which describes the overlap of tree rhizosphere and maize roots (S. Radersma and P.F. Grierson, unpublished).

The first aim of the simulations was to determine if an observed decrease in maize growth near a border line of the tree *Grevillea robusta* was due to either direct competition for P or to indirect effects, via reduced soil water contents (owing to greater water use by the trees). Simulations showed that lower soil water content reduced rates of P diffusion and had a cumulative and negative effect over time on maize root and shoot biomass. Measured relative yields (yield in the zones close to tree lines as a fraction of yield in the zone furthest away from the tree line) were compared with the simulated relative yields arising from the model. Results from the model predicted that a 2.5% decrease in water content reduced P diffusion near the *Grevillea* tree line, resulting in a decrease in crop growth of about 15% (as compared with the maize growing far from the tree line). The iterative and cumulative effect of reduced P diffusion resulted in a further 10% decrease in maize biomass. Thus the total effect of a 2.5% decrease in soil water was a 25% reduction in maize yield. Inclusion in the model of *Grevillea*'s P demand, and a direct water limitation to crop growth, had a negligible effect on predictions of the relative yield levels in the different zones.

A second set of simulations tested whether mobilization of P in the rhizosphere could explain a measured 12–15% increase in maize yield near a line of the small tree *Cassia spectabilis*. As all leaf litter was exported from the field, increased organic matter and P inputs from litter decomposition could not account for the increased growth. In addition, soil water did not change with distance from the tree line in the *Cassia*–maize system. Based on P contents and P demand of the *Cassia*, relative maize biomass levels were predicted to be reduced by about 20% close to the tree line. However, *Cassia* rhizosphere soil and bulk soil did not show significant differences in sequential-P fractions, meaning that *Cassia* takes up as much P as it mobilizes. This was represented in the model by setting

*Continued*

**Box 7.2.** *Continued.*

the *Cassia* P demand to 0. *Cassia*'s mobilization of P, and its transfer to maize, was estimated from measured oxalate concentrations and pH increases in the rhizosphere. It was simulated at a 'minimum' level of effect (oxalate loading 120 mg/kg over the first 10 days and pH 7 over the following 38 days) and a 'maximum' level of effect (oxalate loading 600 mg/kg over the first 10 days and pH 8.5 over the following 38 days). The model predicted that maize yield would increase by between 10% (at 'minimum') and 35% (at 'maximum') near the tree lines, which corresponds well with the observed 12–15% increase in maize yield near the *Cassia* tree line.

From these two simulation examples, it is clear that whereas the rhizosphere effects of one plant may result in a small increase in maize growth, the possible positive effects of some species of trees on P mobilization can be very easily offset by even small reductions in soil water content. These reductions are, in turn, predicted by the model to decrease P diffusion and P uptake by maize, even though they may not be severe enough to cause direct water-deficit problems.

### Conclusions

**1.** The P-fixing capacity of tropical soils can be high and increases with the amount of clay in the profile; tropical soils are also often P depleted.
**2.** Replenishment of soil P capital requires inputs of P from fertilizers or organic plant residues.
**3.** Cycling of organic matter is key to the management of P availability in agroforestry systems.
**4.** Plants and microorganisms can affect P availability by changing rhizosphere pH, by excreting metal-complexing anions or phosphatases, and by affecting ligand reactions and mineralization rates.
**5.** P mobilization and acquisition by some plants, such as white lupin and pigeonpea, can be beneficial to subsequent crops or to intercrops.
**6.** The relative contributions that root exudates and phosphatase activity make to P mobilization and plant uptake in agroforestry systems have not yet been quantified.
**7.** Models of P uptake dynamically simulate the release of nutrients, their movement by mass flow and diffusion through the soil solution phase, and their uptake at the surface of a growing root system.
**8.** Tree–crop competition for soil water has a deleterious effect on growth of maize, which is greater than the positive effects, due to increases in P availability associated with rhizosphere activity of tree roots.
**9.** For some species, P mobilization mechanisms may enhance maize growth if water is not limiting.

### Future research needs

**1.** To quantify fluxes of P, and rates of processes such as P release, due to decomposition and mineralization, when green manure is added to low-P soils.
**2.** To quantify how much P is mobilized by root exudates and to assess critically whether this P is taken up by crop plants.
**3.** To screen both crop and perennial species for selection and breeding, with respect to P release and mobilization, for their potential use in agroecosystems.

# 8 Managing Soil Acidity and Aluminium Toxicity in Tree-based Agroecosystems

Mike T.F. Wong, Kurniatun Hairiah and Julio Alegre

---

**Key questions**

1. What are the main causes of soil acidity in low-input tropical agroecosystems?
2. How can agroecosystems transfer organic alkalinity to ameliorate acidic soils?
3. How do we develop the practical implementation of this knowledge concerning acidity?

---

## 8.1 Introduction

About a third of the total land area of the tropics (about 1.5 billion ha) is strongly acid. In the soils within this area, exchangeable Al occupies more than 60% of the cation exchange sites of the 0–50 cm soil layer. So, Al toxicity is considered to be a major constraint for crop production (Sanchez and Logan, 1992). Aluminium toxicity decreases the efficiency with which scarce plant nutrients and water are used, by interfering with the growth and physiological functions of roots. These strongly acid soils are prevalent in the humid tropics and acid savannahs, and are mostly found in Oxisols, Ultisols and Dystropepts. Soils in these regions commonly have iron oxide to clay ratios of > 0.2, and can fix large amounts of P. Their strong acidity is correlated with low nutrient reserves: their sand and silt fractions contain < 10% weatherable minerals. The capacity of these soils to supply P, K, Ca, Mg and S is limited. In Africa, and elsewhere, depletion of these nutrients, by crop removal, leaching and erosion, exacerbates the problem of low nutrient supply further and causes a decline in soil fertility. Soil fertility decline is considered to be the principal cause of declining per capita food production in Africa. Crop performance under such conditions is limited by deficiencies of such nutrients as N, P, K, Ca, Mg and Mo, a problem that is exacerbated by the poor nutrient uptake efficiency of the Al-affected root system.

Lime is traditionally used to neutralize soil acidity; high rates of application, ranging from 1 to 5 t/ha, are usually needed every few years. However, liming cannot be the sole method used to solve soil acidity in many developing countries, because of low income and because of the low value of farm produce relative to the cost of liming. The per capita income in many countries in Africa is, for example, <US$1 per day. Agriculture is, therefore, mainly practised as a low-input system. An Integrated Soil Acidity Management Strategy needs to be developed in order to provide solutions that will minimize the causes of acidity and treat its effects within the socioeconomic constraints of the regions. Liming is a part of this integrated solution. The technologies involved in liming are well developed, and will not be dealt with here. However, a recent review on the use of lime to treat topsoil and subsurface acidity can be found in Edmeades and Ridley (2003), and a recent report on the use of lime to treat subsoil acidity can be found in Whitten et al. (2000).

In the past, farmers used to abandon degraded land temporarily, by employing a shifting cultivation system typified by slash-and-burn. Land abandonment and the regrowth of natural vegetation allowed soil fertility to be replenished by $N_2$-fixation, atmospheric deposition of plant nutrients and accession of minerals leached into the subsoil or released by mineral weathering (see also Chapter 6, this volume). However, population growth in Africa has resulted in an unsustainable shortening of the soil recuperation period. Tree-based agroecosystems can act as economic mimics of natural systems, by performing some of their ecological functions (such as $N_2$-fixation and cycling of subsoil nutrients). In this chapter, we will: (i) illustrate the methods used to identify and control the main causes of soil acidification, where relevant data exist; and (ii) demonstrate how tree-based systems can be used to treat both the causes and effects of soil acidity.

## 8.2 Identifying the Causes of Acidity in Low-input Tropical Agroecosystems

Soil acidification in tropical agroecosystems is mainly caused by acid production, linked with the cycling of N, C and to a limited extent S. The input of protons causes the soil pH to decrease by an amount that depends on both the amount of protons produced by nutrient and carbon cycling and the pH buffering capacity of the soil. The soil's pH buffering capacity is simply the capacity of the soil to resist pH change. Buffering reactions include the adsorption of protons by variable charge sites on soil mineral and organic matter surfaces and the dissolution of carbonates and oxides. Over a long period of time, the weathering of primary and secondary minerals buffers the soil pH and, in so doing, releases nutrients such as K, Mg, Ca, etc. The soil's buffering capacity is expressed in mol $H^+$/kg soil/pH. Because a large variety of buffer reactions exist, the value obtained is dependent on the method of measurement used. It is important to realize that rapid titrations will only measure pH buffering associated with rapid processes, such as ion exchange and surface proton adsorption.

Sustainable management of an agroecosystem should aim at balancing the inputs of acid with additions of alkalinity from both internal and external sources. External sources of alkalinity include lime and organic matter additions, whereas the most important internal source of the soil's acid neutralizing capacity (ANC) is mineral weathering. In low-input systems, mineral weathering and the uptake of dissolved nutrients by deep-rooted trees help balance acid additions. In natural ecosystems, acidification is minimized because they experience less removal of biomass (in the form of harvested products), possess a greater diversity of plant species, and experience lower leaching losses, both as a result of the rapid immobilization of nutrients and because they have a large proportion of their nutrients stored in biomass. In contrast, in an agroecosystem, sometimes net losses of nutrients and ANC cannot be avoided (as they result from the removal of harvested products). In such a case, the sum of nutrient and ANC losses from product removal, leaching and erosion exceeds the accessions from the atmosphere and from weathering products. The aim is then to minimize such losses and balance the nutrient and ANC losses with inputs into the agricultural system.

The main causes of acidity (which include atmospheric inputs and the processes involved in nutrient cycling) are described in the following sections.

### 8.2.1 Atmospheric sources of acidity

Accession of $H^+$ derived from deposition of ammonium salts, or from acid rain, is small in low-input tropical agroecosystems typified by their remoteness from industrial activity. The main source of acidity in such systems is carbonic acid derived from the dissolution of carbon dioxide. The soil atmosphere contains between 0.15 and 0.65% $CO_2$:

$$CO_2 + H_2O = H^+ + HCO_3^-, \quad pK_a = 6.1 \quad (8.1)$$

Proton input from the above reaction is more important at high soil pH values, since the carbonic acid remains protonated at lower pH values. The acid input from this source can be calculated from estimated drainage, soil pH and rainwater pH values using data tabulated in Helyar and Porter (1989).

### 8.2.2 Nitrogen cycle

Acidification associated with N cycling is an important source of proton input, and occurs as a result of both plant and soil processes. These processes have been reviewed by Bolan *et al.* (1991) with regard to legume-based farming systems, and by Helyar and Porter (1989) with regard to calculating the proton fluxes associated with farming systems in general. The main proton fluxes associated with plant processes result from the uptake of ammonium ($NH_4^+$), nitrate ($NO_3^-$) and, in $N_2$-fixing plants, $N_2$. These processes are illustrated in Fig. 8.1.

The uptake of ionic forms of N, in excess of the accompanying counter ion charge, results in a release of protons or hydroxyl ions. This allows an electrical charge balance to be maintained within the plant. Uptake of ammonium ions, for example, results in the release of 1 mol $H^+$ per mol of $NH_4^+$. The resulting $NH_3$ is assimilated in order to produce amino acids of varying strengths ($pK_a$

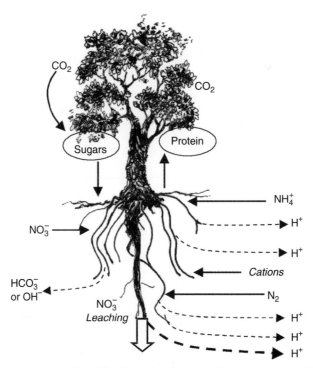

**Fig. 8.1.** Schematic representation of acidification in the root zone due to the uptake of $NO_3^-$, $NH_4^+$, and other cations, and to $N_2$-fixation and nitrate leaching.

values). Strong acids (such as aspartic and glutamic acids) release protons into the plant's cytoplasm; these protons are exported into the rhizosphere in order to maintain the internal pH of the plant cell. The net effect is that plants excrete 1.1–1.2 mol $H^+$ per mol of $NH_4^+$ taken up. Following nitrate uptake, between 0 and 1 mol $OH^-$ per mol $NO_3^-$ is released. This is because the hydroxyl ions can partly be stored in vacuoles by the malate pH buffer system. Charge balance is achieved by the uptake of base cations.

As fixation of $N_2$ does not result in charge imbalance at the root–soil interface no proton flow is associated with this process. Assimilation of fixed $N_2$ into amino compounds results in dissociation of protons according to the $pK_a$ of the compound. Some of these protons are excreted into the rhizosphere, in order to maintain cytoplasmic pH. Tropical legumes tend to produce amino compounds (ureides) with higher $pK_a$ values than those of amino compounds produced by temperate legumes. These ureides are mostly protonated at the pH values of the cytoplasm; hence $N_2$-fixation by such tropical plants has a less adverse effect on soil acidification. Temperate legumes, in contrast, produce strong amino acids with lower $pK_a$ values. In their case, cytoplasmic pH is maintained by $H^+$ excretion into the rhizosphere. Depending on the legume species, between 0.2 and 0.7 mol of $H^+$ is excreted per mol of N fixed. Charge balance is maintained by taking up the equivalent amount of base cations during $H^+$ excretion.

Measurement of the amount by which base cations are in excess of anions in the plant tissue gives the amount of acidity generated in the rhizosphere (Bolan *et al.*, 1991). This acidification of the rhizosphere is not neutralized if the organic matter is allowed to accumulate as plant biomass or soil organic matter or if it is exported from the site of production. It is neutralized by ammonification of organic matter.

Ammonification of organic N to $NH_4^+$ in soil involves deamination of amino compounds and the hydrolysis of urea. In each case, 1 mol of $H^+$ is consumed per mol $NH_4^+$ produced. Nitrification of $NH_4^+$ to $NO_3^-$ releases 2 mol $H^+/mol\ NO_3^-$. Ammonium fertilizers therefore acidify the soil to a greater extent than urea, which benefits from $OH^-$ production during ammonification. Denitrification of $NO_3^-$ to $N_2$ consumes 1 mol of $H^+/mol\ NO_3^-$. This effectively balances acid production during ammonification and nitrification.

Nitrate leaching is an important source of acidification, since the nitrate source is commonly the soil's organic matter, or urea or ammonium fertilizers. The acidity produced during nitrate production cannot be neutralized by uptake or by denitrification if the nitrate produced is leached from the system as a neutral salt solution. The residual soil acidity is equivalent to the amount of nitrate leached, when both are expressed in mol/ha:

$$H^+_{Leach} = \Sigma\ NO^-_{3\ Leach} \quad (8.2)$$

### 8.2.3 Carbon cycle

Soil organic matter, litter and undecomposed plant materials contain anionic functional groups ('alkalinity') that are bound with base cations according to the soil pH and the $pK_a$ of the organic anion. Accumulation of this organic alkalinity in soil, or its export from the site of production, results in acidification. Alkalinity is released from organic matter to the soil when the organic anion is decarboxylated or protonated. The amount of alkalinity contained in plant materials can be determined by ashing or can be estimated from the material's total base cation content. In the case of soil organic matter, it can be calculated from the soil organic matter content and soil pH by relating the surface charge of organic matter to pH (Helyar and Porter, 1989).

### 8.2.4 Sulphur cycle

The amount of S cycling occurring in the ecosystem is about a tenth of the amount of N cycling. Sulphur is also less prone to leaching than N; hence, the contribution S cycling makes to the proton budget of tropical agroforestry ecosystems is likely to be small, unless elemental S is used.

## 8.2.5 Rate of soil acidification

Soil acidification occurs as a result of natural, pedological processes due to external input of carbonic acid from the atmosphere and from internal acidity production (which results from organic matter and from nutrient cycling). Anthropogenic activities increase the rate of acidification by interfering with such cycling processes during land management. The rate of soil acidification (AR, mol $H^+$/ha/year) can be estimated for each soil layer from the rate of change in soil pH and the pH buffer capacity of the soil layer:

$$AR = dpH/dt \times pH_{bc} \times BD \times V \quad (8.3)$$

where $dpH/dt$ is the rate of pH decline in pH units per year, $pH_{bc}$ is pH buffering capacity in mol $H^+$/kg soil/pH unit, BD is bulk density in kg/m$^3$ and $V$ is volume of the soil layer in m$^3$.

The soil pH buffer capacity is seldom measured on a routine basis. It can be estimated from other more readily available soil properties such as organic matter and clay content (Aitken et al., 1990). This estimate of the acidification rate can be used to assess the contribution that various acidifying processes make to the total acidification rate. An example is given below.

## 8.2.6 Case study: sources of acidity

As far as we can ascertain, there exists no published rate of soil acidification for low-input tropical agroecosystems matched with proton budgets, which allows the causes of acidification to be identified. Such information is vital if we are to progress from the treatment of the effect of acidification (liming) to the development of an Integrated Soil Acidity Management Strategy that includes treating the causes of acidification.

Previously unpublished results, from a well-documented, long-term field experiment carried out by ICRAF at Yurimaguas, in the Amazon basin of Peru, have allowed acidification rates to be calculated and proton sources to be identified. The soil at the site is a Typic Paleudult, and annual rainfall is 2200 mm. The site's initial vegetation was a 10-year-old secondary forest fallow that was cleared in 1985. Land was cleared using the slash-and-burn technique, except for a continuous cropping treatment, which was cleared using a bulldozer. The cropping system treatments applied until 1995 were as follows.

1. Shifting cultivation with cropping of rice, cassava and plantain until 1987 (when the land was left to natural regrowth).
2. High-input cropping of maize and inoculated soybeans. The area received 1 t/ha lime. Maize received 100 kg N, 43 kg P and 100 kg K/ha. The soybean crops received 40 kg N, 35 kg P and 85 kg K/ha.
3. Low-input cropping with acid-tolerant rice, cowpea and a leguminous cover crop. The cover crop was given 10 kg P/ha as rock phosphate.
4. Multistrata agroforestry system using rice and cowpea (*Vigna unguiculata*) grown with trees such as tornillo (*Cedrelinga catenaeformis*) for timber; peach palm (*Bactris gasipaes*) for fruits; guaba (*Inga edulis*) for firewood and fruits; and araza (*Eugenia stipitata*) for fruit. Native coffee was planted in 1993. The crops received no fertilizer and a cover crop was used for weed control and to provide N inputs.
5. Peach palm agroforestry with rice and a legume ground-cover crop.

Replicate soil samples were taken for detailed chemical analysis at four depth intervals (0–15, 15–30, 30–50 and 50–100 cm) down to 1.0 m. Bulk density was measured using intact soil cores. The pH buffer capacity of each soil layer was estimated from its organic carbon and clay content (Aitken et al., 1990). The change in soil pH and the input of lime and ash of a known composition for the time period 1985 to 1995 allowed the acidification rate for the whole profile to be calculated (Table 8.1). Inputs of protons from acidifying processes were calculated in order to determine the sources of acidification.

In all treatments, urea was used as the N fertilizer. If urea is used, only nitrate leaching needs to be accounted for in the proton budget shown in Table 8.1. Urea is rapidly

**Table 8.1.** Rates of acidification and sources of acidity under contrasting cropping systems at Yurimaguas, Peru. The values are in kmol H$^+$/ha/year for the 0–100 cm layer.

| Cropping system | Total profile acidification rate | Sources of acidity | | | | | Estimated NO$_3^-$ leaching |
| --- | --- | --- | --- | --- | --- | --- | --- |
| | | Soil organic matter loss | Plant removal | Bicarbonate loss | Accounted acidity inputs | Acidity not accounted | |
| High-input cropping | 8.8 | −1.0 | 2.0 | 0.5 | 1.5 | 7.3 | 6.4 |
| Low-input cropping | 1.4 | −1.3 | 0.8 | 0.0 | −0.5 | 1.9 | 1.9 |
| Shifting cultivation | 0.8 | −1.2 | 1.0 | 0.0 | −0.2 | 1.0 | 1.2 |
| Multistrata agroforestry | 2.2 | −0.3 | 1.1 | 0.0 | 0.8 | 1.4 | 1.1 |
| Peach palm agroforestry | −4.0 | −0.5 | 0.7 | 0.0 | 0.2 | −4.2 | 1.5 |

hydrolysed in soil to produce ammonium N. This process releases 1 mol OH$^-$/mol N. The hydroxyl ions are neutralized either when ammonium N is taken up by plants or when it is volatilized as ammonia gas. Subsequent nitrification of ammonium N results, however, in the net release of 1 mol H$^+$/mol N. This acidity is neutralized by plant uptake, production of alkaline plant materials or denitrification. Calculation of soil acidification rate therefore only needs to take account of the leaching of nitrate since the loss of plant alkalinity is accounted for separately. Similar processes and proton flows occur following ammonification of soil organic N. The acidity input resulting from removal of harvested plant materials was calculated as the product of the biomass removed and its base cation content. The soil organic matter content decreased in all treatments over the 10-year cropping period, and the associated release of alkalinity was calculated using the method described by Slattery et al. (1998).

The range of soil acidification rates for the cropping systems in Yurimaguas is of the same order of magnitude as the rates reported for several tropical agroecosystems (Wong and Swift, 2003). Inputs of acidity as a result of bicarbonate leaching ('bicarbonate loss') only occurred in the limed high-input treatment (Table 8.1), but did not occur in unlimed treatments. The acidity inputs from changes in soil organic matter, plant removal of base cations and bicarbonate loss accounted for only a small proportion of the total acidification profile. The peach palm agroforestry treatment appears to have accessed alkalinity that was not accounted for by measurements made in the 0–100 cm profile.

Nitrate leaching, unfortunately, was not measured at Yurimaguas and the acidity input from nitrate leaching was not accounted for directly. Leaching was estimated indirectly, using data collected from a similar site in southern Nigeria. The site had a soil, climate (with an annual rainfall of 2420 mm) and history of land use that was similar to the Yurimaguas site. In Nigeria, nitrate leaching was measured using large tension-drained monolith lysimeters on a freshly cleared Typic Paleudult; such measurement showed that a third of the nitrate derived from the soil and from urea was leached (Wong et al., 1992). The acidity input, due to this nitrate leaching under high-input maize and rice, was 10.3 kmol H$^+$/ha/year. It therefore seemed reasonable to estimate leaching in Yurimaguas by assuming that 30% of fertilizer and soil mineralized N was leached from the high-input system. This percentage decreased to 25% for other systems. The estimated proton input derived from the estimate of nitrate leaching is a good match for the unaccounted acidity input. As was the case in southern Nigeria, nitrate leaching in this high-rainfall environment appears to be the principal cause of acidification. Nitrate leaching was followed by the removal of harvested materials as the second most important cause of soil acidification.

As expected, leaching was greatest under the treeless high-input and low-input cropping systems, as compared with the tree-based systems. Management practices aimed at minimizing leaching and export of plant materials will minimize soil acidification at Yurimaguas and in similar agroclimatic regions (such as southern Nigeria). The experiment run at Yurimaguas suggested that tree-based systems lower the rate of nitrate leaching more effectively than treeless systems. The short-term gain in alkalinity from soil organic matter loss that occurred in Yurimaguas is not sustainable over the longer term, since this decreases the pH buffering capacity and the acid neutralizing capacity of the systems. In addition, a loss of soil fertility is associated with organic matter loss. Tree-based systems in general maintain the soil organic matter content better than treeless systems.

Other soil acidifying processes may be more important in other agroclimatic regions. In southeast Australia, for example, the accumulation of soil organic matter under legume-based pastures is the principal cause of acidification (Williams, 1980). Research is clearly needed to identify the causes of soil acidification in other tropical agroecosystems, in order to develop socially, economically and environmentally appropriate management options to treat these causes.

## 8.3 Transfer of Organic Alkalinity in Agroecosystems

### 8.3.1 Use of undecomposed plant materials

Liming is a well-known method for the mass transfer of alkalinity. It involves the excavation, processing, transport and application of the material to acid soils. In terms of the distribution of alkalinity, more subtle enrichment occurs at the landscape scale. In many agroecosystems, spatial variability in the distribution of alkalinity offers the opportunity to use plant materials to transfer alkalinity to acid sites. In West Australia's cropping zones, for example, at the farm scale (~2000 ha), soil pH varies from 5.5 to 8.5 due both to solute leaching from higher parts of the landscape and the accumulation of carbonates in lower parts of the landscape (Wong and Harper, 1999). A similar accumulation of alkalinity often occurs in the lower profile of duplex soils. Such soils are typified by marked textural contrasts between the lighter topsoil and the heavier subsoil. In wetter environments, where alkalinity may not accumulate in the landscape, mineral weathering acts as a source of subsoil alkalinity by consuming protons. Local acidification of alkaline sites, or of weathering minerals by plants, allows alkalinity to be accumulated in plant materials. Plant materials, and their associated alkalinity, could then be removed from the sites in a sustainable manner, provided that the critical load for acid input is not exceeded.

Numerous laboratory experiments have recorded increased soil pH, decreased Al saturation and improved conditions for plant growth as a result of the addition of plant materials to acid soils. Such plant materials also supply base cations, such as Ca, Mg and K. The total concentration of these base cations, expressed as equivalents of the cations per kilogram of the materials, is closely related to the ability of the plant materials to neutralize soil acidity (Wong et al., 2000). The base cation content of the plant materials can be measured or estimated from published tables (e.g. Drechsel and Zech, 1991) and from dedicated databases (e.g. Organic Resource Database, Palm et al., 2001). Another direct way of estimating the ability of plant materials to neutralize soil acidity is by determining the ash alkalinity of the plant material, measured by means of the titration of its ash (Noble et al., 1996). Examples of soil pH values achieved by the 14-day incubation of an Oxisol and an Ultisol with pruning materials from seven agroforestry tree species with a base cation content 'b' ($cmol_c/kg$) are as follows (Wong et al., 2000):

Oxisol: $pH = 5.61 - 1.58\ b + 0.77\ b^2$,
$r^2 = 1.00$ \hfill (8.4)

Ultisol: $pH = 4.23 - 0.74\ b + 0.62\ b^2$,
$r^2 = 0.95$ \hfill (8.5)

The intercepts of Equations 8.4 and 8.5 were close to the initial soil pH values.

Several mechanisms contribute to an increase in soil pH (Wong and Swift, 2003). An initial rapid increase in pH may occur due to the complexation of protons by organic anions. The protonation of organic anions depends both on their $pK_a$ and pH values and on soil pH. This process is independent of biological activity. In addition, microbial decarboxylation of organic anions consumes one mol H+ per mol of carboxylate; this further contributes to the pH effect (Yan et al., 1996). The decarboxylation reaction can be illustrated using calcium oxalate as an example (Natscher and Schwertmann, 1991):

$$Ca(COO)_2 + \tfrac{1}{2} O_2 + 2H^+$$
$$= 2CO_2 + H_2O + Ca^{2+} \quad (8.6)$$

Both proton adsorption and decarboxylation depend on the organic anion content of the plant materials. This can be determined from their base cation content or ash alkalinity. Due to ammonification, plant materials high in N give rise to an additional transient increase in soil pH. This pH benefit is reversed once the ammonium ions are nitrified, since the combined ammonification plus nitrification steps result in a net release of protons (Pocknee and Sumner, 1997). The rate of pH reversion is dependent on the pruning materials used. For example, when used in an incubation experiment, those tree prunings that had the highest N content (3.8% N, from *Leucaena leucocephala*) were associated with the most rapid reversion to lower pH values. Furthermore, those with the lowest N content (1.8% N, *Grevillea robusta*) were associated with the slowest reversion (Pocknee and Sumner, 1997). Attempts to use a polyphenols to nitrogen ratio as a measure of likely decomposition rates (Palm and Sanchez, 1991) did not, however, give a good relationship between the ratio and soil pH (Wong et al., 2000).

The nitrification-induced pH reversion measured in the laboratory (Pocknee and Sumner, 1997) may not necessarily be of the same magnitude in the field. This is because nitrate uptake and denitrification can generally be expected to neutralize the acidity produced by the ammonification–nitrification steps. Nitrate accumulation is unlikely to occur in topsoil under many situations in the tropics because of leaching. Reversion would occur according to the amount of nitrate leached. A long-term accumulation of nitrate can often be detected in deep soils following such leaching events (Wong et al., 1990b), and deep-rooted trees can recycle this source of nitrate (see Chapter 6, this volume).

The effect on soil pH of decomposing plant materials, due to protonation of organic anions and decarboxylation, is considered to be long lasting and provides a window of opportunity for crop growth under conditions that normally involve high levels of Al toxicity (Wong et al., 1995; Pocknee and Sumner, 1997). For practical purposes, it can be assumed that the magnitude of the liming effect of plant material addition is equal to its total base cation charge.

There is one field experiment we know of that reported the benefits of a lateral transfer of alkalinity achieved by pruning pure stands of agroforestry trees and applying the pruned materials to pure stands of maize and beans. The experiment was carried out in collaboration with ICRAF on an Oxisol at Karuzi in Burundi (Wong et al., 1995). The maize and bean crops were given adequate fertilizer and the response to application of 3 and 6 t/ha of the pruned material was attributed to amelioration of soil acidity and lowered Al saturation. These rates of addition appear feasible in agroforestry systems. Plants with high biomass production and high ash alkalinity or total base cation charge will be more efficient in transferring alkalinity. Leguminous materials are particularly useful in this respect, because they generally have high ash alkalinity and the added benefit of also fixing $N_2$.

### 8.3.2 Recycling of alkalinity through waste materials

Because of their value as animal feed, it is often not practical to use prunings directly as a soil ameliorant. This is especially true for leguminous plant materials, which are rich in proteins and minerals. In such a situ-

**Box 8.1.** Case study: aluminium detoxification using prunings – pot studies.

Several plant materials, with different total base cation contents but with similar lignin : N ratios (~10) and polyphenols : N ratios (~0.5), were tested in pot experiments using Ultisol soil from Lampung, Indonesia. Four plant species, with a range of total base cation (K+Ca+Mg) contents, were chosen for the pot experiment: *Flemingia congesta* (total cations, 36 cmol$_c$/kg), *Gliricidia sepium* (53 cmol$_c$/kg), *Peronema canescens* (72 cmol$_c$/kg), and *Chromolaena odorata* (100 cmol$_c$/kg). The rate at which pruning materials were added to the soil was equivalent to 15 Mg/ha.

Plant materials with a higher total base cation content increased soil pH and decreased exchangeable Al. At the 10th week of incubation, *Gliricidia* had the strongest effect in terms of increasing soil pH and decreasing monomeric Al concentration. *Flemingia* gave the weakest effect. However, the pH of incubated soil varies temporally, and this ranking changed with incubation time. The longer-term effect of prunings is increasingly determined by their base cation contents and by N mineralization rates.

Figure B8.1 offers, for comparison, results from a previous pot experiment (Hairiah *et al.*, 1996) that an application of 15 Mg/ha *Gliricidia* biomass is able to suppress monomeric Al to the same extent as an application of 90 Mg/ha of *Melastoma* biomass or 15 Mg/ha *Peronema* biomass. The total cation concentration of *Gliricidia*, however, was 50% lower than that of *Melastoma* (105 cmol$_c$/kg) and 25% lower than that of *Peronema* (72 cmol$_c$/kg). This finding is very useful for practical purposes, as the period of Al amelioration provides a window of opportunity for crop growth. In practice, the production of 15 Mg/ha of plant biomass (for application to the soil) should be possible under field conditions.

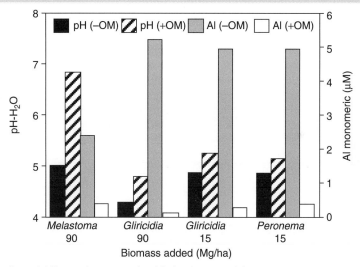

**Fig. B8.1.** Effect of different plant materials added (+OM) to soil from Lampung (in Sumatra, Indonesia) on soil pH and monomeric Al concentrations, at the 7th week of incubation. (−OM = no addition of prunings, 90 = 90 Mg/ha biomass and 15 = 15 Mg/ha biomass). Adapted from Hairiah *et al.* (1996).

ation, the manure of the animals fed with the prunings can be used to transfer alkalinity. Consumption of agricultural produce by rural and urban communities also results in waste materials that could be recycled in order to minimize pollution and the removal of nutrients and alkalinity from the land.

Animal manure and waste materials are normally composted before use, in order to lower the weed, pest and disease load and to produce a more pleasant low-odour material. The composted product is characterized by the fact that it is relatively stable and will not further decompose.

**Box 8.2.** Case study: aluminium detoxification using prunings – a field study from Lampung, Sumatra, Indonesia (Hairiah *et al.*, 1996).

Soil samples were collected from a long-term (9-year-old) hedgerow intercropping experiment run by the BMSF (Biological Management of Soil Fertility) project in North Lampung. Five tree species were tested: *Calliandra calothyrsus*, *Leucaena leucocephala*, *Gliricidia sepium*, *Peltophorum dasyrrachis* and *Flemingia congesta*. Plots without hedgerow trees were used as control plots. Trees were pruned during the growing season, and all pruned biomass was returned to the plot as mulch. The soil samples were collected at 0–5 and 5–15 cm depths, and analyses included soil pH and concentrations of base cations and exchangeable Al.

Soil organic carbon (C-org) content increased as a result of continuous biomass addition of about 8.5 Mg/ha/year. The soil organic matter content (represented by C-org) was positively correlated with soil pH and with the effective cation exchange capacity (ECEC; Fig. B8.2), which was due to increased Ca and Mg contents in soil. However, with increasing C-org, levels of exchangeable Al decreased. Under these field conditions, where addition of plant residues was more realistic than in pot experiments, the concentration of monomeric Al was negatively correlated with soil pH.

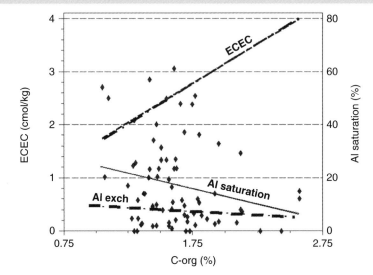

**Fig. B8.2.** Al saturation, exchangeable Al and effective cation exchange capacity (ECEC) as function of soil organic C content (C-org) for a long-term hedgerow intercropping and mulch-transfer experiment in North Lampung, Indonesia (at 0–15 cm depth).

The composted product contains humic substances, which confer pH buffering and metal binding properties. Materials derived from coal waste also contain humic substances that may be beneficial in terms of treating acid soils. Plant residue composts, urban waste compost, animal manures and coal-derived organic products have all been used to increase soil pH, to decrease Al saturation and to improve conditions for plant growth under laboratory conditions (Wong and Swift, 2003). Farmyard manure has been used successfully, under field conditions, to alleviate Al toxicity in an Oxisol in Burundi and to improve the yield of maize and beans grown in that Oxisol (Wong *et al.*, 1995).

Several mechanisms have been proposed to explain the increase in soil pH that was recorded on the addition of these organic materials to acid soils (Wong and Swift, 2003). The most important mechanism is proton adsorption by humic substances contained in the materials. These humic substances have functional groups, such as carboxyl groups, that are able to consume or release protons according to their $pK_a$ and pH values and the pH of the surrounding solution. Titration of the organic materials down to pH 4.0 determines the proton consumption capacities of these organic materials. The proton consumption capacities of such organic materials are closely related to their ability to increase soil pH. For example, laboratory incubation of an Oxisol from Burundi, an Ultisol from Cameroon and a Spodosol from Sumatra with these organic materials resulted in increased soil pH in all three cases. The pH of the incubated soils was linearly related to the proton consumption capacities (x, $mol_c/kg$) of the organic materials used (Wong et al., 1998). The regression equations for the pH of the incubated soils were:

Oxisol: pH = 4.51 + 0.29 x,
$r^2$ = 0.999         (8.7)

Ultisol: pH = 4.11 + 0.41 x,
$r^2$ = 0.989         (8.8)

Spodosol: pH = 4.06 + 0.61 x,
$r^2$ = 0.991         (8.9)

The constants of the regression equations represent the pH values of the control soils. The slopes were determined by soil buffer capacity, the Oxisol being most strongly buffered and the Spodosol least strongly buffered. These findings allow predictions to be made of the final pH that are universally valid across soils; it also provides an effective means of addressing the issue of variability in the composition of humified organic materials used in the treatment of soil acidity (Wong et al., 1998). The increased soil pH, which results from the addition of the organic materials, leads to a corresponding increase in cation exchange capacity and to a decrease in Al saturation.

An additional benefit of organic additions is that a decrease in Al solubility can be expected at a given pH value when soil pH is increased with the use of humified organic matter. Decreased Al solubility is expected to result in lower activity of Al at any given pH value < 5.0. The opposite, an increase in Al solubility, is observed when lime is used, because of the formation of freshly precipitated Al hydroxide, which is more soluble (Helyar et al., 1993). In the case of lime, and due to higher Al solubility, higher pH values would be required to decrease Al activity compared with the use of organic materials. Another disadvantage associated with the use of lime is that it has a limited effect on subsoil acidity. The fulvate-type components of the humified materials are expected to be more effective in increasing subsoil pH than lime because they are more mobile.

Depending both upon how well waste materials have been composted and upon their C : N ratios, additional proton fluxes may occur due to ammonification and nitrification of organic N and of urea. Again, this cycle of pH increase (ammonification) and decrease (nitrification) may not be important under field conditions, due to nitrate uptake and denitrification. Animal waste high in N (such as chicken manure) may result in acidification due to nitrate leaching. However, composting with plant residues that have wide C : N ratios would solve this problem. It is expected that the addition of Ca and Mg, contained in a medium of organic matter, would further alleviate the effect of Al phytotoxicity (Kinraide and Parker, 1987).

### 8.3.3 Root exudation of organic materials

The principal cause of acid soil infertility is Al phytotoxicity. The negative log of the active concentration of Al in soil solution is generally directly proportional to soil pH. This means that the concentration of Al increases by a factor of 10 for each pH decrease of 1 unit. Phytotoxicity normally occurs when Al concentration exceeds 5 µM. The terminal 5 mm of roots (root apices) are particularly sensitive; in acid-resistant plant varieties, root apices respond to Al toxicity by exuding simple biochemical compounds into the rhizosphere (Wong and Swift, 2003). These organic compounds detoxify Al by rapidly binding it in stable organic complexes. This organically bound Al is less readily taken up by roots and is, therefore, less phytotoxic. This mechanism of Al detoxification by plants has been well documented with regard to cereals and grain legumes (Kochain and Jones, 1996). Cultivars differ considerably in their ability to detoxify Al in this way. Thus there is genetic-based variability in resistance to Al, which can be exploited when matching crops to an environment.

Such cultivars rapidly release citrate or malate in response to root exposure to Al, and these form complexes with Al in the rhizosphere. The response is specific to exposure to toxic Al. However, low pH values, low phosphate concentration and the presence of other metals fail to trigger such a response. The amounts of citrate or malate released are Al-dose dependent and are proportional to the external concentration of Al (when Al concentrations range between 0 and 200 µM; Jones, 1998). But the release of citrate or malate does not always lower the rhizosphere pH. With regard to malate, for example, the cation that accompanies the release of malate from wheat roots is $K^+$ (and not $H^+$); hence, the release of malate does not lower rhizosphere pH (Ryan et al., 1995). However, as has been explained above, decarboxylation of malate would increase soil pH. Comparatively less is known about the importance of rhizosphere exudation of organic matter in conferring acid tolerance to agroforestry tree species. In contrast to the rhizosphere effect, detoxification of Al by organic matter complexation also occurs within the tissue of resistant plant species.

Phosphorus deficiency is a problem commonly associated with acid soils. Release of organic acids in the rhizosphere also helps to alleviate P deficiency (see also Chapter 7, this volume). An example of this is given by trees species of the genus *Banksia*, which are commonly found as remnant vegetation on farms in areas of Western Australia with acid sandy soils. The proteoid (cluster) roots of these trees release citrate and malate into the rhizosphere (Grierson, 1992). In banksias, the purpose of these anions is, mainly, to overcome phosphate deficiency, since Al phytotoxicity is not prevalent in these sandy soils. Citrate and malate can both dissolve and desorb unavailable forms of phosphate, by transforming them into soluble forms. Dissolution occurs as a result of cation binding (e.g. the removal of Ca, Fe and Al from insoluble P minerals). Desorption occurs as a result of the exchange of the P ligand with citrate or malate.

Organic anions, derived from the decomposition of plant residues and humic substances contained in composted materials, also have a similar effect to that of citrate and malate, in terms of increasing P availability. In most tropical agroforestry systems, P cannot be captured from the subsoil due to low concentrations of P and low subsoil root densities. Soils in these systems are often P deficient, due to low initial concentrations and to P depletion as a result of crop removal. Phosphorus is therefore always needed as an external input. Many rock-P sources in Africa are of low reactivity in their natural forms, and are not suitable for direct application without additional processing, such as acidulation or compaction with soluble P fertilizers (Buresh et al., 1997). Incorporating rock phosphate into the composting process could increase available P, because of the formation of organic anions from the compost that are capable of chelating Ca, Fe and Al from low-grade rock-P. For example, Singh and Amberger (1998) found a high level of production of organic acids during the composting of wheat straw, which resulted in high rates of dissolution of low-grade rock-P.

Some plants, such as the Mexican sunflower (*Tithonia diversifolia*), have high tissue P concentrations, and so composts derived from them can be used to alleviate soil P deficiencies. The Mexican sunflower is a more effective source of plant-available P than are residues such as those derived from maize stover. Apparently, this is because the sunflower has a higher concentration of inorganic P and a faster decomposition rate than the maize residues (Buresh et al., 1997). Unfortunately, however, composts derived from *T. diversifolia* have a high Ca content, and this limits its potential use in promoting the dissolution of phosphate rock, because Ca inhibits the dissolution process (Smithson et al., 1999).

> **Key points**
>
> 1. Aluminium phytotoxicity is the main problem with acid soils.
> 2. Some plants detoxify aluminium by secreting organic anions (such as citrate and malate) to bind aluminium in non-toxic forms.
> 3. Selection or engineering for aluminium tolerance allows us to match plants to the soil conditions.
> 4. Secretion of organic anions also improves phosphate availability in acid soils, through dissolution and desorption of phosphate. Plant residues and organic waste composts have a similar effect.

## 8.4 How Do We Implement This Knowledge to Manage Acidity?

Measuring acidification rates, and accounting for the rates obtained, by assessing the contributions made by soil and plant processes, offers major opportunities in terms of identifying and remedying the causes of acidification in farming systems. In high-rainfall ecosystems similar to that at Yurimaguas, the control of nitrate leaching is expected to have a major impact in decreasing the rate of soil acidification. Incorporating trees into the farming system can achieve such control. These trees are expected to decrease drainage and nitrate concentration in drainage water due to water and nutrient uptake. At Yurimaguas and in southern Nigeria, the source of a large proportion of the nitrate leached was mineralization of soil organic matter. This was, presumably, due to soil disturbance during clearing and cropping. In addition to nitrate leaching, it is generally true that removal of plant materials will cause acidification at the site of removal. On acid soils, such removal should be limited to the useful harvestable products; residues must be left on site or recycled back to the site. We currently do not have enough data to carry out similar analyses for, and so identify the causes of soil acidification in, other low-input tropical agroecosystems.

The effect of undecomposed plant materials (e.g. the addition of plant residues) on soil pH is relatively well understood. The main processes involved have been identified, and consist of the protonation and decarboxylation of organic anions and the ammonification and nitrification of organic N. Additional knowledge of these processes could be gained by quantifying their rate, so that the temporal pattern of soil pH can be explained. Knowledge gained thus far is based, almost exclusively, on laboratory and glasshouse experiments. There has been little effort made to extend and apply this knowledge in the field. Our knowledge, so far, suggests that plants with high rates of biomass production and a high base cation content will be more effective in increasing soil pH. The caveat is that plant materials with low C : N ratios may acidify the soil, due to N mineralization followed by nitrate leaching.

When applying prunings in order to increase soil pH in the field, one has to consider additional factors. For example, prunings taken from *Gmelina* trees have a high base cation content, but are unsuitable for hedgerow intercropping since the trees outcompete the crops (Wong *et al.*, 1997). This tree species should perform well in a lateral base transfer and mulch system that can also incorporate animals and the recycling of manure. An increase in soil pH caused by addition of prunings will be to the detriment of the site from which the plant materials were collected. This strategy for the treatment of soil acidity should only be considered within the context of farming systems where spatial variability in alkalinity offers the opportunity to harvest this alkalinity for redistribution to acid sites. It is also feasible that deep-rooted perennial plant species (such as lucerne and agroforestry tree species) might be used to access alkalinity derived from mineral weathering. There is a need to field-test the potential for using plant materials to treat acidity in agroecosystems.

The effects of recycled, waste organic material of plant, animal and industrial origins (such as coal waste) on soil pH are well understood. The aim, in future, should be to recycle as much of this waste material as possible, in order to minimize potential pollution hazards, and to return alkalinity, nutrients and organic matter to the land. The knowledge base regarding such undertakings is based almost exclusively on laboratory and glasshouse studies. Again, there is a need here to test under field conditions the potential benefits offered by the use of these materials.

The greatest potential for the use of these materials is offered by the development of an Integrated Soil Acidity Management Strategy for tropical agroecosystems. In addition to the use of lime, the strategy would aim to identify and control sources of acidity in farming systems. It would also seek to use organic materials to transfer organic alkalinity to acid soils. The first step towards the development of such an integrated approach would be the development, at the village scale, of a research and demonstration model. This could be used to test the sustainability of such a system under real-life conditions, and so could take into account socioeconomic and infrastructural constraints. The management strategy will be a major support for the low-input production systems commonly practised in the tropics.

**Conclusions**

1. An Integrated Soil Acidity Management Strategy is required for low-input tropical agroecosystems.
2. Such a strategy should identify and treat the causes of soil acidification.
3. Tree-based agroecosystems can play an important role in this strategy by providing tree prunings that are rich in base cations and that can be used to transfer alkalinity within the system. Such farming systems can mimic natural ecosystems.
4. Recycling of organic waste material is critical in acid soil management.

**Future research needs**

1. To determine proton budgets and the sources of, and solutions to, soil acidification.
2. To develop laboratory-derived knowledge on the use of undecomposed plant materials (residues), waste organic materials and of root exudation for field implementation.
3. To construct village-scale models to optimize the cycling of nutrients, alkalinity and carbon in the ecosystem.

## Acknowledgements

The research work of Mike Wong presented in this chapter was funded by the UK Department for International Development (DFID), UK, for the benefit of developing countries, as part of project R4754, administered by the Forestry Research Programme. The views expressed are not necessarily those of DFID.

# 9 Uptake, Partitioning and Redistribution of Water by Roots in Mixed-species Agroecosystems

Mark Smith, Stephen S.O. Burgess, Didik Suprayogo, Betha Lusiana and Widianto

---

**Key questions**

1. How is the water budget of a cropping system modified by trees?
2. What controls uptake and partitioning of soil water in species mixtures?
3. What is bidirectional flow?
4. How are partitioning and the bidirectional flow of water measured and modelled?

---

## 9.1 Introduction

The fate of rainfall on any agroecosystem can be described in terms of a water budget. For a crop monoculture, transpiration ($E_t$) is balanced against precipitation ($P$) by evaporation directly from the soil ($E_s$), evaporation of rainfall intercepted by canopy surfaces ($E_i$), surface runoff ($R$), drainage out of the rooting zone ($D$) and the change in storage of soil water ($\Delta\theta$), which is written:

$$P = E_t + E_s + E_i + R + D + \Delta\theta \quad (9.1)$$

Modification of the water budget results when species are mixed together. In a simultaneous agroforestry system, transpiration is split into tree ($E_t^t$) and crop ($E_t^c$) components, as is interception ($E_i^t$, $E_i^c$) (Wallace, 1996). Changes to microclimate and belowground interactions between the tree and crop root systems cause modification of $E_s$, $R$, $D$ and $\Delta\theta$. We can therefore write a water budget for the tree–crop mixture as

$$P = E_t^t + E_t^c + E_i^t + E_i^c + E_s^* + R^* + D^* + \Delta\theta^* \quad (9.2)$$

where the asterisks (*) denote spatial averages that integrate values for the tree and crop components. The term $E_t^t + E_t^c$ represents a productive use of rainfall, as transpiration enables the production of biomass by photosynthesis. All remaining terms to the right-hand side of the equation are not productive, as they do not add to biomass production. Runoff and drainage may supply water that can be used to provide valuable services elsewhere on the farm or in the wider community and ecosystem.

A common aim of mixed agroecosystems, at least in environments where availability of water limits farm productivity, is higher efficiency of rainfall utilization – more biomass is produced per unit rainfall. Rainfed agriculture in dryland environments is, typically, strikingly ineffective at making productive use of rainfall. A model-based analysis of water use by millet (*Pennisetum glaucum*) grown on low-input farms in Niger, for example, concluded that a mere 4–9% of available water was used by the crop for transpiration (Röckstrom *et al.*, 1998). In the case of agroforestry, integration of trees into cropping systems can theoretically increase productive use of rainfall by reducing non-productive use of water. Trees may arrest runoff, reduce drainage and cut soil evaporation by the shading of bare soil, which saves water for use by plants if the reduction in $E_s$ exceeds $E_i^t$.

Trees are thus able to increase rainfall utilization in cropping systems by trading the non-productive use of water for productive uptake by plants. A danger associated with using trees to modify the water balance in favour of plant uptake is that *competition* can occur and the trees deprive the crop of water. Competition affects the growth and survival of plants when one neighbour gains an advantage over another because it is able to acquire more of a resource that is required for growth and is in limited supply (Anderson and Sinclair, 1993). For agroforestry to succeed, therefore, competition between trees and crops should be avoided. Cannell *et al.* (1996) expressed this as the central biophysical hypothesis for agroforestry. They stated that, in the case of water: '... benefits of growing trees with crops will occur only when the trees are able to acquire resources of water ... that the crops would not otherwise acquire.'

The latter criterion can be satisfied if there is *complementarity* in water use by trees and crops. In those cases in which water uptake by the root systems of trees and crops occurs from spatially discrete sources or at discrete times, water use is complementary and productive use of water can be enhanced without there being any negative impacts of competition. 'Over-yielding' (where the productivity of agroforestry exceeds the additive yields of tree and crop monocultures) can result from complementarity and increased resource-use efficiency.

Bidirectional flow of water in root systems adds complexity to below-ground interactions in species mixtures. There is now strong evidence that reversal of flow in roots causes redistribution of soil water where the root network spans soil horizons with contrasting water potentials. In agroforestry, bidirectional flow could facilitate the coexistence of trees and crops or increase the competitive advantage of trees, depending on whether water is transferred into or out of the crop rooting zone.

The key to determining how competition for water, complementarity and reverse flow phenomena affect the outcome of agroforestry or other species mixtures is understanding the partitioning of water between coexisting root systems. Partitioning is driven both by the processes that control uptake by each species and by bidirectional flow. Insight into these processes is available from theory and measurement. Ultimately, modelling of partitioning is required to enable the simulation of system performance. Our objective here is to relate evidence of competition, complementarity and bidirectional flow to the modelling of partitioning by reviewing the principles of water uptake by plants. These principles are examined on the basis of both relevant theory and measurements, drawn principally from agroforestry research.

## 9.2 Competition and Complementarity for Water Use in Mixed-species Systems

### 9.2.1 Impacts on productivity

In mixtures of trees and crops, production from component species is determined by their success at capturing and effectively utilizing essential resources that limit growth (Squire, 1990). In seasonally dry environments, where rainfall is variable and frequently deficient, a lack of available water commonly limits growth. Biomass production ($B$, g/m²) in an agroforestry system is then a function of cumulative transpiration by the trees and crop ($\Sigma E_t^t$ and $\Sigma E_t^c$, mm) and their water use efficiencies ($e_w^t$ and $e_w^c$, g/mm) (Black and Ong, 2000):

$$B = e_w^t \Sigma E_t^t + e_w^c \Sigma E_t^c \qquad (9.3)$$

Water-use efficiencies vary with atmospheric humidity deficit and between $C_4$ and $C_3$ species. However, a major benefit derived from combining trees and crops in water-limited environments is a reduction in non-productive losses from the water budget (i.e. an increase in $\Sigma E_t^t + \Sigma E_t^c$). This benefit is weakened or lost if, in violation of the hypothesis of Cannell et al. (1996), acquisition of water by trees reduces $\Sigma E_t^c$ because of competition.

The mechanistic basis of competition for water in agroforestry was demonstrated by Govindarajan et al. (1996) and McIntyre et al. (1997). Their experiments in a semiarid region of Kenya showed that, where seasonal rainfall was insufficient to recharge soil below the crop rooting zone, the grain yields of maize grown in alley cropping systems with *Senna spectabilis* or *Leucaena leucocephala* were reduced by between 39% and 95%. Without available water below the crop rooting zone, uptake by the hedges deprived the maize of much of the water required for growth. In such cases, higher water use by trees causes increased suppression of crop yields (Fig. 9.1).

Combinations of trees and crops are most successful where competition is avoided and resource use is complementary. A classic example of temporal complementarity in water use in agroforestry occurs in the parkland system, which combines dispersed *Faidherbia albida* trees with crop production in West Africa. *F. albida* exhibits 'reverse phenology' and is leafless during much of the cropping season. The trees and crop therefore have discrete periods of demand for water, ensuring minimal competition for water with understorey crops (Sanchez, 1995; Roupsard et al., 1999). Spatial complementarity for water uptake was demonstrated by Smith et al. (1997b), who showed that when the roots of windbreak trees had access to ground water, competition with adjacent crops for water from the crop rooting zone was minimal, a phenomenon associated with improved crop yields.

### 9.2.2 Importance and implications of bidirectional flow

Flow of water through root systems from one soil layer to another has been called 'hydraulic redistribution' (Burgess et al., 1998). Transfer of water upwards, because of uptake from wetter soil at depth, reverse

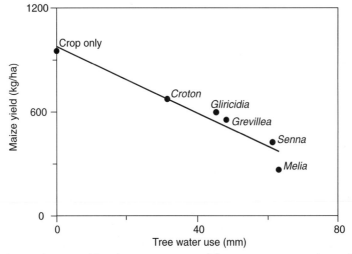

**Fig. 9.1.** Correlation of maize yields with water use among different tree species in adjacent linear plantings in Machakos, Kenya, during a growing season with low rainfall (250 mm). Redrawn from Ong et al. (1999).

flow in lateral roots and subsequent efflux in drier soil near the surface, is termed 'hydraulic lift' (Caldwell et al., 1998). The reverse process, in which water is transferred from wetter soil near the surface into drier soil beneath, was termed 'downward siphoning' by Smith et al. (1999b).

There are uncertainties over both the quantities of water transferred by hydraulic redistribution and its impact on the soil water balance. Where gradients in water potential between soil layers are steep, there is some evidence suggesting that the proportion of uptake diverted by flow reversal in roots can be as high as 20% (Emerman and Dawson, 1996; Smith et al., 1999b), though such high rates probably occur for only short periods. Burgess et al. (2001a) determined that 28 mm of water was transferred downward by roots over a period of 28 days when there was 96 mm of rainfall. Simulations suggest that the contribution to transpiration of reabsorption of water following hydraulic distribution is very small over seasonal timescales, but can be in the order of 20% on some days (Ryel et al., 2002).

Both hydraulic lift and downward siphoning have implications for belowground interactions, especially in agroforestry where component species commonly have contrasting maximum rooting depths. For shallow-rooted crops growing in drying soils adjacent to trees, hydraulic lift by the tree roots could provide a source of water and hence facilitate crop growth. This would require that the root system of the crop is established to an extent that would allow it to out-compete the trees for the water being made available. Alternatively, efflux from the tree roots would have to exceed uptake, which might occur, for example, if water potentials in the tree were consistently elevated because of the tree's access to ground water.

Downward siphoning of water by trees would tend to exaggerate their competitive advantage over crops. The water content of subsoil has been found to increase measurably because of hydraulic redistribution through the taproots of trees (Burgess et al., 2001a), suggesting that water storage in the topsoil is concomitantly reduced. Impairment of crop productivity because of reduced availability of water may result. Under waterlogged conditions, however, downward siphoning could have the benefit of moving excess water out of the topsoil and rewetting very dry subsoil (Burgess et al., 2001a). Additionally, downward siphoning by trees may facilitate the growth of tree roots through dry soil layers, enabling access to ground water.

## 9.3 Partitioning of Water by Plant Root Systems: Theory

### 9.3.1 Principles of plant water uptake

Uptake of water by plants is driven by gradients in water potential along a pathway that links soil, roots, foliage and atmosphere (Fig. 9.2). Water potential declines continually from soil to atmosphere along the flow path. Evaporation within the substomatal cavities of leaves creates tension in a continuous column of water that connects leaves to root tips. Water moves into roots when water potentials are lower in roots than they are in the soil.

Relationships between flow and the drop in water potential ($\psi$) along the uptake pathway can be represented as an Ohm's law analogue. The hydraulic properties of each segment of the flow path are modelled as a transport coefficient termed the hydraulic conductance, $k$ (which is the inverse of the resistance to flow). In the simple linear model depicted in Fig. 9.2, flow in all parts of the pathway is equal to the rate of evaporation from the leaves ($E$), and is related to gradients in potential between the soil ($\psi_s$), root surface ($\psi_{rs}$), base of the stem ($\psi_b$) and leaf ($\psi_l$), as described by Newman (1969):

$$E = (\psi_s - \psi_{rs})k_s = (\psi_{rs} - \psi_b)k_r$$
$$= (\psi_b - \psi_l)k_{sh} \quad (9.4)$$

where $k_s$, $k_r$ and $k_{sh}$ are hydraulic conductances for the soil, root system and shoot, respectively. Shrinkage of roots in very dry soil can create an interfacial resistance to uptake, causing a sharp decline in $k_s$.

When expressed per unit root length, $k_r$ becomes root hydraulic conductivity ($\kappa_r$). As

## 9.3.2 Partitioning in species mixtures

Equation 9.5 can be written for uptake by each species within a mixed stand of vegetation. In agroforestry or intercropping systems, partitioning of soil water is thus determined by the integrated effects of differences between species in root distributions, root hydraulic properties and plant water potential. From Equation 9.4, the latter is dependent on the hydraulic architecture of the plant and the evaporative flux at the leaves, which is determined by above-ground microclimate and stomatal behaviour.

Three hypothetical scenarios can be constructed to illustrate how the principles of water uptake by plants control partitioning between coexisting root systems, at least over short timescales. Partitioning also depends on relative rates of root growth (and dieback) over longer periods. The following statements are true for a soil layer with uniform $\psi_s$ containing both tree and crop roots:

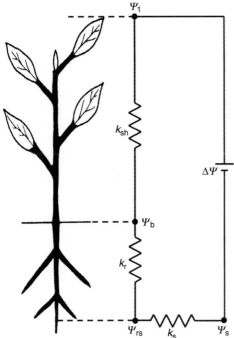

**Fig. 9.2.** The Ohm's Law analogy for water uptake by plants. Water flow is driven by differences in water potential between the soil ($\psi_s$) and foliage ($\psi_l$). In this simple linear model (Equation 9.4), hydraulic conductances for the soil ($k_s$), root ($k_r$) and shoot ($k_{sh}$) connect in series to intermediate water potentials at the root surface ($\psi_{rs}$) and base of the stem ($\psi_b$).

- With equal root densities and equal root water potential (say $\psi_b$, Equation 9.5), partitioning of water is determined by the relative values of $\kappa_r$ for the trees and crop.
- Where the trees shade the crop, partitioning is determined by the relative values of $\psi_b$, and therefore by canopy-level microclimate and stomatal opening, if root densities and $\kappa_r$ are equal.
- With equal $\psi_b$ and $\kappa_r$, but with one root system being more sparse than the other, water is partitioned in proportion to relative values of $L_v$.

Competition could result from each of these scenarios if partitioning favoured one species and placed a neighbouring species at a disadvantage because of insufficient availability of water. Complementary use of water could result if the root distributions of the two species were vertically stratified, or if periods of demand for water by each species were discrete.

defined here, $\kappa_r$ combines the components of conductivity for radial transport across the root, from the root surface to xylem, and axial transport to the base of the stem. Soil hydraulic conductivity ($\kappa_s$) can also be expressed per unit root length (Rowse et al., 1983). The rate of uptake of water (S) by a root system of root length density $L_v$ and occupying soil volume V is then

$$S = \frac{(\psi_s - \psi_b)}{(1/\kappa_s + 1/\kappa_r)} \cdot L_v V \qquad (9.5)$$

Unless roots are very sparse, S is most limited by $\kappa_r$ at soil water contents above the wilting point. Uptake from soil with water potential $\psi_s$ is therefore dependent on the water potential within the plant, the hydraulic conductivity of the root system and the density of the root network.

## 9.3.3 Mechanisms for bidirectional water flow in roots

Efflux of water from roots into soil is possible where water potential is higher in roots than in the surrounding soil. This results in the reversal of flow in the root. Rather than flowing towards the stem, efflux results in flow towards the root tips. Water efflux is driven by gradients in either osmotic or hydrostatic potential. Osmotic gradients can cause efflux of generally small amounts of water. Under hydrostatic gradients, efflux can occur in dry soil if water potential remains higher in the root than it is in the soil, because of hydraulic connections to other parts of the root system present in wetter soil at a higher potential. As water potentials in the plant are highest when the evaporative flux from leaves is close to zero, efflux is most common at night. Efflux from roots occurs principally from young, relatively unsuberized roots (Caldwell et al., 1998).

## 9.4 Measurement of Water Uptake and Bidirectional Flow

Quantification of water uptake and partitioning is important for studies of competition and complementarity in mixed-species agroecosystems. In monoculture stands, it can be sufficient to estimate uptake by difference after measurement of the water budget terms in Equation 9.1. The method is limited by challenges in measuring drainage; a further limitation in plant mixtures is that it is not possible to distinguish uptake by individual species from the water budget. More direct measurements of water use are therefore needed in studies of partitioning in intercropping or agroforestry.

### 9.4.1 Measurement of sap flow in trees and crops

The most direct method available for the quantification of uptake of water by individual plants is measurement of the rate of sap flow in stems (Smith and Allen, 1996). Several methods are available (Table 9.1) that use heat as a tracer for sap movement. All are suitable for use on woody stems, though each is most suited to a specific range of stem sizes. However, only stem heat balance gauges are non-invasive and can be used on herbaceous stems. Uptake of water can be partitioned in plant mixtures by simultaneous measurement of sap flow in each species (Fig. 9.3). Sap flow thus provides a powerful means for the direct quantification of partitioning of water in agroforestry or intercropping. Because of limitations on the number of instruments that can be operated at one time, attention must be given to spatial scaling of sap flow rates from plant to stand, in order to give rates of uptake for unit land area. Scaling must account for heterogeneity in rates of uptake and the spatial distribution of plants (Smith and Allen, 1996).

### 9.4.2 Measurement of sap flow in roots

Sap flow can be measured in, and compared among, individual roots in order to provide insight into uptake from different zones in the soil. For example, the contribution made by deep roots to the uptake of water has been evaluated by the comparison of sap

**Table 9.1.** Methods of sap flow measurement used on trees and crops. For use on tree roots, the ability to measure flow in both directions is required because of the possibility of bidirectional flow. A general review of methods is given by Smith and Allen (1996).

| Operating principle | Method | Crops | Trees | Diameter range (mm) | Tree roots | Reference |
|---|---|---|---|---|---|---|
| Heat balance | Constant power | √ | √ | 2–125 | √ | Sakuratani (1981) |
| | Constant temperature | √ | √ | 2–125 | | Weibel and Boersma (1995) |
| | Trunk sector | | √ | > 120 | | Cermak et al. (1984) |
| Heat pulse | Compensation | | √ | > 30 | | Green and Clothier (1988) |
| | Heat ratio | | √ | > 30 | √ | Burgess et al. (2001b) |
| Empirical | Thermal dissipation | | √ | > 40 | | Granier (1987) |

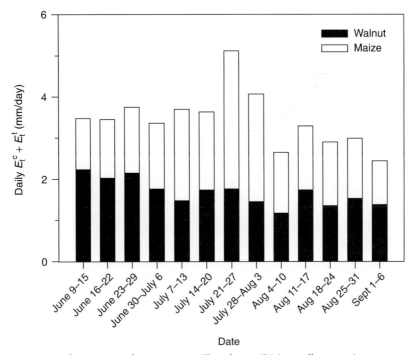

**Fig. 9.3.** Partitioning of transpiration between trees ($E_t^t$) and crop ($E_t^c$) in an alley cropping system combining maize and walnut (*Juglans nigra*) in mid-western USA, measured using sap flow techniques. Redrawn from Jose et al. (2000) after scaling to the area occupied by the tree–crop combination.

flow in lateral roots and vertical sinker or tap roots. The installation of sap flow sensors on herbaceous roots is generally not possible, because they are, typically, too fine and fragile. However, installation on woody roots is easily accomplished using the same procedures used for stems. Care must be taken in selecting the method used, as bi-directional flow can only be detected using the heat ratio and constant-power heat balance methods (Table 9.1). Other methods are used on roots, but there is then a risk that reversal of flow is masked by limitations in instrumentation. An example of flow reversal in roots is shown in Fig. 9.4.

### 9.4.3 Assessment of sources of water used by trees and crops

Variation in the natural abundance of the stable isotopes of water ($^2H$ and $^{18}O$) in soil profiles, and in rainfall and groundwater, can provide a means of tracing the sources of water used by plants. Changes in the isotopic composition of water do not occur when it enters roots or during its flow through roots or suberized stems. Thus, the isotopic composition of water in xylem reflects the source of uptake. If multiple sources of water are exploited, the isotopic composition of sap is a blend of the sources used. In practice, the isotopic ratios $^{18}O/^{16}O$ or $^2H/H$ are compared for water from the soil, water table and samples of stems or branches (Dawson, 1993). Isotopic ratios of water in samples are measured using a mass spectrometer after the recovery of water from tissue and soil by distillation or by using an equilibration method (see Smith *et al.*, 1997b).

Isotopic tracing of water sources has been used in a number of settings to provide qualitative information about the partitioning of water in plant mixtures. At semiarid sites in Niger, isotopic tracing showed that water acquisition by *Azadirachta indica* trees in windbreaks and adjacent millet was verti-

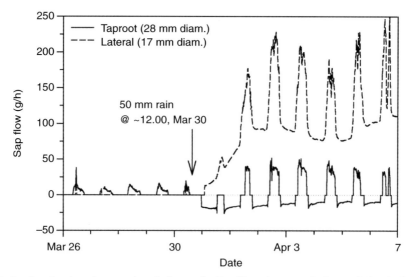

**Fig. 9.4.** Sap flow in a lateral root and vertical root of a *Grevillea robusta* tree before and after the first rainfall after the dry season, measured using constant-power heat balance gauges. Positive flow was towards the trunk and negative flow was towards the root tips. Rainfall on dry soil created a gradient in water potential between the topsoil and subsoil, causing reverse flow in the taproot during the night (Smith *et al.*, 1999b).

cally stratified only where ground water was accessible to the trees' roots (Smith *et al.*, 1997b) (Fig. 9.5). Similarly, Roupsard *et al.* (1999) found vertical stratification of water uptake at a site where *Faidherbia albida* trees accessed ground water. A discrete source of water available only to trees is thus a precondition for spatial complementarity with regard to water, at least when trees are combined with annual crops.

An awareness of the following is of critical importance: an essential requirement for tracing water sources using this approach is that differences in the abundance of stable isotopes must exist between water sources or within the soil profile (Burgess *et al.*, 2000). When the distribution of isotopes in the profile results in non-unique isotopic ratios among depths and sources, so-called pulse–chase experiments may prove helpful. In this approach, a small amount of water with a distinct isotopic ratio is applied to the soil. Percolation of this labelled water occurs after subsequent rainfall events. Sequential sampling of the isotopic profile in the soil and sap allows an assessment to be made of the depth of water uptake by plant roots.

## 9.5 Control of Water Partitioning

It was concluded from Equation 9.5 that partitioning of soil water between coexisting species is determined by the combined effects of differences between them in terms of root distributions, root hydraulic properties and the relationship between plant water potential and evaporative flux from the canopy. Consequently, investigation of the mechanisms controlling partitioning requires the evaluation of the role played by all three attributes in determining uptake by each species.

### 9.5.1 Root distributions

Methods available for the measurement of root length densities are described in detail by Smit *et al.* (2000b). Results from studies of root distributions in mixed agroecosystems are reviewed in Chapter 4 of this volume.

### 9.5.2 Plant water potentials, microclimate and leaf conductances

Differences in the driving force for uptake of water between species growing in combi-

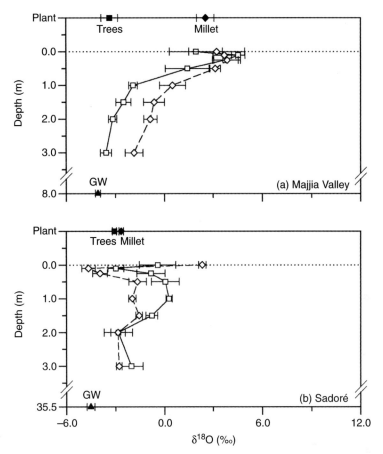

**Fig. 9.5.** Isotopic ratios for $^{18}O$ ($\delta^{18}O$) for the sap of *Azadirachta indica* trees in windbreaks (■), adjacent millet (♦), ground water (GW) (▲) and soil water beneath the trees (□——□) and millet (◊---◊) during a dry spell in the rainy season at (a) the Majjia Valley and (b) Sadoré, in Niger. Trees used water from below the crop rooting zone only at the Majjia Valley site, where groundwater was accessible to tree roots (Smith *et al.*, 1997b).

nation can be determined by comparison of their water potentials. The water potentials of leaves or small shoots are most readily measured after excision using a pressure chamber (Pearcy *et al.*, 1991). The same technique can be used to determine the water potential of stems if measurements are made on non-transpiring leaves attached to the stem. Leaves can be wrapped in plastic or aluminium foil for this purpose. If suckers are present near the base of a tree, this approach can be used to measure water potential in the trunk or root collar. Stem psychrometers provide an alternative method of measuring stem water potentials, but they have limited utility under field conditions. Routine measurement of water potentials in roots or at the surfaces of roots is not possible using currently available techniques.

Water potentials decline (become more negative) as rates of transpiration increase; partitioning of water between species in mixtures is thus dependent in part on above-ground control of transpiration. Evaporation from leaves is driven by solar radiation, with energy use for evaporation determined by humidity deficits in the atmosphere, the efficiency of aerodynamic exchange and stomatal opening. The integrated effects of these variables on evaporation from single leaves or canopies are commonly modelled using the Penman–Monteith equation (Monteith and Unsworth, 1990). Combining species in intercropping or agroforestry typically changes the microclimate, causing shading or a reduction in wind speed for example, which results in altered water use by each species. Instruments and methods used in monitoring microclimatic variables in agroforestry were reviewed by Brenner (1996).

In addition to a dependence on microclimate, partitioning of water among species is also dependent on plant responses to microclimate. Plants use a complex control system to reduce evaporation from foliage when the root system is unable to acquire water at the required rate. Loss of water from leaves is controlled by variation in the aperture of stomatal pores, modelled as a stomatal conductance. Many plants also shed leaves to control water use if dry conditions are prolonged. Control of water use by trees in agroforestry can similarly be achieved by shoot pruning, although feedback between stomatal opening and water potential means that reduction in water use may not be proportional to the leaf area removed. Such relationships and stomatal responses to microclimate can be measured in the field using porometers or by the measurement of leaf gas exchange using portable infra-red gas analysers, with leaf areas being measured directly, by destructive sampling, or using indirect methods (Hall et al., 1993).

### 9.5.3 Root hydraulic properties

The hydraulic properties of roots have received less attention from researchers than above-ground controls of transpiration, root distributions or water movement through the soil itself, despite the importance of $\kappa_r$ in Equation 9.5. As a result, uptake of water from soil has commonly been modelled using empirical functions of root length density and above-ground demand for water. When applied to species mixtures, there is an implicit assumption in this approach that the root systems of different species have equivalent hydraulic properties. There is evidence to the contrary, however, as root hydraulic conductivities reported by Steudle and Heydt (1997) were an order of magnitude lower for a range of temperate tree species than they were for herbaceous species, at least when expressed per unit surface area of root. There is, therefore, a suggestion that, in general, water uptake is more rapid for herbaceous roots than for tree roots under equivalent potential gradients.

Most techniques used to measure root hydraulic conductances are suited only to use in the laboratory with pot-grown or solution-grown plants. Two methods are available for use in the field, however. Assuming steady-state flow, conductances can be calculated using Equation 9.4 from measurements of transpiration rates, soil water potential and water potential at the base of the stem. Provided the soil is moist, the error associated with assuming that $\psi_{rs} = \psi_s$ will be small. An alternative method uses a high pressure flow meter (HPFM) developed by Tyree et al. (1995). This device forces water into severed roots under hydrostatic pressure at known rates of flow. Water moves in the opposite direction to that which it would normally (i.e. towards the root tips). Water pressure is increased rapidly over a period of 1–2 min and recorded with flow rate. The hydraulic conductance of the root is given by the slope of the linear portion of the plot of flow rate against pressure. Close agreement has been found between results obtained using these two field methods for both trees and crops. Care must be taken when interpreting measurements of hydraulic conductance in roots, because the pathway for uptake, and hence conductance, is different for flow that occurs under osmotic and hydrostatic gradients in water potential (Steudle and Heydt, 1997).

A HPFM was used to compare the conductivity for roots of maize and of the tree Grevillea robusta grown in the field in semi-arid Kenya (Smith and Roberts, 2003). On the basis of unit root length, mean $\kappa_r$ was $1.88 \times 10^{-7}$ kg/s/MPa/m for G. robusta and $1.25 \times 10^{-7}$ kg/s/MPa/m for maize. This difference was not significant, suggesting that the assumption of hydraulic equivalence between root systems may hold for some combinations of species. The contradiction that exists between this result and the data from Steudle and Heydt (1997) demonstrates, however, that more measurements of this type are needed for a wide variety of species, especially tropical trees.

## 9.6 Modelling of Water Uptake in Mixed Agroecosystems

### 9.6.1 Uptake partitioning

The principles of water uptake by plants provide the basis for mechanistic models of the partitioning of soil water between species in mixtures. Such models thus incorporate control of uptake by aboveground evaporative demand, plant water potential, root distributions and root hydraulic properties. Unless root lengths can be considered constant, models must also account for the effects of root growth. Sillon et al. (2000) developed a model of partitioning in agroforestry with mechanistic features. In this model, the daily soil water balance is modelled for a two-dimensional soil profile containing tree and grass roots. Extraction of water by roots is determined by a 'sink term' for each species based on Equation 9.5, with a single conductivity term combining $\kappa_s$ and $\kappa_r$. Potential evaporation is partitioned between species in proportion to fractional light interception, with actual evaporation being dependent on a stress function related to leaf water potential. Equations in the model are solved iteratively, to ensure convergence of the calculated uptake and transpiration.

Simpler approaches to modelling the partitioning of soil water are often required, however, because of limitations such as the lack of available data for model parameterization and the sheer complexity of dynamic models of growth and resource use in mixed vegetation. In common with other models of intercropping and agroforestry, the HyPAR model (Mobbs et al., 1998) partitions available soil water in proportion to transpirational demand and root length densities for trees and crops in each soil layer. This approach successfully enables simulation of interspecific competition for water, but would result in reduced accuracy if coexisting root systems have markedly different hydraulic properties. In such cases, differences among species may be implicitly reflected in the empirical coefficients that are applied to uptake functions.

An intermediate approach is used in the WaNuLCAS model (van Noordwijk and Lusiana, 2000). As in Sillon et al. (2000), the soil water balance is modelled in two dimensions and partitioning is controlled by the combined effects of transpirational demand, root distributions and root hydraulic conductivities. Parameterization of the transport equations for soil water is achieved using pedotransfer functions (Box 9.1). Calculations of water uptake by roots are simplified by using an empirical function to relate actual and potential transpiration, thus eliminating the requirement for an iterative solution of the uptake equations. Examples of the output from WaNuLCAS are shown in Box 9.2.

---

**Box 9.1.** Pedotransfer functions.

Lack of data on soil hydraulic properties is commonly a limitation on the use of models of the soil water balance and water uptake, particularly for practical, non-experimental applications. The relations between volumetric soil water content ($\theta$), pressure head ($h$) and soil hydraulic conductivity ($K$) can be measured in the laboratory or field, but only with specialized equipment and expertise. Consequently, simplified methods of estimating $\theta$-$h$-$K$ relationships greatly broaden the applicability of soil water models. An important innovation in methods for quantifying soil physical properties has been the development of pedotransfer functions (PTFs). These are sets of equations that enable the prediction of $\theta$-$h$-$K$ relationships from much more readily measured soil data, such as percentages of clay and silt, organic matter content and bulk density. A review of the determination and applicability of PTFs is given by Wösten et al. (2001).

**Box 9.2.** Simulation of water uptake in a hedgerow intercropping system using WaNuLCAS.

The WaNuLCAS model was used to simulate water uptake by trees and crop in a *Peltophorum dasyrrachis*–maize hedgerow intercropping system in Lampung, Sumatra. Mean annual rainfall for Lampung is 3100 mm. Simulations were made of a second cropping season (March–May) commencing after the end of a poor rainy season in which only 20% of normal rainfall was received. The effects that variation in tree and crop root length densities have on the uptake and partitioning of water were simulated. The system was assumed to have no nutrient limitations.

Results from the simulations are shown in Fig. B9.1. Reduction of tree root lengths had little effect on seasonal uptake in the monoculture tree stand but, in the tree–crop mixture, caused a small shift in partitioning in favour of the crop (Fig. B9.1a). The small influence that changes in root length had on water uptake by the trees suggests that the decline in $\psi_b$ required to maintain $S$ (Equation 9.5) for truncated root systems was not sufficient to limit uptake severely. This suggests water uptake may be more sensitive to tree root length in systems where the trees are not pruned, as higher potential transpiration would require lower $\psi_b$. The simulations also showed that reductions in crop water uptake in both the monoculture and hedgerow system were more substantial under similar relative changes in maize root densities (Fig. B9.1b). Uptake by the trees only partially compensated for reduced uptake by the crop, resulting in a decline in productive water use by the system (Equation 9.3).

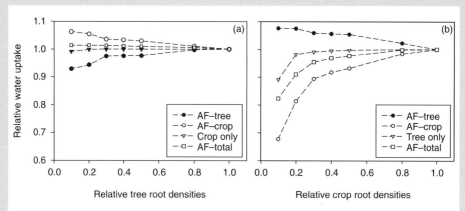

**Fig. B9.1.** Effects of root length densities on water uptake by the trees and crop in a *Peltophorum*–maize hedgerow system, as simulated using WaNuLCAS. Water use is expressed relative to its value at maximum root length density.

There is emerging evidence that use of a simple parameterization for root hydraulic conductivity may result in the effects of root architecture on patterns of water uptake being overlooked. Doussan *et al.* (1999) used an intricate model of root conductance and architecture to contrast uptake by herbaceous and woody root systems. The higher transport capacity of large woody roots near the base of the tree resulted in preferential uptake from soil close to the tree. Uptake by the herbaceous root system was more uniform. On the basis of theory, therefore, partitioning of water in tree–crop mixtures may depend to some extent on root system architecture, in addition to hydraulic conductances. Representation of this phenomenon may become a feature of future developments in the modelling of soil water partitioning in agroforestry.

## 9.6.2 Bidirectional flow and soil water redistribution

Physically based models of root function offer an advantage over more empirical approaches when the simulation of bidirectional flow in root systems is of interest. Flow of water from one soil layer to another through roots can be simulated if root systems are represented mathematically as a network of conductances that connect regions of a plant or of soil with differing water potentials. Reverse flow, from root to soil, is simulated when root water potential in any layer is higher than soil water potential, with the rate of flow being determined by the potential gradient and by the conductances of the flow path. WaNuLCAS (van Noordwijk and Lusiana, 2000) provides a capability for modelling of bidirectional flow using these principles, with the assumption that conductances for flow in either direction are equivalent. However, as reversal of flow entails movement of water from one root to another, flow paths for conventional and reverse flow are not exactly equivalent, suggesting that this assumption requires testing.

## 9.7 Summary and Conclusions

Combining species in agroecosystems can reduce non-productive loss of water and enhance total uptake by plants, resulting in increased biomass production. One danger associated with the mixing of species is competition for water. Unacceptable loss of crop yield can occur if trees deprive a crop of water. To succeed in water-limited environments, therefore, water use by trees and crops should be complementary: uptake by trees and crops should be from spatially discrete sources or should occur at discrete times.

Partitioning of water between coexisting root systems can be conceptualized using the fundamental principles of water uptake by plants. Partitioning of water therefore depends on the combined effects of differences between species in terms of the distributions, hydraulic properties and water potentials of roots. Inequalities in root distributions, transpirational demand and root hydraulic conductivities may each result in partitioning favouring one species over a neighbouring species. Further complexity in below-ground interactions arises from bidirectional flow in roots, which is observed when root systems span soil layers with different water potentials. Flow reversal in roots causes redistribution of soil water, with hydraulic lift potentially facilitating the coexistence of trees and crops, but with downward siphoning enhancing the competitive advantage of trees.

Sap flow methods provide a powerful tool for the direct measurement of tree and crop water use and therefore for the measurement of partitioning. Some of these methods have been used on tree roots to contrast uptake by deep and shallow roots. Qualitative insights into the vertical stratification of uptake by tree and crop roots can also be obtained from an analysis of variation in the natural abundance of deuterium or $^{18}O$ among water sources. Sap flow methods have, additionally, enabled direct quantification of bidirectional flow for the first time; this has resulted in the consideration of this interesting phenomenon as an ecological process with potentially important impacts, rather than as a theoretical oddity. A full understanding of its effects awaits further research.

The mechanistic basis for partitioning can be investigated by the study of root distributions, physiological and microclimatic control of transpiration, and root hydraulic conductivities. Knowledge of the processes controlling partitioning in agroforestry has advanced in the last decade, and has enabled the modelling of water use in mixtures to move beyond simplistic dependencies on root length distributions. However, there remains a need for new data that would enable parameterization of these models, especially in relation to root hydraulic properties.

Modelling has a key role to play in ensuring that recent progress on research into below-ground interactions and water results in real improvements in the management of complex agroecosystems. In water-limited environments, strategies for controlling competition and promoting complementarity in water use are imperative in agroforestry. However, the system-wide impacts of modifications made to management strategies can be hard to predict without reliable models. Advances in our ability to model uptake and the partitioning of water by competing root systems should, therefore, make the task of developing and promoting more sustainable systems of land use easier. Ultimately, however, future models must enable analysis of economic trade-offs caused by the modification of water budgets at the catchment scale, in order to relate improvements in agroecosystem management to better livelihoods for rural communities.

### Conclusions

1. Trees can modify the water budgets of cropping systems to increase rainfall utilization, trading non-productive use of water for productive uptake by plants.
2. Partitioning of soil water in species mixtures is determined by the integrated effects of differences in root distributions, root hydraulic properties and canopy-level control of transpiration; approaches to modelling uptake and partitioning are based on representations of these effects, with varying levels of simplification.
3. Bidirectional flow occurs when root networks span gradients in soil water potential, causing reversal of flow in roots and transfer of water from wetter to drier soil.
4. Sap flow methods provide a powerful tool for direct quantification of partitioning and bidirectional flow; qualitative assessment of vertical stratification in uptake in species mixtures is possible where differences occur in the isotopic composition of water from different sources.
5. Competition for water in agroforestry is severe where rainfall is too low to recharge soil below the crop rooting zone, unless a water table is accessible to tree roots.
6. Spatial complementarity in water use requires both deep sources of water and differential tree and crop root distributions; deep-rooted trees alone are not sufficient.
7. Pruning, the use of species with 'reverse phenology', and sequential planting in agroforestry systems permit temporal partitioning of uptake; however, there is a need to ensure that high water use in one phase of the production cycle does not reduce water availability in subsequent phases.

### Future research needs

1. To further investigate bidirectional flow: what quantities of water are involved, what are its impacts in real systems and what are the physiological controls on flow reversal in roots?
2. To collect data on root hydraulic properties for field-grown tropical trees and crops, and to elucidate relationships between root hydraulic conductances and root system architecture.
3. To quantify economic trade-offs from changes in the water budget and to develop methods for the scaling-up of these trade-offs from plot to catchment and community.

# 10 Catching and Competing for Mobile Nutrients in Soils

Georg Cadisch, Peter de Willigen, Didik Suprayogo, Deena C. Mobbs, Meine van Noordwijk and Edwin C. Rowe

---

**Key questions**

1. How many roots does a plant need for the adequate uptake of water and nutrients?
2. Do soil macropores enhance or decrease leaching of nutrients?
3. Can below-ground competition between plants be predicted from the root length density of each plant?

---

## 10.1 Introduction

The mobility of nutrients in soil is of prime importance with respect to plant nutrition and nutrient losses from a system. Nutrient movement in soils is closely linked to water dynamics and nutrient depletion in the rhizosphere (i.e. plant demand) as well as to the properties of the soil and the nutrient itself. The plant components of ecosystems significantly influence water movement, and hence nutrient utilization and losses, by altering infiltration, runoff, water usage and drainage. Particular emphasis has been given, in the past, to the effect of agroecosystems and their management on the occurrence of macropores and their potential effect in terms of reducing/increasing leaching of nutrients (van Noordwijk et al., 1991a; Dierolf et al., 1997). Deep-rooted trees, crops and pastures have the potential to reduce leaching by acting as a 'safety net' and intercepting nutrients moving through the soil profile (Rowe et al., 1999). Soil properties, such as cation and anion exchange capacity, further act as an inherent nutrient safety net (Mekonnen et al., 1997; Suprayogo et al., 2002). Introducing more diverse cropping systems and mixed-species systems may enhance nutrient capture and recycling efficiency. This is particularly true where species in mixed systems have complementary, rather than competitive, nutrient resource acquisition strategies. Multispecies systems may better exploit underutilized resources or may exploit new resource niches (Cadisch et al., 2002a). Evidence of root plasticity in response to competition is emerging (Gathumbi et al., 2002a; Chapters 4 and 5, this volume), which may lead to a better exploitation of the available resources in mixed-species systems (Schroth, 1998).

There is still great uncertainty as to how many roots per unit soil volume (i.e. fine root length density) are needed to exploit or intercept leaching nutrients (van Noordwijk and Brouwer, 1991; Cadisch et al., 1997). Chapter 9 discusses, in detail, soil–plant interactions with regard to water, and simplified ways to describe soil hydraulic properties (pedotransfer functions), which determine, to a great extent, the mobility of nutrients in the soil solution. Here, we describe factors determining nutrient transport in, and losses from, soils, and investigate ways to describe how plants compete for nutrients or act as safety nets, with examples from recent research.

## 10.2 Nutrient Mobility in Soil Solutions

### 10.2.1 Mobility of nutrients in uniform soil conditions

The basic transport equation for a substance in a porous medium reads:

$$\frac{\partial S}{\partial T} = -\nabla \cdot F + U \qquad (10.1)$$

where $S$ is the bulk density of the substance (mg/cm$^3$), $T$ is time (days), $F$ is the flux (mg/cm$^2$/day), $\nabla$ is the gradient operator (cm$^{-1}$) and $U$ is the production or consumption (mg/cm$^3$/day). This equation states mathematically that the rate of increase or decrease of a substance in a small volume of soil is due to the difference of in- and outgoing flows over its surface plus possible production or consumption in the volume. If, for instance, the substance in question is nitrate, it can be produced by mineralization and subsequent nitrification or consumed by denitrification. The flux of nutrients can, at least within the scope of this book, be assumed to consist of two components – a convective component ($F_C$) and a diffusive component ($F_D$):

$$F = F_C + F_D \qquad (10.2)$$

The convective flux ($F_C$) is brought about by transport of water, and the resulting flux is proportional to the concentration of the nutrient in that water:

$$F_C = VC \qquad (10.3)$$

where $V$ is the volume flux of the water (cm/day) and $C$ is the concentration of the nutrient in water (mg/cm$^3$). The diffusive flux ($F_D$) is assumed here to be given by Fick's first law (i.e. flux is proportional to the gradient of the concentration):

$$F_D = D\nabla C \qquad (10.4)$$

where $D$ is the diffusion coefficient (cm$^2$/day). The convective flux, or mass flow, is the dominant mechanism in leaching, whereas the diffusive flux is often the most important process when transport in the vicinity of the root is being considered. Substitution of Equations 10.2, 10.3 and 10.4 into 10.1 results in:

$$\frac{\partial S}{\partial T} = \nabla \cdot VC + \nabla \cdot D\nabla C + U \qquad (10.5)$$

This then is the general equation that is used to describe transport in the soil to the root.

### 10.2.2 The inherent 'safety net' of soils

The total amount of a nutrient in the inorganic form consists of the amount in solution, which is mobile, and the amount in the solid phase, which is considered immobile. Here, the immobile form will be assumed to consist of two pools: a pool, $Q$ (mg/cm$^3$), which exchanges at a fast rate with the pool in solution, and another pool from which transfer to the solution is so slow that it can be ignored for our purposes. Thus:

$$S = Q + \theta C \qquad (10.6)$$

where $\theta$ is the volumetric water content of the soil (cm$^3$/cm$^3$). It will be assumed here that at any time there exists a unique functional relationship – the adsorption isotherm – between $Q$ and $C$, symbolized by:

$$Q = f(C) \qquad (10.7)$$

Substitution of Equations 10.6 and 10.7 into Equation 10.5 yields an equation with only $C$ as the dependent variable:

$$[f'(C) + \theta]\frac{\partial C}{\partial T} = \nabla \cdot D\nabla C - \nabla \cdot VC + U \qquad (10.8)$$

In the following equations, we will neglect the production/consumption term $U$, which reflects mineralization and plant uptake, as it is dealt with below (Equation 10.16). In Equation 10.8, the role of the soil's chemical properties is given by the derivative of the adsorption isotherm with respect to concentration, as shown on the left-hand side of Equation 10.8. The soil's physical properties and conditions manifest themselves via their influence on the transport parameters $V$ and $D$. Water is both the carrier of solutes (reflected in the role of the flux $V$) and the medium through which diffusion takes place. The diffusion coefficient in soil is very much dependent on water content and can be calculated as (Nye and Tinker, 1977):

$$D = D_0 f_l \theta \quad (10.9)$$

where $D_0$ is the diffusion coefficient in water and $f_l$ the impedance factor, a function of water content. Equation 10.8 assumes special forms, of which some examples follow, depending on subsequent assumptions. If we consider first the case of leaching without root uptake (consumption), then one usually assumes transport in a vertical direction only:

$$[f'(C) + \theta]\frac{\partial C}{\partial T} = \frac{\partial}{\partial z} D \frac{\partial C}{\partial z} - \frac{\partial VC}{\partial z} \quad (10.10)$$

In ecosystems, plant roots play an important role in nutrient transport, in terms of plant consumption and depletion of nutrient stocks, and hence leaching. An obvious and widely used assumption, when transport to a root is considered, is that of the cylindrical form of the root, which leads to the choice of expressing Equation 10.8 in cylindrical coordinates rather than in rectangular coordinates as was done in Equation 10.10. When tangential and vertical gradients are assumed to be negligible, Equation 10.5 (not considering leaching) assumes the form:

$$[f'(C) + \theta]\frac{\partial C}{\partial T} = \frac{1}{R}\frac{\partial}{\partial R} DR\frac{\partial C}{\partial R} - \frac{1}{R}\frac{\partial RV_R C}{\partial R} \quad (10.11)$$

where $R$ is root radius (cm) and $V_R$ is volume flux to the root (cm/day). When transport is dominated by diffusion, and the mass flow component can be neglected, Equation 10.11 simplifies to:

$$[f'(C) + \theta]\frac{\partial C}{\partial T} = \frac{1}{R}\frac{\partial}{\partial R} DR\frac{\partial C}{\partial R} \quad (10.12)$$

provided that $D$ is independent of $C$ and the adsorption isotherm is linear, i.e.

$$f(C) = K_a C \quad (10.13)$$

(where $K_a$ is the adsorption coefficient (cm$^3$/cm$^3$)); then Equations 10.10, 10.11 and 10.12 are linear equations, which can be solved analytically by classical mathematical techniques. If, however, the adsorption isotherm is non-linear (e.g. in the case of phosphate), the equations have generally to be solved by numerical methods. When the diffusive flux can be neglected with respect to the mass flow (e.g. in the case of leaching) and adsorption is linear, substitution of Equation 10.13 into Equation 10.10, and some rearranging, leads to:

$$\frac{\partial C}{\partial T} = -\frac{1}{K_a + \theta}\frac{\partial VC}{\partial z} \quad (10.14)$$

This shows that the flux of a dissolved nutrient is retarded by the retardation factor $N$:

$$N = \frac{1}{K_a + \theta} \quad (10.15)$$

For nutrients with a non-linear adsorption isotherm (e.g. phosphate) the situation is more complicated. The retardation here depends on the concentration itself and the shape of the adsorption isotherm (see also Chapter 7, this volume).

Retardation factors (Table 10.1) can significantly reduce leaching of nutrients. The effectiveness of this inherent 'safety-net' function of soils in reducing mineral N leaching in tropical cropping systems was investigated by Suprayogo *et al.* (2002), using the WaNuLCAS model (Box 10.1; van Noordwijk and Lusiana, 1999). Simulation of different cropping systems in humid tropical conditions in Sumatra, Indonesia (Box 10.2; Suprayogo *et al.*, 2002) suggested that the ammonium retardation mechanism reduced leaching of mineral N by between 5% and 19% (Table 10.2). Effective retardation increased when crops and trees were present, due to their direct uptake of mineral N, and as an indirect result of their uptake of soil water, which decreased net drainage through the rooting zone. When the retardation factors in the model were activated, the simulated leaching

**Table 10.1.** Diffusion and adsorption coefficients of nutrient ions.

| | Diffusion coefficient (cm$^2$/day) | | Adsorption coefficient ($K_a$) (cm$^3$/g)[f] |
|---|---|---|---|
| | Water (25°C)[a] | Soil[a] (range) | |
| Nitrate | 1.6 | $10^{-1}$–$10^{-2}$ | 0.03–0.17[b,d] (Ultisol) 0.07–0.28[c] (Acrisol) |
| Ammonium | n/a[e] | n/a | 1.5–1.8[b] (Ultisol) |
| Phosphorus | 0.8 | $10^{-3}$–$10^{-6}$ | n/a |
| Potassium | 1.7 | $10^{-2}$–$10^{-3}$ | n/a |

Source: [a]Jungk (1991); [b]Suprayogo *et al.* (2002); [c]Wong *et al.* (1987). [d]pH dependent; [e]n/a = not available; [f]by taking into account soil bulk density $K_a$ can be expressed per volume as in Equation 10.13.

---

**Box 10.1.** WaNuLCAS (Water Nutrient Light Capture in Agroforestry Systems).

WaNuLCAS (van Noordwijk and Lusiana, 2000; Fig. B10.1) offers a good compromise between spatial and process-orientated complexity. The model is particularly suited to the evaluation of spatial interactions (such as above-ground shading and competition for water and nutrients), which may occur over a range of distances in mixed-species systems. The resource-capture framework for modelling plant growth used here is based on shoot and root biomass, allocation to leaf and root area index (LAI and RAI, respectively) and its spatial distribution (based on 'architecture'), and capture of light, water and nutrients.

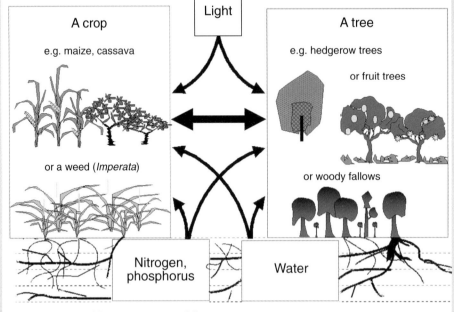

**Fig. B10.1.** Scope of the WaNuLCAS model.

*Continued*

**Box 10.1.** *Continued.*

**Water balance:** Upon infiltration, a 'tipping bucket' model is followed for wetting subsequent layers of soil, filling a cascade of soil layers up to their effective 'field capacity'. Field capacity is estimated from the water retention curve using pedotransfer functions (see Chapter 9). Soil evaporation depends on ground cover (based on the LAI of trees and crops) and soil water content of the topsoil; soil evaporation then stops when the topsoil layer reaches a water potential of −16,000 cm.

In WaNuLCAS 2.0, a simple representation of by-pass flow is added and there is an option for a dynamic simulation of macropore structure. The user can define an initial saturated hydraulic conductivity value that differs from the default value predicted by the pedotransfer value. During the simulation, the value will tend to return to this default value (at a rate determined by the S_KStructDecay parameter). Water uptake by the plants is driven by their transpirational demand, within the possibilities determined by root length densities and soil water content in the various cells to which a plant has access.

**Leaching** of N (and P) is driven by percolation of water through the soil and by the average nutrient concentration in soil solution. The latter is derived from the inorganic nutrient stock, the soil water content and the apparent adsorption constant ($K_a$). Macropore flow bypasses the soil solution contained in the soil matrix. A multiplier $N\_BypassMacro\ i[Zone]$ is used in the leaching equation, to describe preferential macropore flow.

**Nutrient uptake:** A target N content is contrasted with the current nutrient content, to derive the 'Nutrient deficit'. The N deficit can be met either by atmospheric $N_2$-fixation or by triggering an increased nutrient demand for uptake from the soil. Potential nutrient uptake ($U_{ijk}$) from each cell ($ij$) by each component ($k$) is calculated from a general equation for zero-sink uptake (de Willigen and van Noordwijk, 1994) on the basis of the total root length in that cell, and is allocated to each component in proportion to its effective root length:

$$U_{ijk} = \frac{L_{rv,ijk}}{\sum_k L_{rv,ijk}} \frac{\pi D_0 (a_1 \Theta_{ij} + a_0) \Theta_{ij} H_{ij} N_{stock,ij}}{(K_a + \Theta_{ij})[-\frac{3}{8} + \frac{1}{2}\ln \frac{1}{R_0 \sqrt{\pi \sum_k L_{rv,ijk}}}]} \quad \text{(B10.1)}$$

where $L_{rv}$ is root length density (cm/cm³), $D_0$ is the diffusion constant for the nutrient in water, $\Theta_{ij}$ is the volumetric soil water content, $a_1$ and $a_0$ are parameters relating effective diffusion constant to $\Theta_{ij}$, $H$ is the depth of the soil layer, $N_{stock}$ is the current amount of mineral N per volume of soil, $K_a$ is the apparent adsorption constant and $R_0$ is the root radius. Real uptake $S_{ijk}$ is derived after summing all potential uptake rates for component $k$ for all cells $ij$ in which it has roots. Total uptake will not exceed plant demand. The definition of 'demand$_k$' is based on the current biomass and a target nutrient concentration appropriate for that biomass (van Noordwijk and van de Geijn, 1996), *minus* a fraction derived from atmospheric $N_2$-fixation.

events in the crop-only and in the tree-based systems agreed reasonably well with the actual, measured, leaching values of 6.5 and 2.0 g N/m², respectively (Suprayogo, 2000).

Whereas adsorption of ammonium in this plinthic Acrisol significantly decreased the simulated N leaching, the much smaller nitrate adsorption coefficient (Table 10.1) had only a minor effect in reducing leaching. This was exacerbated by the shallowness (0.8 m) of the effective rooting depth of the investigated and simulated soil due to the plinthic subsoil. Thus, anion (and cation) adsorption may more effectively reduce leaching in deep soils with less intensive rainfall. Nitrate adsorption coefficients increase with increasing soil depth. This is associated with: (i) the decrease in soil pH with depth, because of the soil's pH-dependent variable charge (which is more positive at depth); and (ii) the often

**Table 10.2.** Effect of including the adsorption coefficients for $NH_4^+$ and $NO_3^-$ (Table 10.1) in the simulation of leaching of mineral N below 0.8 m over 1 year, in different tropical cropping systems, with fertilizer additions of 30 kg N + 60 kg N/ha, on a Plinthic Acrisol in the humid tropics. (Adapted from Suprayogo et al., 2002.)

| Cropping system[a] | Adsorption coefficients included in simulation | | | |
|---|---|---|---|---|
| | $-K_{aNH_4} -K_{aNO_3}$ [b] | $-K_{aNH_4} +K_{aNO_3}$ | $+K_{aNH_4} -K_{aNO_3}$ | $+K_{aNH_4} +K_{aNO_3}$ |
| | Mineral N leaching in g/m² (% change in parentheses) | | | |
| −C−H | 11.8 (0) | 11.8 (0.1↓) | 11.3 (4.7↓[c]) | 11.3 (5.0↓) |
| +C−H | 10.7 (0) | 10.6 (0.8↓) | 8.7 (18.9↓) | 8.6 (19.9↓) |
| +C+H | 8.3 (0) | 8.2 (0.6↓) | 6.9 (16.8↓) | 6.8 (17.4↓) |

[a] −, without; +, with; C, maize–groundnut rotation; H, *Peltophorum dasyrrachis* hedge; [b] $K_{aNH_4}$ and $K_{aNO_3}$, adsorption coefficients for $NH_4^+$ and $NO_3^-$; [c] ↓ = % decrease.

increased clay content at depth in highly weathered soils. An important factor is also the decrease in organic matter content (SOM) that occurs with soil depth. This is because, with decreasing SOM, the number of negative charges (CEC) decreases and the number of anion exchange sites (AEC) increases (see also Chapter 6, this volume). Mekonnen et al. (1997) found large accumulations of nitrate under maize monoculture systems in a Lixisol in Western Kenya (see also Chapter 6, this volume). This may be explained by the great subsoil depth in that Lixisol (>4 m), which, when compared with the Acrisol investigated in Sumatra (<1 m rooting depth), may have greatly increased the 'safety-net' opportunities.

Nitrogen can be leached not only in mineral form but also in a dissolved organic form (DON). The presence, in soil solutions, of DON in concentrations greater than either $NH_4^+$ or $NO_3^-$ has increased scientific interest in the form and function of this pool of N. Studies by Bhogal et al. (2000) and Murphy et al. (1999a) concluded that leached DON could be an important route of N loss (e.g. under arable and grassland systems DON was found to comprise 10% and 20% of total N leached, respectively). Even larger amounts of DON were found to exist in the soil profile and were not lost into drains or watercourses (Murphy et al., 1999b), probably due to effective adsorption mechanisms. However, in South American forests, DON losses into the rivers accounted for >70% of the N budget (van Breemen, 2002).

### 10.2.3 Mobility of nutrients in heterogeneous soil conditions

Understanding nutrient movement under field conditions necessitates a better knowledge of the flow of water in aggregated soils. Water flow depends on pore distribution and pore size and hence on soil structure and texture. Well-structured soils often exhibit macropore water flow, allowing percolating water to pass rapidly through the soil without displacing the resident soil water contained in micropores (Beven and Germann, 1982); there is thus little nutrient movement out of the soil matrix. Within macropores (Fig. 10.1), dissolved nutrients move downwards rapidly with low rates of adsorption. The result may be increased soil water drainage with increased or reduced leaching of nutrients. The latter will depend on the ratio of nutrients present in macropores to nutrients present in the soil matrix (i.e. whether the nutrients derive from surface-applied fertilizer and are contained in the macropore solution, or whether they derive from mineralization in aggregates and so are contained in the micropore solution).

Macrofauna (such as earthworms and termites), through their burrowing, directly increase the occurrence of macropores (Lavelle et al., 1992). There is also evidence that tree roots change the pore-size distribution in soils. Van Noordwijk et al. (1991a) found evidence of substantial macropore flow in channels developed from decaying tree roots on a previously forested site.

Fig 10.1. Decaying roots (indicated by organic-rich (black) material in pore) provide macropore channels for root growth, water and nutrient flows.

tant in the formation of water-stable aggregates than incorporated foliage litter is (Chapter 11, this volume; Cadisch et al., 2002a). On a plinthic Acrisol, field hydraulic conductivity decreased more rapidly with decreasing soil water content under maize monoculture than it did under hedgerow intercropping systems (Fig. 10.2; Box 10.2), implying that the presence of trees resulted in a greater proportion of larger pores. These results, relating to internal drainage, thus suggest that hedgerow intercropping systems can cause greater 'bypass flow' through macropores than monoculture cropping systems can. Indeed, at a distance of 0.4 m from the hedgerow trees, the average macropore flow after 5 min saturation comprised 12% (± 4%) of the total flow under hedgerow intercropping systems, and 8% (± 3%) of the total flow under monoculture (Suprayogo, 2000). The challenge in dealing with macroporosity is the often large heterogeneity found in fields.

Several models simulate salt movements in aggregated soils by using a 'two-region' approach, e.g. convection and dispersion transport through pores or cracks and diffusion movement inside the micropores (van Genuchten and Dalton, 1986). A simplified approach was incorporated in the WaNuLCAS Version 2.0 agroforestry model (Box 10.1) in order to describe the flow of water through macropores that bypasses the soil solution contained in the soil matrix.

Equally, living root systems can contribute to improved soil structure and hence can enhance soil hydraulic properties. Recent evidence suggests that roots are more impor-

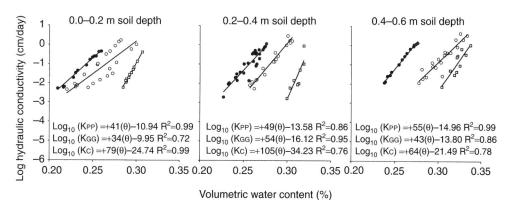

Fig. 10.2. Relationship between field soil hydraulic conductivity and soil water content of a plinthic Acrisol under different cropping systems: (●) *Peltophorum dasyrrachis* and (○) *Gliricidia sepium* hedgerow intercropping systems and (□) maize monoculture. Adapted from Suprayogo (2000).

> **Box 10.2.** Mixed tree–crop cropping systems in Lampung, Indonesia.
>
> Safety-net functions of trees, and competition between trees and crops, were evaluated and simulated in an experiment that was carried out at the Biological Management of Soil Fertility (BMSF) project site of Universitas Brawijaya (Malang, Indonesia)/ICRAF/Wye College (UK)/PT Bunga Mayang (North Lampung, Sumatra) (4°31′S, 104°55′E), on a field with a gentle slope (4%). Annual rainfall is about 2500 mm distributed between December and June and the soil is a coarse loam Acrisol (Ultisol) with a clayey plinthic, Al-rich layer beyond 60–80 cm (van der Heide *et al.*, 1992).
>
> Six cropping systems were established in 1985–1986. Those relevant to this chapter were a system with no hedgerows ('monocrop'), hedgerow intercropping systems using *Peltophorum dasyrrhachis* (P-P) and *Gliricidia sepium* (G-G), and a system in which hedgerows of the latter two species alternated (G-P). Hedgerows were 4 m apart and trees were 50 cm apart within the row. Hedgerows were pruned two to four times per year depending on light competition. Average yields of prunings were (1993–1999) 8.4 t/ha for G-G and 7.2 t/ha for P-P with production of *G. sepium* occurring mainly during the rainy (crop) season while *P. dasyrrachis* grew better during the fallow (dry) period.
>
> Two crops were generally grown per year, maize (December–February) followed by groundnut (March–June). KCl (60 kg K/ha) and triple super-phosphate (60 kg P/ha) were applied to each crop. Mean maize/groundnut yields (t/ha) over four cropping seasons (between 1993 and 1997) were 2.3/0.3 (monocrop with 90 kg N/ha), 1.2/0.2 (G-G, 0 N), 1.8/0.5 (P-P, 0 N) and 1.7/0.4 (G-P, 0 N). For comparison, in the 1998/99 cropping season, equal N fertilizer treatments (90 kg N/ha) yielded 2.1 (monocrop), 1.3 (G-G), 2.6 (P-P) and 1.9 kg/ha (G-P) of maize grain on average.

Sensitivity analysis of the impact of macroporosity on a hedgerow intercropping system (Box 10.2) indicated that the amount of vertical and lateral drainage increased by around 44% and 58%, respectively, when all flow occurred in macropores, as compared with a situation where no macropores were present (Fig. 10.3). Additionally, runoff was significantly decreased, which led to potentially important reductions in soil erosion and associated nutrient and organic matter surface losses. The model simulations demonstrated that the presence of macropores could proportionally reduce N leaching (by up to 30%) per unit of water drainage. However, in high intensity rainfall conditions, increasing the proportion of macropore flow increased total vertical and lateral leaching of mineral N by almost 15% and 24%, respectively, which was associated with increased drainage due to reduced runoff losses. Thus, the effects that macropores have on nutrient flows have to be viewed in the context of the whole system and offset against reduced erosion losses.

The movement of $^{15}$N-labelled fertilizer and organic residues added to this system (measured by Rowe, 1999) suggested that, proportionally, there was greater lateral movement of mineral fertilizer N than there was of N derived from the mineralization of organic residues. This led to the hypothesis that fertilizer N applied in high-rainfall areas becomes mainly distributed in macropores, whereas N derived from residues is associated more with the matrix soil, i.e. in micropores and at exchange sites, as was also proposed by Haynes (1986). This is particularly likely in those cases where residues have been incorporated into the soil. N released from surface-applied residues, however, is more likely to behave like fertilizer N. In comparison with mineral fertilizer N, a lower level of leaching of organic-derived mineral N is sometimes observed to occur, and this has also been associated with the temporary immobilization of N in the microbial biomass (Schroth, 2001). However, less-controllable N-release patterns from organic residues may lead to larger N losses in other systems. Currently, there are no data available on the distribution (between macropores and the soil matrix) of N derived from different sources that could be used to test the hypothesis that leaching losses of mineral fertilizer N are greater than those of N derived from organic residues.

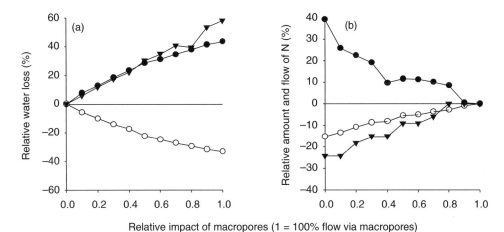

**Fig. 10.3.** WaNuLCAS sensitivity analysis of the importance of bypass flow (0–1 = proportion of total flow as macropore flow) on (a) vertical drainage (▼), runoff (○) and lateral drainage (●) relative to result of soil with no macropores; (b) mineral N stock at 0–0.8 m soil depth (●), vertical N leaching (○) and lateral N movement (▼) relative to result with all flow via macropores in a 1-year cycle of a *Peltophorum dasyrrachis* hedgerow system intercropped with a maize–groundnut rotation at the Lampung site. Adapted from Suprayogo, 2000; Box 10.2).

## 10.3 Catching Nutrients in Single-species Stands

Plants mainly take up nutrients in mineral form, via either passive or active pathways. Recent studies suggest that some plants can also exploit dissolved organic N (DON) directly or via ericoid- or ecto-mycorrhizal associations. Raab *et al.* (1996) showed that the alpine sedge *Kobresia myosuroides* was able to take up glycine from nutrient solutions at higher rates than it could take up $NH_4^+$ or $NO_3^-$. This may be an adaptation that allows the plant to survive when inorganic N is severely limited (such as in cold upland situations) or a strategy to avoid competition for scarce resources. Turnbull (1995) showed evidence of organic-N uptake by *Eucalyptus*. However, the general direct availability of low- or high-molecular-weight compounds to plants remains largely unknown. Thus, most of the currently available plant–soil models only consider nutrient uptake via the inorganic route.

### 10.3.1 Regulation of nutrient uptake processes

De Willigen and van Noordwijk (1987) derived an equation that described the fractional depletion ($f$) of a mineral nutrient as a function of demand, root length density, root length, water content, its diffusion coefficient and water uptake (Equation 10.16). Fractional depletion is defined as the fraction of the available amount that can be taken up by the root system at a rate corresponding to the demand. The equation was derived under the assumptions that the roots being considered were parallel, regularly distributed, and all of the same length and activity:

$$f = 1 - \frac{A}{LD} \frac{R_0^2}{C_i} G(\rho, v) \qquad (10.16)$$

where $A$ is the demand of the plant (mg/cm²/day), $L$ the root length (cm), $R_0$ the root radius (cm), $D$ the diffusion coefficient (cm/day) and $C_i$ the initial concentration (mg/cm³). $G(\rho, v)$ is a function of the root length density and of transpiration:

$$G(\rho, v) = \frac{\rho^2}{4(v+1)} \left\{ \frac{\rho^{2v}-1}{v} - \frac{v+3}{v+2} \right\} \qquad (10.17)$$

In Equation 10.17, $\rho$ is a dimensionless parameter given by:

$$\rho = \frac{1}{R_0 \sqrt{\pi L_{rv}}} \qquad (10.18)$$

where $L_{rv}$ is the root length density (cm/cm³), and $v$ is the dimensionless transpiration:

$$v = -\frac{E_T}{4\pi LDL_{rv}} \quad (10.19)$$

where $E_T$ is the transpiration (mg/cm$^2$). The diffusion coefficient is calculated as in Equation 10.9 where the impedance is a linear function of water content calculated, following Barraclough and Tinker (1981), as:

$$f_l = a_1\theta + a_2 \quad (10.20)$$

where $a_1$ and $a_2$ are parameters relating effective diffusion constant to $\theta$.

From Equation 10.16 it can be seen that the fractional depletion is complete (i.e. $f = 1$) when the diffusion coefficient is infinite, that is, when transport through soil is not limiting. For a given root length density, transpiration rate and diffusion coefficient, the fractional depletion increases with decreasing demand, decreasing adsorption and decreasing root radius. For certain nutrients, like phosphorus, it is further necessary to consider the symbiotic mycorrhizal hyphal network as an extension of the plant's root length, which can increase $L_{rv}$ to several times that of a non-mycorrhizal plant. The above discussion pertains to a root system in soil where no other gains or losses of nutrients occur. The next section considers competition between root uptake and leaching.

### 10.3.2 Critical root length density for intercepting leaching nutrients: root safety net

The root length density required to remove mineral N from the soil is small in relation to that required for less mobile ions, such as phosphate. However, in the dynamic soil environment, if nitrate is not taken up quickly by a plant it may be lost through leaching or may be taken up by other plants; therefore, large root length densities may be needed to acquire N rapidly (Cadisch *et al.*, 1997). We used WaNuLCAS (Box 10.1) to determine how efficiently tree roots intercept leaching nitrogen in a case study in Lampung (Box 10.2). The model runs simulated the December to March high-rainfall period, without N fertilizer application. The trees were pruned on the same day as the crop was planted, with prunings being added to the soil. The safety-net zone (i.e. the zone below the crop root zone, which is explored by tree roots) was, in the simulation, defined as the 60–100 cm soil layer, in order to reflect restricted root subsoil exploration in this soil due to an aluminium-rich plinthic subsoil.

Simulations made using the WaNuLCAS model suggested that only around 20% of leaching N would be intercepted by tree roots in the 60–100 cm soil layer at a density of 0.5 cm/cm$^3$ (Fig. 10.4). Measured tree root length densities at this depth were in the range 0.005–0.015 cm/cm$^3$, sufficient to intercept less than 5% of leaching N according to this simulation. The safety-net zone beneath crop roots may, however, be considerably thicker than this, especially at early stages of crop growth or in soils that allow deeper rooting. WaNuLCAS sensitivity analyses were used to examine the effects of root length density, safety-net layer thickness, rainfall and proportion of N derived from N$_2$-fixation on safety-net efficiency (*SNE*). This was defined as follows:

$$SNE = 100 \times \frac{TN_{upt}}{Leach_{out} + TN_{upt}} \quad (10.21)$$

where *SNE* is safety-net efficiency (%), $TN_{upt}$ is tree N uptake from the safety-net layer (mg/cm$^2$) and $Leach_{out}$ is N leached beneath safety-net layer (mg/cm$^2$).

Increasing root length density within the safety-net layer from 0 to 2 cm/cm$^3$ led to an approximately linear increase in the efficiency of safety-net interception. However, further increases, above 2 cm/cm$^3$, did not result in a larger proportion of leaching N being intercepted. This probably represents the situation in a field, where even a dense mat of roots would not intercept N leaching during periods of low tree demand, or during heavy rainfall events when residence time is short. Smaller amounts of rainfall achieved greater interception efficiencies, and a plateau level of efficiency was reached at a smaller $L_{rv}$ than was the case in simulations with more rain. Simulations demonstrated the importance of safety-net layer thickness, and hence the maximum rooting depth of trees. With a safety-net layer 4 m thick, the interception efficiency achieved with a root length density of 0.02 cm/cm$^3$ (15%) approached that

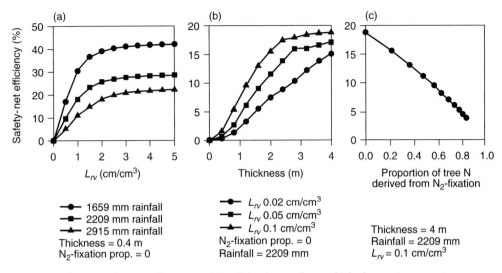

**Fig. 10.4.** Effects on safety-net efficiency within a *Peltophorum dasyrrachis* hedgerow intercropping system (Box 10.2), as simulated by WaNuLCAS (Box 10.1), of (a) tree root length density ($L_{rv}$) in the safety-net layer and rainfall, (b) thickness of this layer and $L_{rv}$, and (c) proportion of N derived from $N_2$-fixation. The safety-net layer was assumed to begin at 60 cm depth. Adapted from Rowe (1999).

achieved with a root length density of 0.1 cm/cm³ (18%). The maximum depth of tree roots will be limited if soil conditions become more adverse with depth, but tree rooting depths of >20 m have been reported for tropical forests (Chapters 4 and 6, this volume), and safety-net layers of this thickness are certainly possible. However, even a thick layer with a high root length density does not intercept all of the leaching N. Saturation of the plant safety net occurs because of asynchrony between leaching and tree N demand. A reduction in tree N demand reduces safety-net efficiency; for example, satisfying a proportion of demand by $N_2$-fixation results in an approximately linear reduction in safety-net efficiency. Using $^{15}N$ injections at depth and the $^{15}N$ natural abundance technique to estimate $N_2$-fixation, Rowe *et al.* (1999) confirmed that the non-fixing legume *Peltophorum dasyrrachis* took up more N (42 kg N/ha) from beneath the main crop rooting zone than the $N_2$-fixing legume *Gliricidia sepium* (21 kg N/ha) (see also Chapter 6, this volume). In summary, safety-net efficiency is limited by the residence time of nutrients in the safety-net layer (e.g. rainfall intensity, layer thickness, retardation factors and water consumption by the trees), the tree root length density within this layer, and the N demand of the tree. Given that the latter factor has a large impact on safety-net efficiency, ensuring that plant demand is not restricted by deficiencies of other nutrients (e.g. phosphorus) or by environmental conditions is a matter of paramount importance (Giller *et al.*, 2002).

## 10.4 Competing for Nutrients in Soil Solution

### 10.4.1 What strategies do plants have for dealing with competition?

The evaluation of the balance of positive and negative interactions in mixed-species systems remains a major challenge. Van Noordwijk *et al.* (1998a) showed both that negative interactions dominate in many simultaneous hedgerow intercropping systems, and that fast-growing $N_2$-fixing species may not necessarily be the best trees to use in hedgerow intercropping systems. Competition for nutrients (and light) is a dominant factor in evaluating the efficiency of systems for resource utilization. Zhang *et al.*

(2001) estimated that a third of the positive effects of intercropping in wheat–maize and wheat–soybean could be attributed to better soil exploration. Root distribution of the component species has often been interpreted as a direct indication of competition and complementarity in mixed cropping systems, including agroforestry (van Noordwijk et al., 1996; Huxley, 1999; Wahid, 2001). Cadisch et al. (2001) have shown that the fallow species *Sesbania sesban* and *Crotalaria grahamiana* react directly to competition by altering their rooting pattern (root plasticity), in order to improve their exploitation of the available nutrient resources. However, such dynamic responses on the part of plants are not considered in most simulation models. Further examples of plant strategies for responding to competition are discussed in Chapters 4, 5, 6 and 17 of this volume.

### 10.4.2 How can we model plant competition for nutrients?

Although low root length densities may be sufficient for the uptake of mobile nutrients such as nitrate in a situation without competition (or one with intraspecific competition in a monoculture of identical plants), the relative competitive strength of plants in a situation with shared access to a soil layer is thought to be proportional to both their respective root length densities *and* plant nutrient demand. Nutrient acquisition is affected by the presence of competition, that is: (i) nutrient depletion by the other plants reduces nutrient availability; (ii) potential nutrient uptake per unit root length is reduced by the presence of other roots since the size of the soil cylinder depleted per average root decreases with increasing total root length density; and (iii) reduced soil water content reduces nutrient mobility. The severity of these effects can be off-set by plants that increase the availability of nutrients in their rhizosphere and/or increase soil water content in nutrient-rich soil layers through the process of 'hydraulic lift' (see Chapter 9, this volume). Plants can indirectly increase their competitiveness by forming associations with mycorrhizas, which act as extended roots and hence increase apparent root length density, although this involves some carbon costs (Chapter 14, this volume).

In WaNuLCAS (Box 10.1), competition for water and nutrients is based on sharing the potential uptake rate for both – based on the combined root length densities ($L_{rv\,tot}$) – on the basis of relative root length multiplied by relative demand ($A$): (see bottom of page) where *PotUpt* gives the potential uptake rate (mg/cm$^2$/day) for a given root length density ($L_{rv}$). This description ensures that uptake by species $k$ is: (i) proportional to its relative root length density if demand for all components is equal; (ii) never more than the potential uptake by a species in a monoculture with the same $L_{rv}$; and (iii) not reduced if companion plants with a high root length density have zero demand (e.g. a tree just after pruning). At this stage, this procedure is applied to four species (e.g. three trees plus a crop or weed in each zone), but the routine can be expanded to include a larger number of interacting plants.

The outcome of modelled competition depends not only on the description of $L_{rv}$ and demand ($A$) but also on the sequence in which they are applied in different soil layers or nutrient patches. For example, in the multispecies agroforestry model HyPAR 4 (Mobbs et al., 2001), partitioning of resources starts at the soil surface layer (Box 10.3). Any unfulfilled demand in the upper soil layer is added to the demand from the next soil layer. This enables deep roots to extract more water and nutrients if necessary. The extracted nutrient is then partitioned to trees and crop in proportion to their demand. In future versions of HyPAR, the uptake routines will be modified in such a way that

$$PotUpt(k) = \min\left[\frac{L_{rv}(k) \times A(k) \times PotUpt(L_{rv\,tot})}{\sum_{k=1}^{n}(L_{rv}(k) \times A(k))}, PotUpt(L_{rv}(k))\right] \qquad (10.22)$$

> **Box 10.3.** Below-ground competition in HYPAR v4.
>
> HYPAR (Mobbs *et al.*, 2001) models the growth of trees and crops in agroforestry systems. The tree growth components within HYPAR are based on those in the HYBRID model (v3.0) whereas the crop components are based on the tropical crop model PARCH. The version of the model described here (HYPAR v4) runs continuously from year to year, allowing several annual crop seasons to be studied, with one or two crops per year.
>
> **Below-ground interaction**
> Every simulation day, the tree and crop components of the model independently calculate the optimum uptake (or demand) of water and N by the plants. The actual uptake is the minimum of the demand, the available resource and a maximum uptake rate in each soil cell. The maximum uptake rate for water (mm (water)/mm soil/day) is
>
> $$X_{i,pot} = \left(\frac{aw_i}{aw_{fc}}\right)^2 \sqrt{\frac{\rho_i}{\rho_{max}}} U_{max} \quad (B10.2)$$
>
> where $aw_i$ is the available water in layer $i$, $aw_{fc}$ is the available water when the soil is at field capacity, $\rho_i$ is the total (tree and crop) root length density in the layer $i$, $\rho_{max}$ is the maximum root length density and $U_{max}$ is the maximum uptake rate of water per unit depth of soil. The maximum uptake rate for N by the trees and crop roots together is
>
> $$U_{i,N} = 0.07\big((1-\exp(-0.09[N_i]))\sqrt{\frac{\rho_i}{0.25\rho_{max}}} \frac{aw_i}{0.5\sqrt{aw_{fc}^2}} \quad (B10.3)$$
>
> where $[N_i]$ is the nitrate concentration in the layer $i$.
> If there is sufficient of a resource to meet the demand of all the trees and the crop, then the full amount is removed from the soil and there is no direct competition. If the sum of the demands in any soil cell is greater than that available, or if the combined extraction exceeds a maximum rate, then competition for resources takes place in that cell.
>
> **Competition**
> **Water:** Starting at the soil surface layer (in each plot independently), the available uptake is compared with the total demand. If the supply is limited, then all of the available water is removed from the soil and any unfulfilled demand is added to the demand for the next soil layer. This enables deep roots to extract more water if necessary. The extracted water is partitioned to trees and crop in proportion to their demand. When competition for water occurs, the crop suffers stress. The tree responds the following day through the effect of a change in soil water potential on stomatal conductance.
> **Nitrogen:** Starting at the soil surface layer (in each plot independently), the available N uptake is compared with the demand. If there is sufficient, then the full amount is removed from the soil and added to the tree and crop internal storage pools prior to reallocation. If the supply is limited, then all of the available N is removed from the soil and apportioned to each tree and the crop in proportion to their demand. Unfulfilled demand in any one layer is passed down to the layer below, allowing extraction from depth where possible.

extraction of water will take place preferentially in wetter layers, rather than from the soil surface downwards (G. Lawson, Edinburgh, 2002, personal communication).

In a way similar to WaNuLCAS, HYPAR (Mobbs *et al.*, 2001) represents competition for water and nitrogen in a grid of soil columns with layers of variable depths (Box 10.3). We used HYPAR to predict the reduction in crop yield through time and space as trees grow and rooting density, and hence competition, increases (Fig. 10.5).

HYPAR also portrays the course of nitrate uptake by the tree and crop components at different depths through the growing season (Fig. 10.6). Close to the tree (Fig. 10.6a), crop growth is very poor (0.07 t/ha) and has little effect on tree N-uptake in an established system. Further from the tree (Fig. 10.6b), the moderately successful crop (2.8 t/ha) has roots that penetrate to layer 8 and which completely out-compete tree roots for nitrate in the latter half of the growing season. Beyond the tree rooting

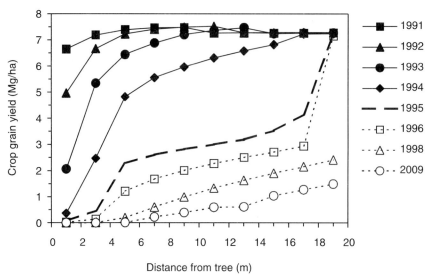

**Fig. 10.5.** Simulated yield of sorghum (using HyPAR, Box 10.3) for a site in Ghana with annual rainfall assumed to repeat an identical seasonal pattern for 20 years (1502 mm), in a sandy soil (80% sand, 3 m depth), with soil water and nitrogen conditions stabilized for a 10-year 'spin-up' period of bare soil prior to 'planting' a generic *Eucalyptus* tree of 4.8 cm dbh and 8 m height. At the end of 20 years this tree had grown to 23 m height and 45 cm diameter. Radius of the rooting zone had grown from 8 m to 16.5 m. Initial nitrogen is assumed to be equally distributed through the top 40 cm depth and to decline exponentially below this.

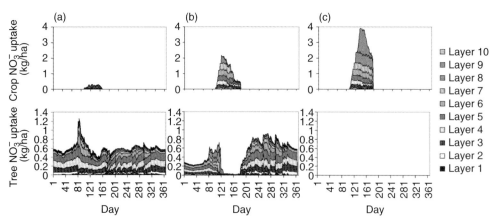

**Fig. 10.6.** Simulated uptake of nitrate (using HyPAR, Box 10.3) by the tree and crop during 1995, adjacent to the tree (a), 9 m from the tree (b) and 19 m from the tree (c). Layer 1 is the surface layer, and deeper layers are thicker (2, 4, 10, 14, 16, 19, 21, 34, 80 and 100 cm respectively – a total of 3 m). The spatial pattern of crop yield for this year is shown in Fig. 10.5.

zone (Fig. 10.6c), a much higher crop yield is predicted (7.2 t/ha), with nitrate supplies down to layer 9 being accessed by the crop roots in the latter half of the growing season. The nature of this competition for nitrogen is greatly affected by the parameters assumed for the exponential decline in tree and crop root density with depth and with distance from the tree. The importance of making field measurements to obtain realistic crop and root density profiles is clearly illustrated.

Using WaNuLCAS, a sensitivity analysis for a tropical hedgerow intercropping system (Box 10.2) showed that total root length per unit soil area and relative distribution of tree root length density with depth may have complex effects on crop performance (Fig. 10.7). In situations where trees have the majority of their roots in the topsoil, negative effects via competition for water and nutrients reduce crop performance (van Noordwijk and Cadisch, 2002). The magnitude of these competitive effects depends on the total tree root length density. A higher total root length (i.e. higher absolute root lengths in both topsoil and subsoil) can have a moderate, positive effect on maize yield even with a large proportion of roots in the topsoil. Tree root systems with 20% or more of their roots in the subsoil had consistently positive effects on the crop: the higher the total root length, the more positive the impact on maize (Fig. 10.7). The resulting safety-net action of the trees increased the amount of N being recycled in the system and hence improved maize growth in systems where the N supply was limiting. The positive effects the trees had, via improved N recycling, would be expected to increase with time (current results show effects averaged over the first 2 years).

The model's output also suggested that there is a delicate balance between the positive and negative effects of trees. For example, at high total root length, the tree root systems with 60% of their roots below the topsoil led to slightly higher maize yields than those trees with more roots (up to 100% of their roots) in the subsoil. Although this effect is too small to be recognized in the field, the differences in N uptake in crops associated with trees that have 0% of their roots in the topsoil and crops associated with trees that have 40% of their roots in the topsoil arise during dry spells in the cropping season. During these dry spells, the sparse crop roots in the deeper soil layers have slightly more N available in situations where the trees forage partly in the topsoil. Sensitivity analysis of the model thus shows that tree root

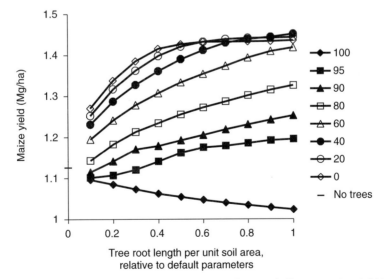

**Fig. 10.7.** Competition vs. complementarity: WaNuLCAS simulations of effects on maize yield in a hedgerow intercropping system (Box 10.2), when relative distribution of tree roots with depth, as well as total amount of tree roots, are varied independently. Whereas the 'default' tree root system had 21.5% of its roots in the top layer, a series of data was made that had 0–100% of its roots in the top layer and the remainder allocated to the deeper layers in proportion to the root length densities of the default case. For each of these root distributions, the total amount of roots was varied from 0.1 to 1 times the default, while maintaining the relative values. Adapted from van Noordwijk and Cadisch (2002).

length density below the main crop root zone may have complex and partially unexpected effects on crop performance in situations where negative effects via competition for water and positive effects via improved N supply vary in intensity during the growing season.

The impacts of competition on competing plants differ between the initial phase of near-exponential growth (where a setback in current growth affects future growth as well) and the crop's closed-canopy stage of linear growth (when only current growth rates are affected). This means that negative early effects of competition cannot normally be compensated for by positive effects later on in the growth period. Interactions of positive and negative effects, acting at different spatial and temporal scales, require the use of simulation models that can keep track of such cumulative effects. As mentioned above, plants may respond to competition by exhibiting a differential response to the exploitation of underutilized resource patches in the field (e.g. root proliferation within nutrient-rich zones). However, most models do not include such effects. Additionally, leguminous plants will satisfy an unsatisfied N demand by switching their source of N – from soil N resources to atmospheric N, via their ability to fix $N_2$ in association with rhizobia (see Chapter 13, this volume).

### 10.4.3 Can competition be managed by managing demand?

Managing the distribution of tree nutrient uptake by designing systems where tree and crop roots are in separate regions of the soil is not the only way to reduce tree-crop competition. Tree nutrient uptake can be manipulated by pruning (see also Chapter 17, this volume). Removing shoot material instantly reduces transpiration demand and thus the rate of nutrient uptake through mass flow. Active nutrient uptake is also likely to be reduced since available root and stem carbon can be used for constructing new shoots. There may thus be potential for the use of pruning to reduce nutrient uptake by intercrop trees during critical periods for crop growth, such as at establishment or pollination. On the other hand, shoot pruning reduces water consumption by the trees and hence may increase leaching; so, the net benefit for the crop is smaller than that which would be expected from reduced nutrient uptake by the tree.

The effect of pruning on soil nutrient demand is exacerbated due to remobilization of internal resources. This is illustrated by the delay, relative to stem regrowth, in uptake of soil N following a pruning event (Rowe, 1999; Fig. 10.8). In this study, hedgerow intercropping trees (Box 10.2) showed no significant soil N uptake during the period 5–14 days after pruning, even though some shoot regrowth occurred during this time. Shoot activity and root activity may not always coincide: tree roots may contain substantial reserves of nutrients and carbohydrates, allowing shoot regrowth to take place with little nutrient uptake from the soil. Conversely, tree roots may take up nutrients during periods of shoot inactivity, an effect noted in temperate ecosystems where fruit tree roots took up N during the late autumn and winter when shoots were inactive (Tromp, 1983). The degree to which plants can uncouple root and shoot phenology depends on their carbohydrate and nutrient storage capacity. In WaNuLCAS, such delays in nutrient uptake can be simulated by changing the tree pruning time relative to the crop sowing date (Fig. 10.9). The optimal time for pruning to minimize nitrogen leaching whilst maximizing crop yield was, thereby, predicted to be at, or just before, the time of sowing.

Reducing tree competition depends on timing management operations to prevent strong competition occurring before the crop has become established. Having strongly growing trees adjacent to crops inevitably leads to competition, and management should aim at keeping regrowth small during the cropping period. The beneficial effects of trees will be maximized by allowing strong and rapid tree growth to occur during other stages of the crop cycle (i.e. in

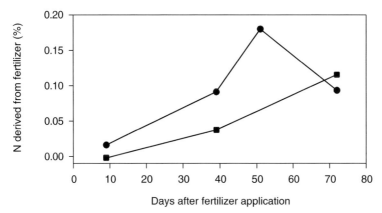

**Fig. 10.8.** Nitrogen uptake by *Gliricidia sepium* trees pruned either 4 (■) or 25 (●) days before fertilizer application in a hedgerow intercropping system in Lampung (Box 10.2), measured using $^{15}$N. Trees pruned 25 days before application (only) were given an additional pruning 51 days after application. Adapted from Rowe (1999).

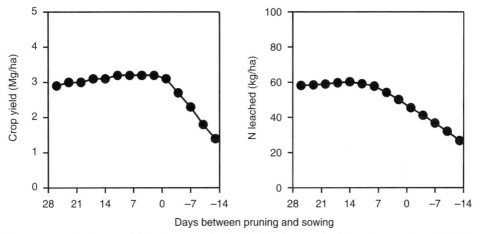

**Fig. 10.9.** Simulated maize yield and nitrogen leached during the growth of the maize crop in a *Gliricidia sepium* hedgerow intercropping system in Lampung (Box 10.2), as affected by the interval between pruning and sowing. Day 0 is pruning day and negative numbers are days after pruning. From Rowe (1999).

the later stages of crop growth and during any fallow periods). However, tree pruning management is labour intensive, and is particularly unwelcome when it coincides with labour demands for other activities. Labour demands make hedgerow intercropping barely economically viable (Whitmore et al., 2000) and are also one reason why farmers in central American coffee plantations exchange traditional legume shade trees for timber trees with less pruning requirements (Tavares et al., 1999).

### 10.4.4 Environmental modifiers

In a series of WaNuLCAS simulations that excluded P limitations on crop or tree growth, a gradual shift from water- to N-limited growth conditions was obtained (Fig. 10.10) by applying multipliers to the daily rainfall records for Lampung for a hedgerow intercropping system involving maize and groundnut (grown in rotation) and *Peltophorum dasyrrachis* or *Gliricidia sepium* hedgerows to which N fertilizer (90

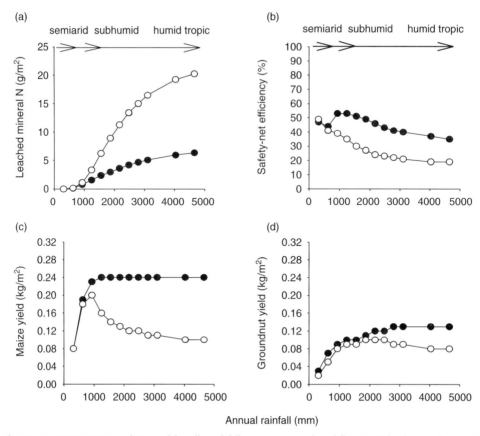

Fig. 10.10. WaNuLCAS simulations of the effect of different amounts of rainfall in (●) *Peltophorum dasyrrachis* and (o) *Gliricidia sepium* hedgerow intercropping systems on (a) amount of mineral-N leached (g/m$^2$), (b) safety-net efficiency (%), (c) maize yield (kg/m$^2$) and (d) groundnut yield (kg/m$^2$). From Suprayogo (2000).

kg/ha) was applied (Box 10.2). The simulated rainfall represented semiarid (< 1000 mm annual rainfall), subhumid (1000–2000 mm) and humid (> 2000 mm) areas. No leaching of mineral-N was observed when rainfall was below 620 mm, as there was no net water drainage (Fig. 10.10a). Leaching of mineral-N was greatly increased when rainfall was above 900 mm, particularly in the system that contained the N$_2$-fixing tree *Gliricidia*. This was due to additional N inputs arising from biological N$_2$-fixation, as well as to the faster rate of mineralization that occurred with the high-quality *Gliricidia* prunings, as compared with the polyphenol-rich *Peltophorum* prunings (Handayanto et al., 1994, 1997). Thus N$_2$-fixing systems appear to be more 'leaky' than low-input systems without legumes. The N safety-net efficiency decreased with increasing rainfall above 900 mm (Fig. 10.10b). The maximum maize yield in the *Peltophorum* system was obtained above 1200 mm of rainfall, but in the *Gliricidia* system, the maximum yield was obtained at 930 mm of rainfall (Fig. 10.10c). The observed decrease in maize yield in the *Gliricidia* hedgerow system above 1000 mm of rainfall was due to the fast growth of *Gliricidia* trees and hence to a reduction in the amount of light available to the maize. However, this was not the case in the *Peltophorum* system due to the different canopy shape of *Peltophorum* trees and lower tree biomass production at the time of maize growth.

The maximum yield of groundnut, grown as a second sequential crop, was obtained above 2000 mm of rainfall in both *Peltophorum* and *Gliricidia* systems, mainly as a result of the fact that growth was restricted by water deficiency towards the end of the rainy season at lower rainfall levels (Fig. 10.10d). Thus, in the presence of hedgerow trees, water can be the growth-limiting factor on 5–10% of the days in the cropping season, even with an annual rainfall of 2000 mm. Cannell *et al.* (1998) suggested that below 800 mm of rainfall the trees in an agroforestry system have a low productivity, which is not sufficient to compensate for the loss in the yield of associated grain crops that results from competition for light and water. Their results also suggested that between 800 and 1000 mm of rainfall, total productivity of trees may compensate for the loss in crop grain yield, with only a small increase in total site productivity. Moreover, their results confirmed that above 1000 mm of rainfall, total productivity can be increased by the presence of trees. The simulated results do not take into account the effects of trees that result in improved microclimate conditions for crops, or the case of trees with inverse phenology (such as *Faidherbia albida*), which may alter the range of rainfall conditions in which simultaneous agroforestry systems can be considered useful.

### 10.4.5 Risk management

Because of the complexity of interactions in mixed-species systems, many farmers have resorted to monocrop systems. Moreover, mixed-species systems may not yield more than the best monocrop species unless cropping density can be increased with the introduction of a new species (Gathumbi *et al.*, 2002b); that is, the additional species must acquire resources that the crop would not otherwise acquire (Cannell *et al.*, 1996). Apart from improved resource utilization, risk management is another important factor to consider in designing new systems (Yachi and Loreau, 1999; Loreau, 2000).

Diseases, pests and adverse weather conditions (e.g. drought or flooding) commonly affect plant production in tropical regions. In fact, management of risk or stress(es) may in many cases be of greater importance than resource utilization. Resource capture may thus become limited by risk factors via a reduced resource demand. Equation 10.23 describes the effect of reduced demand on the potential nutrient uptake of each species. Thus, the impact of stress events on resource capture of the system will depend on which species is most affected and what its proportional demand for resources is. On the other hand, there may be some compensatory growth ($\varepsilon$) on the part of the more stress-tolerant species due to underutilization of resources and space e.g.

$$PotUpt = \varepsilon \times \sum_{k=1}^{n} PotUpt(k)stress \qquad (10.23)$$

where:

$$\varepsilon = (1-s) \times \sigma \qquad (10.24)$$

where $s$ is an integrated stress factor (0–1) denoting the reduced nutrient demand and $\sigma$ is a factor (0–1) depicting the compensatory growth potential of the more tolerant species. Thus, mixed-species systems are most likely to have an advantage over highly productive monocrops under conditions of increased environmental stress and when the species mixture constitutes a better buffered system (Fig. 10.11). Van Noordwijk and Ong (1999) used a slightly different approach to assess risk reduction in diverse systems based on average expected yield (mean) and its variance due to environmental stress. They suggested that risk reduction through increased diversity is least effective against 'disasters' and most effective (leading to the lowest system variance) when a small number of strong competitors are present, which vary primarily in relation to tolerance to specific biotic stress factors.

## 10.5 Conclusions

The process of catching mobile nutrients in soils cannot be described simply by mapping root distributions in soils, as it depends to a

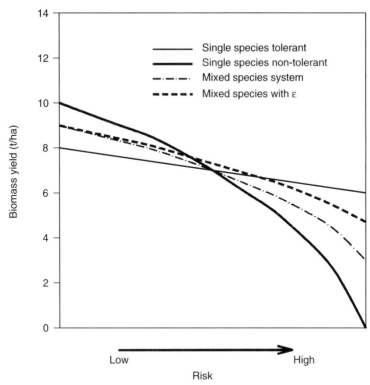

**Fig. 10.11.** Effect of environmental factors (weather, pests) on advantages of a mixed-species stand vs. monocrops of different susceptibility to stress, without and with compensatory growth ($\varepsilon$) capabilities.

large extent on plant nutrient and water demand and residence time of the nutrient in the rooting zone. Soils have inherent safety nets, in the form of cation and anion retardation mechanisms. However, the impact of retardation factors for nitrate appears to be small in relation to the overall effects of the cropping system on water dynamics and soil structure. More diverse systems, with intense root activity as well as increased soil cover and recyclable organic material, favour the formation of macropores, which have the potential to reduce leaching of nutrients due to a fast bypass flow and limited exchange with nutrients in micropores. However, in high-intensity rainfall areas, macropore flow results in increased drainage, which may or may not result in greater leaching depending on the (mostly unknown) distribution of nutrients in macropores/micropores. The consequent reduction in runoff may decrease losses of soil, organic matter and nutrients via erosion, and thus there should be an overall net benefit to the system.

With increased environmental concerns, more emphasis is given nowadays to the nutrient use efficiency of systems and, in particular, to their safety-net or filter function. Safety nets are strongly dependent on the presence of sufficient $L_{rv}$ at lower depths and hence species selection is crucial. Given that demand has a large impact on safety-net efficiency, ensuring that plant demand is not restricted by other nutrient (e.g. phosphorus) deficiencies or by ill-adapted species, is of paramount importance. Recycling of the nutrient resources captured in the safety net has the potential to increase crop productivity in nutrient-limited systems, but we have to ask why a species should preferentially explore less fertile lower soil depths when nutrient-rich patches exist mainly in the topsoil. This points to the need to search for less-competitive, deep-rooted species for use in simultaneous mixed species systems, rather than fast-growing $N_2$-fixing species (which are commonly perceived to be the best for such systems). However, root distribution is governed by genotype × environment interactions, and competition, with the resulting root plasticity, may be necessary for creating the complementarity in soil exploration that is necessary for efficient resource use, as noted by Schroth (1998). Simulations and existing data also suggest that $N_2$-fixing systems appear to be more 'leaky' than non-fixing ones. Future challenges include the evaluation of how the safety-net functions of systems at the landscape scale can help to reduce nutrient losses (pollution) into stream waters, whilst benefiting those systems at a lower position in the watershed.

Mixed-species systems remain a challenge, due to their complex interactions. Thus benefits must be sought beyond resource capture advantages (e.g. risk management, etc.) for them to become viable alternatives to farmers. The development of mechanistic agroforestry models has greatly advanced our understanding and prioritization of complex interactions in multispecies systems. However, for site-specific recommendations to be developed, a major effort towards the calibration of such models is needed, in order to initialize soil conditions and plant performance, because these are not determined only by the parameters described in the model.

---

**Conclusions**

1. There is no universal critical root length density ($L_{rv}$) for efficient nutrient interception, since this is determined mainly by the mobility of the nutrient in question, and by plant demand and net drainage.
2. Systems that promote the formation of macropores may reduce leaching of mineral N, but the associated reduction in erosion appears to be equally important.
3. Soil anion retardation factors have a smaller impact on nutrient retention than does the type of cropping system.
4. Low-input legume-based systems appear to be more N 'leaky' than similar systems without $N_2$-fixing legumes.
5. Potential options for managing nutrient interception by perennial components include pruning (timing and severity), species selection (slow/fast growth), and arrangement in relation to lateral flows of nutrients.
6. The balance between competition and safety-net efficiency in mixed-species systems appears to be better achieved by slow-growing, deep-rooted and well-adapted species than by fast-growing $N_2$-fixing legumes.

---

**Future research needs**

1. Investigation of the partitioning of nutrients and roots between matrix soil and macropores.
2. Studies on the effect of management options in mixed-species systems on temporal changes in demand.
3. Studies to determine whether changes in $L_{rv}$ distribution due to competitive plasticity will have a major effect on plant performance or filter/safety-net efficiency.

---

## Acknowledgements

This publication is an output from projects funded by the UK Department for International Development (DFID, NRSP and FRP) for the benefit of developing countries. The views expressed are not necessarily those of DFID.

# 11 Below-ground Inputs: Relationships with Soil Quality, Soil C Storage and Soil Structure

## Alain Albrecht, Georg Cadisch, Eric Blanchart, S.M. Sitompul and Bernard Vanlauwe

---

**Key questions:**
1. What is the magnitude of below-ground inputs?
2. Where is carbon stored?
3. What are the impacts of increased soil C on soil functions and biodiversity?
4. Is there a relationship between soil C and plant productivity?

---

## 11.1 Introduction

The impact that different land use systems have on the functioning of agroecosystems depends largely on the amount and quality of inputs provided by the system, on the properties of the soils themselves, and on climate and management. Organic resources play a critical role in the maintenance of soil organic matter and nutrient cycling in most smallholder farming systems in the tropics. Past research has mainly focused on the influence of above-ground inputs on soil properties and functions, presumably because these are much easier to measure. However, it is increasingly recognized that the importance of below-ground plant inputs has been underestimated (McNeill et al., 1997; Cadisch et al., 2002b). Their importance has been highlighted further with the increasing interest in soil carbon (C) sequestration and issues related to global climate change. In this chapter, we will investigate the relationships that exist between below-ground inputs and soil chemical and physical properties and soil C storage. We will then assess the link between improved soil C content and soil erodibility, soil charges, soil mineral nutrient supply and, ultimately, plant productivity.

## 11.2 Magnitude of Below- Versus Above-ground Plant Inputs

Globally, about 120 Gt C/year are fixed by plants during photosynthesis, of which about half is subsequently lost through plant respiration. Furthermore, a large proportion of the remaining plant-fixed C is lost during the decomposition of dead organic matter through heterotrophic respiration.

The amount of annual-plant residues or litter recycled every year generally increases with decreasing latitude (Jenkinson, 1981). The amount of biomass available for recycling in a given climate depends primarily on the duration of plant growth; this varies between short- and long-duration crops, and with the age of fallow systems (Table 11.1). Plant species, adaptation, $N_2$-fixing ability (see Chapter 13, this volume) and soil fertility are factors that further modify the speed of biomass accumulation, particularly in the short term. Whereas well-fertilized crops often accumulate biomass faster in the short term (e.g. 10–20 t/ha in 4–6 months, of which around half is often removed in the harvest), woody fallow systems often produce more recyclable biomass in the medium to long term (> 40 t/ha in 2 years) with no or lower fertilizer inputs (see also Chapter 4, this volume).

Roots are an important part of the C balance, because they transfer large amounts of C into the soil (Table 11.1). Depending on rooting depth, a considerable amount of C is stored below the plough layer and is, therefore, better protected from disturbance, which leads to longer residence times in the soil. With some trees having rooting depths of more than 60 m, root C inputs can be substantial, although the amount declines sharply with soil depth (Chapter 4, this volume). Although most of the biomass of the roots of annual crops consists of fine roots (< 2 mm in diameter) a large proportion of the below-ground biomass of tree roots consists of coarse roots (> 2 mm diameter; see Table 11.1 and Chapter 4, this volume). Fine roots of both trees and crops have a relatively fast turnover (days to weeks; van Noordwijk et al., 1998c), but the lignified coarse roots decompose much more slowly and may thus contribute substantially to below-ground C stocks (Vanlauwe et al., 1996).

Roots are characterized by lower tissue N and polyphenol concentrations and a higher lignin content than the recyclable above-ground materials (Table 11.2). Root N concentration often falls below the critical content of 1.8–2.5%, i.e. the threshold that determines whether there will be net N mineralization or immobilization (Palm et al., 2001). As such, roots appear to contribute more to the maintenance of soil organic

**Table 11.1.** Above- and below-ground biomass production of species used in improved fallows and pastures.

| | Duration (months) | Biomass (t/ha) | | | Root : shoot ratio | Reference |
|---|---|---|---|---|---|---|
| | | Above ground | Below ground | Fine roots | | |
| Crotalaria grahamiana | 8 | 6.1 | 4.8 | 4.0 | 0.79 | Boye (2000) |
| Crotalaria grahamiana | 12 | 8.5 | 2.7 | n/a | 0.32 | Ndufa (2001) |
| Calliandra calothyrsus | 12 | 21.0 | 7.0 | n/a | 0.33 | Ndufa (2001) |
| Cajanus cajan | 12 | 8.5 | 3.9 | n/a | 0.46 | Ndufa (2001) |
| Senna spectabilis | 12 | 7.0 | 4.8 | n/a | 0.69 | Ndufa (2001) |
| Sesbania sesban | 12 | 14.2 | 7.3 | n/a | 0.51 | Ndufa (2001) |
| Tephrosia vogelii | 12 | 10.8 | 4.0 | n/a | 0.37 | Ndufa (2001) |
| Crotalaria grahamiana | 18 | 24.7 | 10.9 | 6.4 | 0.44 | Ndufa (2001) |
| Crotalaria paulina | 18 | 19.8 | 13.6 | 3.7 | 0.69 | A. Albrecht 2001, unpublished |
| Tephrosia candida | 18 | 31.0 | 33.2 | 3.6 | 1.07 | A. Albrecht 2001, unpublished |
| Calliandra calothyrsus | 22 | 27.0 | 15.5 | 2.8 | 0.57 | Nyberg (2001) |
| Sesbania sesban | 22 | 36.9 | 10.8 | 2.4 | 0.29 | Nyberg (2001) |
| Grevillea robusta | 22 | 32.6 | 17.7 | 2.8 | 0.54 | Nyberg (2001) |
| Eucalyptus saligna | 22 | 43.4 | 19.1 | 2.4 | 0.44 | Nyberg (2001) |
| Brachiaria decumbens | 12 | 5.6 | 2.7 | n/a | 0.48 | Cadisch et al. (1994a) |

n/a, data not available.

**Table 11.2.** Shoot and root qualities of various plant species used in improved fallows and as cover crops in western Kenya (A. Albrecht, 2001, unpublished data).

| Species | Plant part | N (%) | Lignin (%) | TEP[a] (%) | N/lignin ratio | N/TEP ratio |
|---|---|---|---|---|---|---|
| Calliandra calothyrsus | Shoot | 2.56 | 12.33 | 9.94 | 0.21 | 0.26 |
|  | Root | 1.37 | 18.39 | 10.43 | 0.07 | 0.13 |
| Gliricidia sepium | Shoot | 3.44 | 8.60 | 2.07 | 0.40 | 1.66 |
|  | Root | 2.36 | 22.22 | 0.13 | 0.11 | 18.84 |
| Leucaena leucocephala | Shoot | 3.91 | 9.94 | 6.23 | 0.39 | 0.63 |
|  | Root | 1.88 | 22.28 | 3.11 | 0.08 | 0.60 |
| Senna siamea | Shoot | 2.31 | 10.30 | 4.60 | 0.22 | 0.50 |
|  | Root | 1.65 | 30.60 | 0.71 | 0.05 | 2.32 |
| Tephrosia vogelii | Shoot | 2.47 | 9.28 | 4.92 | 0.27 | 0.50 |
|  | Root < 2[b] | 2.35 | 15.05 | 0.56 | 0.16 | 4.20 |
|  | Root > 2[c] | 0.73 | 12.65 | 1.40 | 0.06 | 0.52 |
| Tephrosia candida | Shoot | 3.63 | 15.55 | 1.74 | 0.23 | 2.09 |
|  | Root < 2 | 1.90 | 23.48 | 1.25 | 0.08 | 1.52 |
|  | Root > 2 | 1.07 | 16.07 | 0.48 | 0.07 | 2.23 |
| Crotalaria grahamiana | Shoot | 3.14 | 6.05 | 3.00 | 0.52 | 1.05 |
|  | Root < 2 | 1.70 | 23.66 | 1.04 | 0.07 | 1.63 |
|  | Root > 2 | 0.87 | 15.49 | 0.83 | 0.06 | 1.05 |
| Crotalaria paulina | Shoot | 3.74 | 13.53 | 3.70 | 0.28 | 1.01 |
|  | Root < 2 | 1.70 | 21.08 | 1.07 | 0.08 | 1.59 |
|  | Root > 2 | 0.80 | 16.87 | 1.24 | 0.05 | 0.65 |
| Tithonia diversifolia | Shoot | 2.57 | 11.96 | 3.43 | 0.21 | 0.75 |
|  | Root | 1.20 | 13.65 | 2.39 | 0.09 | 0.50 |
| Desmodium uncinatum | Shoot | 3.44 | 9.43 | 3.12 | 0.36 | 1.10 |
|  | Root | 1.03 | 8.42 | 0.77 | 0.12 | 1.34 |
| Glycine max | Shoot | 3.67 | 7.24 | 1.01 | 0.51 | 3.65 |
|  | Root | 0.94 | 26.25 | 0.44 | 0.04 | 2.15 |

[a] Total extractable polyphenols, [b] roots < 2 mm in diameter, [c] roots > 2 mm in diameter.

matter and to improved soil structure than to plant nutrition (Risasi et al., 1998). Indeed, Urquiaga et al. (1998) observed initial immobilization of N on incubation of legume and grass root samples. Their double exponential equation of $CO_2$ evolution predicted that between 43% (in the case of *Stylosanthes guianensis*) and 62% (in the case of *Brachiaria decumbens*) of root C would never be decomposed under laboratory incubation conditions. For a more comprehensive overview of resource quality and its relationship with nutrient dynamics, see Palm et al. (2001) who developed an organic resource database that contains information on organic resource quality parameters (including macronutrient, lignin and polyphenol contents of fresh leaves, litter, stems and/or roots) of almost 300 species found in tropical agroecosystems.

## 11.3 C Storage Dynamics and Determinants

Simple models of the decomposition of organic matter C ($C$) use single exponential decay functions to assess changes in soil C stocks over time, e.g.:

$$\frac{\delta C}{\delta t} = fA - kC \tag{11.1}$$

where $A$ = the amount of C added (from residues), $f$ = the fraction of $A$ that decomposes to become soil-C each year, $k$ = the fraction of soil-C decomposed each year and $C$ = the organic soil C pool (Jenkinson, 1981). Thus, C-storage dynamics are strongly determined by the amount of C added and by its quality. Sitompul et al. (1996) estimated that in a humid tropical ecosystem 8 t C/ha needed to be added in residues every year to compensate for $CO_2$-C respiration losses from soil organic matter. The exponential model above indicates that there exists an equilibrium in soil organic C ($C_e$) under continuous soil management (Jenkinson, 1981):

$$\frac{\delta C}{\delta t} = 0 \quad --> C_e = \frac{fA}{k} \quad (11.2)$$

The turnover time ($t$) is defined as the time required for the mineralization of an amount of organic carbon equal to the amount in the soil at equilibrium or in the residue:

$$t = \frac{fA}{kfA} = \frac{1}{k} \quad (11.3)$$

Using this model (Equation 11.3) Jenkinson and Rayner (1977) evaluated a cropping system of continuous wheat for 100 years at Rothamsted, UK, with 1.2 t/ha/year C input, a humification factor ($f$) of 0.33 and an equilibrium topsoil-C content of 26 t/ha. This resulted in a $k$ value of 0.015 (this value represents the fraction of soil-C decomposed each year) and an average soil C turnover time of 66 years. A major assumption of such a simplified model (Equation 11.3) is that all the soil C and the residue-input C have the same availability with regard to microbial utilization. However, soil organic matter and organic inputs are heterogeneous. Thus, the rates of turnover of the different pools can vary greatly, e.g. the age of fulvic acid (as measured by radiocarbon dating) was 420 years whereas that of humin was 2400 years in a temperate soil (Jenkinson and Rayner, 1977). Similarly, large variations in C dynamics in different soil fractions have been recorded (Cadisch et al., 1996; Magid et al., 2002). Evidence from physical soil fractionation studies suggests that the half-life of such fractions in tropical soils is relatively rapid, varying between 8 years (for 50–2000 µm fractions) and 22 years (0–2 µm fractions) (Feller and Beare, 1997).

Young (1997) suggested that the conversion-loss fraction (when litter is converted into humus) of roots (0.67, i.e. humification factor $f$ = 0.33) is less than that of aboveground residues (0.85). Paustian et al. (1997) account for more recalcitrant materials (such as roots) by using the lignin/N ratio as a factor controlling decomposition, and also by partitioning C into more stable soil organic matter pools. These approaches, which take into account root quality, increase soil C accumulation and reinforce the importance of roots in the maintenance of soil organic matter. Urquiaga et al. (1998) observed a double exponential decay function for the decomposition of the roots of tropical grasses and legumes with apparently very long turnover times for the recalcitrant fraction. This may account for some of the C accumulation observed in tropical grassland systems (Fisher et al., 1995; Fearnside and Barbosa, 1998). Thus, more recent models use multi-compartment approaches for both the residue part and soil organic matter pools (Paustian et al., 1997); these can include one, or several, soil microbial biomass pools (Jenkinson et al., 1987).

Estimations of C stocks need to account for the amount of C contained in the whole soil profile. For example, Fisher et al. (1994) found that the majority of soil C was situated below the top 20 cm in an improved grassland on a deep Oxisol. According to the International Panel on Climate Change (IPCC, 2000), the total terrestrial C stock at any time ($t$) is equal to the product of the area of each land-use system ($A_i$) and the C stock value ($C_i$) associated with that land use, i.e.

$$C_t = \sum_{i=1}^{n} A_i, t C_i, t \quad (11.4)$$

Hence, both changes in area as well as in above- and below-ground C stocks have to be considered when assessing the impact of land-use changes. Globally, around 2000 Gt of C are stored in soils and detritus, while only about 500 Gt are located in the vegetation (Batjes, 1996). The soil's C content is the result of the history of organic inputs to the soil, and the rates of decomposition (Equation 11.2). Inputs and decomposition are determined by the inherent properties of the soil, and by management factors and site factors. Thus, soil C storage will be controlled by soil texture (Fig. 11.1), which will control the physical protection of C from decomposers. In addition, soil C is also controlled by the amount and pattern of rainfall

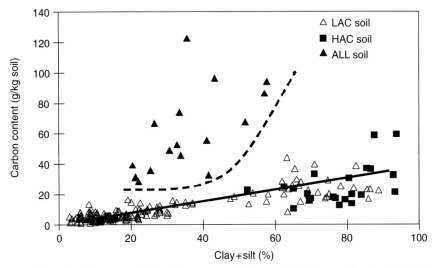

**Fig. 11.1.** Relationship between soil clay+silt content and clay type (LAC, low activity clay; HAC, high activity clay; ALL, allophane) with soil organic carbon content. Redrawn from Feller *et al.* (2001).

and by temperature; these factors control microbial activity and plant productivity.

In soils at, or close to, equilibrium (e.g. native grasslands and forests) there is often a close relationship between the amount of clay and silt and the amount of organic matter in soils (Fig. 11.1, and Hassink, 1997), for example:

$$C = a + b(\%clay + \%silt) \quad (11.5)$$

where *a* and *b* are constants. The greater soil C observed in heavier soils is attributed to the stabilization (e.g. chemical protection) of organic matter by clay. Clay–humate complexes are thought to form when humates are adsorbed to clay by polyvalent cations such as $Ca^{2+}$ and $Fe^{3+}$, and by association with hydrous oxides (ligand exchange or anion exchange via positive sites that exist on iron and aluminium oxides). Positive sites on sesquioxides will not exist in soil with pH>8, and so clay–humate complexes are less likely to form in highly alkaline soils. A special case appears to be allophane soils, which have specific properties and which showed a weaker relationship between soil C and clay + silt content (Fig. 11.1). Our available dataset did not show a clear separation between low (1:1) and high (2:1) activity clay types. However, a recent compilation by Six *et al.* (2002) suggests that low activity soils, which include a wide range of tropical soils, have a slightly lower C-protection capacity. The amount of C in the clay + silt fraction found in well-managed, undisturbed grasslands or forest is defined as the C storage capacity of soils.

C stocks under natural vegetation have often been used as a reference for the C storage potential of a soil in a given climate. With changes in land use and management, soil C often declines in relation to levels in the natural vegetation (Fig. 11.2). The difference between current and potential C storage can be expressed as the C saturation deficit (van Noordwijk *et al.*, 1997):

$$C_{satDeficit} = (C_{ref} - C_{org})/C_{ref} \quad (11.6)$$

where $C_{ref}$ is a reference soil C level representative of a forest soil of the same texture and pH, and $C_{org}$ is the current C stock. It is often difficult to find an adequate natural reference site and so van Noordwijk *et al.* (1997) developed an equation based on depth, texture, pH and soil type, based on a large dataset from Indonesia, to estimate $C_{ref}$. The saturation deficit not only depicts the potential amount of C that can be stored, but also influences the speed of C accumulation (i.e. the closer a soil is to its maximum

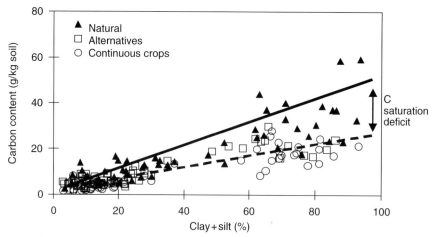

**Fig. 11.2.** Influence of land-use system and soil clay+silt content on soil organic carbon content. Redrawn from Feller et al. (2001).

potential C storage, the slower the C accumulation, as proportionally less C becomes protected). Trenbath (1989) developed a model for shifting cultivation in which the rebuilding of soil C depends on the C saturation deficit, that is:

$$\frac{dC}{dt} = \frac{(C_{ref} - C_t) \times C_t}{(C_{ref} - C_t + RC_{ref})} \quad (11.7)$$

where $R$ is the half-recovery time of the implemented system. This suggests that protection of soil C becomes less as the soil reaches its C storage capacity.

In systems where fire is frequently used, charcoal formation leads to a long-term increase in storage of inert C, even though only a small fraction (about 2%) of the plant biomass is normally converted into charcoal by such fires. Indeed, Cadisch et al. (1996) found that charcoal persisted in the light organic fraction of a grassland soil for 18 years after conversion from rainforest in Brazil. As charcoal is biologically inert, slash-and-burn systems may actually lead to increased soil C stocks where the vegetation is allowed to fully recover. Recent estimates in Australia suggest that, on average, between 20% and 30% of total topsoil C was charcoal in systems prone to fire (Skjemstad et al., 1999). Thus, further investigations in this direction are warranted, particularly given the link between soil C sequestration (e.g. into an inert pool) and global climate change.

Erosion is a major uncertainty in our current assessment of C dynamics. Whereas fine organic particles containing organic matter are preferentially eroded and hence transported from the field, they may not necessarily be lost from the global system. Redistributed C may be deposited in fields at lower elevations, or buried in freshwater and marine sediments (in which case it may become protected from decomposition processes). It is estimated that, globally, 0.2 Gt C are deposited in sediments every year and that the amount of C in fossils and carbonate mineral is > 6000 Gt C (IPCC, 2000). However, erosion losses definitely lead to a local depletion of soil fertility.

## 11.4 How can Soil Organic C Stocks be Increased by Below-ground Inputs and Activities?

Land management drives the availability of above- and/or below-ground inputs in relation to productivity and to outputs (export). In the past, most studies have focused on the impact that above-ground inputs have on soil organic matter dynamics. Here, we examine the impact that below-ground inputs have on soil organic matter and structure in particular.

The impact that roots and macrofauna activity have on soil organic C (SOC) was investigated by Chevallier et al. (2001) in a young pasture of *Digitaria decumbens* on a Vertisol in Martinique. Using combinations of herbicides and nematocides, as well as above-ground plant harvest (involving the removal of biomass), it was possible to separate the effect of roots and earthworms on soil properties (Fig. 11.3). After 4 years, the SOC had increased substantially in the topsoil (0–10 cm depth): SOC values were 14 mg C/g soil (in the 'no plants-no earthworms' treatment, 'P0E0') and 24 mg C/g soil (in the 'with roots-no earthworms' treatment, 'P+E0'). This effect was due to the root inputs alone. However, the effect of roots declined strongly with depth, and below 30 cm the 'with roots-no earthworms' treatment had no significant impact on SOC. The combined effect of roots and earthworms led to the highest total SOC storage, which was mostly due to the increased SOC at greater soil depth in that treatment. This effect was associated with two events: first, the burying activity of earthworms and second, the increase in water-stable aggregates (WSA) (Fig. 11.3).

In order to distinguish between the effect of above-ground inputs and that of below-ground inputs, Ndufa (2001) injected fallow trees in Western Kenya with $^{15}$N, via the stem, and imposed the following treatments at tree harvest: (i) above-ground woody biomass, foliage and litterfall biomass removed from the microplot – to assess the contribution of roots; and (ii) above-ground $^{15}$N-labelled biomass (from the labelled microplot in (i) above) applied to an unlabelled microplot – to assess the impact of foliage on soil structure. Subsequent soil fractionation after the maize harvest suggested that, in these tree-based systems, a large proportion (31–39%) of below-ground N became protected in water-stable meso- and macroaggregates, whereas around 20% was associated with the clay + silt size fraction (Table 11.3; Cadisch et al., 2002b).

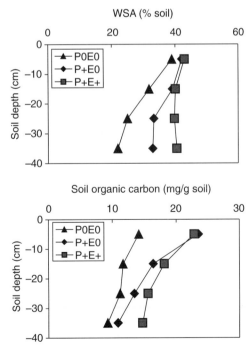

**Fig. 11.3.** Impact of root inputs and earthworm activity on water-stable aggregates (WSA) in soil, and on soil organic carbon, evaluated using herbicides and nematicides. Adapted from Chevallier et al. (2001). P0E0, exclusion of roots and earthworms; P+E0, root effect only; P+E+, root and earthworm impact.

**Table 11.3.** Role of roots (estimated using stem $^{15}$N labelling) and foliage in formation of water-stable aggregates at 6 months after fallow harvest. Adapted from Ndufa (2001).

| N source | $^{15}$N recovery (% of $^{15}$N recovered) | | |
| --- | --- | --- | --- |
| | WSA[a] >20 μm | Free OM[b] >20 μm | Clay+silt |
| Root | 39 | 39 | 22 |
| Foliage | 31 | 49 | 20 |

[a] Water-stable aggregates; [b] organic matter.

Interestingly, roots/below-ground inputs contributed most to soil-structure building, with 39% of root-derived $^{15}$N being found in water-stable aggregates 6 months after fallow harvest. This contrasts with the fate of $^{15}$N-labelled foliage litter, of which only 31% was found in aggregates. However, a larger proportion of the foliage litter (49%) was found in free organic matter fractions. This confirms the important role played by foliage inputs with regard to soil fertility (e.g. N mineralization), whereas, in contrast, roots play a larger role in soil structure formation.

## 11.5. Effects of SOC Increases on Different Soil Properties

### 11.5.1 Soil structure improvement

Well-structured soils contain a number of aggregate types. These have different stabilities (e.g. water-stable aggregates and aggregates obtained from dry–wet sieving) and are of different sizes (micro- to macroaggregates). We place particular emphasis on water-stable aggregates (WSA), since they are used as indicators of soil structural properties. Improvement of physical properties (such as water-holding capacity, compaction control, infiltration and erosion) is often closely linked to improvements in WSA. Soil aggregation is primarily controlled by the amount of clay and the type of clay mineralogy. Significant amounts of WSA will only exist in soils with clay contents of more than 10%. For heavy clay soils, Feller *et al.* (1996) showed that the nature of the clay type will determine the range of water-stable macroaggregates (WSA > 200 μm). For 1:1 clay (e.g. kaolinite) soils, such as Oxisols, WSA contents are high, even when the soils have a low SOC content (Fig. 11.4a). The large amount of WSA in such soils is attributed to the presence of sesquioxides ($R_2O_3$), which are strong and permanent binding agents of the 1:1 clay particles. For 2:1 clay (e.g. montmorillonite or illite) soils, such as Vertisols that develop on volcanic rocks, there are no permanent binding agents, and so the soil is very sensitive to dispersion. Thus, increases in SOC cause large improvements in WSA in Vertisols, but a have a weaker effect in 1:1 clay soil types (Fig. 11.4a).

Data from western Kenya (A. Albrecht, unpublished) suggest that, in comparison with a continuous maize system, short-term (6–15 month) legume fallows can significantly increase the amounts of WSA in the 1:1 clayey soil found in that area (Fig. 11.4b). The average increase in WSA was about 14%, 8% and 5% in the 0–5, 5–10 and 10–20 cm soil layers respectively, at the end of the fallow period, before incorporation of the above-ground material. At the same time, there was an increase of 25% in SOC under the fallows in the 0–5 cm layer; no treatment differences were observed in the lower soil layers. The data thus suggest that the root activity of short-term legume fallows can significantly increase the amount of WSA. In accordance with the observations of other authors, these data also suggest that soil aggregation has to be improved first by below-ground biotic activity (e.g. of roots and macrofauna), in order to enable an increase in SOC: WSA provide a mechanism for the protection of SOC and hence the reduction of carbon mineralization. In the case of the surface soil layer (0–5 cm), the respective roles

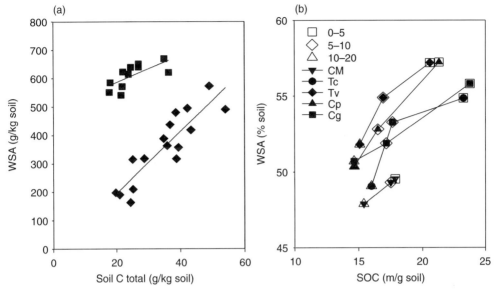

**Fig. 11.4.** Effect of soil carbon on the occurrence of water-stable aggregates (WSA). (a) Relationship between total soil C and WSA, for two clay types (1:1 clay ■; 2:1 clay ◆); (b) relationship between soil organic carbon (SOC) and WSA at different depths, under improved fallows of four different species (Tc, *Tephrosia candida*; Tv, *T. vogelii*; Cp, *Crotalaria paulina*; Cg, *C. grahamiana*) and under continuous maize (CM); note that WSA response is only shown for 5–10 cm (A. Albrecht, unpublished).

played by above- and below-ground inputs with regard to SOC increases are difficult to separate. However, the joint increase of SOC and WSA allows us to suspect that below-ground activities have the preponderant impact in this improvement of the soil.

### 11.5.2 Agents of soil aggregation: bacteria, fungi and macrofauna

WSA are closely linked to SOC content, but soil structure improvement is also controlled by below-ground biological activities. Fungal hyphae and roots have physical binding capacities (Miller and Jastrow, 1990; Degens, 1997). Rillig *et al.* (2002) showed that glomalin (a soil protein produced by arbuscular mycorrhizal fungi (AMF)) was much more effective, in terms of its direct effect, in cementing soil particles and in aggregate formation than were AMF hyphae themselves, in terms of their direct effect. Macroaggregates always contain more microbial biomass per unit soil mass than microaggregates (Angers and Caron, 1998) and there is a trend for microbial diversity to be higher in macroaggregates (Lupwayi *et al.*, 2001). Bacteria with specific functions are often found in aggregates, especially exopolysaccharide-producing species (Achouak *et al.*, 1999), and these further contribute to the binding of particles. SOC and organic inputs are a resource for these below-ground microbes, as well as soil macrofauna; hence their positive effect is decreased if soil degrades and SOC declines. In multispecies systems, diversity of inputs may lead to greater below-ground biodiversity, but Wardle and Lavelle (1997) stated that litter diversity does not necessarily result in predictable changes. The relationship between soil organisms and SOC is not a simple one, because of the multiple functions of the living soil components. For example, earthworms are SOC consumers but they are also 'engineers' able to build stable soil structures; therefore, they can protect SOC in aggregates from mineralization (Fig. 11.3). Earthworms are more effective in 1:1 soils, in terms of building soil structure, than they are in the 2:1 soils, which have a tendency to swell and shrink – a characteristic that reduces the impact of earthworm activity (Blanchart et al., 2002). Earthworm casts, after drying, in kaolinitic soils are an important source of WSA (Shipitalo and Protz, 1988; Blanchart et al., 1997) and will control soil physical properties (Blanchart, 1992). However, a suitable balance between 'compacting' species (e.g. Polypheretima elongata and Millsonia anomala) and 'decompacting' species (e.g. Hyperiodrilus africanus, small eudrilid species in Western Africa) should be preserved. This is because, when compacting species become dominant, and the diversity of soil invertebrates decreases, soil degradation can occur, as was reported by Chauvel et al. (1999) in a study of Amazonian pastures.

### 11.5.3 SOC and soil erodibility control

Increased SOC content is also associated with the occurrence of larger WSA, which result in decreased soil dispersibility and thus lead to a decrease in the amount of soil that is susceptible to detachment and transport by runoff. Thus, SOC controls soil erodibility. Simulating rainfall on 1 m² plots with different SOC contents and three different land uses (pasture, bare soil – simulating no-tillage practices – and a freshly tilled surface), A. Albrecht et al. (unpublished data) found that the impact of SOC on soil erosion was most important in the most-erodible surface treatment, i.e. the freshly tilled surface (Fig. 11.5). When the soil was protected by vegetation, or when soil roughness was non-existent (bare soil), SOC had no effect on soil erodibility.

### 11.5.4 SOC and soil charges

In weathered tropical soils, soil organic C contributes significantly to the soil's cation exchange capacity (CEC), due to the low charge density of 1:1 clay minerals. Typical values for the CEC of soils dominated by kaolinite and amorphous oxides range from 2 to 6 $cmol_c/kg$ (Gallez et al., 1976).

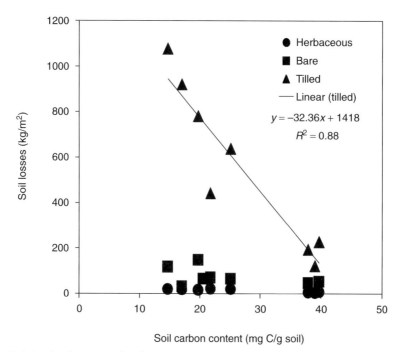

**Fig. 11.5.** Relationship between soil carbon content (0–5 cm) and soil losses from different land use systems on a Vertisol in Martinique (A. Albrecht, unpublished).

Permanent charges in such soils are generally only 1–2 $cmol_c$/kg, illustrating the importance of the CEC due to variable charge, which is primarily associated with oxides and soil organic matter (SOM). A substantial amount of information is available on the contribution of the SOM pool to the total CEC at different soil pH values for tropical soils (Stevenson, 1982; Oades et al., 1989). As SOM is not a homogeneous pool, different fractions may contribute to soil CEC to differing degrees.

The residue quality of organic inputs is usually associated with short-term C and N dynamics, but little is known about the influence that the composition of organic inputs has on the CEC of the SOM pool. Such knowledge could, potentially, generate a tool that could be used to manage the CEC pool of weathered soils. One study (Oorts et al., 2000) examined the contribution that different SOM fractions (of differing particle size) make to the CEC of a tropical soil, and how this was influenced by organic matter inputs of different biochemical composition. The effect of inputs from *Leucaena leucocephala*, *Dactyladenia barteri*, *Afzelia africana*, *Pterocarpus santalinoides* and *Treculia africana* were studied in a 16-year-old arboretum established on a Ferric Lixisol. It was found (Oorts et al., 2000) that the CEC of the fractions smaller than 0.053 mm was inversely related to their particle size: clay (< 0.002 mm) > fine silt (0.002–0.02 mm) > coarse silt (0.02–0.053 mm). The clay and fine silt fractions were responsible for 85–90% of the CEC of the soil. Whereas the charge on the clay fraction was largely governed by pH, the charge on the silt fractions was strongly related to the quality of organic inputs. Organic inputs with a high C/N and lignin/N ratio produced fine- and coarse-silt-sized SOM fractions with the highest charge density, suggesting that inputs of slowly decomposing organic residues show promise in terms of increasing the CEC of highly weathered soils (Fig. 11.6). In another trial (Oorts et al., 2002), which used decomposition tubes, the development of charge in the top 10 cm of soil over a period of 2 years was assessed, in relation to residue quality (using *Afzelia*, *Dactyladenia*, *Gmelina arborea*, *Leucaena* and *Treculia*). After 23 months, total soil C contents ranged between 3.8 and 5.3 g C/kg soil, and CEC values at pH 5 (= average soil pH) ranged between 1.9 and 2.5 $cmol_c$/kg soil. Fine-silt C contents ranged between 18.3 and 26.5 g C/kg fine silt, and CEC values at pH 5 varied between 5.3 and 8.9 $cmol_c$/kg fine silt. Fine-silt fractions again reflected the differences between the treatments most clearly, indicating that the lowest-quality residues (such as those from *Treculia* and *Dactyladenia*) resulted in the largest CEC values, and the largest C contents (Table 11.4). This indicates that even a single addition of these residues enhances charge characteristics significantly, and for a significant length of time.

## 11.6 Impacts of SOC Increases on Plant Productivity

It is commonly reported in the literature that soil organic matter is closely linked to plant productivity. However, a better understanding of such relationships has only recently emerged. Soil organic matter provides plant-available mineral N through decomposition and mineralization processes, and hence directly influences plant nutrition. As for charge build-up, the N-supply capacity of different SOM fractions may vary considerably: this depends partially on the age of the organic material in the fractions and partially on the degree of physical protection of these materials from decomposition. An example of this was given by the results of a microplot experiment with $^{15}$N-labelled high-quality *Leucaena leucocephala* leaf residues, and low quality *Dactyladenia barteri* leaf residues, which was established in Ibadan, Nigeria. Highly significant ($P < 0.001$) relationships between residue-derived N (RDN) present in the soil particulate organic matter (POM) fraction, and uptake of RDN by maize, indicated the high availability of RDN in the POM fraction (Vanlauwe et al., 1998). The weak relationships between RDN in smaller particle size classes (SOM < 0.053 mm) and uptake of RDN by maize, indicated the lower availability of N in the finer SOM fractions.

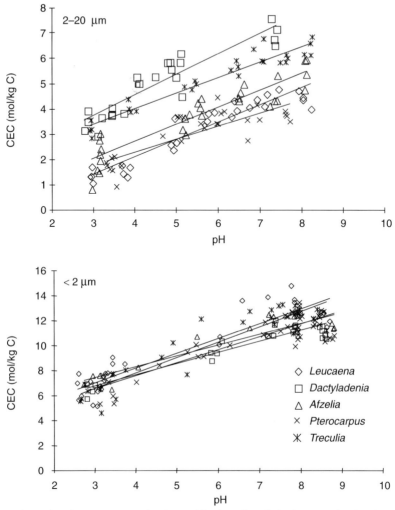

**Fig. 11.6.** Relationships between CEC and soil pH of the fine silt and clay fraction of soil under tree species having leaves with varying resource quality, sampled from a 16-year-old arboretum established on a Ferric Lixisol in Ibadan, Nigeria. Leucaena, *L. leucocephala*; Dactyladenia, *D. barteri*; Afzelia, *A. africana*; Pterocarpus, *P. santalinoides*; Treculia, *T. africana*. Redrawn from Oorts et al. (2000).

In another set of trials, the impact that organic inputs had on total SOM- and POM-N contents in relation to soil type, and also the relationships between sources of N and maize N uptake, were assessed in alley-cropping trials in the West African moist savannah (Vanlauwe et al., 1999). The initial POM-N content varied between 50 and 160 mg N/kg soil. The average proportion of soil N belonging to the POM pool ranged between 9% and 29%. This was significantly related to the annual N inputs from maize stover and prunings, when averaged over the different alley-cropping treatments. The relative change in POM-N content between the 2 years of sampling was about twice the relative change in total soil N content. This suggests that N incorporated in the POM is relatively labile, compared with the N incorporated in the other SOM fractions. A highly significant linear relationship ($R^2 = 0.91$) was also observed between the total N

**Table 11.4.** Carbon contents and CEC of the whole soil, fine silt and clay fractions of the different treatments after 23 months decomposition in the field. After Oorts *et al.* (2002).

| Treatment | Whole soil | | Fine silt | | Clay | |
|---|---|---|---|---|---|---|
| | C (g/kg) | CEC at pH 5 (cmol$_c$/kg) | C (g/kg) | CEC at pH 5 (cmol$_c$/kg) | C (g/kg) | CEC at pH 5 (cmol$_c$/kg) |
| Control | 3.85 | 2.17 | 18.2 | 5.79 | 28.5 | 22.6 |
| *Afzelia africana* | 3.82 | 1.94 | 18.7 | 5.31 | 27.6 | 21.1 |
| *Dactyladenia barteri* | 4.67 | 2.19 | 24.4 | 8.08 | 27.9 | 21.9 |
| *Gmelina arborea* | 4.63 | 2.30 | 20.6 | 6.46 | 28.6 | 20.6 |
| *Leucaena leucocephala* | 3.76 | 2.08 | 21.3 | 6.50 | 27.1 | 21.3 |
| *Treculia africana* | 5.25 | 2.53 | 26.5 | 8.94 | 26.7 | 21.9 |

uptake by the shoots and roots of 7-week-old maize and the POM-N content found in a set of West African savannah soils under various land uses (unfertilized continuous maize cropping, unfertilized and fertilized alley cropping with maize, *Gliricidia sepium* tree fallow, natural fallow; see Fig. 11.7 and Vanlauwe *et al.*, 2000). This relationship was not observed in three soils from the humid forest zone with high clay contents, presumably as a result of physical protection of the POM pool (Fig. 11.7).

Data from Carsky *et al.* (1998) showed the relationship between land management and SOC content and maize grain yield (Fig. 11.8). The relationship shows the importance of resource allocation, i.e. management by farmers can induce an increased SOC in fields close to their homes (because they provide them with a larger proportion of farm inputs). A higher SOC will sustain greater crop production through improved soil fertility (in terms of nutrient supply and water retention) but also provides other services as mentioned above. This example also shows that there are opportunities/niches at the farm scale for increasing SOC by appropriate management of distant plots.

**Conclusions**

1. Below-ground inputs in improved fallows can be substantial, amounting to as much as 30 t/ha in 18 months.
2. Soil C sequestration in low- and high-activity clay soils is strongly related to their clay + silt content, but this is not the case in allophane-dominated soils (Andosols).
3. Water-stable aggregates provide physical protection for C and reduce soil erodibility. Their formation is enhanced by root and faunal activity, minimal tillage practices and relates strongly to soil-C content in 2:1 clay soils (e.g. Vertisols); however, the impact of earthworms is larger in 1:1 clay (highly weathered) soils.
4. Increased soil C not only enhances the activity of the decomposer community, it also increases the number of charges in the soil (i.e. cation exchange capacity) and hence nutrient retention. Inputs of slowly decomposing organic residues show promise with regard to increasing the CEC (particularly in the fine- and coarse-silt-sized SOM fractions) of highly weathered soils.
5. Direct relationships between soil C (or the C content of soil fractions, e.g. particulate organic matter) and plant productivity are evident, and on farm these are often related to the distance of the plot from the house.

**Future research needs**

1. A better quantitative and temporal (turnover) assessment of below-ground inputs.
2. Assessment of the stability of sequestered soil C.
3. Quantification of inactive pools (e.g. charcoal) in soils.
4. A better understanding of the impacts of biodiversity on soil functions.
5. Investigation of the potential for exploitation of niches at the farm scale in order to increase soil organic C by appropriate management of more distant plots.

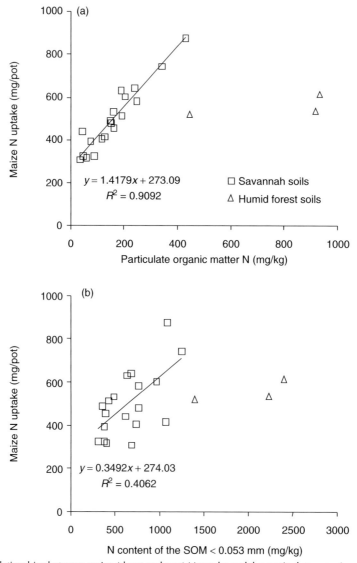

**Fig. 11.7.** Relationships between maize (shoot and root) N uptake and the particulate organic matter N content (a) and between maize N uptake and the N contained in the SOM <0.053 mm (b) for a series of West African savannah soils. The humid forest soils (Mbalmayo and Ebolowa soils) were excluded from the linear regression analysis. Source: Figs 1(a) and 1(b), Vanlauwe et al. (2000) © Springer, reproduced with permission.

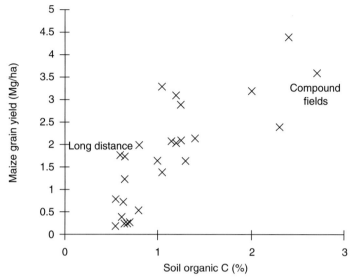

**Fig. 11.8.** Relationship between soil organic carbon content and maize grain yield for fields at varying distances from the farmer's house. Adapted from Carsky *et al.* (1998).

# 12 Soil–Atmosphere Gas Exchange in Tropical Agriculture: Contributions to Climate Change

Louis V. Verchot, Arvin Mosier, Elizabeth M. Baggs and Cheryl Palm

---

**Key questions**

1. Which 'greenhouse gases' are produced and/or consumed in terrestrial agroecosystems as part of the nitrogen and carbon cycles?
2. Which soil processes are responsible for the production, emission and consumption of these gases?
3. What contributions do tropical land-use change and managed agricultural soils make to atmospheric loading?

## 12.1 Introduction

The objective of this chapter is to summarize the body of knowledge concerned with the effects of agriculture on greenhouse gas (GHG) loading in the atmosphere. Before we approach the heart of this matter, we will first review several concepts relevant to greenhouse gases and the role of atmospheric gases in producing the greenhouse effect. We will then consider the soil processes responsible for the emission and consumption of GHGs. Finally, we will review what is known about the contribution of agricultural soils to atmospheric GHG loading, focusing on the effects of land-use change and agricultural management.

## 12.2 Greenhouse Gases

Several gases play an important role in the greenhouse effect. Of these, water vapour is the most abundant, comprising 1% of the atmosphere by volume. Other gases are present in the atmosphere in very low concentrations. These gases are collectively referred to as '*trace gases*', a term that refers to their presence in *trace* amounts (in a parts-per-million order of magnitude). It is these trace gases that are increasing and producing the radiative forcing that is responsible for climate change. The trace gases that are important in the enhanced greenhouse effect are carbon dioxide ($CO_2$), methane ($CH_4$), nitrous oxide ($N_2O$), halocarbons (including chlorofluorocarbons such as $CFCl_3$, $CFCl_2$) and ozone ($O_3$;

see Fig. 12.1). Because each of these gases has a different atmospheric lifetime and a different heat absorption capacity per molecule, the warming effect they have on the atmosphere differs between the gases. This difference is expressed as the global warming potential (GWP) of the gas. For example, a molecule of $CO_2$ resides, on average, for 100 years in the atmosphere, a molecule of $N_2O$ for around 120 years and a molecule of $CH_4$ for around 9 years. GWP is expressed on a per molecule basis relative to $CO_2$, which is assigned a GWP of 1. In fact, $N_2O$ has a GWP of 310 and $CH_4$ has a GWP of 15.

Since our subject is the contribution that agriculture makes to increased atmospheric loading of greenhouse gases, the rest of this chapter will only consider three greenhouse gases upon which agriculture has an important effect: $CO_2$, $N_2O$ and $CH_4$. For reasons that will become apparent, our discussion of $N_2O$ will be expanded to include another N oxide, nitric oxide (NO).

### 12.2.1 Carbon dioxide

Historical climate data show a tight coupling between global temperatures and atmospheric $CO_2$ concentrations. $CO_2$ concentrations have increased by 31% since 1750. The present concentration is now much higher than it has been in the past 400,000 years, and is probably higher than it has been in the last 20 million years. In the 1800s, and until the 1960s, the principal source of increased $CO_2$ in the atmosphere was the deforestation that accompanied the expansion of agriculture. In the 1800s, this deforestation was concentrated in Europe and North America. Recently, as agricultural areas shrink in the temperate zone and forests regrow, temperate areas have become a sink for atmospheric $CO_2$. A 'sink', in the context of this chapter, is any process that removes a GHG from the atmosphere. Today, expansion of agriculture in tropical regions is the principal source of increased atmospheric $CO_2$ from deforestation, contributing around 25% of the annual $CO_2$ increase occurring in the atmosphere.

Modelling efforts began in the 1960s, to try to estimate the effects that the increase in GHG concentrations will have on the global climate. Results of these early analyses suggested that the expected increases in atmospheric temperature could be of an order of magnitude such that global climate systems would be perturbed. More recent refinements made to these models have only reinforced these initial predictions, and have given better spatial resolution to the estimates of change.

### 12.2.2 N oxides

Two important N oxides are produced in agricultural soils: nitrous oxide ($N_2O$) and nitric oxide (NO). The atmospheric concentration of $N_2O$ has increased rapidly in recent times, from a preindustrial concentration of ~ 275 p.p.m. to 320 p.p.m. today. Industrial emissions were blamed for the increasing atmospheric burden of this gas, but we now recognize that both natural and anthropogenic emissions are largely biogenic (i.e. arising from living organisms). Table 12.1 shows the magnitudes of the different sources of $N_2O$ to the atmosphere and shows that emissions from agricultural soils are by far the most important anthropogenic source of this gas.

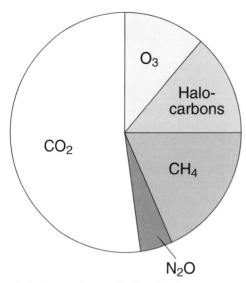

**Fig. 12.1.** Relative contribution of different gases to anthropogenic radiative forcing from increased GHG concentration (Turco, 1997).

Table 12.1. Global atmospheric nitrous oxide budget (Kroeze et al., 1999).

|  | Tg N/year |
|---|---|
| Sources |  |
| *Natural* |  |
| Terrestrial and aquatic ecosystems | 9.6 |
| *Anthropogenic* |  |
| Agriculture | 6.2 |
| Biomass burning | 0.6 |
| Energy | 0.9 |
| Industry | 0.3 |
| Total sources | 17.6 |
| Sinks |  |
| *Stratospheric decomposition* | 10.9 |
| Annual $N_2O$ accumulation | 6.7 |

The other important N oxide, NO, is not a greenhouse gas, but is a strong oxidant in the chemistry of the atmosphere. The global budget shows that tropical soils are also an important source of NO, and account for 25–35% of global NO emissions (Holland et al., 1999). Reactions of NO with $OH^-$ and with other gases can result in the production of $O_3$. As mentioned above, tropospheric $O_3$ is a greenhouse gas, but it is also a toxin. As levels increase, there is growing concern about the negative impacts of this gas on human health, agricultural productivity and ecosystem health. Experimental evidence suggests that NO emissions increase in fertilized systems and that high agricultural emissions contribute to $O_3$ problems in rural areas. There is also good reason to believe that atmospheric photochemistry in tropical regions is sensitive to NO emitted from soils.

Although these two gases behave very differently in the atmosphere, there is a well-established rationale for considering them together: they are both by-products of the same microbial processes in soils – nitrification and denitrification. Several biochemical and abiotic pathways exist in soils for the production of these gases (Fig. 12.2). NO production appears to be more closely associated with nitrification, whereas $N_2O$ appears to be more closely associated with anoxic soil conditions and denitrification. Our current understanding of N oxide production in soils suggests that there are two levels of regulation of the production of these gases in soils. The first level of regulation is the rate of N cycling that determines the total amount of N oxides produced (NO + $N_2O$). Thus, with high nitrification rates, high levels of these gases are produced. The second level of regulation is the soil water content, which controls the oxidation/reduction status of the soil and consequently determines the relative importance of NO and $N_2O$ as the gaseous end products of these processes. In aerobic soils with oxidizing conditions, the more oxidized gas (NO) is the predominant gas emitted, whereas in wet soils, where reducing conditions prevail, the more reduced gas ($N_2O$) is produced in relatively higher quantities.

### 12.2.3 Methane

The atmospheric concentration of the GHG methane has doubled since preindustrial times. The current concentration is ~ 1780 ppbv; and, until recently, the concentration was increasing. The rate of increase in the concentration of atmospheric $CH_4$ slowed from about 15 ppbv/year in the 1980s to near zero in 1999. Since 1990, the annual rate of $CH_4$ increase in the atmosphere has varied between less than zero and 15 ppbv. The reasons for this change are not known.

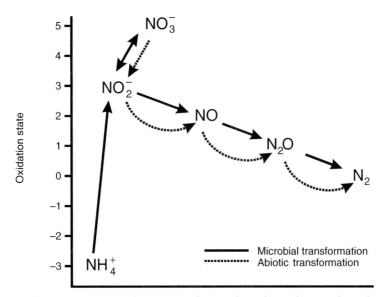

**Fig. 12.2.** N transformations in terrestrial ecosystems showing the pathways for N-oxide production during nitrification and denitrification. Adapted from Davidson (1991).

Soils both produce and consume $CH_4$. The net soil–atmosphere $CH_4$ flux is the result of the balance between the two offsetting processes of methanogenesis (microbial production) and methanotrophy (microbial consumption). Methanotrophy is a process that involves the oxidation of $CH_4$ by bacteria; $CH_4$ is the C source, or the electron donor, in the respiration reaction. Methanotrophs and methyltrophs are all obligate aerobes; the biochemical process requires a monooxygenase enzyme and, therefore, requires molecular $O_2$. Methanotrophy is the dominant process in upland soils. In these soils, oxidation generally exceeds production and there is a net uptake by the soil of $CH_4$ from the atmosphere. Methanotrophy is also an important process in wetland soils at the aerobic soil–water interface. One estimate of the importance of soil methanotrophy suggests that as much as 50% of the $CH_4$ produced in soils and sediments is consumed therein (Reeburgh et al., 1993).

Methanogenesis is the process of microbial production of $CH_4$ in anaerobic sites, as a result of fermentation. The major pathways of $CH_4$ production in anaerobic conditions involve: (i) the reduction of $CO_2$, with $H_2$, fatty acids, or alcohols being the hydrogen donors; and (ii) the transmethylation of acetic acid or methyl alcohol by methane-producing bacteria. Methanogenesis is an important process in wetland soils and rice paddies; these systems are usually sources of $CH_4$ for the atmosphere. But methanogenesis can also occur in upland soils, inside soil aggregates where anaerobic 'microsites' occur.

In the global $CH_4$ budget, upland tropical soils are a net sink (Table 12.2). In fact, soils worldwide are the largest biotic sink for atmospheric $CH_4$, consuming 15–45 Tg annually (Table 12.2), a rate that is of the same order of magnitude as the rate of $CH_4$ accumulation in the atmosphere during the 1990s. Thus, any significant change made to the soil $CH_4$ sink could alter the net biosphere–atmosphere flux and alter the atmospheric accumulation rate of this potent greenhouse gas. Evidence is rapidly accumulating that suggests that land-use change and expansion of agriculture has significantly reduced the strength of the soil sink.

Several factors are known to contribute to the spatial and temporal variability of soil emissions and consumption in landscapes; these include soil carbon contents, substrate quality (in terms of polyphenol and lignin

**Table 12.2.** Global atmospheric $CH_4$ budget. Note: the values are appropriate for much of the 1990s, as atmospheric accumulation has declined over the past few years (Mosier *et al.*, 1998).

|  | Tg $CH_4$/year |
|---|---|
| **Sources** | |
| *Natural* | |
| Wetlands | 100–200 |
| Termites | 10–50 |
| Oceans | 5–20 |
| Fresh water | 1–25 |
| $CH_4$ hydrate | 0–5 |
| *Anthropogenic* | |
| Coal mining, natural gas and petroleum industry | 70–120 |
| Rice paddies | 20–150 |
| Enteric fermentation | 65–100 |
| Animal wastes | 10–30 |
| Domestic sewage treatment | 25 |
| Land fills | 20–70 |
| Biomass burning | 20–80 |
| **Sinks** | |
| *Atmospheric (tropospheric plus stratospheric) removal* | 420–520 |
| *Removal by soils* | 15–45 |
| Annual $CH_4$ accumulation | 28–37 |

contents, N concentrations and C : N ratios, etc.), temperature, moisture, soil diffusivity, microbial activity, pH and N availability. Some generalizations can now be made. In well-drained soils where $CH_4$ concentrations are at ambient atmospheric levels, gas-phase transport or diffusion of $CH_4$ limits soil uptake and the effect of temperature is weak or non-existent. In poorly drained soils, when $CH_4$ concentrations greatly exceed atmospheric concentrations, oxidation reactions can become saturated so that enzyme activity, rather than gas-phase transport or gas–water exchange, limits oxidation. In this case, soils emit $CH_4$ and there is a parabolic relationship between the rate of production and temperature.

Other factors have been shown to play important roles in controlling $CH_4$ consumption. Fertilizer experiments have shown that high levels of N availability reduce $CH_4$ oxidation, and laboratory culture experiments have demonstrated that $NH_4^+$ and $NO_2^-$ reduce the activity of methanotrophic enzymes and inhibit the growth of methanotrophs. Two mechanisms appear to be at work here. First, use of N fertilizers reduces $CH_4$ consumption because the enzymes that oxidize $CH_4$ also oxidize $NH_4^+$ and thus the two compounds compete for enzymes in the soil. Therefore, in fertilized systems, increased availability of $NH_4^+$ reduces $CH_4$ oxidation. Secondly, most land-use change is accompanied by soil compaction. Since the rate at which $CH_4$ diffuses into the soil is often the rate-limiting process, compaction reduces $CH_4$ oxidation in soils.

## 12.3 Trace Gases and Land Use

The soil C pool is estimated to be in the order of 1300–1500 Gt C, which is about twice the pool extant in terrestrial plant biomass and three times the atmospheric pool. Approximately 32% of these soil C stocks occur in tropical soils. The effects of land-use change and agriculture on soil C and the soil-to-atmosphere flux of $CO_2$ have been fairly well documented, summarized, and modelled (Davidson and Ackerman, 1993; Davidson and Trumbore, 1995;

Trumbore et al., 1995; Woomer et al., 2000). Soil C losses associated with $CO_2$ emissions, following conversion from forest to row-crop agriculture, generally vary in the order of 20% to 30% of the original C stocks: these losses generally occur within 20 years of deforestation.

The response of pastures is somewhat less consistent than that of cultivated lands, and land conversion from forest has mixed effects on soil C stocks. In some studies C stocks have been shown to increase under pasture, whereas in others C stocks have been shown to decrease. In other studies, management effects determine whether pastures gain or lose C relative to the forests. In Rondônia, Brazil, soil C stocks increased in the top 50 cm of the soil profile by about 20 t/ha over 80 years. Fisher et al. (1994) have also suggested that C stocks increase significantly in soils under very productive pastures to a depth of 2 m. Some of these differences result from the fact that different authors studied changes in soil C to different depths. However, we can conclude that, generally, soil C stocks increase in the surface layers of the soil (10–20 cm) in pastures.

Most studies look at surface soil layers, but in many areas of the tropics, where soils are deep, the C stock below a depth of 1 m may be greater than the stock above a depth of 1 m. Studies that look at deeper soil C suggest that overall stocks may decrease following land-use change, but high spatial variability and sampling difficulties often lead to inconclusive results. However, in eastern Amazonia, Trumbore et al. (1995) showed that 13% of the soil C at a depth of 8 m was from recent inputs and was cycling rapidly. Thus, if the land-use systems that replace forests do not maintain deep soil C inputs, significant soil C pools could be lost.

As noted above, the soil C story is fairly well understood and summarized. Chapter 11 of this book deals with the importance of soil C, management of soil organic matter and potential methods of increasing soil organic carbon (SOC) in agricultural systems. The objective for the remainder of this chapter is to focus on N oxides and on $CH_4$, to summarize the accumulating body of knowledge on the contribution of agriculture to the increasing atmospheric burden of these gases, and to examine some promising approaches to modelling the fluxes of these gases from agricultural soils.

### 12.3.1 Natural vegetation

Information on, and understanding of, trace gas fluxes from soils in the humid tropics has increased substantially over the past 10 years. A growing number of estimates of annual $N_2O$ emissions from humid tropical forests are emerging; secondary forests generally tend to have lower fluxes than primary forests. For example, Verchot et al. (1999) found that $N_2O$ and NO fluxes in a 20-year-old secondary forest in eastern Amazonia were less than half the emissions of a primary forest. Estimates of the annual $N_2O$ fluxes from Amazonian forests are remarkably similar between sites, suggesting that primary forests in this region emit about 2 (± 0.5) kg $N_2O$-N/ha year. Younger sites on volcanic soils (e.g. Costa Rica) are generally more N rich than the Amazonian forests and have higher $N_2O$ emissions (Reiners et al., 1994). The older secondary forests of Puerto Rico have lower $N_2O$ emissions, in the order of 0.5 kg $N_2O$-N/ha/year (Erickson et al., 2000). Only two of the $CH_4$ studies in the tropics measured fluxes in secondary forests. Keller and Reiners (1994) found that secondary forests were sinks of a magnitude similar to that of primary forests. Verchot et al. (2000) found that secondary forests were weaker sinks than primary forest, consuming 50% less $CH_4$.

### 12.3.2 Biomass burning

The first step in the conversion of forests into agricultural uses typically involves biomass burning in some form. For example, shifting cultivation requires that the forests be cut down, the logging debris and unwanted vegetation burned, and the land farmed for several years before it is left fallow to re-vegetate. Savannah and rangeland biomass is often burned to improve livestock forage. Agricultural residues are also burned in the field, to return nutrients to the soil or to reduce the regeneration of shrubs on rotational fallow lands. Such agriculture-related burning may account for 50% of the biomass burned annually. Estimates indicate that 8700 Tg of biomass, and between 1% and 5% of the world's land area, are burned per year. Because of incomplete combustion, $CH_4$ and N-oxides are emitted. Annual global $CH_4$ emissions from biomass burning are approximately 39 Tg; emissions from tropical forest clearing for agriculture, savannah burning, and agricultural crop residue burning account for about 19 Tg $CH_4$/year (Andreae and Merlet, 2001).

During combustion, the N in end groups, open chains, and heterocyclic rings of organic compounds can be converted into gaseous forms, such as $NH_3$, NO, $N_2O$, $N_2$ and HCN. These compounds are liberated at all temperatures at which smouldering and combustion occur. Early estimates suggested that ~ 8 Tg $N_2O$-N/year are emitted into the atmosphere as a result of biomass burning. However, recent work suggests that these emission factors were too high and that $N_2O$ is probably only a minor product of biomass burning. We now believe that the contribution of biomass burning to $N_2O$ emission may be less than 1 Tg/year. Over the longer term, however, biomass burning increases the abiotic production of NO in soils, and perhaps that of $N_2O$, by removing vegetation and thereby decreasing $NO_3^-$ uptake. As $NO_3^-$ accumulates in these soils, there is a greater potential for chemodenitrification and thus for N oxide production. For example, Verchot et al. (1999) found that NO emissions increased greatly following the burning of Amazonian pastures. Otter et al. (2001) indicate that biogenic emissions of NO from southern African savannahs exceed those from biomass burning and industrial emissions from the area.

### 12.3.3 Pasture formation

Most studies considering the effects that conversion of forest into pasture have on trace gases have been conducted in Latin America. Little work has been done in natural grasslands. A few studies have examined the effects of deforestation and pasture creation on $CH_4$ sink strength, and their conclusions suggest that the conversion of primary forest to pastures results in decreased net $CH_4$ uptake in soils and, in many instances, in the conversion of a sink into a source (Table 12.3). There is great variation in $CH_4$ consumption rates in forests; soil texture is a major determining factor. In humid tropical forests, fine texture soils consume 1.5–2.0 kg/ha/year, while medium and coarse texture soils consume >4.0 kg/ha/year. In all studies pastures were, for the most part, sources during the wet season and sinks during the dry season. The exception to this rule was found in some young pastures in eastern Amazonia, which act as sinks all year long. Only two of the studies in tropical pastures derived from forests measured $CH_4$ fluxes in abandoned pastures. Keller and Reiners (1994) found that the abandoned pastures in Costa Rica

Table 12.3. $CH_4$ consumption rates (kg/ha/year) in forest, old pasture and young pasture for sites in Latin America, with soil texture and rainfall information for the study sites.

| | $CH_4$ consumption rates (kg/ha/year) | | | | |
|---|---|---|---|---|---|
| Site | Primary forest | Old pasture | Young pasture | Rainfall | Soil texture |
| Guacimo, Costa Rica[a] | −4.0 | −1.0 to 5.3 | −1.1 to 2.8 | 4000 | Medium |
| La Selva, Costa Rica[a] | −4.8 to 4.4 | −2.5 to 1.5 | – | 4000 | Medium |
| La Selva, Costa Rica[b] | −4.6 | 0.8 | – | 4000 | Medium |
| Paragominas, Brazil[c] | −2.5 to −2.1 | −1.3 | −6.2 to 1.1 | 1800 | Fine |
| Rondônia, Brazil[d] | −5.9 to −3.4 | 1.0 to 12.0 | −0.8 to 3.4 | 2200 | Coarse |

[a] Keller et al., 1993; [b] Keller and Reiners, 1994; [c] Verchot et al., 2000; [d] Steudler et al., 1996.

were sinks of a magnitude similar to that of primary forests. However, Verchot et al. (2000) found that degraded and abandoned pastures in eastern Amazonia were very strong sinks, consuming 50% more $CH_4$ than primary forest sites. The sources of the differences between these two studies lay in the dry season uptake rates. Costa Rica was a much wetter site and had a shorter and wetter dry season.

Conversion of tropical forest into pasture generally results in decreased rates of N cycling and a change in the nature of soil inorganic-N pools, from nitrate-dominated in forests to ammonium-dominated in pastures, as nitrification rates decrease. These shifts in N cycling have implications for the N oxide emissions of soils.

Luizão et al. (1989) first raised the possibility that the conversion of forests into pastures in the tropics may be partially responsible for the current increase in atmospheric $N_2O$ burden. These authors intensively sampled $N_2O$ fluxes in a young (3 years old) and a primary forest. They found that annual emissions from the pasture exceeded forest emissions by a factor of 3. Luizão et al. (1989) also sampled four additional pastures once during the rainy season (the pastures represented a chronosequence of sites, ranging in age from 3 to 10 years), and found that two of the pastures had fluxes that greatly exceeded the forest fluxes. Extrapolating the increased flux, they estimated that tropical deforestation contributed around 1 Tg of $N_2O$-N to the atmosphere annually. However, they urged caution because the pastures they used were not necessarily representative of the broad diversity of soil, management and climatic conditions that exist within the tropics. Both Keller et al. (1993) working in Guacimo, Costa Rica, and Melillo et al. (2001) working in Rondônia, Brazil, found elevated $N_2O$ emissions in young pastures, but significantly lower emissions in older pastures. In the Costa Rican sites, emissions were elevated in pastures that were less than 10 years old, whereas in Rondônia, elevated emissions lasted for only 2 years. Verchot et al. (1999) found no increase in $N_2O$ production following land clearance in eastern Amazonian sites; on the contrary, pastures at two sites had lower emissions than forests. They also found no consistent trend associated with pasture age for young pastures on their first rotation. Thus, Luizão et al. (1989) may have captured a transitory situation of elevated emissions in the 3-year-old pasture, as was found by both Keller et al. (1993) and Melillo et al. (2001). Results from intensive sampling by Coolman (1994), at a site near that used by Luizão et al. (1989), showed that emissions from three abandoned pastures, which underwent a short rotation (6 years) and were then abandoned for 3 years, were only slightly higher than those at a comparable upland primary forest site.

Soil emissions are also cited as an important source of NO in global and regional budgets. The body of data that exists concerning the effects that tropical land-use change has on this source is smaller than the body of data that exists for $N_2O$, but the studies presented in Table 12.4 suggest that changes in NO fluxes mirror changes in $N_2O$ fluxes on an annual basis. Therefore, it seems reasonable to expect that effects of tropical land-use change may be a transient increased soil source following conversion from forest to pasture, but over the long term there will be a decreased soil source of this gas relative to the forest.

### 12.3.4 Row crops

Non-$CO_2$ gas fluxes from agriculture have been poorly studied in general: this is particularly true for tropical systems. However, the processes that produce N oxides and consume $CH_4$ fluxes are the same across different ecosystems and climatic zones. Thus, mechanistic studies conducted in various ecosystems and in the temperate zone will allow us to draw some inferences for tropical systems; however, the picture is still incomplete. As a result, it is difficult to fully assess the factors that regulate trace gas fluxes (e.g. crops, weather patterns, soils and management) in tropical upland agricultural systems.

The first general conclusion we can draw is that agricultural soils often have higher N

Table 12.4. Differences in N-oxide emission rates (kg N/ha) in forest, old pasture and young pasture sites in the neotropics.

| Site | Primary forest $N_2O$ | Old pasture $N_2O$ | Young pasture $N_2O$ | Primary forest NO | Old pasture NO | Young Pasture NO | Reference |
|---|---|---|---|---|---|---|---|
| Guacimo, Costa Rica | 6.1 | 1.8 to 10.5 | 34.1 to 51.7 | 4.8 | 0.9 to 1.1 | 3.9 to 8.8 | Keller et al. (1993) |
| La Selva, Costa Rica | 3.5 to 7.9 | 0.9 to 2.6 | – | 0.8 | 1.1 | 0.1 to 0.5 | Keller et al. (1993) |
| La Selva, Costa Rica | – | – | – | 0.9 | 0.2 | – | Keller and Reiners (1994) |
| Manaus, Brazil | 1.9 | – | – | – | – | – | Luizão et al. (1989) |
| Manaus, Brazil | 1.4 | – | 1.6 | – | – | – | Coolman (1994) |
| Rondônia, Brazil | 2.02 1.70 | 1.15 to 1.65 | 3.09 to 5.13 | – | – | – | Melillo et al. (2001) |
| Paragominas, Brazil | 2.6 to 3.3 | 0.1 to 0.3 | 0.5 to 1.7 | 1.5 | 0.5 to 0.7 | – | Verchot et al. (1999) |
| Luquillo, Puerto Rico | 0.6 to 1.7 | 0.5 to 0.7 | – | 0.1 to 0.4 | 0.1 to 0.2 | 0.2 to 1.7 | Erickson et al. (2000) |

oxide emissions and lower $CH_4$ consumption than soils under native vegetation. The principal reason for this is that fertilization increases the N-oxide flux from soils. Fertilizer application stimulates the microbial processes of nitrification and denitrification, which results in increased gas fluxes. Goreau and de Mello (1988) showed that N fertilization increased $N_2O$ emissions 15-fold in cowpeas grown in Amazonia. Crill et al. (2000) found that fertilizer applications of 122 kg N/ha to maize in Costa Rica increased $N_2O$ emissions threefold, from 0.5 to 1.8 kg/ha per season. Thus $N_2O$ losses amounted to 1.4% of the fertilizer application. Weitz et al. (2001) found similar results in fertilized sites in Costa Rica, where $N_2O$-N losses amounted to between 0.2% and 2.3% of N fertilizer applications in a maize system. Increased $N_2O$ fluxes following fertilization have been observed in many other systems, including temperate agriculture, fertilized forests, tree plantations and grasslands.

Fertilization also increases soil NO fluxes. Sanhueza (1997) recorded emissions between 3.3 and 3.7 kg N/ha/year in cereal crops in Venezuela. Ortiz-Monasterio et al. (1996) found that NO emissions increased from 2.7 to 6.3 kg/ha/year following the fertilization of an irrigated wheat system in Mexico. Observations in temperate systems reinforce the results of these tropical studies.

$CH_4$ uptake is inhibited by fertilization, primarily through enzymatic competition with $NH_4^+$. We know of no studies that systematically isolate the effect of N fertilization in tropical systems, but work in temperate systems suggests that this mechanism is sufficiently robust to expect that this is the case in tropical systems. Hütsch (1996) showed that long-term N fertilization in Germany decreased the ability of soils to consume $CH_4$. These observations are also supported by observations in other systems. For example, Castro et al. (1994) found that fertilization of pine plantations in Florida decreased $CH_4$ uptake by a factor of between 5 and 20. Steudler et al. (1989) found that N fertilization decreased $CH_4$ uptake by 33% in a temperate forest, and Mosier et al. (1991) found that fertilization of grasslands decreased $CH_4$ uptake by 65%. It should be noted that several studies have reported no N fertilizer effect on $CH_4$ uptake rates by soils, so some questions about this mechanism remain. Schimel and Gulledge (1998) suggested that the response of $CH_4$ consumption to fertilization varied from site to site and that such variations were due to differences in soil microbial communities. Visscher et al. (1998) suggested that cation exchange capacity was responsible for much of the site-to-site variation, as soils with high cation exchange capacity bound $NH_4^+$ and reduced its inhibitory effect.

A second general conclusion can be drawn: fertilization and weather interact strongly to affect the $N_2O$ flux. Heavy rainfall soon after fertilization or the application of fertilizer to wet soils stimulates $N_2O$ emissions. Both Crill *et al.* (2000) and Weitz *et al.* (2001) observed that the highest $N_2O$ fluxes were associated with very wet soils, where the water-filled pore space of the surface soil was between 80% and 99%, and that fertilization and soil moisture were the dominant regulators of $N_2O$ flux. Rainfall has a comparatively small impact on $N_2O$ fluxes in unfertilized soils. Mechanistic studies that elucidate the effects of the interaction of weather events and fertilization on NO fluxes are lacking, and it is not possible to draw any conclusions at the present time. The best guess that can be offered at this point is that NO emissions decrease during wetting events following shortly after fertilizer application, when $N_2O$ dominates the soil N-oxide flux.

Another generalization that can be made is that, in comparison with tilled systems, conservation and reduced tillage systems generally increase $N_2O$ emissions and increase the fixation of C by decreasing $CO_2$ emissions. Tillage may also decrease the oxidation rate of atmospheric $CH_4$ in aerobic soils. The method of incorporation in tilled systems has been shown to affect the magnitude and pattern of emissions, presumably by varying the supply of organic C and N to microorganisms and by changing the soil moisture/aeration status around the incorporated material. Baggs *et al.* (2003) found emissions of $N_2O$ to be up to seven times higher from no-till treatments than those from conventionally tilled treatments. This was attributed to the creation of anaerobic conditions under the mulch in the no-till treatment, with localized concentrations of mineralized C and inorganic fertilizer $NO_3^-$ being conducive to denitrification.

Several studies have found rates of microbial activity to be higher in no-till soils than in ploughed soils. Doran (1980) found that denitrifier populations were up to 43 times greater, and nitrifier populations up to 20 times greater, in the surface 7.5 cm of no-till soils under maize residues. $CH_4$ consumption is also very sensitive to cultivation, which lowers uptake rates following tillage. Several authors have attributed this effect to increased $NH_4^+$ availability following cultivation, but soil physical factors may also play a role.

## 12.3.5 Rice production

Rice production is an important part of tropical food production, particularly in Asia. The area of rice harvested has increased by 75% since 1935. About 90% of the world's harvested area of rice paddies is located in Asia, and about 60% of this is located in India and China. In the period 1995–1997, the global rice production area was approximately 154 million ha, to which 15.4 Tg of N was applied as fertilizer. Because of the unique nature of rice production, we will treat it separately from other row crops. With typically flooded soils and relatively high N input, there is a potential for high emission of $CH_4$ during flooded periods and high $N_2O$ emissions during non-flooded periods.

During the past 15 years, a large number of field studies have quantified $CH_4$ emissions from rice fields during the growing season (e.g. Wassmann *et al.*, 2000). These, and a host of other studies, have shown that emissions are affected by several factors related to both natural conditions and to crop management. In reviewing the numerous field studies, however, it seems justified to distinguish between primary factors that determine the level of emissions (water regime, organic amendments, soil characteristics and climate), and secondary factors that modulate emissions within a smaller range (i.e. the selection of rice cultivars, use of sulphate fertilizers, etc.). Given the broad scope of this chapter, the following discussion will focus on the primary factors.

Permanent flooding favours the formation of large amounts of methane, whereas even short periods of soil aeration significantly reduce emission rates. An unstable water supply is a condition generic to rainfed rice, so that this production system is generally characterized by lower emissions of methane than irrigated rice. In vast parts of equatorial Asia, rainfed rice suffers from dry periods either at the beginning or at the end of the growing season, an occurrence that reduces the overall emissions by ~ 50%. However, ample and evenly distributed rainfall may create soil conditions comparable to irrigated rice in some rainfed systems; for example, in eastern India emissions were similar between irrigated and rainfed rice (Adhya et al., 2000). Permanent flooding of rice fields over the entire annual cycle is found in some remote parts of Central China, and leads to extremely high emissions of approximately 900 kg $CH_4$/ha/year (Khalil et al., 2000). Consistent flooding throughout the growing season, an occurrence that is relatively common during the wet season crop in large areas of South-East Asia, also leads to high emissions. Numerous field studies under this type of flooding indicate that emissions range from less than 100 up to 500 kg $CH_4$/ha.

In many rice-growing regions of China, the flooding of the fields is interrupted by short drainage periods in the middle of the growing season. Although the reduction effect varies considerably at different locations and in different seasons, in comparison to permanent flooding, this local practice reduces emission rates by between 20% and 40% in most cases. In northern India, irrigation has to compensate for high percolation losses, so that frequent flooding causes pulses of oxygenated water to replenish the flood water as frequently as once a day (Jain et al., 2000). Methane emissions from this type of irrigated rice field are generally below 30 kg $CH_4$/ha (Jain et al., 2000).

The second management factor determining the level of emission rates is the quantity of organic inputs. Traditional agriculture in China uses relatively large amounts of manure, leading to high emission rates. The decline of this practice over the last few decades has subsequently led to a major reduction in the methane source strength of Chinese rice fields. In addition to exogenous organic material, such as animal manure, the management of crop residues (i.e. straw, stubble and roots) influences methane emissions. Incorporation of rice straw into the soil generally stimulates emissions, but the incremental effect depends on the timing of the straw application. The practice of straw addition is rather unpopular among those farmers who have access to other fertilizers, so the plant parts remaining in the field represent the only input of organic material into the soil. Under these conditions of low organic inputs, even the height of the stubble can have an impact on emissions.

The comparison of emissions under identical crop management and measurement protocols clearly demonstrates the strong influence that natural environmental factors have on $CH_4$ emission. In incubation studies, rice soils showed a wide range of $CH_4$ production potentials that may be responsible for the pronounced variation of in situ emission rates found even within small areas (Wassmann et al., 1998). Climate can also act as a natural determinant of methane emissions. In northern China, rice is grown in one crop that experiences a very low temperature at the late phase of the growing season, which brings emissions down to very low levels. This temporary impediment to emissions occurs during a period of the growing season that represents the bulk of methane release in other rice-growing regions. Thus, in comparison with a tropical climate, the temperature regime of this temperate climate causes a significant reduction in the cumulative emission of methane over the entire season.

Sass (1994) reviewed $CH_4$ emission studies in China, India, Japan, Thailand, the Philippines and the USA in order to estimate global $CH_4$ emissions from rice cultivation. He combined the data on total area of rice paddies with published flux estimates. The rice-growing areas in the countries considered represented 63% of the total world's rice paddy area, and resulted in an estimated annual $CH_4$ emission of 16–34 Tg. Extrapolating these estimates to the world gave a global range of between 25.4 and 54 Tg/year.

New emission models and Geographic Information System (GIS) databases are being used to narrow down the enormous uncertainties implied in recent estimates of methane source strengths. One approach, combining a GIS database on rice ecosystems with data on soils and weather, allowed the computation of national source strengths under different crop management scenarios (Matthews et al., 2000). The baseline scenario, assuming no addition of organic amendments and permanent flooding of the fields during the growing season, yielded $CH_4$ emissions of 3.73, 2.14, 1.65, 0.14 and 0.18 Tg $CH_4$/year for China, India, Indonesia, the Philippines and Thailand, respectively.

Relatively few studies have quantified annual fluxes of $CH_4$ and $N_2O$ in rice-based cropping systems. In many rice-based agricultural areas, one or two rice crops and an upland crop are grown. Between cropping periods, there are fallow times when no crops are grown. Of the few studies that have quantified $CH_4$ and $N_2O$ emissions through whole annual cropping sequences, most notable are those conducted at the International Rice Research Institute (IRRI), in the Philippines, on rice–fallow–rice–fallow and rice–fallow–wheat–fallow cropping sequences. In these studies, automated chamber systems were employed, which permitted several flux measurements per day at each location throughout several years of measurement (Bronson et al., 1997a,b; Abao et al., 2000). For the cumulative amount of $CH_4$ and $N_2O$ (expressed as g $CO_2$ equivalents/m² per season) emitted during each rice production (Bronson et al., 1997a) or fallow season (Bronson et al., 1997b), see Table 12.5. The GWP of $CH_4$ emissions generally exceeded the GWP of $N_2O$ from continuously flooded rice. Incorporation of rice straw increased $CH_4$ emissions during rice cropping, but had little effect on $N_2O$ emissions. When rice fields were drained, particularly at the midtillering stage of rice crop development, $CH_4$ emissions substantially decreased while $N_2O$ emissions increased. A side-by-side comparison of the effect of midseason drainage on trace gas emissions showed that total GWP was little changed, in comparison with continuously flooded rice. The total GWP for $CH_4$ and $N_2O$ combined was 1040 g $CO_2$ equivalents/m² for the drained plots compared with 800 for the continuously flooded plots. In terms of GWP, the increase in $N_2O$ emissions more than offset the decrease in $CH_4$ emissions due to field drainage.

Tsuruta et al. (2000) found similar trends in which $CH_4$ fluxes were relatively high during intermittently flooded cropping and $N_2O$ emissions were low. The opposite was the case during the fallow period. Over the ~ 480 observation periods it became clear that $N_2O$ is an important part of rice-based agriculture's GWP. Where two rice crops were continuously flooded, GWP totalled

**Table 12.5.** Seasonal $CH_4$ and $N_2O$ emissions expressed as GWP in terms of $CO_2$ equivalents from rice field fertilized with urea or urea plus rice straw (Bronson et al., 1997a,b) under different management regimes. All plots were fertilized with 200 kg urea-N/ha for dry season and 120 kg N/ha for wet season in four equal split applications at plot harrowing, midtillering, panicle initiation and flowering stages. A second set of treatments consisted of application of rice straw (5.5 Mg dry straw/ha with 32 kg N/ha) and urea-N (dry season: 160 and wet season: 80 kg/ha). Drainage was conducted at panicle initiation or at midtillering stages. Two fields were continuously flooded and never drained. Measurements are for wet or dry seasons, with season length indicated in parentheses. Data are also presented for a short and a long fallow.

| Crop | Season (days) | Drainage timing | Urea fertilized | | Urea + rice straw | |
|---|---|---|---|---|---|---|
| | | | $N_2O$ | $CH_4$ | $N_2O$ | $CH_4$ |
| Fallow | (46) | – | 832 | 4.2 | 1420 | 7.7 |
| Rice | Dry (111) | Panicle initiation | 122 | 296 | 78 | 5520 |
| Fallow | (36) | – | 2340 | 4.6 | 2350 | 5.5 |
| Rice | Wet (98) | None | 179 | 907 | 81 | 9840 |
| Fallow | (89) | – | 263 | 98 | 492 | 93 |
| Rice | Dry (97) | Midtillering | 870 | 168 | 225 | 5340 |
| Rice | Dry (97) | None | 353 | 441 | 38 | 9340 |

~ 5800 g of $CO_2$ equivalents/m², with about 70% of this figure being contributed by $N_2O$. During the three rice-growing periods, the total GWP was ~ 2300 g $CO_2$ equivalents/m², with ~ 72% being contributed by $CH_4$ emissions.

### 12.3.6 Agroforestry

Agroforestry systems are those systems that incorporate a tree component in the production system. This would include tree-based systems and systems that combine trees with row crops or livestock. Improved fallow systems that aim to restore soil fertility, fodder plantations, windbreaks, and riparian forest management are all examples of agroforestry.

Tropical tree-based cropping systems (e.g. banana, cocoa, coconut, coffee, oil palm and rubber) covered approximately 16 million ha in 1995. Although little trace gas emission data exist for such crops, field studies have been conducted in banana (Veldkamp and Keller, 1997) and papaya (Crill et al., 2000) plantations in Costa Rica and in a rubber plantation in Sumatra, Indonesia (Tsuruta et al., 2000). Veldkamp and Keller (1997) measured $N_2O$ and NO emissions from two soil types (Andosol and Inceptisol) within a 400 ha banana plantation in a humid tropical area of Costa Rica. Emissions were higher from the Inceptisol (18.3 and 20.5 kg N/ha/year for $N_2O$ and NO, respectively) than from the Andosol (4.7 and 10.4 kg N/ha/year for $N_2O$ and NO, respectively).

Few measurements have been made of trace gases in complex agroforests, but they have begun to be made on a small scale in the Alternatives to Slash-and-Burn Program that spans the humid tropics. In slash-and-burn systems in southern Sumatra, forest is cleared and replaced by a multistorey rubber agroforest system. This 'jungle rubber' system is characterized by a relatively high density of rubber trees, in which other useful fruit and timber trees are interplanted. The system is established through a complex succession of production stages, involving the planting of crops and trees for commercial and domestic products. When mature, this system has a forest-like structure. Measurements comparing these systems with indigenous forests in Jambi province, in southern Sumatra, gave somewhat equivocal results for $CH_4$ fluxes. A primary forest site dominated by dipterocarp (*Dipterocarpus crimtus*) and mahogany (*Swietenia macrophylla*) was a strong sink for atmospheric $CH_4$, whereas a forest dominated by mahogany and *Scaphium macropodum* was a net $CH_4$ source (Fig. 12.3). Soils were a weak sink during the logging and burning phase, and sink strength was only slightly stronger under the jungle rubber plantation. $N_2O$ emissions were easier to interpret, with higher emissions being associated with disturbance (Fig. 12.3). The emissions were moderate in the primary forest sites, higher in the logged sites, low following the burning phase of regeneration, and low in the rubber plantation. Although a lack of replication and a lack of detailed site descriptions make it difficult to draw broad generalizations about the effects of the conversion of

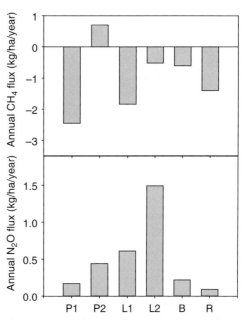

**Fig. 12.3.** Annual $CH_4$ and $N_2O$ fluxes in a slash-and-burn agricultural system in Sumatra (Tsuruta et al., 2000). Data are for primary forest (P1, P2), logged forest (L1, L2), a burned site (B) and jungle rubber agroforest (R).

forest into tree-based agricultural systems, we do note a transient increase in $N_2O$ emissions associated with increased N availability at these sites (Ishizuka *et al.*, 2000). For $CH_4$, diffusion appeared to be the main factor controlling $CH_4$ fluxes in these soils. Scaling these and other results up to the landscape level, using a spatial database of land cover derived from satellite imagery, Prasetyo *et al.* (2000) estimated that, between 1986 and 1992, land-use change in Jambi province resulted in the loss of 50 million t of C from above-ground biomass. Annual greenhouse gas emissions from soils increased by 4.3 million t, in the case of $CO_2$, and 256 t, in the case of $N_2O$, while the soil $CH_4$ sink decreased by 183 t/year. These soil GHG emissions equate to 4.4 million t of C equivalents per year, with $N_2O$ accounting for 0.18% of this radiative forcing and $CH_4$ accounting for <0.01%.

Agroforestry systems were also the subject of a long-term experiment established in 1985 in the Peruvian Amazon (Palm *et al.*, 2002). The study was conducted at the Yurimaguas Experimental Station in the Peruvian Amazon. The area has a long-term annual average temperature of 26°C and an average annual rainfall of 2200 mm. The experiment provided six land management systems, from which $N_2O$ and $CH_4$ fluxes were compared over a year and a half of measurement. At the beginning of the experiment, a 10-year-old shifting cultivation forest fallow was slashed and burned, according to local practice, and five land management systems were established. The land management treatments included two annual cropping systems ('high' and 'low' input), two tree-based systems (a multistrata agroforestry system, and a peach palm tree plantation), as well as a secondary forest fallow control. There was no primary forest nearby to serve as the control. The shifting cultivation forest fallow that was left undisturbed, to serve as a control treatment, was approximately 22 years old when the trace gases were sampled. Results of this study showed that high-input agriculture was a net source of $CH_4$, whereas the other systems were sinks (Fig. 12.4). The fallow control was the strongest sink. The high and low input agriculture systems were greater $N_2O$ sources than the agroforestry systems, and the fallow control was an even lower source. No NO measurements were made in this study.

In agroforestry systems designed to restore or maintain soil fertility, trees are often grown in the fields, or nearby, and tree litter is used as a green manure. $N_2O$ flux is very much dependent on the quality of the plant litter that is produced and incorporated into the soil (Millar, 2002). Greater emissions have been recorded following incorporation of residues with low C : N ratios, such as those of legumes, than have been recorded after incorporation of material with high C : N ratios, such as cereal straw (Kaiser *et al.*, 1998; Baggs *et al.*, 2000). Baggs *et al.* (2001) showed that $N_2O$ production in a controlled-environment experiment is influenced by the polyphenol content of the agroforestry prunings and by their ability to

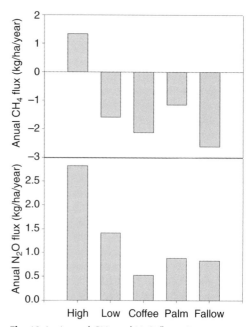

**Fig. 12.4.** Annual $CH_4$ and $N_2O$ fluxes in row crop and tree-based agricultural systems in Yurimaguas, Peru (Palm *et al.*, 2002). Data are for high- and low-input cropping systems, shade coffee and peach palm plantations and a 22-year-old tree fallow.

bind proteins. Emissions from high-quality *Gliricidia sepium* prunings (C:N 12, total extractable polyphenols (TEP) 1.3%, protein binding capacity (PBC) 22 µg bovine serum albumin (BSA)/mg) were significantly higher than those from *Calliandra calothyrsus* (C:N 13.6, TEP 3.5%, PBC 317 µg BSA/mg) or from *Peltophorum dasyrrachis* (C:N 20.0, TEP 3.9%, PBC 245 µg BSA/mg) prunings, due to the rapid release of N from the *Gliricidia* (Fig. 12.5). Emissions after the addition of *Calliandra* and *Peltophorum* leaves were significantly lower than from the *Gliricidia* treatment, despite similar C:N ratios. The lower $N_2O$ emissions and mineralization from the former prunings can be attributed to the higher polyphenol contents and higher protein binding capacities of *Calliandra* and *Peltophorum*.

In improved fallow systems in western Kenya, total $N_2O$ emissions over 34 days, following incorporation of *Sesbania* residues (2 kg $N_2$O-N/ha), were higher than they were following the incorporation of *Macroptilium atropurpureum* and natural fallow residues (Fig. 12.6). A flux of 7.2 g $N_2$O-N/t/ha/day was measured in the *Sesbania* treatment on the first day after incorporation. This result was attributed to the rapid release of N from this high-quality (high N, low lignin) residue.

### 12.3.7 Effects of combined organic and inorganic N applications on $N_2O$ emissions

It is commonly believed that combining organic matter with an inorganic fertilizer will increase synchrony and reduce losses, by the redistribution of inorganic N into organic forms. Studies have generally looked at organic inputs of a lower quality, such as crop residues. Trade-offs exist between the possible reductions in yields associated with the use of organic materials and the greater potential nutrient losses associated with the use of inorganic nutrients alone. Is it possible that high-quality organic materials could reduce losses of inorganic N without considerably reducing yields?

Janzen and Schaalje (1992) found that fertilizer N losses were twice as large when green manure plus fertilizer was applied to barley. Their interpretation was that green manure promoted high levels of nitrate and available carbon in the soil, enhancing denitrification. Losses were reduced with the use of smaller, repeated applications of green manure, again implying that the use of high-quality green manures as a partial substitute for, rather than addition to, inorganic fertilizer N may increase nutrient use efficiency. Xu *et al.* (1993a,b) found that large losses of N occurred (25–41% of the N added from *Leucaena leucocephala*

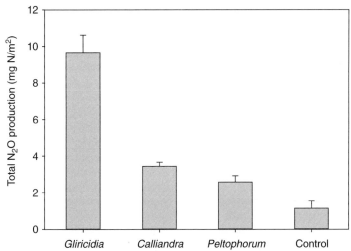

**Fig. 12.5.** Cumulative emissions of $N_2O$ from soil, during the 23 days following the incorporation of prunings from three agroforestry tree species: *Gliricidia sepium*, *Calliandra calothyrsus* and *Peltophorum dasyrrachis*.

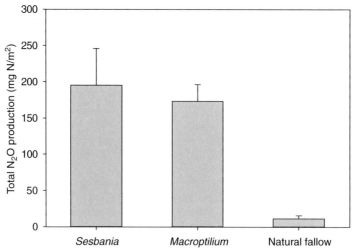

**Fig. 12.6.** Cumulative emissions of $N_2O$ during the 36 days following incorporation of residues from two improved fallow tree species, *Sesbania sesban* and *Macroptilium atropurpureum*, and the natural fallow vegetation in western Kenya.

prunings), which they attributed to denitrification. Losses were greater when materials were incorporated rather than surface applied. Although these three studies did not compare the losses from fertilizer alone, they do indicate that losses from high-quality organics alone can be quite high. Additionally, Ganry *et al.* (1978) showed that incorporation of a low-quality organic matter (straw) with fertilizer could result in large losses of fertilizer N, through denitrification. These studies indicate that N losses can be quite large from both organic and inorganic sources, contrary to the popular belief that the application of organic sources will result in fewer losses.

Other studies exist that make it difficult to make broad generalizations. Azam *et al.* (1985) reported that applications of ammonium sulphate combined with *Sesbania aculeata* residues reduced total losses of fertilizer N by up to 30%, by increasing microbial immobilization, but increased the contribution of *Sesbania* N to N losses, as a result of increased mineralization of *Sesbania* N. Baggs *et al.* (2003) found that combined organic and inorganic additions to temperate soils had an interactive effect on $N_2O$ emissions that was either positive or negative, depending on the quality of the residue applied and the cultivation technique employed. This interaction was positive in a no-tillage rye system, but negative in the conventionally tilled rye system. Negative interactions such as this indicate the potential that management has to reduce $N_2O$ emissions from such systems.

## 12.4 Conclusion

In comparison with temperate systems, few measurements have been made of trace gas fluxes in tropical agricultural systems, but the underlying processes that produce or consume these gases are the same, regardless of the system or the climate. An understanding of this allows us to draw some broad generalizations about the nature of the impact agriculture has on atmospheric GHG loading. However, pinning down the magnitude of this impact is more difficult. In general, tropical agriculture contributes to atmospheric loading of both $CH_4$ and N oxides. Management significantly influences both the magnitude of N oxide and $CH_4$ emissions. Management also affects the magnitude of $CH_4$ consumption in upland systems. As tropical agriculture intensifies, and as inorganic fertilizer use becomes more widespread, the contribution that tropical agriculture makes to increased GHG loading will surely increase.

The increasing intensification of agriculture in the tropics will most certainly increase N-oxide emissions, as N fertilization must increase, for many crops, to sustain production. The increase in N-oxide emissions, however, may be of the same order as that expected in temperate crop production, and thus would be smaller than some have predicted. From the studies presented above, comparison of $N_2O$ emissions from temperate agriculture indicates that $N_2O$ emissions from tropical cropping systems are not, as they are often quoted to be, higher than agricultural systems in temperate zones. Even the very high N-oxide emissions noted by Veldkamp *et al.* (1997) fall within the range of emissions from temperate agricultural soils. The potential certainly exists for large emissions from tropical systems, but the timing of rainfall and competition between plants and soil microbes for soil N appear to limit N-oxide emissions. As a result, appropriate management in high-input agricultural systems could limit N-oxide emissions. It is likely that such considerations would also increase fertilizer use efficiency.

One must bear in mind the fact that the generalizations made in this chapter and our current understanding of tropical agroecosystems are based on very limited information. Few studies exist for most tropical agroecosystems outside those that measure $CH_4$ from flooded rice fields. Since trace gas fluxes vary by orders of magnitude in time and space, depending on land use and management, soils, rainfall and nutrient input, it is clear that more information, representing the major land uses and agricultural systems and encompassing seasonal and annual cycles, is needed. Additionally, the fluxes of the suite of gases $CO_2$, $CH_4$, $NO_x$, $N_2O$ and possibly volatile organic carbon compounds (VOCs) need to be quantified simultaneously in ways that further our understanding of the impact not only of changes in land use, but also of changes in agricultural management. This information could then be used to help address issues of crop production, water quality, and air quality across local and regional scales.

---

**Conclusions**

**1.** Although measurements of trace gas fluxes in tropical agricultural systems are few compared with temperate systems, the underlying processes that produce or consume these gases are the same, regardless of the system or the climate.
**2.** In general, tropical agriculture, even at current low levels of fertilizer use, contributes to atmospheric loading of both $CH_4$ and N oxides. However, management can significantly influence both the magnitude of N oxide and $CH_4$ emissions and the magnitude of $CH_4$ consumption in upland systems.
**3.** Intensification of agriculture in the tropics, with increased use of N fertilizer, will most certainly increase N-oxide emissions; however, this increase will probably be no greater than that experienced in the temperate zones.

---

**Future research needs**

**1.** As trace gas fluxes vary by orders of magnitude in time and space (depending on land use and management, soils, rainfall and nutrient input) it is clear that more information (encompassing both seasonal and annual cycles) is needed on the major land-use and agricultural systems.
**2.** The fluxes of the suite of gases $CO_2$, $CH_4$, $NO_x$, $N_2O$ and possibly volatile organic carbon compounds (VOCs) need to be quantified simultaneously in ways that further our understanding of the impact had not only by changes in land use but also by changes in agricultural management.
**3.** The information obtained from studies concerned with the above two points should be used to help address issues of crop production, water quality, and air quality across local and regional scales.

# 13 Benefiting from $N_2$-Fixation and Managing Rhizobia

Paramu L. Mafongoya, Ken E. Giller, David Odee, S. Gathumbi, S.K. Ndufa and S.M. Sitompul

---

**Key questions**

1. Do all legumes fix atmospheric $N_2$?
2. Do all legumes have to be inoculated? How can you decide in a specific case?
3. How can genotype × environment interactions be managed to increase the amount of $N_2$ fixed?
4. How much $N_2$ is fixed by different legumes in different systems?
5. What is the fate of fixed N in cropping systems?

---

## 13.1 Introduction

Nitrogen is an element key to soil fertility and the development of sustainable food production systems. Although other elements, such as phosphorus (Chapter 7, this volume), potassium, magnesium and micronutrients (Chapter 6, this volume), are of course essential for plant growth, in most situations it is the N supply that determines crop yields. Soil N contents are maintained by the natural processes of biological $N_2$-fixation and atmospheric deposition, as well as by the addition of organic manures and mineral fertilizers (Giller et al., 1997b). Plant-available N is easily leached or volatilized from the soil, but soil organic matter (SOM) also contains N, which can contribute to a sustained N supply. This chapter considers the amount of N contributed by biological $N_2$-fixation, the fate of the fixed N, and the necessity for inoculating with $N_2$-fixing bacteria.

In the tropics, $N_2$-fixation contributes to agricultural production in three areas: firstly, in crops in which the $N_2$ fixed is translocated directly into the harvested product, be it grain, fruit, fodder or timber; secondly, in cases when the N fixed goes into fodder and is used in animal production; and thirdly, by contributing to the maintenance and replenishment of soil fertility.

Legumes may increase the productivity of other crops when incorporated into agroecosystems, either as intercrops or as a crop within a rotation. In intercropping systems, legumes contribute protein-rich grain, and also assist in maintaining soil fertility by fixing their own N. Grain legumes are often intercropped with cereals, for instance, common bean (*Phaseolus vulgaris*) is often intercropped with maize in Central and East Africa, and cowpea (*Vigna unguiculata*) is often intercropped with sorghum or millet in the Sahel. The N contribution made by legumes in intercropping systems is linked to the decay of crop residues, roots, nodules and fallen litter.

In rotational systems, legumes can contribute a substantial amount of N to subsequent crops. However, the amount of N that grain legumes contribute in cereal rotations is often small, because much of the N is removed in the grain. But, when compared with green manure or tree legumes, grain legumes are often much more attractive to farmers, because they contribute directly to the household food supply. The residual benefits legumes make to the cereal crops that follow them can often be greater than would be expected from the amount of N in the residues at harvest, indicating that N contributions made by fallen leaves and below-ground residues may be significant (Kasasa et al., 1999). Other rotational effects may also be important, such as pest control and the improvement of soil physical properties. Since 1996, soybean has expanded rapidly as a smallholder crop in Zimbabwe, because it has a good market value (Mpepereki et al., 2000). Another major contribution grain legumes can make to soil fertility is the provision of cash to buy fertilizers.

Although herbaceous legumes clearly have the potential to improve soil fertility, their actual use is surprisingly infrequent (Giller, 2001). In most cropping systems, only small amounts of organic residues are contributed by green manures, which limits the role they can play in maintaining soil fertility. Woody (tree) legumes generally produce biomass at a greater rate than herbaceous legumes. The yield response of crops following improved fallows containing tree legumes normally depends on biomass and N accumulation within the fallows (Szott et al., 1999). Given that biomass accumulation peaks after about a year in most herbaceous legumes, woody legumes are more effective in fallows of more than 1 year, since they allow more biomass to be accumulated. Much of the biomass and N produced may, however, be exported as wood from the field. In many systems, legume fallows of 6 months in duration accumulate insufficient N to produce yield responses beyond one subsequent crop. Longer duration fallows, of 2–3 years, involve the accumulation of larger quantities of N, and provide a residual yield effect sufficient to benefit two or three subsequent crops (Sanchez, 1999b; Szott et al., 1999).

## 13.2 Nodulation and $N_2$-Fixation in the *Leguminosae* Family

The *Leguminosae* (or, more correctly, the *Fabaceae*) is one of the largest families of flowering plants, and contains a large number of $N_2$-fixing species. However, not all legumes can nodulate and fix $N_2$. Surprisingly, few reports of legume nodulation have been published since the seminal work of Allen and Allen (1981). Some of the new additions to the corpus can be found in the paper by de Faria et al. (1989), which uses a more up-to-date legume taxonomy. Collation of these data shows that more than 39% are the genera have still not been examined for nodulation, and the data for a further 3% are ambiguous. Many of these genera include woody shrubs and trees with great potential for exploitation in agroforestry systems. In general, the taxonomy of the *Leguminosae* acts as a good guide as to which trees are $N_2$ fixers.

The *Papilionoideae* is generally regarded as a subfamily in which nodulation is widespread (Sprent and Parsons, 1999). However, there are several well-documented cases of non-nodulation. Of the 677 genera, 165 remain to be checked for nodulation. Most of those with no reports of nodulation are herbaceous plants, small shrubs and a number of important tree genera, which urgently need to be examined. One such important genus is *Pterocarpus*. There are consistent reports of its nodulation in Africa, but there are no reports of nodulation in Brazil, despite extensive searches and inoculation experiments (de Faria et al., 1989).

The *Mimosoideae* subfamily only has 66 genera; but, of these, 15 have not been checked for nodulation. Most of those remaining to be checked are located in tropical West Africa, though there are some significant groups in Argentina and Paraguay. It has been reported that at least six genera lack the ability to nodulate (de Faria *et al.*, 1989; de Faria and de Lima, 1998). One of the genera for which data are ambiguous is *Parkia* (Allen and Allen, 1981).

Out of the 256 genera in the *Caesalpinioideae*, 84 remain to be checked for nodulation. Most of these are in West Africa, tropical South America and Central America (Sprent and Parsons, 1999). Although it is likely that many of these do not nodulate, a number of important genera in this subfamily (e.g. the tree genus *Erythrophleum*) do nodulate. Another interesting genus is *Parkinsonia*, which was cited by Halliday and Nakao (1982) as nodulating. However, recent evidence by Sprent and Parsons (1999) indicates that it is unable to nodulate.

The state of our knowledge on nodulation in woody legumes has increased in recent years. Even when a genus can nodulate, it may not do so under certain conditions. The task of checking remaining genera for their ability to nodulate under different conditions remains incomplete.

It is worth mentioning here that $N_2$-fixing symbioses are also formed between certain non-legume trees and actinomycetes (*Frankia* spp.). These are termed 'actinorrhizal symbioses' (Giller, 2001). The most important of these non-legume trees are the *Casuarina* spp. from Australia, which are used, throughout the tropics, for soil stabilization, as windbreaks and to produce poles and fuelwood. However, actinorrhizal symbioses have been poorly studied in agroforestry.

## 13.3 Rhizobial Classification

The rhizobia that nodulate legume trees have received little research effort, in comparison with those that nodulate grain legumes. The taxonomy of rhizobia has developed rapidly since the advent of phylogenetic methods for bacterial classification based on sequence analysis of the 16S rRNA gene (Young, 1992). The older classification methods were largely based on the legume host range of the rhizobia. Although this approach was recognized to be severely flawed early on (Wilson, 1944), the new method of classification reveals the true extent of such problems, in that nodulation ability is closely related to the nodulation genes carried by the rhizobia, rather than to their evolutionary similarity (Young and Haukka, 1996). Such nodulation genes are generally carried on transmissible plasmids in rhizobia and, in some cases, on transmissible 'symbiotic islands' of chromosomal DNA. Rhizobia far apart on the phylogenetic tree may thus carry the same nodulation genes and have a very similar host range for nodulation and $N_2$-fixation.

Research undertaken in the last 10 years has revealed some surprising overlaps between rhizobia that can nodulate legume trees and herbaceous and grain legumes. The best documented case is that of *Rhizobium* sp. NGR234, which has been shown to effectively nodulate legumes from 112 genera, including members of the three different subfamilies of the *Leguminosae* (Pueppke and Broughton, 1999). Another broad-host-range rhizobia that nodulates trees is *Rhizobium tropici* (Martinez-Romero *et al.*, 1991), which nodulates *Leucaena* spp. and *Phaseolus vulgaris*. Slow-growing rhizobia are all considered to be members of a single genus (*Bradyrhizobium*) whereas the fast-growing rhizobia have been split into several genera. Five genera of fast-growing rhizobia are currently recognized: *Rhizobium*, *Azorhizobium*, *Sinorhizobium*, *Mesorhizobium* and *Allorhizobium*. These genera should be regarded with caution, since these classifications are currently being revised. Some trees are reported to nodulate only with slow-growing rhizobia (Dreyfus and Dommergues, 1981), but the vast majority examined nodulate with fast-growing rhizobia. Recent studies of rhizobia that nodulate legume trees have resulted in the description of the new species *Sinorhizobium terangae* and *S. saheli*, isolated from nodules of *Sesbania* and *Acacia* spp. in Senegal (de Lajudie *et al.*, 1994), and *Mesorhizobium plurifarium* (de Lajudie *et al.*, 1998), *Sinorhizobium arboris* and *S. kostiense* (Nick *et al.*, 1999), isolated from nodules of *Prosopis* and *Acacia* species in Sudan and Kenya, respectively.

Although it is widely accepted that legumes are generally highly specific in terms of their rhizobia, this view has largely been based on experience gained of temperate grain and pasture legumes. When a much broader range of legumes is examined, including many of the tropical legumes, it becomes clear that promiscuity for nodulation is actually the 'normal' condition. Generally, specificity occurs only in limited situations, when legumes are taken away from their normal geographic range or habitats (Giller, 2001). A detailed study of rhizobia that nodulate fast-growing legume trees in soils from across the tropics showed that species like *Leucaena leucocephala*, *Calliandra calothyrsus* Messn. and *Gliricidia sepium* were nodulated by a wide range of rhizobia, which guaranteed their nodulation success in most soils (Bala and Giller, 2001). *Sesbania sesban* (L.) Merr. was an exception for which indigenous rhizobia were not generally found in soils, except in the case of those from low-lying areas in Africa typical of the habitat of this short-lived tree.

## 13.4 Quantification of $N_2$ Fixed by Different Legumes

### 13.4.1 Methods of measuring $N_2$-fixation

Several methods are used to estimate fixation rates. These include acetylene reduction assay, the N difference method, the N solute method, isotope dilution techniques and the natural $^{15}N$ abundance method. This chapter does not review these methods, and readers are encouraged to refer to the excellent review written by Giller (2001).

Previously, N accumulation was used to indicate $N_2$-fixation rates, but could not be used to actually distinguish $N_2$-fixation from the plants' ability to scavenge soil N. Hence, previous estimates of $N_2$-fixation were often exaggerated. There is no single 'correct' way of measuring $N_2$-fixation. No one technique will provide an accurate measure of $N_2$-fixation for all legumes grown in any soil under diverse environmental conditions. Each technique has its own unique advantages and limitations. Problems encountered in measuring $N_2$-fixation are magnified in trees, as compared with grain and pasture legumes. The roots of trees may penetrate very deeply into the soil (see Chapter 4, this volume). They may, therefore, take up soil N from pools with $^{15}N$ signatures very different from those from which reference plants obtain their N, a fact that introduces errors into $N_2$-fixation estimates calculated using the natural $^{15}N$ abundance method. The wide variety that occurs in the rooting patterns of different tree species makes it difficult to select suitable reference crops. Similarly, because of the difficulty inherent in labelling soil evenly, to a sufficient depth, using $^{15}N$ fertilizer, estimates based on the enrichment method are likely to be flawed. Nevertheless, in the last decade the likely range of $N_2$-fixation rates has been quantified for a range of agroforestry trees using the $^{15}N$ natural abundance method (Table 13.1; Gathumbi *et al.*, 2002a).

A large amount of information exists concerning the amount of $N_2$ fixed by grain and pasture legumes (Giller, 2001). Most $N_2$-fixation studies have, however, been conducted on research stations, and thus may not reflect the situation in farmers' fields, where a restricted supply of nutrients and water may limit $N_2$-fixation. Most of the legumes used for food, forages, or in agroforestry systems are $N_2$ fixers. A review of all the data available on $N_2$-fixation shows that grain legumes derive between 50% and 98% of their N supply from $N_2$-fixation (equivalent to about 80–200 kg N/ha under non-limiting conditions). Forage legumes can derive between 70% and 90% of their N supply from $N_2$-fixation (the equivalent of 60–380 kg N/ha) and tree legumes can derive between 14% and 100% of their N supply (70–270 kg N/ha) from $N_2$-fixation (Giller, 2001). This clearly shows that the amount of $N_2$ fixed varies

**Table 13.1.** Net N contribution of 9-month-old monoculture legume fallows to the overall soil N economy. N derived from $N_2$-fixation was estimated using the natural $^{15}N$ abundance method using a range of non-$N_2$-fixing reference plants. Adapted from Gathumbi et al. (2002a).

| Fallow species | Above-ground biomass N (kg/ha) | N source | | Soil (kg/ha) | N off-take (wood) (kg/ha) | N balance[d] (kg/ha) |
|---|---|---|---|---|---|---|
| | | $N_2$-fixation (kg/ha) | (%) | | | |
| Crotalaria | 177 | 142 | 80 | 35 | 28 | 149 |
| Tephrosia | 150 | 100 | 67 | 50 | 33 | 117 |
| Pigeon pea | 148 | 91 | 62 | 57 | 38 | 110 |
| Sesbania | 100 | 52 | 52 | 48 | 36 | 64 |
| Siratro[a] | 145 | 64 | 44 | 81 | 0 | 145 |
| Calliandra | 55 | 24 | 44 | 31 | 9 | 46 |
| Groundnut | 18 | 8 | 44 | 10 | 0 | 18 |
| S.E.D.[b] | 24.0*** | 20.1*** | | 12.6*** | 4.9*** | 25.7*** |
| CV (%)[c] | 31.3 | 41.4 | | 38.9 | 33.7 | 39.3 |

[a]*Macroptilium atropurpureum*; [b]SED, standard error of the difference between treatment means (***$P = 0.001$); [c]CV, coefficient of variation; [d]Amount of total above-ground N yield minus N off-take in wood.

widely among legumes and among different genotypes of the same legume, across environments in which legumes are grown. It is difficult to make useful generalizations, as the $N_2$-fixation ability of different legumes is affected by a variety of factors. The amounts of $N_2$ fixed by grain, pasture and green-manure legumes are reviewed in more detail by Giller (2001). This review will focus, therefore, on tree legumes and, in particular, on knowledge of their $N_2$-fixation rates obtained in the last decade.

### 13.4.2 How much $N_2$ can trees fix?

Estimates of $N_2$-fixation by legume species are, to a large extent, site specific. For instance, recent work on a Kandiudalfic Eutrudox in western Kenya showed that pigeon pea (*Cajanus cajan* (L.) Millsp.), *Calliandra calothyrsus* and siratro (*Macroptilium atropurpureum* (DC.) Urb.) fixed approximately 62%, 46%, and 43% of their N requirement (Gathumbi et al., 2002a), as compared with the corresponding percentages of 65%, 14%, and 92% reported for the same species, respectively, in Australia and India (Peoples and Craswell, 1992). In Indonesia, *Sesbania sesban* in pure stands was estimated to fix about 84% of its N requirement, as compared with the estimated range of 46% to 59% reported in Kenya. This variation is due to site factors, the tree genotype used and effectiveness of the microsymbiont.

In a study of $N_2$-fixation in field-grown herbaceous/shrub/tree legume species, *Crotalaria grahamiana* was found to be highly dependent on $N_2$-fixation (75–83%), whereas only about 54% of the total N in *Calliandra calothyrsus* was found to be derived from fixation (Gathumbi et al., 2002a). The amount of fixed N contained in the above-ground biomass of 9-month-old fallows using these species ranged from 8 to 142 kg/ha (Table 13.1). Slow growth and/or poor establishment of some species at early growth stages apparently led to a low demand for N relative to the supply from the soil N pool, as indicated by the positive correlation between total N uptake and the proportion derived from $N_2$-fixation (Cadisch et al., 2002a). It is hypothesized that, since the soil N pool remains the same, species with a greater N demand must obtain a greater proportion of their N from fixation. This implies that legumes have substantial control over the rate of $N_2$-fixation, and can fix sufficient $N_2$ to make up the shortfall in mineral N supply.

The contribution that $N_2$-fixation makes to the overall N balance of a soil can be summarized in the following simple expression, modified from Peoples and Craswell (1992):

$$\text{Net N-balance (kg N/ha)} = Nf - Nw \quad (13.1)$$

where $Nf$ = proportion of N from $N_2$-fixation × total fallow N (kg N/ha) and $Nw$ = wood N (kg N/ha) exported.

Assuming that all the above-ground foliage biomass was retained in the soil, Niang et al. (1996) estimated that improved fallows in western Kenya could recycle between 18 and 149 kg N/ha to crops planted following a 9-month fallow. However, after testing these systems in the field, and after taking into account the amount of N exported in the wood, the N contributions were smaller than expected.

Mixing legume species has the potential to increase above-ground and below-ground resource acquisition, due to complementarity of acquisition niches (see Chapter 6, this volume) and compensatory gains resulting from the synergistic plant interactions. The net gain is primarily governed by the degree of inter- and intraspecific competition during the growing period, as a result of the establishment vigour of the different species and the availability of below-ground and above-ground resources. The hypothesis that increased competition for soil mineral N in species mixtures results in increased $N_2$-fixation in mixed fallows has been tested using a Kandiudalfic Eutrudox soil in western Kenya. The results obtained indicated that this hypothesis does not generally hold true (Table 13.2). $N_2$-fixation in mixtures increased only when the mixture out-performed the single-species fallows, as in the case of the Sesbania + pigeon pea mixture (where both the proportion and amount of $N_2$ fixed were larger in the mixed system). In the other mixtures, Sesbania suffered from competition exerted by the fast-growing Crotalaria, due to its slow rate of establishment. Hence, total biomass production in these mixtures was less than in sole-Crotalaria fallows, resulting in less demand for soil N and, hence, reduced $N_2$-fixation (Cadisch et al., 2002a). Evidence that growth in mixtures increases the proportion of N derived from $N_2$-fixation (but not necessarily the amount of $N_2$ fixed) has been observed when legumes were grown in the vicinity of non-fixing crops, e.g. in maize–bean intercrops (Giller and Cadisch, 1995) and hedgerow intercropping (Hairiah et al., 2000a).

**Table 13.2.** Comparison of $N_2$-fixation in 9-month-old single and mixed-species legume fallows. Adapted from Cadisch et al., (2001a).

|  | N derived from atmosphere | | | N derived from soil |
|---|---|---|---|---|
|  | (%) | (kg N/ha/species)[b] | (kg N/ha/system)[c] | (kg N/ha/system)[c] |
| Sesbania | | | | |
| Sesbania alone | 30 | 12[c] | 23 | 61 |
| Sesbania + Crotalaria | 14 | 2 | 46 | 87 |
| Sesbania + siratro[a] | 29 | 12 | 22 | 53 |
| Sesbania + pigeon pea | 44 | 10 | 108 | 49 |
| Crotalaria | | | | |
| Crotalaria alone | 50 | 57[c] | 113 | 117 |
| Sesbania + Crotalaria | 36 | 44 | 46 | 87 |
| Pigeon pea: | | | | |
| Pigeon pea alone | 62 | 36[c] | 71 | 48 |
| Pigeon pea + Sesbania | 73 | 98 | 108 | 49 |

[a] Macroptilium atropurpureum; [b] kg N/ha/species, estimated N yield of monoculture at same density as in mixture, i.e. monocrop yield/2; [c] kg N/ha/system, actual yield of all legume components in the field, per area.

## 13.5 Managing Environmental Constraints to Increase $N_2$-Fixation

### 13.5.1 Genotype × environment ($G^2$ × E) interactions

Apart from restricted N supply, environmental limitations that adversely affect plant growth and vigour generally decrease the amount of $N_2$-fixation occurring in legumes, although the symbiosis is sometimes more sensitive to such constraints than other aspects of plant growth (Giller, 2001). $N_2$-fixation is sensitive to nutrient deficiencies, in particular P deficiency (which may restrict nodule formation if acute). Molybdenum deficiency influences $N_2$-fixation directly, as Mo is a component of the nitrogenase enzyme. $N_2$-fixation is thought to be more sensitive to drought stress than other processes (such as photosynthesis), although the evidence is somewhat equivocal. A further interesting feature of the legume–rhizobium symbiosis is that sensitivity to stress may be expressed through the bacteria, the legume host or the formation of the symbiosis itself. Legumes are invariably more sensitive to salinity than rhizobia (Sprent, 1984), but rhizobia are more sensitive than their hosts to heavy metal pollution. Large differences in sensitivity to stresses (such as aluminium toxicity in soil) are found among rhizobial strains and among legume hosts.

If the environment (E) cannot readily be altered by farmers, then we must maximize our exploitation of ecological adaptation in terms of genotype × environment interactions ($G^2$ × E, where $G^2$ refers to the genotype of both legume $G_l$ and its microsymbiont $G_r$). A tremendous variety of grain, pasture or tree legumes and their rhizobia remain to be exploited. Much variation in tolerance to climatic and edaphic stresses has been demonstrated. In this section we will discuss promising management options intended to improve $N_2$-fixation by the direct selection of rhizobia in soil or by the screening and breeding of legumes for increased $N_2$-fixation.

### 13.5.2 Effects of pruning on nodulation and $N_2$-fixation

Legume trees are often pruned severely in agroforestry systems, to provide fodder or foliage for soil amendment. It is well established that defoliation causes nodule senescence (Witty and Minchin, 1988) and that pruning, or the browsing of trees by animals, causes temporary decreases in the rates of $N_2$-fixation. Re-establishment of $N_2$-fixation depends on the formation of new nodules; this can sometimes occur rapidly, as legumes often harbour latent infections in young roots that can develop when N demand in the plant is large. However, researchers found that the pruning of *Erythrina poeppigiana* resulted in complete mortality of nodules, coupled with a lag of 10 weeks before active nodules were re-formed (Nygren and Ramirez, 1995). Nygren (1995) highlighted the danger that nodulation might be completely suppressed if trees are pruned too frequently. Nodule senescence associated with defoliation of trees is a mechanism by which N is made available to plants growing in close proximity; however, the amounts of N released in this way are likely to be small, as total nodule biomass is a small fraction of the total N in the plants.

The influence that periodic pruning has on nodulation is not well documented (Sanginga *et al.*, 1990). Snoeck (1996) observed that when the aerial parts of *L. leucocephala* were cut there was a substantial decrease in the quantity of nodules. Furthermore, after defoliation, it took almost three months for the quantity of nodules and the level of $N_2$-fixing activity to return to pre-defoliation levels. In a more detailed study, Kadiata *et al.* (1997) observed that when plants were defoliated once at 16 months after planting, nodule dry weight decreased by 12%, as compared with non-defoliated plants. However, in plants that were defoliated once at 12 months after planting, nodule dry weight decreased by as much as 54%.

### 13.5.3 Effects of soil phosphorus and pH on $N_2$-fixation

Aluminium toxicity (which increases with decreasing soil pH) and phosphorus deficiency are recognized as major factors affecting legume growth and nodulation. There has been some debate as to whether acidity affects plant growth or the associated rhizobia responsible for nodulation and $N_2$-fixation. Acid-tolerant rhizobia strains (such as TAL 1145) have been selected (Halliday and Somasegaran, 1983) but these have not been any more effective than non-acid-tolerant strains when used in the field.

Phosphorus fertilization and liming of soil can also affect N supply via symbiosis; but, the results obtained have not been very clear thus far. A combined application of lime and phosphorus in three soils in Nigeria significantly enhanced growth of *L. leucocephala* (Duguma *et al.*, 1988). That there is high demand for phosphorus by *L. leucocephala*, as a requirement for both normal growth and nodulation, is supported by more recent work by Brandon and Shelton (1997a,b). Another important aspect of this problem is demonstrated by the dependence of *L. leucocephala* on mycorrhizal associations for improved P supply (Shepherd *et al.*, 1996). Phosphorus deficiency can be overcome by phosphorus fertilization, but this is often not a practical solution for smallholder farmers. Hence, there exists the need to select species or provenances of trees or grain legumes able to grow in soils that are poor in terms of available phosphorus (see also Chapter 7, this volume). Large differences in growth and P-use efficiency were observed among 11 accessions of *L. leucocephala* by Sanginga *et al.* (1991). These differences were crucial at early growth stages, and suggest that the selection of *L. leucocephala* accessions tolerant to low phosphorus deserves further research. Such a research effort could also be extended to other tree legumes and herbaceous legumes. Studies by Sanginga *et al.* (1995) have shown that *Gliricidia sepium*, a species widely used in agroforestry systems, does not need large, available concentrations of soil P to fix $N_2$. However, little is known of the P requirements of other tree species important in agroforestry systems. Mycorrhizal associations, and the production of organic acids that can bring sorbed P into solution, are proposed as the main mechanisms by which P-use efficiency may be increased (Chapters 7 and 14, this volume). The variability of these attributes in different legume species and provenances is hardly known, and is another area in need of further research.

### 13.5.4 Effect of available soil nitrogen on $N_2$-fixation

The inhibitory effect of plant-available N on nodulation and $N_2$-fixation is common in most symbiotic $N_2$-fixing associations. For example, $N_2$-fixation by well-nodulated *L. leucocephala* was reduced by 50% by the application of 40 and 80 kg/ha of fertilizer N in two experimental treatments (Sanginga *et al.*, 1988). The extent of the inhibition is highly variable, and reflects the tolerance to N application of both the host legume and the rhizobial strain (Sanginga *et al.*, 1995). As N accumulates in the soil, dependency on $N_2$-fixation declines, though it may resume if soil N levels subsequently decline. A study by van Kessel *et al.* (1994) demonstrated that $N_2$-fixation declined steadily over time as N accumulated in a *Leucaena* plantation. In fact, $N_2$-fixation ceased within 6 years of plantation establishment. If this observation reflects the general situation with tree legumes, agronomic solutions need to be devised that maintain $N_2$-fixation rates. One such strategy would be the exploitation of the variation between species or accessions and the selection and breeding of genotypes that have greater tolerance to soil N, as suggested by Sanginga *et al.* (1992) for *Gliricidia sepium*.

### 13.5.5 Nematodes and rhizobia: interaction effects on $N_2$-fixation

Many of the tree legumes used in improved in fallows are susceptible to attack by nematodes (see Chapter 15, this volume). Root-knot nematodes (*Meloidogyne javanica*) have been observed to have negative effects on nodulation in several legumes (Taha, 1993), leading to decreased plant vigour. Recent studies in western Kenya have demonstrated the occurrence of negative interactions between *S. sesban*, in terms of nodulation, and nematodes (Desaeger and Rao, 2000). The potential for soil N replenishment by $N_2$-fixing legumes can therefore be greatly reduced by root-knot nematodes. It is thus necessary for screening programmes for fallow tree legumes to be implemented, to examine their ability to fix $N_2$ and nodulate in the presence of nematodes. Desaeger and Rao (2001a) demonstrated interesting differences in tolerance to nematode infection between two provenances of *S. sesban* ('Kisi' and 'Kamega'). Kisi demonstrated higher productivity than Kamega, partly because of a greater tolerance to root-knot nematode and greater nodulation in the presence of nematode infection. This striking difference suggests that there is scope for selecting tree legumes that can effectively nodulate in the presence of nematodes.

## 13.6 The Need to Inoculate with Rhizobia

As mentioned above, the productivity of agricultural plants is often discussed in terms of the G × E interaction. In dealing with legume–*Rhizobium* symbiosis, the interaction is second order, i.e. $G^2$ × E, where $G^2$ refers to the genotypes of the legume ($G_l$) and its microsymbiont ($G_r$). In order to effectively harness the potential of this symbiosis, we must fully understand these multiple interactions.

It has been proposed that rhizobial genotypes are differentially adapted to soil conditions and that it is this adaptation, rather than the relationship with the host, that governs outcomes related to competition or persistence (Howieson *et al.*, 2000). Expressed in terms of the G × E formula, it can be contended that symbiotic effectiveness ($G^2$) is unimportant to the persistence and nodulation success of rhizobia relative to adaptation of the rhizobia to soil and climate ($G_r$ × E). An alternative hypothesis is that the soil population of rhizobia may be enriched with strains that are effective on resident legumes as a result of the superior growth of effectively nodulated plants and the subsequent release of rhizobia from their nodules. In this case $G^2$, rather than $G_r$ × E, is important to rhizobial population development. Some legume species may facilitate this process through their capacity to select effective rhizobial strains from a mixed soil population (Robinson, 1969). There appears to be little evidence to support this in the tropics, although populations of rhizobia may be increased by compatible crops in the field (e.g. Andrade *et al.*, 2002).

There is considerable potential for the improvement of $G^2$ in suboptimal tropical legume symbioses. An understanding of $G^2$ × E interactions should lead to progress in the optimization of $N_2$-fixation. The conceptual framework (Fig. 13.1 and Table 13.3) developed by Howieson *et al.* (2000) can be used to study these multiple relationships. This scheme proposes three relatively common scenarios, leading to research options that arise when investigating $N_2$-fixation and legume nodulation. For each scenario there exist a number of research options that could be used to improve $G^2$:

- Scenario 1, where the soil contains a large population of variably effective rhizobia that cause reduced $G^2$.
- Scenario 2, where the soil contains a small population of variably effective rhizobia that cause reduced $G^2$.
- Scenario 3, where the soil does not contain rhizobia capable of nodulating with the host legume of interest and hence inoculation is required.

In a review of rhizobium inoculation in tropical cropping systems, Date (2000) felt that inoculation would be beneficial in the following scenarios:

- When the legume of interest, or another legume with related symbionts, has not been grown previously on the land.
- When the legume grown previously on the land was poorly nodulated.
- In a rotation when a legume follows a non-leguminous crop.
- In the reclamation of severely degraded land.

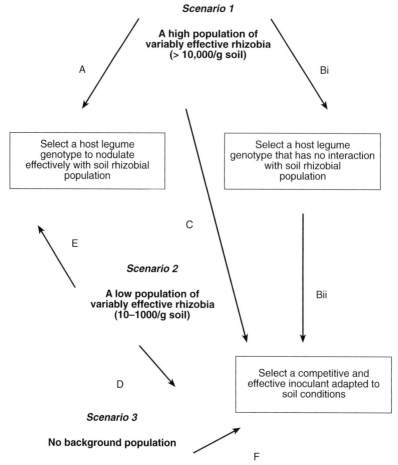

**Fig. 13.1.** A schematic representation of strategies to improve $N_2$-fixation in legumes through the selection of either the host or rhizobial genotype (from Howieson et al., 2000). For letters in the figure see Table 13.3 and text for detailed explanations.

**Table 13.3.** Examples in which the application of research pathways A–F (given in Fig. 13.1) have been successful with tropical legumes and have resulted in improved symbiotic $N_2$-fixation.

| Pathway (Fig. 13.1) | Legume | Reference | Location |
|---|---|---|---|
| A, E | Glycine max (promiscuous) | Mpepereki et al. (2000) | Africa |
| | Tephrosia vogelli | P.L. Mafongoya (unpublished) | Africa |
| | Crotalaria spp. | | |
| | Cajanus cajan | Mapfumo et al. (2000) | Zimbabwe |
| | Leucaena, Calliandra, Gliricidia | Bala and Giller (2001) | In tropics |
| Bi, Bii | Glycine max (specific) | Mpepereki et al. (2000) | Zimbabwe |
| C | Arachis hypogaea | Date (2000) | Australia |
| | Calliandra calothyrsus | Date (2000) | Australia |
| D | Desmanthus spp. | Date (1991) | Australia |
| F | Glycine max (specific) | Bushby et al. (1983) | Australia |
| | Acacia crassicarpa | P.L. Mafongoya (unpublished) | Zambia |
| | Sesbania sesban | Bala and Giller (2001) | Across the tropics |

The presence of nodulating legumes is an obvious indicator of the presence of an indigenous rhizobial population. Variations occur in the abundance of indigenous rhizobia, and are mainly driven by the presence or absence of an appropriate nodulating legume host (Odee et al., 1995). Indigenous rhizobial populations exceeding $2.4 \times 10^5$/g of soil have been reported for *S. sesban* and *Acacia* species in their naturalized habitats (Odee et al., 1995). Similarly, Bala (1999) showed that the presence of a particular rhizobial species or genotype is determined by the occurrence of the appropriate legume host. Roughley et al. (1995) showed that an established indigenous *Bradyrhizobium japonicum* population declined from $1.32 \times 10^5$ to $6.40 \times 10^2$/g of soil in the fifth year after the last soybean crop. Within this period, three crops of rice and one crop of triticale ($\times$ Triticosecale) had been sown, alternating with 6-month fallow intervals. In our own current work with improved fallow legumes in western Kenya (Table 13.4), we have established that populations of indigenous rhizobia can build up from undetectable levels to $2.5 \times 10^5$ *Sesbania*-rhizobia/g of soil in 18 months. The same work shows that rhizobia associated with fallow legume species (e.g. *Tephrosia candida* and *Crotalaria paulina*) are naturally abundant in these soils, indicating a recent history of legume hosts with similar rhizobial affinities.

### 13.6.1 Inoculation response: what is the critical size of the indigenous population?

Many studies have demonstrated the futility of inoculation with an introduced strain when populations of indigenous rhizobia are large. It is generally impossible to get an inoculation response in cases where indigenous populations are $> 10^3$ rhizobia/g of soil (e.g. Singleton and Tavares, 1986; Thies et al., 1990; Turk et al., 1993). However, this is not always a valid threshold and can be influenced by other factors, such as the effectiveness and competitiveness of both the inoculant strain and the indigenous rhizobia. For example, Turk et al. (1993) reported no significant increase in above-ground biomass following inoculation of *Acacia auriculiformis* grown on a soil with an indigenous rhizobia population of $< 10$/g of soil. By contrast, the response of *Acacia mangium* grown on a soil with an indigenous rhizobia population of $> 10$ rhizobia/g of soil was significant (see Table 13.5). Brockwell et al. (1995) argued that inoculation should not be used in soils with large numbers of indigenous rhizobia. Instead, attempts should be made to strategically manipulate the effective components of the populations. However, there may also be large populations of ineffective rhizobia; hence inoculation with a competitive and effective inoculant rhizobium strain will be necessary.

Table 13.4. Influence of natural fallow or fallow legume species (grown for 18 months), and two seasons of maize + bean intercropping, on size of indigenous rhizobial populations in soils in western Kenya (D.W. Odee et al., unpublished).

| | | MPN of rhizobia/g soil for species[a] | | |
|---|---|---|---|---|
| Site | Treatment | Cp | Tc | Ss |
| Teso, sandy soil | Site species | $4 \times 10^4$ | $2.4 \times 10^5$ | $2.5 \times 10^5$ |
| | Maize + beans | $1 \times 10^3$ | $4.5 \times 10^4$ | 0 |
| | Natural fallow | $2 \times 10^3$ | $4.5 \times 10^4$ | 0 |
| Kakamega, clayey soils | Site species | $4 \times 10^4$ | $4 \times 10^4$ | 0 |
| | Maize + beans | $3 \times 10^3$ | $4 \times 10^4$ | 0 |
| | Natural fallow | $2 \times 10^4$ | $4 \times 10^4$ | 0 |

[a]Cp, *Crotalaria paulina*; Tc, *Tephrosia candida*; Ss, *Sesbania sesban*.
MPN, most probable number.

**Table 13.5.** Inoculation response of legumes in relation to the size of indigenous rhizobial population in some tropical soils.

| Site | Species | No. of rhizobia/g soil | Inoculation response | Reference |
|---|---|---|---|---|
| Haiku, Hawaii | Glycine max | 0 | Significant[a] | Singleton and Tavares (1986) |
| | Vigna unguiculata | $1 \times 10$ | Not significant[a] | |
| | Phaseolus vulgaris | $1 \times 10^2$ | Not significant[a] | |
| Keahua, Hawaii | Phaseolus lunatus | 7.2 | Significant[b] | Turk et al. (1993) |
| | Leucaena leucocephala | $3 \times 10^3$ | Not significant[b] | |
| | Acacia auriculiformis | 8.6 | Not significant[b] | |
| | Acacia mangium | 41.3 | Significant[b] | |
| Muguga, Kenya | Calliandra calothyrsus | $1 \times 10^4$ | Not significant[c] | D.W. Odee et al. (unpublished) |

Inoculation response on the basis of: [a]total shoot N; [b]seed (grain legumes) and above-ground biomass; [c]shoot dry weight.

## 13.7 Fate of Fixed Nitrogen

### 13.7.1 Recovery by crops

The main purpose of many agroforestry practices undertaken with the aim of soil fertility replenishment is to provide nutrients, especially N, to the crops. Hence it is pertinent to ask: how much of the N supplied by prunings from agroforestry species is taken up by crops? The recovery of N supplied in leaves after pruning is quite low, generally ranging from 10% to 20% during the first cropping season (Giller and Cadisch, 1995; Palm, 1995). The recovery efficiency of N is affected by the quality of prunings, i.e. their N, polyphenol and lignin contents (Cadisch and Giller, 1997; Mafongoya et al., 1998). These values are similar to the N-recovery efficiencies from residues of annual legumes, but less than the recoveries from inorganic fertilizers. The fate of the remaining 80–90% of the added N has been the subject of debate and experimentation.

In sequential agroforestry systems (such as improved fallows) few attempts have been made to investigate the fate of fixed N. The amount of N released during the course of crop growth may be only 30–60% of the added N (Palm, 1995). The rest of the N will be held in undecomposed material in the soil. Thus, the proportion of released N recovered may be greater than 20%. However, if the amount of N taken up by crops and that remaining in undecomposed material are added up, approximately 60–80% remains unaccounted for, with the larger values corresponding to high-quality materials (Palm, 1995). It can be hypothesized that this N can either be found in SOM or in inorganic N leached beyond the crop rooting zone or that it has been lost through volatilization or denitrification. Nitrogen may also have been taken up by trees in intercropping situations. Based on $^{15}$N labelling of various tree prunings, Haggar et al. (1993) and Ndufa (2001) showed that the majority of legume N often ends up in readily mineralizable fractions of soil organic matter. These fractions did not only consist of the active soil pool (i.e. soil microbial biomass, which contained only between 1% and 5% of the added N) but included unprotected organic matter in various stages of decomposition ('light' fractions, i.e. fractions with a density of < 1.3–1.8 g/cm$^3$). Furthermore, the quantity of light-fraction SOM at the end of the fallow period was found to differ when different leguminous trees were used (Barrios et al., 1997); higher-quality inputs, such as those from S. sesban, were more effective in terms of increasing light-fraction SOM. The amount of N in the light fraction was correlated with maize yield. These studies demonstrate that legume inputs benefit crop production through the long-term build-up of soil organic N, as well as through the direct release of N from decomposing legume residues. The N remaining in undecomposed plant material and SOM fractions has the potential to contribute to the nutrition of subsequent crops. However, Mafongoya et al. (1997) and Cadisch et al. (1998), working with various tree legumes, found that only 3–5% of N was recovered during the three cropping seasons subsequent to the initial application of legume residue inputs. Similar values have been reported for grain–legume residues. These recovery rates were not strongly related to the quality of legume inputs and, hence, the residual effect cannot easily be managed through manipulation of pruning quality.

Nutrient release from SOM is dependent on the biologically active fractions of SOM. SOM fractions have been shown to be sensitive indicators of changes in soil fertility management (Barrios et al., 1997). Physical SOM-fractionation methods based on particle size have been used to separate SOM fractions of different composition and biological fractions. In physical fractionation the soil is separated into sand, silt and clay-sized fractions. The SOM associated with each mineral fraction is then separated by density fractionation into light and heavy fractions. A better understanding of the effect fallows have on the movement of residue N through various biologically meaningful SOM fractions is essential if we are to develop management practices aimed at optimizing N release and SOM build-up. Such information is useful in improving the prediction of the nutrient-supply capacity of tree prunings and plant residues.

It has been suggested that the decomposability (turnover) of organic fractions in soils decreases in the following order: sand > clay > silt content of the soil matrix (Christensen, 1987). Quantification of both the short- and long-term fate of legume N in SOM fractions and of the N supply to the subsequent maize crop is essential when developing management practices for optimizing N-use efficiency and sustainability. Ndufa (2001) found that, at 6 months after fallow residue incorporation, legume-$^{15}$N was distributed throughout all particle-size fractions (Fig. 13.2). However, the relative proportion of legume-N recovered in different fractions was different in *Calliandra calothyrsus*, as compared with *Sesbania sesban*, *Macroptilium atropurpureum* and urea treatments, though the latter treatments exhibited a similar distribution pattern (Fig. 13.2). A large amount of *Calliandra* residue-N was recovered in the > 212 μm organic fraction (OF). By contrast, a large amount of *Sesbania* and *Macroptilium* N was recovered in the < 20 μm fraction (20OMF (mineral organic fraction)). The fate of legume-$^{15}$N in different particle-size classes of SOM was thus greatly influenced by litter quality. Although more *Calliandra* residue-N was found in coarser fractions, its residues released less N. This signifies that it contributed more to the build-up of SOM. By contrast, residues of *Sesbania* and *Macroptilium* decomposed faster and the remaining residue-N became associated (stabilized) with smaller fractions (e.g. clay + silt fractions). Residues with a high lignin and/or polyphenol content may be more effective in the formation and stabilization of particulate SOM, but may not provide sufficient N to crops. By contrast, high-quality residues with a low lignin and polyphenol content or a low C : N ratio may, on the other hand, provide a short-term increase in the labile SOM fraction of C and N but have little effect on the maintenance of SOM. Therefore, both quantity and quality factors should be balanced if SOM improvement and nutrient availability to the crop are to be achieved. The long-term implications of such organic matter distributions require further investigation.

### 13.7.2 Leaching and gaseous N losses

Losses of N through gases such as $N_2O$ and ammonia depend on many soil and climatic factors, such as pH, the presence of soluble C and anaerobic conditions in the soil. Measured in laboratory incubations, loss of N via volatilization from green manures ranges from 5% to 50%, but is usually less than 20% of added N (Costa et al., 1990;

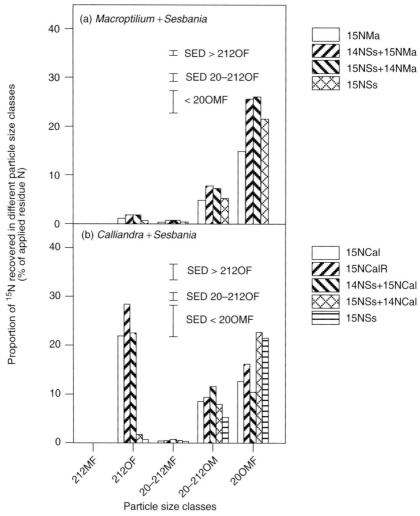

**Fig. 13.2.** Effect of mixing (a) *Macroptilium atropurpureum* and *Sesbania sesban* and (b) *Calliandra calothyrsus* and *S. sesban* residues on $^{15}$N recovery from different soil mineral and organic particle sizes, at Owano in western Kenya, 6 months after residue application. MF, mineral fraction; OF, organic fraction; OMF, mineral organic fraction. 15NSs+14NMa, $^{15}$N *Sesbania*+$^{14}$N *Macroptilium*; 14NSs+15NMa, $^{14}$N *Sesbania*+$^{15}$N *Macroptilium*; 15NMa, $^{15}$N *Macroptilium*; 15NCal, $^{15}$N *Calliandra*; 14NSs+15NCal, $^{14}$N *Sesbania*+$^{15}$N *Calliandra*; 15NSs+14NCal, $^{15}$N *Sesbania*+$^{14}$N *Calliandra*; 15NSs, $^{15}$N *Sesbania*; 15NCalR, $^{15}$N *Calliandra* in *Sesbania* plot (S.K. Ndufa, unpublished data).

Glasener and Palm, 1995). High-quality materials that release N rapidly may lose more N via volatilization (Glasener and Palm, 1995). Few studies have reported denitrification losses in tropical agroforestry systems. In improved fallow systems, where high N inputs are accompanied by high inputs of soluble C, this pathway could be very important and deserves further research (see Chapter 12, this volume). Leaching of nitrates is a possibility when there is a large pool of inorganic N in the soil, such as might occur if N release and demand by crops are asynchronous. Many field measurements have indicated that leaching in both subhumid and humid zones contributes to large losses of N (see Chapter 6, this volume).

## 13.8 Conclusions

Inputs of N from $N_2$-fixation in tropical cropping systems are limited both by the small proportion of legumes actually grown and by restrictions placed on the fixation rate of these legumes by drought and nutrient deficiencies (in particular of P). Inoculation may be necessary for some newly introduced grain legumes (such as soybean), but many other tropical legumes grown to improve soil fertility may not need inoculation. A tremendous amount of work has been done on grain legumes and their inoculation requirements. Major efforts now need to be made to disseminate ready-to-use technologies, in order to increase the adoption of the inoculation technologies by smallholder farmers. Whether inoculating legume seed with rhizobia bacteria will be useful under the conditions faced by African farmers is a matter that needs to be tested. Although many studies have shown that legume inoculation is simple, inexpensive, and highly successful in terms of increasing crop yields, the African experience has shown that the technology is not widely adopted. The issues that need to be addressed to increase adoption rates are:

- lack of knowledge about biological $N_2$-fixation or rhizobia inoculation;
- limited access to rhizobial inoculants when they are needed;
- lack of knowledge concerning the use of rhizobia;
- lack of good seed and good agronomic practices;
- poor soils;
- poor extension advice.

The development of agroforestry practices related to biological $N_2$-fixation has gained momentum in recent years, with the screening of large collections of leguminous multipurpose trees. Most research to date has concentrated on species × rhizobial interactions for effective $N_2$-fixation and, hence, upon the need to identify elite rhizobial strains for these newly introduced tree species. Much of the work on rhizobia–legume compatibility has been carried out under controlled conditions in glasshouses. It is essential that the effectiveness and competitiveness of such strains is determined under field conditions in the target area. We recommend inoculation trials using the method used by Date (1977). This is of paramount importance when new species or provenances are introduced whose rhizobial strain requirements are unknown. It is important to determine the biophysical boundary conditions of different species and their rhizobial requirements.

---

**Conclusions**

1. Use should be made of legumes that do not require inoculation.
2. Use should be made of rhizobia strains that effectively nodulate a wide range of hosts, e.g. *Tephrosia vogelli*, *Tephrosia candida*, *Acacia angusitissima*, *Gliricidia sepium* and *Cajanus cajan*.
3. Use should be made of management practices that build populations of soil rhizobia to eliminate the need for (further) inoculation.
4. Inputs of N from $N_2$-fixation in tropical cropping systems are limited by the small proportion of legumes actually grown, and by restrictions on the fixation rate of these legumes caused by nutrient deficiencies (in particular of P) and drought.
5. Inoculation, based on available technology, may be necessary for some newly introduced grain legumes (such as soybean), but many other tropical legumes grown for soil fertility improvement may not need inoculation.
6. Most research to date has concentrated on the identification of elite rhizobial strains for use in the introduction of new legumes, but the competence of such strains under field conditions in the target area often remains unknown.
7. The applicability of inoculating legume seed with rhizobia bacteria needs to be tested with regard to the conditions faced by African farmers, in order to increase adoption rates.

*Continued*

**Conclusions** *Continued.*

**8.** The development of agroforestry practices using $N_2$-fixating components has gained momentum in recent years with the screening of large collections of leguminous multipurpose trees.

**Future research needs**

1. A better assessment of losses of N in legume-based systems.
2. An improved understanding of rhizobia ecology using molecular techniques.
3. Quantification of the amount of $N_2$ fixed by different legumes under on-farm conditions.

# 14 Managing Mycorrhiza in Tropical Multispecies Agroecosystems

Thomas W. Kuyper, Irene M. Cardoso, Neree Awana Onguene, Murniati and Meine van Noordwijk

---

**Key questions**

1. How do mycorrhizal associations function in the context of a multispecies agroecosystem?
2. What role do mycorrhizal networks play in multispecies agroecosystems, and how important are these networks?
3. Through what mechanisms do the beneficial effects of mycorrhizas in multispecies agroecosystems become manifest?
4. How can mycorrhizal associations be represented in models of nutrient uptake and nutrient cycling?
5. How can mycorrhizal associations be managed in multispecies agroecosystems? Specifically, how can we determine the conditions under which there is a need for mycorrhizal management, rather than mycorrhizal inoculation? Can specific management practices that have a negative impact on mycorrhizal functioning be avoided?

## 14.1 Introduction

'Plants do not have roots, they have mycorrhizas' (Begon et al., 1996). This quoted here in order to stress the well-known, but often neglected fact that most plant species do not simply have roots – they have a mutually beneficial root–fungus association known as 'mycorrhiza'. In this arrangement, the fungus (just like the root system) receives carbohydrates from the above-ground part of the plant in exchange for mineral nutrients. Although mycorrhizas are generally accepted to be of importance, mycorrhizal research remains segregated into a niche, rather than being fully integrated into research on plant ecology and agronomy. However, on the basis of the near universal occurrence of mycorrhizas, we must subscribe to the warning given by Newsham et al. (1995):

> [b]oth ecologists and physiologists need to be aware that the results of experiments on non-mycorrhizal individuals of normally mycorrhizal plants are most probably artefacts.

In this chapter we will therefore consider plant–fungus interactions, placing particular emphasis on multispecies agroecosystems.

Mycorrhizal research has come a long way from its beginnings at the end of the 19th century. Research on tropical mycorrhizas effectively started in 1897, when Janse described some morphological 'curiosities' in the roots of many of the plants in the Bogor Botanical Garden, Indonesia (Janse, 1897). These curiosities turned out to be almost ubiquitous in plant roots. Discussion followed about the nature of such plant–fungus interaction, and it soon became clear that the fungus was (in most circumstances) not harming but, probably, benefiting the plant. Under conditions of high nutrient supply, however, the net benefit for the plant may be zero or even negative (whilst still being positive for the fungus).

As a number of recent reviews of the taxonomic (Cairney and Chambers, 1999; Hibbett et al., 2000; Morton and Redecker, 2001; Schlüßler et al., 2001), ecological (Allen, 1992; van der Heijden and Sanders, 2002), agronomic and silvicultural (Sieverding, 1991; Bethlenfalvay and Linderman, 1992; Gianinazzi and Schüepp, 1994; Pfleger and Linderman, 1994; Gianinazzi et al., 2002) aspects of mycorrhiza are available, we will here focus on the specific challenges of understanding and managing mycorrhizal associations in tropical multispecies agroecosystems, whilst placing emphasis on agroforestry systems. The methods used in mycorrhizal research will not be discussed in detail here (although Box 14.1 provides a very brief overview) and the reader is, therefore, referred to books by Norris et al. (1994), Brundrett et al. (1996) and Varma (1998).

Much research has focused on the life cycle of annual plants and thus on the sequence of events that lead a developing (crop) seed to become a fully mycorrhizal plant that can be harvested. The terminology and methods used when studying mycorrhizas in annual cropping systems have been borrowed from plant pathology – in which discipline, fungi establishing themselves on plant roots are viewed in terms of the potentially strong negative effects they may have on plant performance. The 'borrowing' of terms is evident in that researchers still use the expressions 'mycorrhizal infection of' or 'infection sites' when referring to host plants. Part of this conceptual bias also continues in discussions of whether mycorrhizal fungi can be considered to behave as parasites in those cases where non-mycorrhizal plants outperform mycorrhizal plants (Johnson et al., 1997).

Such a life cycle approach is also possible for perennial plants; but it should not be forgotten that the mycorrhizal fungus is an organism with a long lifespan. Lifespans of more than a hundred years have been reported for individuals of ectomycorrhizal fungi, whilst the lifespan of individual arbuscular mycorrhizal fungi (AMF) could, in principle, be almost indefinite – considering that they have exhibited an asexual life style for over 400 million years. A focus on short-cycle phenomena (such as annual spore formation and the establishment of new colonies as a result of germinating spores) is more relevant to a rotation of crop monocultures than it is to multispecies agroecosystems. This is especially true when long-lived woody plants are included in such systems. Persistence of the fungal mycelium should be the key interest in the latter case. In general ecological terms, management of mycorrhizal fungi in perennial, multispecies agroecosystems should be directed less at conditions where $r$-selected organisms and strategies (high growth rate, smaller-sized individuals, high reproductive output but smaller investment in survival) prevail, and more towards conditions where $K$-selected organisms and strategies (lower growth rate, larger-sized individuals, smaller investment in reproduction but larger investment in survival) prevail (Hart et al., 2001).

The balance between plant and fungus is easily overlooked in such partnerships: most (applied) ecologists look at the symbiosis exclusively from a 'plant's-eye' (phytocentric) view, and forget that a 'fungus'-eye' (mycocentric) view is equally valid. The phytocentric view considers the association between plant and fungus as something that (almost inevitably) results in maximum plant fitness. However, a more realistic approach is to address the question of the extent to which the maximization of plant

**Box 14.1.** Methods for arbuscular mycorrhizal research.

***Identifying mycorrhizal associations.*** Establishing the identity of an arbuscular mycorrhizal (AM) fungus is not easy. The microscopic structures of the mycorrhizal fungus in the plant root allow (to the experienced eye) identification to the level of fungal genus but not to species level, while the external mycelium is similar for almost all arbuscular mycorrhizal fungi (AMF). The taxonomy of AMF is, therefore, entirely based on spore structure. Although spores can be directly extracted from field samples (see below), their subsequent identification is often difficult. Therefore, field soils are often used as the basis for setting up pot cultures of different fungi (based on so-called trap plants). However, different trap plants and different cultivation conditions (temperature, soil pH, etc.) result in different species combinations. The other disadvantages of trap cultures are that: (i) they select for fungi that sporulate prolifically (which probably makes it necessary to have a sequence of traps, which takes a lot of time); and (ii) they sometimes miss fungi that are highly selective for certain plants. With the advent of modern molecular tools, it is now possible to identify the fungi in roots directly. Not all laboratories possess the necessary equipment, however. Another difficulty is that primers used to selectively amplify fungal DNA can be too restrictive (leaving out members of the *Paraglomaceae* and *Archaeosporaceae*) or too inclusive (also amplifying the DNA of fungi that do not belong to the *Glomales*).

***Determining the abundance of mycorrhizal fungi.*** The abundance of mycorrhizal fungi can be based on estimates of the number of fungal spores, the length of the extraradical mycelium in soil, or the extent to which plant roots are colonized (Varma, 1998). Spores can be extracted from the soil by washing and sieving, followed by centrifugation in a sucrose gradient. Fungal hyphae can be extracted from soil and the hyphae of AMF identified, as they are non-septate. From such samples, hyphal lengths can be calculated. Root fragments can be cleared and stained, after which fractional colonization can be assessed. It can be helpful to separate fractional colonization by hyphae, arbuscules and vesicles. Such an assessment of root colonization is difficult for roots that possess dark pigments. An important question we must ask is 'to what extent do different methods yield comparable results?' Efficiency of spore extraction depends on soil texture (spores can stick to aggregates in clayey soils, necessitating the use of a dispersant). Also, the size of the smallest screen of the sieve ultimately determines how many spores will be extracted. For these reasons estimates of spore abundance in similar ecosystems still show a very wide range. Comparisons between different methods can also yield divergent results. Spore extraction could bias the sample towards those species that are prolific spore formers. Onguene (2000) assessed mycorrhizal inoculum potential using three different methods (spore abundance, colonization of a test plant grown in disturbed soil, and colonization of a test plant in intact soil columns). The three methods yielded very similar results (Fig. B14.1). However, such good correlation between methods does not always occur, and the literature on this subject also provides examples where results obtained with different methods were substantially different.

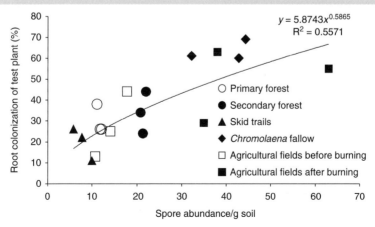

**Fig. B14.1.** Relationship between spore abundance and root colonization in a test tree (*Distemonanthus benthamianus*, Caesalpiniaceae) grown for 12 weeks, in six land-use types in three locations in southern Cameroon. After Onguene (2000).

fitness (or of plant primary production, as this is the farmer's target) is actually traded off against fungal fitness. Kiers *et al.* (2002) recently addressed the question of the extent to which farmers' management practices change the benefits of the mycorrhizal association. They suggested that agricultural practices would result in evolutionary changes in the fungus. More specifically, they predicted that increased use of fertilizer will lead to less effective mycorrhizal fungal genotypes, that crop rotation will prevent dominance of certain fungal genotypes, that tillage will have contrasting effects, and that inoculum addition will increase the chances of roots being colonized by more fungal genotypes, thereby allowing less beneficial ('parasitic') fungal strains to escape the defence mechanisms of the host.

'Superstrains' or 'superspecies' (single species or strains resulting in claimed, though often not proved, superior plant performance) may be insufficiently competitive or insufficiently able to reproduce. However, fungal species that maximize spore formation could do so at the expense of the plant, utilizing its carbon or the nutrients that could otherwise be transmitted to the plant. Strains could be 'superstrains' by virtue of a negative feedback mechanism, whereby the fungus demonstrates higher fitness on host plant A than on host plant B, yet plant B derives more benefit than plant A from that mycorrhizal fungus (Bever *et al.*, 2001). The occurrence of such negative feedback probably explains the observations made by Johnson *et al.* (1992), who noted that monocropping maize or soybean decreased diversity (evenness) of mycorrhizal fungi and led to a gradual decline in yield in these monocropping systems. Crops in a rotation that do not themselves depend on mycorrhiza, can give an important boost to the fungal population and, hence, have a positive effect on a subsequent crop in a rotation. If 'superstrains' are sufficiently competitive, they have the potential, under field conditions, to impoverish mycorrhizal fungal diversity. Plant species diversity and mycorrhizal fungal species diversity are often positively correlated (van der Heijden *et al.*, 1998). Low mycorrhizal species diversity is, in general, characteristic of highly fertilized or disturbed intensive agricultural systems (Johnson, 1993; Helgason *et al.*, 1998). By contrast, natural ecosystems and less intensively managed agroecosystems with a lower disturbance level are often characterized by high(er) diversity (see, however, Box 14.2).

Inoculating the trees at the nursery stage had a small, but statistically significant, positive effect on the fraction of trees that sur-

---

**Box 14.2.** Can mycorrhizal inoculation of trees help transform *Imperata cylindrica* grasslands into productive agroforestry systems?

Large areas of former rainforest in South-East Asia are covered in coarse grassland dominated by *Imperata cylindrica*. Fires tend to block succession to a woody secondary vegetation, and the systems are generally regarded as being degraded. Where such land is (or has become) accessible, it may be economically attractive to try to start an agroforestry land-use system. We must therefore ask: (i) will lands that previously supported mainly one species still host sufficient mycorrhizal inoculum to allow rapid tree growth?; and (ii) is it useful, or even necessary, to provide trees with suitable fungal partners at the nursery stage? Murniati (2002) tried to answer the latter question for four tree species (*Aleurites moluccana, Peronema canescens, Swietenia macrophylla* and *Artocarpus altilis*) using a series of experiments in East Kalimantan (Indonesia). A semicommercial product ('Mycofer') was used as an inoculum containing spores of four species of arbuscular mycorrhiza. Data collected included spore counts, spore identification (with 26 morphospecies of spores being identified in the survey as a whole), and records of the mycorrhization of roots and of the survival and growth of the trees.

In the grassland phase, the spore numbers of arbuscular mycorrhizal fungi were significantly lower in plots at the top of a ridge than they were on the midslope or in the valley. However, the number of morphospecies per 50 g of soil was only slightly lower in the ridge plots. In a total of nine samples, the cumulative number of morphospecies was 14, with an average of 5.6 species per sample.

**Box 14.2.** *Continued.*

During the first 2 years of the site's transformation to agroforestry (with samples being taken after 6 and 24 months; Fig. B14.2) the differences noted between sample positions disappeared, and overall spore numbers declined. In the samples taken after 6 months, the number of morphospecies was found to be slightly reduced in comparison with the number found in the grassland phase (36 samples now provided a total of 36 species, with an average of 3.8 per sample). However, at 24 months both spore numbers and the number of morphospecies identifiable in the samples were markedly reduced (12 samples contained seven species, with an average of 2.5 species per sample). Contrary to the expectations of the researcher, inoculating the trees with a mix of fungi had no positive or negative effect on spore numbers or diversity at 6 or 24 months. Of the four species introduced, only two were found in the soil at 24 months (and none at 6 months), with one dominating the spore numbers collected (the same being true in non-inoculated plots, despite trenching between plots), and one being found in small amounts. The fraction of tree roots that was mycorrhizal was 87 and 85% at 6 months, and 37 or 51% at 24 months, for inoculated and non-inoculated trees, respectively. These differences between treatments at a given sampling time were not statistically significant, but the hypothesis that inoculation would increase mycorrhization could be clearly rejected.

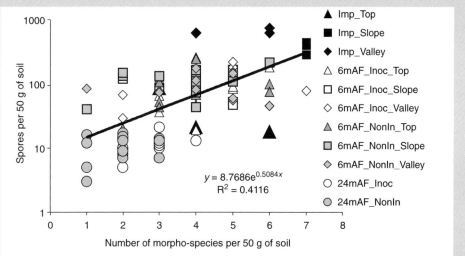

**Fig. B14.2.** Relationship between the number of morphospecies and total AM spore density in 50 g samples of soil taken in three landscape positions (Top = ridge top), in *Imperata cylindrica* grassland (Imp) before planting agroforestry tree species, and in the agroforestry plots (AF), 6 months after planting the trees, with or without inoculation with an AM spore mix (Inoc and NonIn, respectively); 24mAF represents a composite sample from AF plots at all slope positions, 24 months after tree planting. Based on Murniati (2002).

vived the transition from nursery to field (survival after 24 months was 80.0% and 86.5% for inoculated and non-inoculated trees, respectively ($P < 0.05$)). The overall conclusion of the research was that inoculation at the nursery stage is not essential for early tree growth, but it may have some positive effect on early survival. The shifts with time in spore densities and morphospecies composition were found to be substantial, so any observation on a single date may be difficult to interpret as an indication of whether or not inoculum potential is sufficient for a new site. In the grassland phase, the number of arbuscular mycorrhizas (AM) morphospecies clearly exceeded the number of plant species; an increase in plant diversity was accompanied by a reduction in AM spore diversity – but, of course, spore diversity is an incomplete indicator of AM fungal diversity.

On the basis of their morphology (which correlates reasonably well with fungal taxonomy) mycorrhizal associations can be divided into four major types:

- AM – formerly known as vesicular–arbuscular mycorrhiza (VAM);
- Sheathing mycorrhizas – including ectomycorrhizas (ECM), ectendomycorrhizas, arbutoid mycorrhizas, and monotropoid mycorrhizas;
- Ericoid mycorrhizas;
- Orchid mycorrhizas.

The last two mycorrhizal types occur in specific plant families, as their names indicate. AM are formed in the roots or rhizoids of a wide range of plants (mosses and liverworts, ferns, gymnosperms and angiosperms). In fact, the arbuscular mycorrhizal condition can be said to be closest to that of what might be termed 'the original plant'. Several lines of evidence demonstrate that the first primeval plant was AM (Pirozynski and Malloch, 1975; Brundrett, 2002). It might therefore be concluded that non-mycorrhizal plants evolved from AM plants. From such an evolutionary perspective, the phenomenon of non-mycorrhizal plants (Tester et al., 1987) needs to be explained, as it indicates that, under at least some conditions, plant fitness may be increased as a result of excluding the fungi from the roots.

An AM is formed by means of intracellular colonization by aseptate, obligatory symbiotic fungi belonging to the order *Glomales* (or even to an autonomous phylum *Glomeromycota*; see Schlüßler et al., 2001). The other types of mycorrhiza (the sheathing, ericoid and orchid mycorrhizal types) are formed by septate fungi belonging to the *Ascomycota* and *Basidiomycota*, and include many species with above-ground fruiting bodies ('mushrooms'). These other types of mycorrhiza also form their association by means of intracellular colonization; the only exception is the ECM type, in which the colonization of healthy roots is always intercellular (Smith and Read, 1997).

ECM associations are of minor importance in many multispecies agroecosystems, as most ECM plants are woody perennials. The most important ECM food crop is the genus *Gnetum*, which is usually colonized by the highly host-specific fungus *Scleroderma sinnamariense*, easily recognized because of its yellow mycorrhizas and hyphal cords. ECM associations might be important in agroforestry systems in which ECM trees surround (or share) agricultural fields. However, members of only a few families of tropical trees form ECM (with *Caesalpiniaceae*, *Uapacaceae* and *Dipterocarpaceae* being the most important families). Agroforestry systems with a high diversity of trees are, therefore, unlikely to contain a large proportion of ECM trees. The relative importance of ECM in different vegetation types is not very well understood as yet. Wubet et al. (2003) reported the virtual absence of ECM trees in Afromontane forests. In contrast, in the miombo woodlands of southeastern Africa, ECM trees belonging to the *Caesalpiniaceae* (e.g. genera such as *Brachystegia, Julbernardia, Afzelia*) dominate and produce large amounts of fruiting bodies. These mushrooms are edible, and can contribute a substantial amount of protein to the diets of local people, especially at the start of the rainy season when food reserves run low. They can also become a meaningful source of income. However, introducing non-indigenous ECM trees into tropical countries is not without risk, as there are strong indications that introduced species of the ECM tree genera *Pinus* and *Eucalyptus* can increase the rate at which the areas' original soil organic matter is broken down (Chapela et al., 2001). Introduced ECM trees could also harbour poisonous ECM mushrooms that local people are not familiar with. The remainder of this chapter will therefore only consider AM, as this association plays a very large role in seminatural ecosystems and in agroecosystems.

## 14.2 Arbuscular Mycorrhiza

An AM has three important components: the root itself, the fungal structures within the cells of the root (arbuscules, coils, vesicles, intraradical mycelium), and an extraradical mycelium that explores and exploits the soil for nutrients and then transports those nutrients to the root. On the basis of mycorrhizal morphology, the following two types can be recognized (Smith and Smith, 1996): (i) the *Arum*-type, which has distinct intracellular arbuscules and an extensive intercellular phase in the root cortex; and (ii) the *Paris*-type, which has intracellular hyphal coils but lacks an intercellular phase. Intermediate forms also occur, so the distinction is not absolute. Most agricultural plants form mycorrhizas of the *Arum*-type, whereas many tree species form mycorrhizas of the *Paris*-type. Root morphology (and hence the taxonomic identity of the plant) determines which type of mycorrhiza is formed. The functional significance of both types is, however, hardly known. It is striking, though, that plants that lack chlorophyll (and thus which are parasites or saprophytes) and form arbuscular mycorrhizas, have mycorrhizas of the *Paris*-type. The only AM example of interplant carbon transport through a common mycelial network (see below) involves a tree seedling with *Paris*-type mycorrhizas receiving carbon from a plant with *Arum*-type mycorrhizas (Lerat *et al.*, 2002). One is therefore tempted to speculate that the *Paris*-type is correlated with parasitic behaviour on the part of the plant.

AM are ecologically obligate. AMF cannot complete their life cycle in the absence of a host plant and, for that reason, cannot be grown in pure culture. In fact, AM fungi lack the ability to take up and metabolize carbon through the extraradical mycelium (all carbon must hence go through the intraradical mycelium), and it is likely that genes relevant to the carbon metabolic pathway were lost during their long symbiosis with plants. AMF can, however, be grown in Petri dishes in monoxenic cultures with root cultures (Fortin *et al.*, 2002). Under field conditions many, if not most, AM plants are also unable to complete their life cycle in the absence of the fungi, although under specific conditions (absence of competition, addition of nutrients, etc.) they can grow without them. Janos (1996) argued that we should separate the concepts of mycorrhizal dependency (an intrinsic characteristic of plants that has evolved under certain environmental conditions) and mycorrhizal responsiveness (which depends not only on plant species, but also on the identity of the fungal isolate(s) and the abiotic conditions present). However, the concept of mycorrhizal responsiveness (Box 14.3) is known in the older literature as mycorrhizal dependency (Plenchette *et al.*, 1983).

AM fungi are symbionts of a very diverse set of herbaceous plants, shrubs, and trees of temperate and tropical habitats. In most tropical soils, very few woody species of tropical trees are non-mycorrhizal. In French Guyana, 75 species were investigated and were all found to be mycorrhizal. In Korup National Park (Cameroon) 55 out of 56 species investigated were found to be mycorrhizal, and in southern Cameroon this was true of all 97 woody species investigated (Onguene, 2000). Most tropical crops are also strongly dependent on and responsive to arbuscular mycorrhizas (Sieverding, 1991). Only a few families and genera of plants do not generally form arbuscular mycorrhizas; these include *Brassicaceae* (their root exudates are possibly even toxic to AM fungi), *Caryophyllaceae*, *Cyperaceae*, *Juncaceae*, *Chenopodiaceae* and *Amaranthaceae* (although each of these families has some representatives that are usually colonized by AM fungi).

The taxonomic structure of the *Glomales* is depicted in Table 14.1.

The number of species of AM fungi discovered worldwide to date (159) is quite low, especially when we consider that there are probably more than 200,000 plant species that regularly form an arbuscular mycorrhizal association. Individual forest stands or grasslands can harbour between 30 and 50 AM fungal species (Bever *et al.*, 2001), whilst low-input or low-till agricultural systems can harbour up to 15 species (Franke-Snyder *et al.*, 2001; Jansa *et al.*, 2002). The obvious disparity between the number of AM plant species and the number of AM fungal species has traditionally been explained as being the result of a lack of specificity or selectivity on the part of the fungus. Such an explanation is based on the evolutionarily plausible scenario that, in mutualistic symbioses, there is no selection

**Box 14.3.** Mycorrhizal responsiveness.

Mycorrhizal responsiveness (MR) is defined as:

$$MR = (DW_{myc} - DW_{nonmyc}) / DW_{myc} \text{ (Plenchette et al., 1983)}$$

MR is expressed on a dry weight (or C) basis (DW = dry weight). Instead of using carbon as the currency to measure plant response to mycorrhizas, plant phosphorus content of shoots and roots can also be used. This results in the mycorrhizal phosphorus responsiveness (MPR):

$$MPR = (P\text{-content}_{myc} - P\text{-content}_{nonmyc}) / P\text{-content}_{myc}$$

MPR is higher than MR if mycorrhizal plants have higher P concentrations than non-mycorrhizal plants. Use of MPR could give a biased view of plant response, because seed P reserves are often included in the P balance of the plant. Increased P concentrations can also be an artefact of experimental systems in which the non-mycorrhizal plants are P-limited, whereas the mycorrhizal plants are limited by another, unknown, nutrient, but not by P. This leads to luxury P-uptake (for details see Cardoso, 2002).

MR and MPR usually range from 0 to 1 (if the non-mycorrhizal plant fails to grow the MR is 1). If non-mycorrhizal plants outperform mycorrhizal plants, the MR is negative. The value of the MR is plant-species dependent, fungal-species dependent, and soil dependent, which means that comparisons are usually difficult to make. Large-seeded plants often rely on internal reserves for a prolonged time and often give a low MR, even when application of inoculum increases mycorrhizal colonization. MR can easily be measured under laboratory conditions, under which non-mycorrhizal controls can be obtained. Under field conditions, it is rare to find soils that are completely devoid of mycorrhizal inoculum. Application of the concept of MR under field conditions is therefore questionable. Instead, a mycorrhizal inoculation effect (MIE) can be used. This indicates the effect the introduced inoculum has as compared with the inherent field inoculum, and is defined as:

$$MIE = (DW_{inoc} - DW_{uninoc}) / DW_{inoc} \text{ (Munyanziza et al., 1997)}$$

Again, MIE can be expressed on a C basis or on a P-uptake basis. MIE usually varies between 0 (if there is sufficient mycorrhizal inoculum and if field-inoculum quality is good enough) and 1 (if the amount of mycorrhizal inoculum is limiting under field conditions and/or the inoculum is not sufficiently effective). Negative values indicate either mycorrhizal redundancy (whereby costs for the mycorrhizal fungus in terms of carbon or nutrients are higher than mycorrhizal benefit) or that the applied inoculum is less beneficial than the field inoculum. In a comparison of five tree species in South Cameroon, Onguene (2000) noted MIEs of between 0.55 and 0.90 in skid trails, and values that were only slightly lower than these at landings. In an agricultural soil in which seedlings of *Terminalia superba* were grown, MIE was negative after the addition of inoculum collected in a pure stand of the grass *Paspalum conjugatum*. As mycorrhizal colonization increased after the addition of a grass inoculum, these data suggest that inoculum quality (related to selectivity of the inoculum) has an effect and should not be forgotten.

Usually, MIE data are compared after one growing season. However, in perennial agroecosystems, it is important to assess changes in MIE over time and then relate this to changes in the species composition of the inoculum.

Finally, different species of mycorrhizal fungi, or different mixed inocula, can be compared. Such comparisons can be useful for plants grown in pots but, for field conditions, persistence of those fungi is again as important as their initial effects.

---

for host specificity or selectivity. Consequently, a lack of specificity or selectivity was often taken for granted, and the issue not investigated. Until recently, there was also little empirical evidence for specificity between particular fungi and plants – at least at the fungal (morpho-) species level. During the last decade, however, more instances have been noted of selectivity in mycorrhizal associations, with specific combinations of plant and fungal species occurring more often than would be expected to result from chance alone. Previously, the fungi commonly used in experiments were

**Table 14.1.** Taxonomic structure of the *Glomales*. (After Morton and Redecker, 2001.)[a]

| Order | Family | Genus | No. species[d] |
|---|---|---|---|
| *Glomales*[b] | *Archaeosporaceae* | *Archaeospora* | 3 |
| | *Paraglomaceae* | *Paraglomus* | 2 |
| | *Glomaceae* | *Glomus*[c] | 85 |
| | *Acaulosporaceae* | *Acaulospora* | 31 |
| | | *Entrophospora* | 4 |
| | *Gigasporaceae* | *Gigaspora* | 5 |
| | | *Scutellospora* | 29 |

[a] Schüßler *et al.* (2001) proposed a somewhat different classification with a strong inflation in taxonomic rank (one phylum (*Glomeromycota*), four orders, and eight families).
[b] Although *Glomerales* would be grammatically correct, we prefer the use of the well-known name *Glomales*.
[c] *Glomus* is not monophyletic and needs to be split in three groups.
[d] Species number taken from the INVAM website (http://invam.caf.wvu.edu/), but many undescribed species still await formal recognition.

those most amenable to culture conditions (i.e. mostly generalist, *r*-selected species). It is possible that because of this choice of species, selectivity was underestimated and also that the potential for mycorrhizal networks (see below) was overestimated. Many recent studies have found unknown spore types (which researchers have so far not been able to cultivate) or unknown molecular types, based on unique sequence differences obvious between them and known species. Such findings support the idea that some degree of selectivity exists.

## 14.3 Mycorrhizal Functioning in (Multispecies) Agroecosystems

The essential beneficial effects that mycorrhizal fungi have on plants are a result of their ability to absorb nutrients in their inorganic (mineral) form more efficiently than a plant could alone (i.e. in a less costly manner). Therefore, the role that mycorrhizal fungi play in absorbing nutrients is most relevant under conditions of low nutrient availability, such as those commonly found in (sub)tropical (agro)ecosystems (Smith and Read, 1997). Mycorrhizal fungi extend the depletion zones around roots (see Chapter 10, this volume) of elements such as phosphorus (P) and zinc (Zn), which are relatively immobile as a result of their low diffusion rates. Under dry conditions, such as those occurring in semiarid climates, mycorrhizas may also be important in that they enhance nitrate uptake. Due to the much smaller diameter of hyphae (on average 5–10 µm, compared with 10–20 µm for root hairs and 100–500 µm for plant roots) and the large amounts of hyphae in soil, the total absorptive area is greatly increased in comparison with that of roots alone. If, for instance, the length of hyphae (hyphal diameter 10 µm) is 20 times that of roots (root diameter 200 µm) per unit soil volume, the contribution to nutrient uptake made by mycorrhizal hyphae could be similar to that made by roots if the surface area (length × diameter × π) were assumed to be an appropriate basis of comparison (see below).

For AMF, the ratio of hyphal length to root length generally varies between 25 and 250; but much larger ratios (well over 1000) have been found. However, various other mechanisms have been proposed to explain the effects that hyphae have in terms of extending depletion zones. For example, it has been proposed that the hyphae of mycorrhizal fungi may colonize soil pores that are too small for plant roots. It has also been suggested that the kinetic properties of the uptake systems of plants and mycorrhizal fungi may also differ, potentially allowing a closer approximation to 'zero sink' uptake at the primary absorptive surface (see Chapter 10, this volume). However, such differences have not been clearly demonstrated. Finally, it has been claimed that mycorrhizal fungi may be able to access sources of nutrients that are not available to plants (organic phosphorus or nitrogen and sparingly soluble phosphorus). Although it is unlikely that large differences exist in the organic nitrogen and organic phosphorus uptake capacities of plants and mycorrhizal fungi, mycorrhizal fungi might have an advantage over plants insofar as they can better explore the soil, by which means they can more effectively compete with saprotrophic microorganisms. The suggestion that mycorrhizal fungi have access to forms of phosphorus that are (biochemically) inaccessible to plants has not been confirmed under field conditions; however, it has been confirmed in monoxenic cultures, pot systems with well-defined sparingly soluble P-sources, and pot experiments using natural soil containing sparingly soluble P-sources (Cardoso, 2002). The role of mycorrhizas in nutrient uptake models is elaborated below.

The beneficial effects of the AM symbiosis have been attributed to improved phosphorus nutrition. It is also important to identify whether the mycorrhizal symbiosis has other effects that might benefit the plant (e.g. protection against pathogenic fungi, heavy metals or aluminium; better drought resistance; improved soil aggregation). Such beneficial effects have been reported previously; however, in those instances, it was not clear whether said effects were genuine mycorrhizal effects or a result of the improved phosphorus status of the plant. A comparison of mycorrhizal plants and non-mycorrhizal plants with the same P-status has now made clear that said beneficial effects are genuine, independent of P-status, and that AM symbiosis should therefore be considered to be multifunctional (Newsham et al., 1995). As different species of mycorrhizal fungi forage at different distances from the root surface, there is also functional diversity within the role of P-uptake (Jakobsen et al., 2001). It is possible that functional diversity is linked to taxonomic diversity within the Glomales. Boddington and Dodd (1999) have suggested that members of the genus Glomus could be more important to the plant in terms of the provision of P, whereas members of the genus Gigaspora might be more important in terms of the contribution made to soil structure. Our database is, at present, insufficient to address the question of the relationship between the taxonomy and ecological function of the Glomales in more detail.

## 14.4 The Importance of the Mycorrhizal Network

As the mycelium (the network of hyphae) of mycorrhizal fungi is perennial, and grows away from its centre, several plants can be connected by the same fungal individual in a common mycorrhizal network. Such interconnections occur both between plants of the same species and between plants of different species. Interconnecting plants of different species is a logical consequence of the limited selectivity exhibited by (several) AMF. However, as indicated above, lack of selectivity could have been overestimated, and may not be universally true for all species of AMF.

In networks involving more than one plant both nutrients and carbon could be moved from one plant to another. Interplant movement of carbon (a concept known as the 'wood wide web') has attracted a lot of interest and heated debate. The existence of such networks has raised a number of questions:

- Is there gross transport of carbon only (with the benefits being equal to the costs in the case of both partners), or are net quantities of carbon transferred? Carbon labelling one plant (e.g. with $^{14}CO_2$) and then finding that isotope label in another plant only demonstrates gross transport. However, it does leave open the possibility that similar amounts of unlabelled carbon could move in the opposite direction. Double labelling (e.g. using $^{14}CO_2$ to label one plant and $^{13}CO_2$ to label the other) could help resolve this issue.

- If net transport occurs, is it taking place in quantities that are ecologically relevant? Isotope labelling is an extremely sensitive method, whereby even trace amounts of carbon transported can be measured.
- What is the fate of the carbon subsequent to transportation? Does it remain in the fungus (optimizing fungal fitness), or does it end up in the plant (contributing to plant fitness)?

In the case of AMF, data based on double labelling indicate that net transport can occur, though in most cases the quantities involved are small. It also indicates, however, that the magnitude of the flux depends on sink–source relationships. Therefore, in the case of plants whose photosynthetic rates differ because one plant is shaded, transport could be larger. In most cases, however, the carbon transferred remained in the roots, and was not transported to the shoots, suggesting that the carbon is rigorously controlled by the fungus. Recently, Lerat et al. (2002) provided the first evidence that carbon transported through a common AM network ends up in the shoots of the receiver plant. Such reversed carbon transport (from fungus to plant) must also occur in the case of 'saprotrophic' plants that are without chlorophyll and so depend on fungi to obtain their energy for them (Bidartondo et al., 2002).

In light of the above, it is unlikely that a strongly phytocentric version of the concept of mycorrhizal networks (whereby each plant gives as far as it is able and receives according to its needs) can be maintained. However, less plant-oriented, more mycocentric versions of the concept of mycorrhizal networks may instead be applied. Such a concept has been introduced by Fitter (2001):

> To the plants, therefore, the common mycelial network is a club with a variable subscription fee and a range of potential membership benefits; to the fungus, the plants are the potential club members whose subscriptions keep the club afloat.

It remains to be investigated to what extent these mycorrhizal networks are important in multispecies agroecosystems.

## 14.5 Benefits of a Perennial Mycorrhizal System in Multispecies Agroecosystems

The following mechanisms, which all result in a perennial mycorrhizal system, have been proposed to explain how diverse and beneficial mycorrhizal communities are maintained in multispecies agroecosystems.

- Mixtures of plants generally allow a larger diversity of mycorrhizal fungal species to flourish. As the mycorrhizal symbiosis is multifunctional, and different fungal species are likely to fulfil (partly) different functional roles, a mixed-species system gives rise to a larger range of potential benefits for individual plants. Increased mycorrhizal diversity as such (or increased numbers of species of mycorrhizal fungi) has also been shown to increase primary productivity (van der Heijden et al., 1998). In mixtures of plants it may, in principle, be possible to maintain fungal species that greatly benefit a certain plant species but are unable to reproduce on it, as they depend on another plant species for reproduction (see the concept of 'negative feedback', as described by Bever et al., 2001). However, direct evidence of the latter mechanism is, at present, lacking.
- A mixture of plants, especially a mixture of crops and trees such as occurs in agroforestry systems, may exhibit deep rooting, resulting in higher levels of mycorrhizal inoculum at greater depths. This increases the volume of soil in which nutrients can be efficiently taken up (Cardoso, 2002; see Fig. 14.1). In several countries in the West African savannah zone, it has been claimed that a specific cultivar of cassava (*Manihot esculenta*) can restore soil fertility after nutrient depletion as a result of continuous cropping. This cultivar is a slow-growing and deep-rooting landrace, and it would be interesting to investigate whether the claimed beneficial effect is due to mycorrhizal activity in deeper soil layers.
- A mixture of plants, especially a mixture of plants with different growth phenologies, could result in a continuity of hosts over time, thereby allowing mycorrhizal fungi to differentially take carbon from

different plants depending on their photosynthetic activity. Mycorrhizal continuity can also be maintained if weeds or cover crops are established on a field after the major crop has been harvested. Bare fallows, or long periods during which land is kept free from any (mycorrhizal) plant growth in order to conserve soil water, have been related to a plant nutritional disorder in dry areas of Australia. A decline in the levels of the mycorrhizal inoculum involved leads to P and Zn deficiency, causing poor growth in oilseed crops (especially linseed), pulses and cereals grown on clayey soils that are otherwise considered quite fertile. The problem (known as 'long-fallow disorder') can be remedied by applying P and Zn fertilizers. However, it can also be addressed through the use of agricultural practices that provide mycorrhizal continuity (Thompson, 1996).

- A mixture of plants, especially one that includes perennial plants, prevents mycorrhizal mycelium being regularly disturbed. In regularly disturbed agroecosystems dominated by annual crops, where an annual life cycle is imposed on the mycorrhizal fungi, the 'late' establishment of mycorrhizas could be a factor that limits seedling growth. This could result in limited phosphorus uptake by the seedling, which feeds back into a lower growth rate. The important role that mycorrhizal networks play in the early growth of maize has been convincingly demonstrated by Miller (2000). Moreover, early establishment and nodulation of legumes is enhanced in the presence of a mycorrhizal network (Goss and de Varennes, 2002). Mycorrhizal sufficiency in agricultural fields could help young sorghum plants escape, or compensate for, the detrimental effects of witchweed (*Striga hermonthica*; Lendzemo and Kuyper, 2001). However, the exact mechanisms still require further study under field conditions.
- Mixtures of plants allow a continuity of carbon flow and, hence, mycorrhizal activity. This contributes to improved soil carbon sequestration and soil aggregation, and helps prevent soil erosion. Both fungal hyphae and glomalin, a specific glycoprotein produced by AMF, play a very important role in these respects (Rillig and Steinberg, 2002).

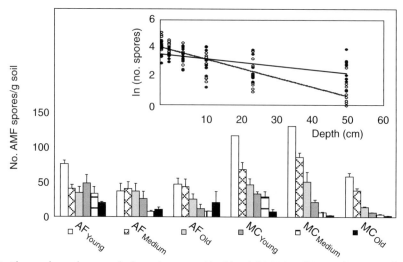

**Fig. 14.1.** The numbers of spores of arbuscular mycorrhizal fungi (AMF)/g soil in agroforestry and conventional coffee systems of different ages. The figure in the top right-hand corner is the regression analysis of the natural logarithm of the number of spores of AMF as a function of soil depth in the agroforestry (AF) system (●) and monocultural (conventional) (MC) system (○). Both intercepts and slope differ significantly between both systems. After Cardoso (2002).

## 14.6 Mycorrhizas in Models of Nutrient Uptake

Most nutrient uptake models for plants are still based on roots rather than on mycorrhizas. The simplest way of incorporating mycorrhizal hyphae into such models is to treat them as 'thin roots', and describe nutrient transport as occurring by diffusion to a small cylindrical sink. In the modelling tradition followed by Nye and Tinker (1977) and Barber (1984), the physiological uptake parameters at the root (or hyphal) surface have to be known to estimate the likely rate of uptake. In the approach taken by de Willigen and van Noordwijk (1987; see also Chapter 10, this volume) the uptake potential is derived on the basis of high estimates of the physiological parameters, in which the root can (approximately) act as a 'zero sink' (i.e. a sink of infinite strength that is able to maintain a concentration of zero at its surface, regardless of the rate of external supply). Actual uptake is taken to be the minimum of the current 'demand' and this uptake potential. Mycorrhizal hyphae will, in this approach, increase uptake potential. However, they will only increase uptake if plant demand cannot be met by the roots alone. The WaNuLCAS model (van Noordwijk and Lusiana, 1999, 2000) of agroforestry systems or other mixed-plant communities includes a representation of mycorrhizal hyphae as part of the 'effective root length' that determines the uptake potential of all plant components in each time step.

In order to understand this approach, we first have to consider how we can best deal with variation in root diameter in a tree or crop root system. As root diameter affects potential uptake rate in a cylindrical zero-sink model, an appropriately derived average root diameter in each layer and zone is needed for the uptake function. Also required is a way to estimate the equivalent effective root length of each component at such a diameter. A number of options exist for making this comparison between roots and/or hyphae of different diameter, and involve the use of the relationship $L \times D^x$ (where L is length, D is diameter and $x$ a parameter to be defined). If we sum roots (of variable D) on the basis of root length (so effectively use $x = 0$) we will probably underestimate the potential contribution of high-diameter roots (de Willigen and van Noordwijk, 1987). If root surface area is used ($x = 1$) the potential contribution made by high-diameter roots will be overestimated in a zero-sink uptake process, where diffusion through soil is the rate-limiting step. If biomass is used ($x = 2$) the result will be even more biased towards high-diameter roots. A comparison of the product of root length and the square root of root diameter (so $x = 0.5$) appears to give the best results (van Noordwijk and Brouwer, 1997), in the sense that, when comparing roots of different diameter on the basis of an equal $L \times D^{0.5}$, the predicted uptake potential is least sensitive to D (Fig. 14.2).

In equation form, the average root diameter for a mix of crops and tree roots of different diameters, as used in the WaNuLCAS model, is: (see bottom of page) where $CLrv$ and $TLrv$ refer to root length densities (cm/cm$^3$) of crop and tree, respectively, and $CDiam$ and $TDiam$ to root diameters.

Based on the above rule for adding roots of different diameter on the basis of the square root of their diameter, we can also get a first approximation of the effects of mycorrhizal hyphae. The total length of hyphae can be derived from the fraction of crop or tree roots that is mycorrhizal and the length of hyphae per unit length of mycorrhizal root. The effective root length ($EffLrv$) can therefore be derived from:

$$EffLrvC_{ij} = LrvC_{ij}\left[1 + \frac{Inffrac.HypLeng.\sqrt{HypDiam}}{\sqrt{RtDiamC}}\right]$$
(14.2)

where the $Inffrac$ parameter indicates the fraction of roots that is mycorrhizal, $HypLeng$ gives hyphal length per unit mycorrhizal root and $HypDiam$ the average diameter of hyphae.

$$RtDiamAV_{ij} = \left[\frac{Rt\_CLrv_{ij}\sqrt{Rt\_CDiam} + Rt\_TLrv_{ij}\sqrt{Rt\_TDiam}}{Rt\_CLrv_{ij} + Rt\_TLrv}\right]^2 \quad (14.1)$$

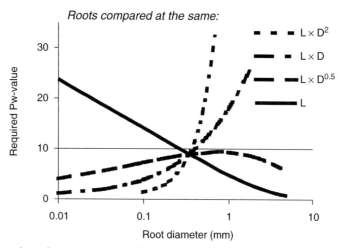

**Fig. 14.2.** Effect of root diameter D on the soil P supply (as expressed in the Pw-value) that is needed to meet the demand of a crop such as wheat when root systems of different diameter are compared at equal length L, root surface area $\pi L \times D$ or volume $0.25 \pi L \times D^2$; the smallest effect of root diameter (the flattest line) exists when root length times the square root of the root diameter ($L \times D^{0.5}$) is used (Van Noordwijk and Brouwer, 1997).

This equation effectively converts the mycorrhizal hyphae into an equivalent length at the diameter of the roots. This option is provided in WaNuLCAS for both crop and tree.

An alternative way of calculating hyphae length entails calculating the total length of mycorrhizal hyphae in the soil (for various depth and spatial zones). These are then assigned, based on proportions, to the various plants sharing the space. Either way, a total length of hyphae plus roots could be estimated for each plant component in each soil compartment.

The inclusion of mycorrhizas in models of the Barber–Cushman type, such as that of Yanai et al. (1995), which considers mycorrhizal hyphae as very thin roots, tends to greatly overestimate actual uptake. It is difficult to incorporate into such models the negative feedback that occurs between saturation of plant demand and a reduction in net uptake. In reality such feedback may include time lags, and the transfer of nutrients from hyphae to roots, and the receipt of a 'feedback signal' in the case of saturation of demand, may take more time than in a situation with roots only.

Whatever the case, root-uptake models that include mycorrhizas on the basis of total hyphal length and transport to cylindrical sinks suggest that the amount of mycorrhizal mycelium found under normal conditions greatly exceeds that needed for plant growth. Such results from models could well be correct, as any consideration of (excess) mycelium outside plant roots should not approach the issue from a phytocentric perspective. Rather, they should be considered from a mycocentric perspective, whereby the large fungal biomass reflects the maximization of the fungus' fitness. A large fungal biomass immobilizes substantial amounts of nutrients that are not accounted for in the plant.

Finally, it should be noted that models such as those above are based on the assumption that nutrient uptake occurs over the whole hyphal surface, analogous to plants where nutrients and water are taken up over the whole surface of the fine roots. Insufficient consideration has been given to the possibility that mineral nutrient acquisition in mycorrhizal fungi occurs only at the hyphal tip or only through specialized structures. It has, however, been suggested that mineral uptake occurs (preferentially) through specific structures, called Branched Absorbing Structures (BAS). These small, bushy structures (resembling arbuscules) form on runner hyphae at regular intervals. They have a small diameter (1.5–3.5 µm) and a relatively short lifespan (approximately 5 weeks), after which they senesce and are closed off by septa (Bago et al., 1998; Bago, 2000).

An important challenge faced by those modelling nutrient uptake through fungal hyphae at the single plant level is, therefore, the need to assess: (i) whether or not preferential sites for nutrient uptake exist; and (ii) how much of the nutrients taken up by the fungus are immobilized in the microbial biomass and, hence, not made available to the plant. It should be borne in mind, however, that phosphorus immobilized in mycorrhizal fungal biomass should, in the long term, be considered to be more available to plants than would be the case if the mineral phosphorus had been fixed to iron and aluminium (hydr)oxides (cf. Cardoso, 2002). When applying these ideas to mixed-plant communities or agroecosystems, attention needs to be directed at the 'rules' that govern the sharing of access to a mycorrhizal network (Box 14.4).

Quantitative simulation models can be used as a tool to test whether, through their interaction with relatively well-known aspects of plant–soil interactions, relatively simple mechanisms are sufficient to explain observed phenomena. Thus far, most models are phytocentric (a result of their agronomic focus), and do not include a perspective that considers the long-term survival of the fungal partner. A number of general principles can be formulated, however, to help manage actual agroecosystems for better 'mycorrhization'.

---

**Box 14.4.** Model approach to sharing access to a mycorrhizal network.

Version 2.2 of the WaNuLCAS model includes a simple option to describe root parasitism. This was inspired by the parasitic trees of the sandalwood family (*Santalaceae*), which provide high-value wood through important forms of agroforestry practised in the drier, eastern parts of Indonesia.

The conventional idea of root parasitism is one in which the 'parasite' steals water and nutrients from the host. In fact, however, it may be more accurate to say that a parasite such as sandalwood 'steals' or 'takes control of' the roots of other plants. The roots of the 'host' then start to function as though they belong to the sandalwood, and will take up water and nutrients as needed by that tree. By 'stealing roots', the sandalwood saves the energy associated with the making of fine roots. However, it is not at all clear who 'pays' the energy costs associated with the maintenance of these fine roots. Probably such roots are not adequately maintained by the sandalwood tree. The parasitic plant, on an evolutionary timescale, faces the following dilemma: if it does not maintain the 'capital stocks' it has 'taken over', the benefit received is short lived; however, if it does maintain said 'capital', the benefit of parasitism compared with a plant that makes and maintains all its fine roots itself may be small.

In a recent survey of sandalwood roots, Wawo (2002) only found parasitic contact in the case of relatively small roots. This suggests two things:

**1.** Sandalwood is not able to make contact with thicker, woody roots, and thus cannot take over major parts of a root system in one go.
**2.** After sandalwood has parasitized a root, the further growth of the fine roots is limited (in terms of both length and girth). Such limiting occurs because the host stops investing resources in the parasitized roots, whilst the sandalwood itself makes no investment.

Looking at the parasitic process in this way enables us to perceive that sandalwood must still need a fairly elaborate root system, in order to constantly find new roots belonging to other plants that it can parasitize.

These concepts can also be applied to multiple access, by various plants, to a network of mycorrhizal fungi in the soil.

## 14.7 Managing Arbuscular Mycorrhizal Associations

Applied mycorrhizal research has often been aimed at the production of inoculum. From a management perspective, however, attention should focus on the identification of: (i) the conditions under which the management of indigenous inoculum is the better approach; and (ii) the conditions under which the application of externally produced inoculum is either desirable or imperative. Decision trees (Fig. 14.3) to allow an answer to that question have been published by Brundrett *et al.* (1996) and Dodd and Thompson (1994). Such decision trees suggest implicitly what is actually confirmed by agricultural and forestry practices: most sites still contain sufficient inoculum. However, if insufficient inoculum is present, judicious management will allow sufficient inoculum to be created and maintained.

In order to manage an indigenous inoculum, it is imperative to know which factors are beneficial or inimical to the inoculum potential, and to the diversity or functioning of mycorrhizal fungi. Beneficial and adverse practices are listed in Table 14.2.

In agroecosystems, plant species selection is very important. In rotational systems, the sequence in which plant species with different mycorrhizal responsiveness are planted could affect the productivity of any one specific crop. Failure is likely to result if a highly responsive species, such as linseed (*Linum usitatissimum*), is planted after either a bare fallow or a crop that negatively affects mycorrhizal inoculum potential, such as the non-mycorrhizal Brassicaceae, which can even poison mycorrhizal fungi (Schreiner and Koide, 1993). Species selection in multispecies agroecosystems is important too, although the question of matching species in terms of fungal selectivity has not been explicitly addressed.

Forest disturbance as a result of commercial logging may reduce or even eliminate mycorrhizal fungi from forest sites. Alexander *et al.* (1992) noted a severe decrease in the levels of AM fungal spores found in a Malaysian forest following heavy logging. Selective logging, however, was found to have had a slightly positive effect. Sites of forest operations (skid trails and landings) in Cameroon also showed a strong decrease in spore numbers and mycorrhizal

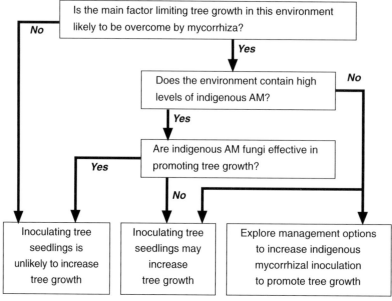

**Fig. 14.3.** Decision tree for use in determining under which conditions inoculation with mycorrhizal fungi or management of mycorrhizal associations is likely to be most successful. After Dodd and Thompson (1994).

**Table 14.2.** Positive and negative influences on arbuscular mycorrhizas by different agricultural management practices. (After Smith and Read, 1997.)

| Management factor | Positive influence | Negative influence |
|---|---|---|
| Plant species | Host species<br>High colonization<br>High spore production<br>High mycorrhizal root length density | Non-host species |
| Bare fallow | None | Reduces populations |
| Pasture | Increased propagule densities | |
| Disturbance | Minimum tillage | Conventional tillage<br>Compaction |
| Management | Organic–biodynamic | Conventional |
| Fertilizer | Drip feeding<br>Slow release<br>Rock phosphate | High applications of soluble P and N |
| Fumigation | None | Reduces propagules |
| Fungicides | Variable effects | Variable effects |
| Low light (glasshouse) | None | Colonization or growth decreased |

inoculum potential (MIP), as assessed by baiting (Onguene, 2000). There was no sign of a recovery over time. Alexander *et al.* (1992) showed that heavy logging decreased MIP by 75%, whereas heavy logging coupled with soil compaction caused a 90% decrease, and heavy logging with subsequent erosion a 95% decrease. Even when followed by a slight recovery, a strong decline in the level of mycorrhizal inoculum may retard secondary succession. Such retardation of secondary succession was reported by Cuenca and Lovera (1992) after bulldozers were used to clear the topsoil of a Venezuelan savannah. If vegetation develops on such sites, it often consists of plants that are not dependent on mycorrhizas. Succession can therefore be arrested by the lack, or the slow build-up, of sufficient inoculum. Under such conditions, successful revegetation may well depend on inoculum addition.

Sieverding (1991) stated that slash-and-burn agriculture has little negative impact on mycorrhizal inoculum. Data from Cameroon (Onguene, 2000) are consistent with this observation, as mycorrhizal inoculum levels were found to be somewhat higher in both agricultural fields and in young fallow, as compared with secondary or primary rain forest. Higher inoculum levels after the onset of shifting cultivation could be due to increased soil surface temperatures and decreased soil moisture after canopy opening (both of which act as triggers for spore formation) as well as to the shift in species composition. Both climatic factors probably force an annual life cycle on mycorrhizal fungi and select for mycorrhizal fungi that have an *r*-strategy that depends less on the mycelium and more on spore production. The importance of higher surface soil temperature and/or lower soil moisture has also been demonstrated by Cardoso (2002; see Fig. 14.1), who found that spore numbers in the topsoil of a conventional coffee field were significantly higher than those in an agroforestry coffee system, whereas the opposite pattern was evident for the deeper soil layers.

Agricultural intensification does not only cause mycorrhizal inoculum to decline through regular disturbance, but also selects for mycorrhizal fungi that are less beneficial to the plant. In The Netherlands, Dekkers and van der Werff (2001) demonstrated that phosphorus fertilizers negatively affect the functioning of mycorrhizal communities. Fifteen years after P-fertilizer treatment ended, mycorrhizal communities in soils that had been loaded with phosphorus fertilizers were still functioning less efficiently than those in unfertilized soil. Corkidi *et al.* (2002) demonstrated, in two semiarid grasslands in North America, that N-fertilization not only altered the balance between the costs and benefits of the mycorrhizal symbiosis, but also shifted the community towards less beneficial mutualists.

Only if mycorrhizal inoculum quantity or quality at a certain site is limiting productivity in agroecosystems should inoculation become an option. Inoculation treatments consist either of a single species inoculum (containing a so-called 'superstrain') or a generic, mixed inoculum ('biofertilizer' in general), based on the truism that any mycorrhizal inoculum is always better than no inoculum. Judicious management or application of such an inoculum for agricultural purposes potentially reduces the need for a phosphate fertilizer. However, it should be emphasized that the comparison between rhizobia (as biofertilizers that deliver nitrogen) and mycorrhizal fungi (as biofertilizers that deliver phosphorus) falls short. Mycorrhizal fungi do not add phosphorus to ecosystems, they only increase a plant's access to this often scarce resource. This distinction has important implications for long-term nutrient balances in agroecosystems. In the short term, however, phosphorus depletion is not likely to become problematic, as the total phosphorus pool in P-fixing tropical soils up to 1 m in depth can easily amount to 6000 kg/ha, a volume that will not be quickly depleted by the annual P removal rate of 5 kg/ha (assuming two harvests per year, of 2.5 t biomass per cropping season, and a biomass P-content of 0.1%). Although, from the plant's perspective, it is immaterial whether the P taken up is derived from a relatively inaccessible pool through the activity of the mycorrhizal mycelium or from a soluble pool after (excessive) fertilizer use, we should not forget that it could make a difference with regard to micronutrients such as Cu and Zn. Fertilizer use could decrease mycorrhizal activity and, hence, result in micronutrient deficiencies (Lambert et al., 1979).

Several single isolates of AMF have been shown to promote the growth of fast-growing tree species in low-nutrient soils (Prematuri, 1995; Setiadi, 1996; Prematuri and Dodd, 1997). Isolates of *Gigaspora rosea*, *Glomus etunicatum*, *Acaulospora scrobiculata* and *Acaulospora* sp. significantly promoted the growth of *Paraserianthes falcataria* and *Acacia mangium* at degraded nickel-mine sites (Setiadi, 1996). Such mine sites are often characterized by surface erosion and, as a consequence, by a low mycorrhizal inoculum potential. Isolates of *Scutellospora weresubiae*, *Glomus manihotis* and *Glomus mosseae* significantly boosted growth of *Pterocarpus indicus*, *P. vidalianus* and *Albizia saman* to levels that were 1.5–3 times higher than the control (Prematuri, 1995; Prematuri and Dodd, 1997). However, the performance of the inoculum over time is as important as initial plant response to inoculum addition. It is therefore necessary to study changes in species composition in terms of the way it is affected by competition with the indigenous species. A commercially acquired inoculum, which results in a high initial benefit but the effect of which does not persist over time, forces the buyer to acquire this inoculum regularly. This creates buyer dependency on the supplier of the mycorrhizal inoculum.

This chapter began by considering the discovery of mycorrhiza a century ago. Though observational techniques, concepts and methods have now greatly improved, many crop plants throughout the world still suffer serious nutrient deficiencies. Better mycorrhization would, at least in the short or medium term, help this situation by allowing more efficient 'mining' of the soil. Finally, it should be stressed that, in the context of multispecies agroecosystems, the concept that better management of the fungal partner in mycorrhization can improve the overall nutrient-use efficiency of the agricultural sector is one that retains potential. However, that potential remains, thus far, unproven.

### Conclusions

**1.** The biology of mycorrhizal associations should be understood from both a 'plant-centric' and a 'fungus-centric' perspective.
**2.** Many crop plants around the world suffer serious nutrient deficiencies: better mycorrhization would, at least in the short or medium term, help them 'mine' the soil more efficiently.
**3.** Mycorrhizal fungi do not add phosphorus to (agro)ecosystems – they only increase a plant's access to this often scarce resource. They therefore differ from rhizobia, which can act as true biofertilizers by delivering nitrogen to the (agro)ecosystem.
**4.** Applied mycorrhizal research has often been aimed at the production of inoculum; however, management of indigenous inoculum may provide more direct benefits.
**5.** Agricultural intensification can not only result in a decline in mycorrhizal inoculum due to regular disturbance, but it can also select for mycorrhizal fungi that are less beneficial to the plant.
**6.** Forest disturbance as a result of commercial logging may reduce or even eliminate mycorrhizal fungi from forest sites.
**7.** Isolates of AMF can promote the growth of fast-growing tree species in low-nutrient soils, especially where these have been severely disturbed (e.g. mine spoils); grasslands, however, may contain a healthy inoculum potential for trees.

### Future research needs

**1.** Better indicators, applicable at the farmer level, of situations where inoculation with mycorrhizal fungi is opportune.
**2.** Better understanding of the persistence of diverse mycorrhizal networks in multispecies agroecosystems as the basis for inoculation, to balance current understanding, which is based on spore counts and identification.
**3.** Mycorrhization of models of plant nutrient and water uptake: comparison of existing quantitative approaches and development of new algorithms.

# 15 Nematodes and Other Soilborne Pathogens in Agroforestry

Johan Desaeger, Meka R. Rao and J. Bridge

---

**Key questions**

1. What are the main factors that govern the build-up of soilborne pathogens?
2. Which strategies can be followed to avoid outbreaks of soilborne diseases?
3. How can agroforestry be a tool in the management of soilborne disease problems?

## 15.1 Introduction

Soilborne organisms (such as plant parasitic nematodes, fungi, bacteria, phytoplasma, protozoa and viruses) are among the most underestimated of the factors that affect plant productivity in tropical regions. Because of their microscopic size and the non-specific symptoms of an infection, these organisms live out of sight and, generally, out of mind of the growers and plant protection workers. Root-knot nematodes are an exception in that they cause distinctive symptoms in the form of root galls, which are sometimes referred to as 'root elephantiasis' by subsistence farmers in central Kenya (Fig. 15.1a). Otherwise, most farmers and extension staff are not able to identify nematodes and other soilborne diseases (Sharma *et al.*, 1997). Moreover, interactions commonly occur between nematodes and other soil pathogens, complicating any quick recognition of the problem and assessment of the damage done. Soilborne plant pathogens affect plants primarily through the infection of roots. These organisms occur as complexes in soils and in plant tissues, the nature of which are generally poorly understood and little quantified. In addition to pathogenic and parasitic organisms,[1] the soil contains a wide range of competitor saprobes, antagonists, beneficial organisms, yeasts, bacteria and nematodes

---

[1] 'Pathogenic' indicates the ability to cause disease, whereas 'parasitic' means that one organism obtains its food from another organism, with or without causing disease.

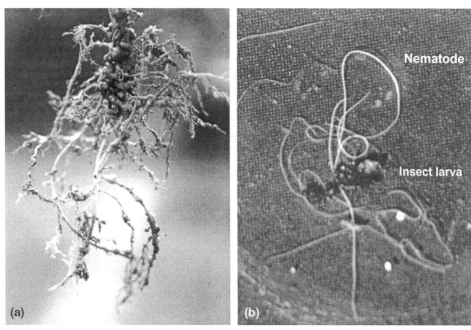

**Fig. 15.1.** Negative and positive nematode associations with *Sesbania sesban*. (a) Root-knot nematode (*Meloidogyne* spp.) causing destruction of the taproot and 'elephantiasis' in *Sesbania* roots. (b) Entomoparasitic nematode (*Hexamermis* spp.), emerging from larvae of a defoliating insect pest (*Mesoplatys ochroptera*) on *Sesbania*.

(Fig. 15.2). The population size of each of these groups is determined by edaphic and environmental factors, as well as by the availability of host roots.

Soilborne plant pathogenic and beneficial organisms are one of the key factors that determine crop health and productivity. The mechanisms that keep these organisms in check are influenced by environmental conditions and by the cropping practices used. Soil management practices greatly affect the dynamics of soil biota in managed ecosystems. While considerable attention has been paid to aspects of soil fertility and water management, research on soil health and its relation to ecosystem productivity has been neglected. Little attention has been paid to less obvious disease problems, where suppression of the causal pathogens may be due to a particular cropping environment and/or to the activity of competing organisms in the soil. Instead of focusing only on individual pathogens, it is important to take a holistic view of the soil environment and examine the total soil fungi and nematode populations, etc. and the ways in which they are affected by changing the management practices.

The growing emphasis placed on agroforestry as a means of producing tree products on farm constitutes a recognition of the need to reduce the pressure being placed on forests and natural vegetation. As a result of the intensive cultivation of selected trees on farms (agroforestry), many pest problems have come to the fore; it is now accepted that unless these pest problems are solved, the potential benefits of improved agroforestry cannot be realized. Not only is research on insects and diseases in tropical agroforestry limited, but the linkage between farmers and extension services in the area of plant health is also poorly developed. There exist certain general misconceptions, which hold that trees have no, or limited, pests and that diversified systems based on trees reduce pests (insects and diseases).

In this chapter, we discuss the factors governing the build-up of soilborne

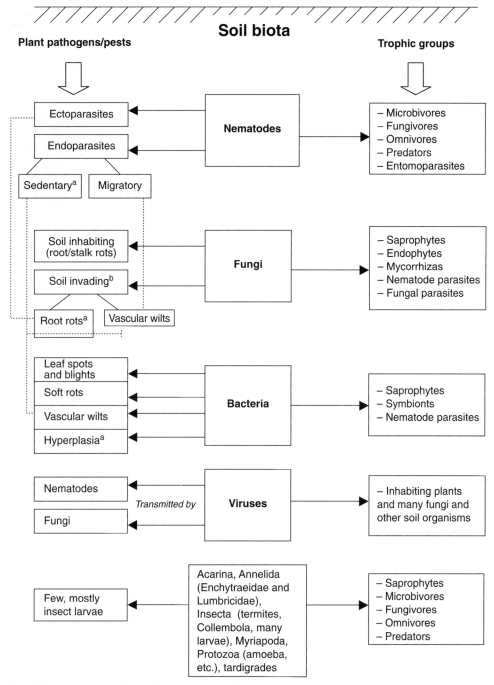

**Fig. 15.2.** Pathogenic and beneficial groups of soil biota and interactions among pathogenic groups (indicated by broken lines). [a]Galls or tumours on roots are usually due to root-knot nematodes (see also Fig. 15.1a); on crucifers, however, they may be caused by clubroot fungus (*Plasmodiophora brassicae*); also crown gall bacteria (*Agrobacterium tumefaciens*) cause galls on the crown (and roots) of fruit seedlings. [b]Most soilborne fungal pathogens are categorized as soil invaders, which indicates their ability to be facultative saprophytes; *Rhizoctonia* and *Pythium* are two important pahogens that are soil inhabitants.

pathogens, and the opportunities that exist for their management in tropical agroforestry ecosystems (see 'Key questions', above). The agroforestry systems considered are rotational systems, tree/crop combinations and complex multistrata systems and home gardens.

## 15.2 Factors Contributing to Soilborne Pests and Diseases

The development and severity of disease in plants can be visualized as a triangle, which is the result of interactions between the host, the environment and the pathogen (Fig. 15.3). The size of the epidemic or the amount of disease is proportional to the sum total of these factors, as long as none of the factors are zero. In the following sections, we will discuss the role of each of these components, namely the importance of agroforestry systems (Section 15.2.1 'Cropping systems'), the impact of soil and climate (Section 15.2.2 'Soil and climate') and the extent and impact of interactions between soil biota in the build-up of soilborne diseases (Section 15.2.3 'Interactions between soil biota').

### 15.2.1 Cropping systems

Cropping systems in tropical regions are generally more diverse and less reliant on chemical inputs than are those in temperate regions. There is also a greater diversity of nematodes and other pests in tropical regions (Luc et al., 1990). Pest outbreaks are considered to be more frequent in the tropics, although crop damage by soilborne pests is usually masked by many other, more visible, limiting factors (Smart and Perry, 1968; Wellmann, 1972). The reasons for the more serious pest problem in the tropics are the generally favourable climatic conditions, the greater pathogenicity of pest species and the more severe disease complexes (Mai, 1986). Table 15.1 lists some of the most common soilborne pathogens in the tropics and the crops and trees that may be affected in different systems.

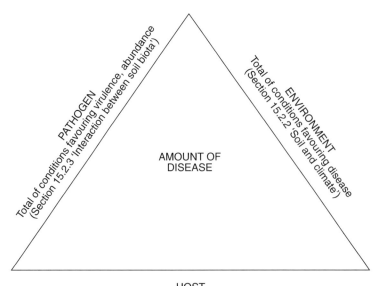

**Fig. 15.3.** The disease triangle. The amount of disease is proportional to the quantities of favourable host, pathogen and environmental conditions converging at a given time and space. After Agrios (1997).

Table 15.1. Common soilborne nematode, fungal and bacterial pathogens on agroforestry trees and shrubs, herbaceous cover crops, and major field crops in the tropics.

| Pest/pathogen | Food and cash crops | Herbaceous cover crops | Trees/shrubs for planted fallows | Trees for boundaries | Trees in croplands and home gardens |
|---|---|---|---|---|---|
| **Nematodes** | | | | | |
| *Meloidogyne* spp. (root-knot nematodes) | Many vegetables, legumes, tubers, coffee and other cash and utility crops | *Desmodium distortum*; *Tithonia diversifolia*; *Vicia* spp.; *Vigna* spp. | *Sesbania* spp.; *Tephrosia* spp. | *Acacia* spp.; *Albizia* spp.; *Faidherbia albida*; *Prosopis juliflora* | *Adansonia digitata*; *Carica papaya*; *Ficus* spp.; *Phoenix dactylifera*; *Psidium guava*; *Vitis* spp. |
| *Pratylenchus* spp. (lesion nematodes) | Cereal crops, root and tuber crops, banana, coffee, tea | *Arachis* spp., forage grasses | *Crotalaria* spp.; *Senna* spp. | *Pinus* spp. | *Hevea* spp. |
| *Radopholus similis* (burrowing nematode) | Banana, citrus, pepper | | | | Palms; *Persea americana* |
| *Rotylenchulus* spp. (reniform nematodes) | Vegetables, cotton, pineapple | *Indigofera hirsuta* | *Cajanus cajan* | | *C. papaya*; *Passiflora edulis* |
| **Fungi** | | | | | |
| *Fusarium* spp. (wilt and rot) | Banana, bean, coffee, cotton, melon, potato, tomato | *Vicia* spp. | *C. cajan*; *Crotalaria juncea*; *Sesbania sesban* | | Palms |
| *Phytophthora* spp. (rots) | Many vegetables, cocoa, citrus, tobacco | *Lupinus* spp. | | *Eucalyptus* spp.; *Pinus* spp. | *P. americana*; *Macadamia* spp.; citrus |
| *Armillaria mellea* (root rot) | Coffee, tea, root and tuber crops | | *C. cajan* | *Acacia* spp.; *Erythrina* spp.; *Grevillea robusta* | *Annona* spp.; *Macadamia* spp.; *Vitis* spp. |
| *Sclerotium rolfsii* (southern blight) | Solanaceous crops, root and tuber crops, legumes, rice | *Mucuna* spp. | *S. sesban* | | *P. americana* and many other fruit trees |
| *Verticillium dahliae* (wilt) | Cocoa, cotton, potato, tomato | | | *Dalbergia sissoo* | Anacardiaceae (mango, cashew, pistachio); *P. americana* |
| **Bacteria** | | | | | |
| *Ralstonia solanacearum* (bacterial wilt) | Solanaceous crops, banana, ginger, groundnut | | | *Casuarina equisetifolia*; *Eucalyptus* spp. | *Annona* spp. |

NB Blank table cells probably indicate lack of knowledge rather than lack of hosts.
Sources: Desaeger and Rao (1999a); Desaeger (2001); Dommergues (1990); Lenné and Boa (1994); Mayers and Hutton (1987); McSorley (1981); USDA (1960); Waller and Hillocks (1997).

### 15.2.1.1 Fallow–crop rotational systems

Planted or 'improved' fallows, which use fast-growing leguminous trees and shrubs (also referred to as cover crops), are being promoted in east and southern Africa, in order to replenish soil fertility in nitrogen-depleted soils and to increase crop yields (ICRAF, 1998). However, a disadvantage of growing *Tephrosia vogelii* and *Sesbania sesban*, two of the most promising species for short-duration (6–12 months) planted fallows, is that they are susceptible to root-knot nematodes (*Meloidogyne* spp.), and markedly increase the nematode's population in the soil (Desaeger and Rao, 1999a, 2001b). Root-knot nematodes (Table 15.1) are by far the most devastating nematode pest in the tropics. Maize (*Zea mays*) yields that follow *S. sesban* or *T. vogelii* fallows are not affected, as maize is a poor host to most isolates of root-knot nematodes, but yields of highly susceptible crops, such as bean (*Phaseolus vulgaris*), are severely reduced.

A number of herbaceous and shrubby cover crops are good hosts to root-knot nematodes (Table 15.1), although some of them, such as *Tithonia diversifolia*, do not show the typical root gall symptoms. *Crotalaria* species effectively suppress populations of root-knot nematodes, and those of most other sedentary plant-parasitic nematodes (Good *et al.*, 1965; Sukul, 1992; Wang *et al.*, 2002), but instead they host lesion (*Pratylenchus* spp.) and spiral nematodes (*Helicotylenchus* spp. and *Scutellonema* spp.). In western Kenya, maize yield reductions of up to 10–50% were ascribed to damage caused by lesion nematodes (Desaeger, 2001).

Very little is known about the importance of soilborne fungal or bacterial diseases in rotational systems involving crops and planted fallows. In western Kenya, *Fusarium oxysporum*, which causes wilt in *Crotalaria juncea* (Hillocks, 1997), is suspected of being responsible for the severe early wilting of a number of *Crotalaria* spp. used as cover crops (e.g. *C. grahamiana*). *S. sesban* experienced wilt caused by *Sclerotium rolfsii* under dry conditions in Hawaii (Evans and Rotar, 1987) and root rot and wilt caused by *F. oxysporum* f.sp. *sesbaniae* in India (Lenné and Boa, 1994). Root rots caused by *Macrophomina phaseolina, Armillaria mellea* and *Ganoderma* spp. may seriously affect the growth of several *Acacia* species (Lenné and Boa, 1994).

### 15.2.1.2 Mixed systems

Multiple cropping systems are still the norm among traditional and subsistence farmers in the tropics, and are estimated to provide as much as 15–20% of the world's food supply (Altieri, 1991). Trees or shrubs can be planted together with crops in different arrangements and for different purposes. They may be dispersed, in order to provide shade to the understorey crop(s); they may be planted along field boundaries, to act as a windbreak or fence; they may be managed as hedgerows, for mulch and/or fodder and for soil conservation; they may be grown to support climbing crops such as beans, betel vine (*Piper betle*) and black pepper (*Piper nigrum*), or under-sown with herbaceous covers, as is the case in plantation crops. Unless the correct choice of tree species is made, there is a danger that they will promote soil pests by serving as alternative hosts. If two or more species in a system have a common pest (or disease), the chances that it will spread and have a severe effect are greater in a mixed system than they are in a rotational system, because species are in close proximity and because there is continuous interaction among the species.

The typical 'under-forest' shrub crops – coffee (*Coffea arabica*), cocoa (*Theobroma cacao*) and tea (*Camellia sinensis*) – are often combined with shade trees of the genera *Erythrina, Albizia, Gliricidia* and *Leucaena*. Shading is an effective insurance strategy against above-ground insect pests, as well as against diseases in cocoa and coffee plantations. Intercropped coffee suffers less damage from coffee rust than pure coffee, as the latter experiences greater physiological stress (Waller, 1984). Below-ground pathogens may also be affected, either by reducing the host's stress and predisposition, or by altering the soil microclimate. In Uganda, solar radiation stress predisposed unshaded cocoa to attack by *Verticillium dahliae* (Palti, 1981). Disease damage will be aggravated, however, if any of the shade trees host the soilborne pathogens that infect the understorey crops. The use of banana as a shade crop in coffee or cocoa could increase infection by *Pratylenchus coffeae*, as banana is a good host for this nematode. In south Asia, the burrowing nematode *Radopholus similis* and the reniform nematode *Rotylenchulus reniformis* may become more damaging when tea is intercropped with crops such as coffee, cloves (*Syzygium aromaticum*) and pepper, all being good hosts for both nematodes (Sivapalan, 1972; Campos *et al.*, 1993).

No major nematode problems were noted on trees employed for hedgerow intercropping, such as *Leucaena leucocephala*, *Calliandra calothyrsus*, *Gliricidia sepium* and *Inga edulis*, but root rots caused by species of *Fusarium* and *Ganoderma* have been reported to kill *Leucaena* in Asia and Australia (Lenné and Boa, 1994). Many soilborne pathogenic fungi of the genera *Armillaria*, *Fomes*, *Ganoderma*, *Verticillium* and *Rosellinia* have been found to spread to alternative hosts, such as coffee, cocoa and tea, from moribund shade trees or tree residues left after clearing (Schroth *et al.*, 2000b). However, *Gliricidia sepium* (also called, because of its toxic seeds, bark, leaves and roots, *mata ratón* in Spanish, meaning 'rat' or 'mouse killer') is free of soilborne diseases, despite the fact that it has been grown widely throughout the tropics. In mixed fallows of *S. sesban* and *G. sepium* in eastern Zambia, there was greater mortality of the *Sesbania*-defoliating beetle *Mesoplatys ochroptera* than in pure *S. sesban* fallows. This was probably due to the beetles feeding on the poisonous *G. sepium* leaves in the mixed stand (Sileshi and Mafongoya, 2000).

Home gardens and multistrata systems typically occur in the humid and subhumid tropics, and resemble the local tropical forest ecosystems. Fruit trees and palms, such as areca nut (*Areca catechu*), coconut (*Cocos nucifera*), oil palm (*Elaeis guineensis*) and peach palm (*Bactris gasipaes*), are often the major components of home gardens. With the exception of the commercially important plantation crops, little research has been done on the soilborne diseases of perennial crops or on other economically valuable tree species.

It is common for taxonomically related plants to share the same pests and diseases. Among plant-parasitic nematodes, the semi-endoparasitic *Tylenchulus semipenetrans* is often found to infect many tropical fruits belonging to the families Rutaceae (*Citrus* spp.), Rosaceae (*Eriobotrya japonica*), Oleaceae (*Olea europea*), and Ebenaceae (*Diospyros* spp.). *Citrus* spp. and *Diospyros* spp. are also good hosts for *Radopholus similis*. Species of *Meloidogyne* and *Rotylenchulus*, on the other hand, commonly infect fruit trees belonging to the Caricaceae, Passifloraceae and Moraceae (*Artocarpus* spp.) (McSorley, 1981). The nematode *Rhadinaphelenchus cocophilus*, which causes red ring disease in coconut, attacks 17 other palm species, including oil palm. This nematode is transmitted by the pantropical palm weevil *Rhynchophorus palmarum*.

*Radopholus similis* is a widespread nematode, and infects many crops, including coconut and areca nut. It is known to infect crops such as betel vine, black pepper, banana (*Musa paradisiaca*), ginger (*Zingiber officinale*) and turmeric (*Curcuma longa*) in the multispecies cropping systems of southern India (Griffith and Koshy, 1993).

*Verticillium dahliae* and *V. albo-atrum* are extremely polyphagous soilborne fungi, and affect many species in home gardens, including fruit trees, such as mango (*Mangifera indica*) and avocado (*Persea americana*), and vegetables, such as aubergine (*Solanum melongena*) and tomato (*Lycopersicon esculentum*) (Palti, 1981). Moreover, the fungus *Sclerotium rolfsii*, as well as root-knot nematodes, attacks both vegetables and fruit crops. Planting orchards on old vegetable land therefore often warrants soil disinfection prior to planting.

### 15.2.2 Soil and climate

Many abiotic factors and physical and chemical soil properties interact with soilborne diseases, and may enhance or reduce the impact they have on economic plants. The susceptibility of crops to pathogens is often greater in the case of those grown on infertile soils than in the case of those grown under fertile soil conditions, and many mild pathogens may cause severe disease under conditions of nutrient stress or aluminium toxicity. The damage caused by lesion nematodes was relatively greater in unfertilized maize than was the case in the fertilized crop (Desaeger, 2001). The increased susceptibility of many solanaceous plants to *Fusarium* wilt, *Alternaria solani* early blight, *Pseudomonas solanacearum* wilt, *Sclerotium rolfsii* and *Pythium* damping-off of seedlings was ascribed to nitrogen deficiency (Agrios, 1997). Excessive nitrogen supply, however, especially in the absence of adequate potassium and phosphorus may reduce crop resistance to (mainly aboveground) pests and diseases, implying the need for careful consideration when planting fallows and using cover crops to input nitrogen into the systems.

Calcium reduces the severity of several diseases, such as the fungi *Rhizoctonia*, *Sclerotium*, *Botrytis* and *F. oxysporum* and the nematodes *Meloidogyne* spp. and *Ditylenchus dipsaci* (stem nematode). However, it increases the severity of black shank disease in tobacco (caused by *Phytophthora parasitica* var. *nicotianae*) and of the common scab in potato (caused by *Streptomyces scabies*; Agrios, 1997). The effect that calcium has on disease resistance seems to be a result of its effect on the composition of cell walls and on their resistance to penetration by pathogens. As most crops have broad pH tolerances, manipulating the soil reaction could potentially be a means of combating certain diseases. Among fungal diseases, *Fusarium* wilt and clubroot of crucifers are more severe in low pH soils, whereas *Verticillium* is more damaging in high pH soils (Palti, 1981). Similarly, in the case of nematodes, lower pH levels increased galling by root-knot nematode (Steinmüller, 1995), but decreased infection by cyst nematodes (Grau, 2001).

The damage potential of many plant-parasitic nematodes is higher in coarse-textured soil than it is in fine-textured soil, because of higher nematode activity as a result of better soil porosity and oxygenation and because of generally lower inherent soil fertility and biological activity. Root rots are most severe in soils with a low organic matter content, poor soil structure and high compaction with inadequate drainage. Soil degradation due to the loss of organic carbon was reported to increase the vulnerability of banana to major pests, including nematodes, causing huge declines in yield (Page and Bridge, 1993). Many tropical soils contain subsurface zones of dense and/or hard materials, such as claypans and hardpans. Such soils result in an abnormal root distribution, with a large portion of the roots being locked up in the upper parts of the profile where most of the pathogens also reside. In high strength soils, aggregation of roots in cracks or voids may lead to inoculum aggregation if the roots become infected and decay. New roots may grow through the channels left by old roots, virtually assuring contact with very large numbers of pathogens. *Phytophthora cinnamomi* devastated *Eucalyptus marginata* in Australia on shallow lateritic soils, as the fungal zoospores were easily dispersed into cracks in the hard layer that had previously been penetrated by the trees' sinker roots (McDonald, 1994).

High temperatures in the tropics provide favourable growth conditions for pathogens, and also induce heat and/or drought stress in host plants, which further increases both their susceptibility to, and rapid development of, disease (Liddell, 1997). Rainfall and soil moisture affect the host–pathogen interaction in several conflicting ways. Symptom expression tends to be more severe under hot and dry conditions, and 'dryland' pathogens, such as *Fusarium* stalk rots and *Macrophomina* charcoal rots, are typical examples of diseases that appear on crops under stress (Palti, 1981). Drought stress weakens the host plant's resistance and tolerance (the former term indicates the ability of the plant to resist pathogen infection and the latter the ability of the plant to withstand infection without apparent damage). Root endoparasites, such as root-knot nematodes, often cause considerably more damage when soil water becomes limiting for plant growth, as they are protected from environmental stress by the root tissue. In western Kenya, root-knot nematodes caused less damage to *S. sesban* on deep and heavy soils under good rainfall conditions (despite high nematode reproduction) than they did in Malawi. There, the same nematodes caused high mortality and poor growth of *Sesbania* on shallow and light soils under drought conditions. Although the semiarid tropical region is infested with hordes of deadly pathogens, the majority of plant pathogens and severity of disease outbreaks is closely correlated with high humidity and rainfall. In fungal diseases, formation, liberation, germination and movement of spores, mycelial growth, and root invasion are always greater under moist conditions. Diseases such as *Sclerotium* and bacterial blight, *Phytophthora* and *Pythium* rots, and ectotrophic root diseases such as *Armillaria* and *Fomes* rots are predominantly found in humid and subhumid agroecological zones (Hillocks and Waller, 1997). Damping-off diseases of seedlings are generally more common in relatively cooler climates.

Few functional linkages between soilborne pathogens and edaphic factors have yet been found (Campbell and Benson, 1994). Still, in spite of its complexity, the spatial and temporal heterogeneity of the soil environment in agroforestry systems presents a unique research opportunity in terms of elucidating interactions that occur between environmental factors and assessing and predicting the impacts of agricultural practices. A great deal of insight could be gained if more quantitative information were to be available on the interactions between soil type, root growth and pathogen epidemiology.

### 15.2.3. Interactions between soil biota

An understanding of the interactions that occur among different groups of soil flora and fauna will help us to manipulate them in a manner that achieves favourable effects. Direct interactions occur when microorganisms compete for space or nutrients, or when one group antagonizes another by producing toxic metabolites. Indirect interactions are mediated through the root system; for instance, one pathogen may increase or decrease the susceptibility of the host plant to another pathogen. Interactions may work either way – there are probably as many organisms favouring pathogens as there are antagonists, and often the quality and degree of the interactions change as soil conditions are altered (Khan, 1993).

#### 15.2.3.1 Synergistic interactions

Several reviews have discussed the role of nematodes in disease complexes, and numerous nematodes have been associated with viral, bacterial and fungal diseases (Khan, 1993; Abawi and Chen, 1998). Nematodes can play different roles in disease complexes; they act as: (i) vectors (e.g. for several viruses); (ii) wounding agents (e.g. *Meloidogyne arenaria* and *Cylindrocladium crotalariae* on groundnut); (iii) host modifiers (e.g. *Pratylenchus penetrans* and *Verticillium dahliae* on potato); (iv) rhizosphere modifiers (e.g. *Meloidogyne incognita* and *Rhizoctonia solani* on tomato and okra); and (v) resistance breakers (e.g. *M. incognita* and *Phytophtora parasitica* on tobacco) (Hussey and McGuire, 1987). Several longidorid and trichodorid nematode species transmit a range of economically important viruses, which cause diseases of fruits, vegetables and ornamentals in temperate climates. However, there is no evidence yet of nematode-transmitted viruses in the tropics, even though tropical Africa is regarded as the place of origin of these nematodes (Hillocks and Waller, 1997). Information on soilborne viruses, and on related mycoplasms in the tropics in general is very scarce, due to the fact that these organisms are difficult to study and also to a lack of adequate resources.

The most common synergistic (or positive) interactions are those that occur between plant-parasitic nematodes and root fungi, such as species of *Fusarium*, *Rhizoctonia*, *Pythium*, *Sclerotium* and *Verticillium*. *F. oxysporum* readily establishes itself in the feeder roots of banana when they are invaded by the nematode *R. similis*, but the fungus has seldom been recovered from nematode-free roots (Blake, 1966). In general, endoparasitic nematodes tend to increase diseases caused by vascular wilt fungi, whereas ectoparasitic nematodes increase infection by cortical rot pathogens

(Hillocks and Waller, 1997; Fig. 15.2). Positive interactions between *Meloidogyne javanica* and *Fusarium* spp. have been observed on pigeonpea (*Cajanus cajan*), coffee and mimosa, as well as on a wide range of annual crops. *Verticillium* species often form disease complexes with species of the *Pratylenchus* and *Globodera* nematodes. Interactions between nematodes and root-rot and wilt-inducing fungi have been reported for at least 45 crops, mainly from the tropics, and have involved over 15 nematode genera and more than 20 fungal genera (Evans and Haydock, 1993; Francl and Wheeler, 1993). Disease complexes between nematodes and bacteria are less common; the best-known example is probably the interaction between the root-knot nematode and *Ralstonia* (*Pseudomonas*) *solanacearum*, which causes bacterial wilt. The latter is widely considered as the most important soilborne bacterial pathogen, and has a broad host range and different pathovars (indicating a subspecies or group of strains that can infect only plants within a certain genus or species).

Soil fungi may reduce plant resistance to pest infection and promote, or reduce (several examples exist of either occurrence), nematode root penetration (Freckman and Caswell, 1985). Occasionally, moreover, non-pathogenic soil-inhabiting microorganisms (such as *Trichoderma* spp.) may become pathogenic when roots are infected by nematodes (Melendez and Powell, 1969). Some authors have questioned the importance of organism interrelationships in disease complexes involving nematodes and pathogens, and have suggested that the perceived synergism may be caused instead by alterations in the abiotic soil environment (Sikora and Carter, 1987). The latter theory stresses the need for more effort to be made to unravel the complex nature of these interactions.

Not much is known about the role played by saprophytic fungi and non-parasitic or free-living (bacterial- and fungal-feeding) nematodes in disease epidemiology. Free-living nematodes are usually dominant over their plant-feeding counterparts in the soils, and are also commonly found inside plant roots as secondary feeders. They play a major role in soil organic matter decomposition and in nutrient cycling processes in the soil, and are most abundant in soils that are rich in organic matter (Sharma and Sharma, 1999). Although free-living nematodes may cause reductions in the populations of pathogenic bacteria and fungi, they may also aid in the dispersal of the same fungi and bacteria, as well as of mycorrhizas and rhizobia (Hillocks and Waller, 1997). As they can also disrupt plant health by interfering with symbionts, some of them need to be considered as facultative parasites. Some fungivorous nematodes have been observed to suppress ectomycorrhizas on pines and endomycorrhizas on many plants, and some bacterivorous nematodes have been observed to inhibit $N_2$-fixation (Freckman and Caswell, 1985; Huang, 1987). However, with regard to symbionts, plant-parasitic nematodes (such as root-knot nematodes) are more damaging than free-living nematodes. Although the majority of interactions are negative (Taha, 1993), nodulation in some *Acacia* spp. was stimulated by root-knot nematodes, possibly by facilitating the entry of the bacteria or by physiological mechanisms that favour the initiation of rhizobial symbiosis (Duponnois *et al.*, 1997).[2] Nodulation of *S. sesban* in western Kenya was greatly reduced by high levels of root-knot nematodes, but was slightly stimulated by low levels of the same nematodes (Desaeger, 2001). However, using a *Rhizobium* strain from northern Kenya, nodulation of *S. sesban* was not reduced even at the highest nematode population (Desaeger, 2001), indicating the need to select *Rhizobium* inoculum appropriate to nematode-infested soils. Alternatively, the result may have been exacerbated by the often low numbers of compatible rhizobia for *S. sesban* in these soils (see Chapter 13, this volume).

---

[2] Root galls and Rhizobium nodules look very similar to the untrained eye, but unlike galls, which are an integral part of the root, nodules are distinct structures and can easily be rubbed off.

### 15.2.3.2 Antagonistic interactions

Antagonistic interactions among soil organisms reduce the risk of soilborne diseases. 'Antagonist' is an umbrella term for parasites, predators, pathogens, competitors, and other organisms (such as rhizobacteria, mycorrhizas, fungal endophytes, bacterial and fungal parasites, nematode-trapping fungi and predatory nematodes) that repel, inhibit or kill pathogens (Table 15.2). The antagonistic potential of soils has been defined as the capacity of a soil ecosystem to prevent or reduce the introduction and/or spread of plant pathogens or other deleterious agents. Considering that more than 90% of soil microorganisms have not been cultured and studied, the potential of antagonistic interactions within the soil is probably greatly underestimated (Sikora, 1992).

Interesting case studies of so-called suppressive soils have been reported, which, typically, show low disease incidence on a susceptible host, even in the presence of adequate inoculum and abiotic factors favourable to the pathogen (Rouxel, 1991). For example, the *Chinampa* soils in Mexico, which date back to the Aztec era, and which are characterized by large inputs of aquatic mud, plant residues and manures, are well known for the suppression of pathogens such as *Pythium* spp. and *Meloidogyne* spp. (Zuckerman *et al.*, 1989; Garcia-Espinosa, 1998). Although low pest incidence in suppressive soils is, in general, linked to

**Table 15.2.** Parasitic and antagonistic potential of soil organisms to kill or inhibit plant-parasitic nematodes and soilborne pathogens.

|  | Target pathogen |
|---|---|
| **Hyperparasites**[a] | |
| Bacterial | |
| *Pasteuria* spp. | Nematodes |
| Fungal | |
| Nematode-trapping: *Arthrobotrys* spp. | Nematodes |
| Nematode-parasitic: *Paecilomyces lilacinus*, *Verticillium chlamydosporium*, *Dactylella oviparasitica* | Nematodes |
| Fungal parasites: *Trichoderma* spp. | Damping-off and root rot fungi (*Rhizoctonia*, *Pythium*, *Fusarium*) |
| Insect parasites: *Beauveria* spp. | White grubs and cockchafers |
| Other | |
| Predatory amoeba, nematodes, tardigrades and mites | Nematodes; fungi |
| Entomopathogenic nematodes | Various insects (adults and larvae) |
| **Antagonists**[a] | |
| Bacterial | |
| *Agrobacterium radiobacter* | *Agrobacterium tumefaciens* |
| Fluorescent *Pseudomonas* spp. | Nematodes; fungi |
| Rhizobacteria, *Bacillus subtilis* | Several fungal diseases |
| Fungal | |
| Non-pathogenic *Fusarium* spp. | *Fusarium* wilts |
| Mycorrhizas: *Glomus*, *Gigaspora*, *Endogone* spp. | Root-knot nematodes |
| *Trichoderma* spp. | Wide range of soilborne and foliar fungal pathogens |
| Others | |
| Actinomycetes (*Streptomyces* spp.) | Fungi |

[a] 'Hyperparasites' use their host directly as food, whereas 'antagonists' act by substrate invasion or modification, or by excreting agents such as antibiotics, siderophores and bacteriocins. The distinction between hyperparasites and antagonists is not always clear (certain agents, such as *Trichoderma virens* may behave in both ways, depending on the pathogen parasitized or antagonized).
Sources: Baker and Cook (1974); Cook and Baker (1983); Stirling (1990); Tjamos *et al.* (1991); Sikora (1992); Copping and Menn (2000).

physicochemical properties associated with the fine texture of these soils (Bruehl, 1987), evidence also points to biological factors, particularly rhizobacteria and species of *Fusarium* and *Trichoderma*, being determinants. The fluorescent pseudomonads, aggressive bacterial root colonizers that sometimes produce antibiotics, are often cited as being the major biological factors.

Many examples exist of fungal hyperparasitism, a well-known case being the use of *Trichoderma* spp. as biocontrol agents against soilborne fungal pathogens (Baker and Cook, 1974). Another example is the delayed development of symptoms caused by fungal diseases in seedlings inoculated with non-pathogenic fungi or hypovirulent strains of the pathogen (Gindrat, 1979). Despite numerous attempts, however, only limited success has been recorded with biocontrol in the field, and most of the agents tested remain as 'Petri-dish antagonists'. The successful introduction and establishment of chosen antagonists or hyperparasites in the soil requires either an empty ecological niche or an abundant food/resource base large enough for there not to be competition with existing soil inhabitants. Both situations are difficult to achieve in the real world. Instead of inoculating the soil with antagonists, the practical option is to stimulate the growth of natural antagonists in the soil through the incorporation of organic matter and other soil amendments (see Sections 15.2.2 'Soil and Climate' and 15.5.5 'Soil Amendments/Mulching'). However, few attempts have been directed at investigating the actual effects of amendments on antagonists.

In addition to the nutritional benefits they offer, rhizobia and mycorrhizas offer protection against certain root diseases. *Rhizobium japonicum* prevented the development of root diseases caused by *Fusarium oxysporum* and *Phytophthora megasperma* on soybean (*Glycine max*) and lucerne (*Medicago sativa*), except when pathogens had already infected these legumes, in which case the pathogens interfered with rhizobial activity (Palti, 1981). When an ectomycorrhizal mantle develops on fine roots, pathogens are rarely able to penetrate such roots, as is the case with *Phytophthora cinnamomi* on pines (Marx, 1975). Also, endomycorrhizas such as *Endogone* spp. and *Glomus* spp. have been shown to give protection against several soil-borne pests and pathogens (Gindrat, 1979).

Competition among different taxa of plant-parasitic nematodes often has a negative effect on at least one of the competitors, especially when their feeding habits are similar. Tomato roots had lower populations of lesion nematodes when root-knot nematodes were present (Estores and Chen, 1972). Mechanisms of competition may include mechanical destruction, physical occupation of feeding sites or induced physiological changes in the host's suitability or attractiveness (Eisenback and Griffin, 1987; Khan, 1993). Nematode interactions are often difficult to explain, and the knowledge base, especially with regard to tropical systems, is small.

A huge amount of information is available on nematodes that parasitize insects (Poinar, 1975; Nickle, 1984). Entomoparasitic nematodes may kill their host by feeding and entomopathogenic nematodes vector a bacterium, which actually causes the insect's death. The entomoparasitic nematode *Hexamermis* sp. was found inside the larvae and adults of the *Sesbania*-defoliating beetle *Mesoplatys ochroptera*, causing their death in Zambia (Kenis *et al.*, 2001; Fig. 15.1b). In Florida, the mole cricket nematode *Steinernema scapterisci* is used in pastures to keep populations of mole crickets, of the genus *Scapteriscus*, under control (Smart *et al.*, 1991).

## 15.3 Strategies for the Management of Soil Pests Based on General Sanitation

The aims of sanitation are to prevent the introduction of the pathogen inoculum into cultivated fields and to reduce or eliminate the inoculum from fields that are already infected. Pathogens can be introduced through seed and through vegetative propagation material, such as cuttings, tubers and seedlings. Vegetatively propagated crops, such as bananas and plantains, ginger, cassava (*Manihot esculenta* Crantz), potatoes (*Solanum tuberosum*), yams (*Dioscorea* spp.), sugarcane (*Saccharum officinarum* L.), taro (*Colocasia esculenta* (L.) Schott) and sweet potatoes (*Ipomoea batatas* (L.) Poir), are frequently infected by pathogens, especially viruses, systemic pathogens and nematodes.

Aphids and whiteflies transmit many seedborne viruses. As these insects become more scarce as altitude increases, traditional potato growers in the Andes, and cassava farmers in the Kenyan highlands, obtain their seed from high-altitude areas where insect-transmitted virus diseases are minimal (Thurston, 1992). Similarly, the seed of different legumes and cucurbits affected by seedborne diseases should be grown and harvested during dry seasons or in arid areas under irrigation, in order to avoid the high scope for disease infection provided by wet weather. The use of disease-free banana suckers, which can be obtained by meristematic tissue culture, is one of the main practices used to reduce nematode damage to banana.

Soil quality, soil water status and choice of shade trees all affect the incidence and severity of soilborne diseases in nurseries. It is advisable not to plant bananas or plantains for shade in a nursery, as they host many nematode pests. Root-knot nematodes are among the most common soilborne pests in tropical nurseries, and account for poor rooting and seedling quality.

Many strategies exist to rid nursery beds of soilborne pathogens: in the tropics they are often based on heat therapy. Burning plant debris, such as dry tobacco stalks, maize stover, rice husk and wood, on the surface of seedbeds is a practice commonly used in Africa to ward off root-knot infestation. To be effective, a sufficiently hot burn, which causes heat to penetrate the soil, is required: this requires the use of wood or woody stalks rather than grass, for example. Root-knot is very effectively controlled in seedbeds by turning soil broken up into a fine tilth at regular intervals during the dry season. Nematodes are killed as they are exposed to high temperatures, solar rays and drying (Bridge *et al.*, 1990; Bridge, 1996, 2000b). Soil solarization, or heating the soil by covering the seedbed with transparent polyethylene, is one of the most practical and efficient means used to reduce soilborne pathogens in nursery beds. Solarization will be effective provided the soil is moist, the plastic is properly sealed at the edges, and the solar radiation is high. Heat therapy is also used to kill nematodes inside the corm tissue of banana suckers by immersing them in hot water (55°C) for 15–25 min (Stover, 1972).

It is extremely difficult to eradicate soilborne pathogens once they are established in the field. Cultural and physical methods, such as crop rotation, use of resistant cultivars, removal or burial of infected plants and crop debris, adjusting crop density, and depth and time of planting, are only effective against certain diseases. However, these measures may not be effective if the pathogen pressure is too high: in such a case they can only be controlled by chemical soil disinfection, generally fumigation.

Chemical control of soilborne diseases is impractical for most small-scale farmers in the tropics, because of cost considerations and a lack of knowledge concerning chemical use. The use of chemicals is, in general, directly related to farmers' economic situations, and very few farmers in the tropics consider ecological considerations to be a reason for not using them. In reality, chemical pesticides are, unfortunately, the preferred option for most farmers, if they are available and affordable, as farmers often (wrongly in many cases) believe that chemicals are the best pest-control solution. Farmers who consider alternatives to pesticides generally expect instant results similar to those provided by chemicals (Bridge, 1998). Apart from their high cost, some of the problems associated with the use of chemicals on small-scale farms in the tropics are a lack of knowledge regarding the use of correct chemicals (farmers often use insecticides to try and control fungal diseases, for example), their own safety, and the correct application rate and frequency. Many of the broad spectrum and highly persistent pesticides, which have been banned in most developed countries, are still being marketed in many developing countries.

Many weed species serve as alternative hosts to soilborne pathogens (such as *Meloidogyne* and *Verticillium* spp.) with or without being seriously affected themselves. If such weeds are not properly controlled, they may reduce the effectiveness of break crops employed in crop rotation to eliminate soilborne pathogens. In western Kenya, *Striga hermonthica*, a parasitic weed in maize, was found to be a good host for root-knot nematodes, which nullifies the effectiveness of maize as a rotation crop used to reduce root-knot nematodes in *Striga*-infested fields (Desaeger, 2001). Planted fallows of *Crotalaria* were more effective in reducing *Meloidogyne* populations in researcher-managed plots than in farmer-managed fields, probably because of poorer weed control in the latter (Desaeger, 2001).

The soil's physical, chemical and biological environments impose restrictions on the dispersal of soilborne pathogens. Unlike windborne pathogens, these do not spread by means of movement at the landscape level; nematodes, for example, only move a few metres per year. However, the exception to this rule are some systemic fungi, which may spread with seeds, and certain facultative soil-inhabiting insects, which may disperse over long distances at the adult stage. Foraging rhizomorphs of *Armillaria* spp. are also capable of spreading over several acres.

Of greater significance is the indirect spread of nematodes and pathogens through water and soil movement (erosion and sandstorms) and human interventions (irrigation and movement of machinery). Nematodes were spread by the wind in the groundnut growing area of Senegal (Baujard and Martiny, 1994) and an entire field of pepper was lost to *Phytophthora capsici* downstream of windblown rain and floodwater in the USA (Bowers and Mitchell, 1990). Within a field, a disease inoculum often accumulates and causes high infection rates in the lowest parts of the field, especially when the field is irrigated. At the landscape level, various pathogens (such as nematodes, wilts and blights) may spread to distant fields by means of irrigation channels, drainage ditches and even rivers. The spread, from isolated foci to a whole plantation, of *F. oxysporum* f.sp. *albedinis*, which causes wilt ('bayoud') in oil palm, was attributed to flood irrigation (Kranz et al., 1977). Flooding can also be advantageous, however, as most parasitic nematodes and fungi such as *Verticillium* and *Fusarium* can be controlled by flooding the soil for at least 2 months (Sumner, 1994). This is particularly interesting in areas where flooding occurs either naturally or as part of the farming system (paddy rice or fish ponds).

Many of the sanitary measures mentioned here do not require a great deal of technology and/or money. Small-scale farmers in the tropics can gain a lot by incorporating the relevant sanitation practices into their farming systems.

## 15.4 The Avoidance Approach to the Management of Soil Pests

### 15.4.1 Crop rotation

Historically, crop rotation has been a major tactic for the control of soilborne pests and diseases. Crop rotations suppress soilborne pathogens if the crops employed in rotation are poor hosts, if they act as trap crops for the pathogens, if they produce toxic or inhibitory allelochemicals (Table 15.3), and/or if they provide niches for antagonistic flora and fauna (Table 15.2). Evidence for utilizing specific crop rotations to ward off soilborne pathogens and enhance productivity can be found in ancient Chinese and Indian literature and the system was used in the pre-Columbian Inca culture, and in Medieval and Renaissance Europe (Rodríguez-Kábana and Canullo, 1992). In Peru, potato cyst nematodes (*Globodera* spp.) have traditionally been managed by rotating potato with other Andean tubers, such as oca (*Oxalis tuberosa*), mashua (*Tropaeolum tuberosum*) and ullucu (*Ullucus tuberosus*). Inca law demanded that potatoes must not be grown on the same land more than once in 7 years (Thurston, 1990). Mashua was recently found to contain isothiocyanates – nematicidal compounds that are commonly found in cruciferous plants and that are related to methyl isothiocyanate, which is the active ingredient of the soil fumigant metam sodium.

**Table 15.3.** Trap and antagonistic crops and organic amendments with potential for control of plant-parasitic nematodes and soilborne pathogens.

| Trap and antagonistic crops | Target pest/pathogen |
|---|---|
| **Cover crops** | |
| *Arachis* spp. (wild groundnut); *Brassica* spp. (mustard); *Cassia fasciculata* (partridge pea); *Crotalaria* spp.; *Macroptilium* spp. (siratro); *Mucuna* spp. (velvetbean); *Pueraria* spp.; *Stylosanthes gracilis* | Root-knot nematodes |
| *Indigofera hirsuta* (hairy indigo) | Root-knot and lesion nematodes |
| *Crotalaria* spp.; *Mucuna* spp. (velvetbean) | Root-knot and reniform nematodes |
| **Flower crops** | |
| *Gaillardia* spp., *Helenium* spp., *Tagetes* spp. (marigold) | Nematodes; *Verticillium* wilt |
| **Oil crops** | |
| *Arachis hypogaea* (groundnut) | Root-knot nematodes |
| *Ricinus communis* (castor); *Sesamum indicum* (sesame) | Root-knot and lesion nematodes |
| **Pasture crops** | |
| *Chloris gayana* (Rhodes grass); *Eragrostis curvula* (weeping lovegrass); *Panicum maximum* (panic grass) | Root-knot nematodes |
| *Cynodon dactylon* ('coastal' bermudagrass) | Root-knot nematodes; *Fusarium* wilt |
| *Sorghum bicolor* × *S. sudanense* (sorghum-sudangrass) | Root-knot nematodes; bacterial wilt; *Striga* |
| **Tree crops** | |
| *Azadirachta indica* (neem) | Root-knot nematodes; various soil pests |
| *Sesbania rostrata* | *Hirschmanniella* spp. (rice root nematodes) |
| **Organic amendments** | |
| Agricultural wastes/residues | |
| Lucerne and cereal straw, cassava peelings, cocoa pods, coffee husks, sugarcane residue, tea waste, tree bark, wood ash, etc. | Nematodes; fungal root rots; *Verticillium* wilt; *Sclerotium* blight |
| Animal wastes | |
| Bonemeal, crab chitin, farmyard manure, poultry manure, etc. | *Phytophthora* root rot; *Verticillium* wilt; *Fusarium* wilt; nematodes |
| Green manure | |
| *Aeschynomene* spp. (jointvetch); *Azolla* spp. | Nematodes |
| Asparagus, clover, crucifers, neem, sudangrass, velvet bean, water hyacinth, etc. | Nematodes; *Pythium* rot; *Fusarium* wilt; fungal root rots |
| Oil cakes | |
| Castor, cotton, mustard, neem, groundnut, sesame, soybean, etc. | Nematodes; fungal root rots; bacterial blight |

Sources: Egunjobi (1985); Saka (1985); Bridge (1987); McSorley *et al.* (1994); Sumner (1994); Thurston (1997); Abawi and Chen (1998); Noe (1998); Gamliel (2000); McSorley (2001); Wang *et al.* (2002).

Besides directly reducing the main pathogen, crop rotation can act against a predisposing pathogen, as is the case in the reduction of *Fusarium* wilt, which results when the predisposing agents (root-knot nematodes) are controlled. A potential exists for exploiting rotations as a control for soilborne diseases. However, the use of such a technique requires the existence of a knowledge base regarding the disease incidence and host status of the component species. Generally, closely related crops and trees are more likely to support the same diseases than unrelated species are (see Section 15.2.1.2 'Mixed systems'). In particular, continuous cropping of legumes, crucifers, cucurbits and other vegetables should be avoided; the rotation of unrelated crops, for instance legumes with cereals, is recommended. Still, this principle is not a guarantee against soilborne diseases, as there are many examples of soilborne diseases that are shared by taxonomically unrelated plant species.

Many soilborne pathogens, such as *Meloidogyne* spp., *Verticillium* spp., *S. rolfsii* and *Armillaria* spp., have wide host ranges and are difficult to control. They require carefully designed cropping systems. By contrast, many cyst nematodes and pathogens that cause bacterial wilts and root-rots are more host specific and are easier to control by crop rotation. In general, although the same nematodes and viruses often affect numerous crops and trees, many fungal and bacterial pathogens are more specific in their host ranges and do not pose so great a threat to other plant species associated or rotated with their primary hosts. Some fungal and bacterial pathogens have a broad host range at the species level, but are host specific at lower taxonomic levels. They are, therefore, subdivided into *pathovars* or *formae speciales*, and are then further divided into races. (*Pathovars* or *formae speciales* are specific to certain plant species, whereas races are specific to a certain variety of a plant species.) *Fusarium oxysporum* is an example of a well-known generalist that is actually highly host specific at lower taxonomic levels.

Both the duration of the fallow phase and the number of 'break crops' that need to be grown in rotation in order to control soilborne diseases depend on the length of time that the pathogen can survive. Cyst nematodes resist disintegration in the soil for long periods, so rotations of 3 or more years may be required. Also, many fungal pathogens (for example, *Fusarium* spp. and *Verticillium* spp.) may survive for several years in the soil. In western Kenya, planted fallows of *S. sesban* and *T. vogelii* require a one-season rotation with pure maize, instead of the traditional maize–bean intercrop, in order to avoid root-knot nematode damage to susceptible beans (Desaeger and Rao, 2000). Table 15.3 gives several cover crops and pasture grasses that can be used, in crop rotations, to reduce root-knot nematodes and other soilborne plant pathogens (either by acting as trap crops or through some other mechanism).

Many *Crotalaria* species have proved to be excellent rotation crops for the control of root-knot nematodes throughout the world (Wang *et al.*, 2002). Cassava, pineapple, sweet potato, sugarcane, tomato and bitter orange are good rotation crops for banana, for which *Radopholus similis* (a burrowing nematode) is the main parasite (Loos, 1961; Luc *et al.*, 1990). Although these crops may increase root-knot nematodes, banana is not affected greatly by root-knot nematodes, except in a few special production areas outside normal growing regions (Bridge, 2000a; Gowen and Quénéhervé, 1993). *Phaseolus aureus* (mung bean), *Vigna mungo* (black gram), *Vigna unguiculata* (cowpea), *Sesamum indicum* (sesame) and the cover crop *Indigofera hirsuta* (hairy indigo) were reported to be good rotation crops for the control of *Pratylenchus* spp. in rice (Bridge *et al.*, 1990).

Much of the information available regarding crop rotations is highly site-specific. The effects of environment and season on the effectiveness of rotations are poorly understood, as are the effects that rotations have at a regional level and on non-target pests. An example of a rotation that aggravated a non-target pest is that of a sorghum rotation, for managing root-knot nematodes, which increased problems with wireworms in a subsequent potato crop (McSorley, 2001). There is no such thing as a 'miracle plant', and whether a certain crop or tree species is 'good' or 'bad' when used in a rotation depends on its proper use. *S. sesban* fallows, for instance, increase the risk of root-knot nematode damage to susceptible crops, but they also act as a false host to the parasitic weed *Striga* (*Striga asiatica* and *S. hermonthica*) and deplete its seed in the soil (Gacheru *et al.*, 2000). Therefore, *S. sesban* should be considered an excellent rotation crop for maize in East Africa in terms of soil fertility replenishment as it reduces *Striga* infestation of maize. Although it increases the number of root-knot nematodes present, the latter do not significantly affect maize in the area.

### 15.4.2 Plant tolerance/resistance

Developing plant material that is tolerant/resistant to soilborne diseases is a continuing process, although some progress has been made in the case of some crops (Waller and Hillocks, 1997). Resistance within a species is mainly against highly specific pathogens, such as the different *Fusarium* wilts; resistance against generalists, such as root-knot nematodes or *Sclerotium rolfsii*, is less common. However, alternative crops can be chosen that have high levels of resistance even to these generalist pathogens.

Increasing genetic uniformity in our major food crops is a dangerous trend, especially with regard to major disease outbreaks. Of the 3000 or so plant species that humans have used for food, about 150 have entered into world commerce; today humans are fed primarily by only about 15 plant species (Thurston, 1992). Therefore, at the very least it seems, maintaining the genetic diversity of these plant species appears crucial, as this would offer some kind of insurance against large-scale crop failures. The disastrous potato blight epidemic that occurred in Ireland in the 1840s, and which was caused by the introduced soilborne fungus *Phytophthora infestans*, was favoured by the genetic uniformity of the crop (Bezdicek and Granatstein, 1989).

Many local races or cultivars show remarkable tolerance to certain soilborne pathogens. Although these races are not necessarily high yielding under optimal conditions, they yield some harvest even under the worst conditions. As many as 50 different cultivars of potato are grown by Andean farmers in South America, a figure that should be compared with the use of only four main cultivars in the USA (Altieri, 1991).

Several comprehensive reviews have been written on the mechanisms, genetics and breeding behind resistance or tolerance (Robinson, 1976; Lamberti *et al.*, 1983). In addition to traditional breeding methods, genetic engineering is becoming more and more important, especially with regard to resistance against pests and diseases. The potential of biotechnological advances is vast. However, the possible risks (such as the creation of new weeds, the amplification of existing weeds and harm to non-target species) should not be disregarded. Biosafety measures (including against certain biocontrol agents) should be put in place to restrict or prevent the spread and introduction of pests of plants and plant products (Schumann, 1991).

## 15.5 The Confrontational Approach to the Management of Soil Pests

### 15.5.1 Mixed systems

Mixed systems, involving plant species that host different pathogens, provide a continuous food source for the reproduction of these pathogens and may aggravate disease incidence. On the other hand, lower density of host species and mutual competition among species in mixed systems reduces the chance for pathogens to increase to damaging levels. The presence of permanent hosts in the system ensures continuous food for predators and antagonists of the pathogen. The control strategy of frequent disturbance of pest and disease populations in crop rotations is, to some extent, substituted for by the strategy of increased stability and internal control mechanisms used in mixed systems (Schroth *et al.*, 2000b).

### 15.5.2 Tree–crop intercrops

Considerable documentary evidence exists to suggest that the reduction of both above-ground and below-ground insect pests is greater in annual intercrops than it is in pure crops (Altieri and Liebman, 1986). This is usually explained by the lower resource concentration for the pest and an increased abundance of predators and parasitoids due to greater availability of alternative food sources and suitable microhabitats in annual intercrops (Risch, 1981). In addition, the trap-crop principle is often cited as a mechanism of reduced pest attack: one component in the system attracts the pest and serves as a trap or decoy, preventing the infection of the host species. The secretion, by one of the species of the system, of harmful substances into the rhizosphere may be detrimental to the pathogen of the other species. This is the mechanism put forward to explain the fact that the occurrence of *Fusarium* wilt is lower in pigeonpea intercropped with sorghum than it is in pure pigeonpea (ICRISAT, 1984). The nematicidal activity of root exudates has been shown for certain plants, such as marigolds (*Tagetes* spp.) and neem (*Azadirachta indica*).

Antagonistic (trap or pesticidal) crops have been fairly well documented thus far (Table 15.3). Interplanting of neem (*A. indica*) seedlings in chickpea (*Cicer arietinum*) or mung bean reduced the incidence of root-knot nematodes (Narwal, 2000). Guinea arrowroot or topinambour (*Calathea allouia*) has been reported to be antagonistic to *Meloidogyne* spp. Its preference for shade also makes it an interesting species for association with trees, as with coffee in Puerto Rico (Noda *et al.*, 1994).

Cover crops are often sown in plantation crops and orchards as a means of suppressing weeds and providing grazing for cattle. It is also common to grow agricultural crops between commercial tree crops, especially during the establishment phase. Such systems offer great opportunities for managing the soilborne diseases of the plantation crops, either through the growing of trap crops, or through the stimulation of antagonists and predators of the pathogen. In Florida and California, cover crops, such as vetch, clovers, grasses and forage groundnut (*Arachis pintoi* and *A. glabrata*), are planted inside vineyards and citrus orchards in order to control nematodes and weeds (Porazinska, 1998). Also *Crotalaria* species (such as *C. spectabilis* and *C. ochroleuca*) have been successfully used to control sedentary plant-parasitic nematodes in peach and banana orchards. In Nicaragua, *A. pintoi* and *Desmodium ovalifolium*, respectively, reduced the populations of *R. reniformis* and *M. incognita* when grown together with coffee (Herrera and Marbán-Mendoza, 1999). Carpets of creeping legumes stimulated the action of saprophytes and antagonistic microflora, which worked to eliminate *Armillaria mellea* root rot inside infested stumps of rubber trees (Liyanage, 1997).

### 15.5.3 Multistrata complex systems and home gardens

Home gardens in West Java, and the Chagga multistorey gardens on Mt Kilimanjaro, in Tanzania, have been in use for centuries without any major disease problems being apparent. High species diversity, combined with the individual care of each plant, generally results in a minimum incidence of insects and pathogens. Soils in these systems contain highly diverse biota, which reduce the risk of any one soil pathogen becoming predominant. It is generally accepted that a close relationship exists between species diversity and performance stability, although merely increasing diversity will not necessarily increase the stability of all ecosystems. One of the goals of pest management should be the identification of those elements of diversity that should be retained or added and those that should be eliminated (Nickel, 1972). The incidence of *R. similis* in multispecies systems in southern India decreased when coconut and areca nut palms were interplanted with cocoa, in comparison with when they were interplanted with banana, black pepper and cardamom (Griffith and Koshy, 1993).

An important aspect of multistrata systems is the physical, non-specific effect that a plant species may have on pest and disease incidence – as opposed to the more obvious, specific biological effects. Although physical effects on microclimate (for example increased shade, moisture, or other physical factors) would mainly be significant in terms of the effect they have on above-ground insects and pathogens, biological activity below ground could also be affected by means of changes in the moisture, structure and porosity of the soil environment. Mortality of *Acacia mangium* trees due to species of the fungal pathogen *Rosellinia* was greater in pure stands than in mixed agroforestry systems, a fact attributed to the wider spacing and faster growth of the trees in the mixed agroforestry systems (Kapp and Beer, 1995).

### 15.5.4 Multispecies fallows

Natural fallows, which have a mixed vegetation, are less likely to experience damaging levels of soilborne pathogens. Desaeger and Rao (2000) found that short-duration (1 year) natural fallows did not increase populations of the parasitic nematodes *Meloidogyne* and *Pratylenchus* to damaging levels in western Kenya. Long-duration fallows (10 years or more) were found to decrease plant-parasitic nematodes in western Kenya (Kandji *et al.*, 2001) and to increase nematode species diversity in Senegal (Pate, 1998). Therefore, increasing species diversity in planted fallows, in order to mimic the functions of natural fallows, is a technique that can limit the build-up of pathogens and their potential to damage susceptible crops in a rotation. Use of such a practice would increase the flexibility that farmers have in terms of choosing crops for rotation. However, practising rotations strictly for the purpose of controlling soilborne diseases may conflict with farmers' preferences for certain crops and with the suitability of certain crops in relation to the soil and climate.

Bean crops that followed mixed improved fallows of *S. sesban* + *C. grahamiana* and *T. vogelii* + *C. grahamiana* did not experience yield losses due to root-knot nematodes (Desaeger and Rao, 2001b). This result should be compared with the result gained when bean followed pure *S. sesban* or *T. vogelii* fallows. Inclusion of *C. grahamiana* (a poor host for root-knot nematodes) in the fallows had reduced the build-up of root-knot nematode and had increased the populations of lesion and spiral nematodes, as well as the populations of other less-pathogenic nematodes. Similarly, the presence of the weakly pathogenic spiral nematode, *Helicotylenchus dihystera*, reduced the pathogenic impact of the nematode community in millet (*Pennisetum typhoides*; Villenave and Cadet, 1997). Not all mixed fallows guarantee the reduction of pests and diseases; quite the opposite may occur if the component species happened to host the same pathogen. An *S. sesban* + *T. vogelii* mixed fallow resulted in very high root-knot nematode damage to bean in rotation, in comparison with pure fallows of the respective species (Desaeger and Rao, 2001b).

Multispecies fallows may have other advantages, such as greater biomass production and greater resilience against environmental stresses than monospecies fallows (Khanna, 1998). Another advantage of multispecies planted fallows is that better synchrony exists between crop nutrient demand and mineralization of plant residues, especially when the foliage of component species has different chemical characteristics (Mafongoya *et al.*, 1998).

### 15.5.5 Soil amendments/mulching

Maintaining high soil organic matter, by regularly incorporating organic materials, farmyard manure, crop residues and composts, usually improves a plant's ability to withstand pathogens (Linford *et al.*, 1937; Palti, 1981). Several mechanisms may be responsible for the suppressive effect of organic matter on root pathogens:

1. Germination and lysis of propagules.
2. Competition for nutrients.
3. Release of toxic compounds, such as sulphur-containing volatiles and high concentrations of ammonia (Stirling, 1990).
4. Stimulation of antagonists, or parasitic or predacious biological control agents, e.g. chitin amendments increase populations of nematophagous fungi.
5. Interference with inoculum dissemination, e.g. mulch reduces soil splashing and the spread of bean web blight (Galindo *et al.*, 1983.
6. Modification of soil environment (temperature and moisture).

Some examples of soilborne pathogens that can be controlled by using soil amendments are given in Table 15.3.

Organic materials and mulches offer many agronomic benefits: lowering soil temperature, maintaining soil moisture, protecting the soil against erosion, providing nutrients and organic matter to the soil, improving soil structure and reducing weeds. Probably because of this, they are seldom used with plant health as the primary consideration. Mulches may have a negative effect on crop health, as in certain cases they increase the incidence of plant diseases and shelter other pests, such as insects, slugs and rodents, as well as venomous snakes. Fresh plant material may initially be colonized by pathogens rather than by saprophytes. The application of green manures with a low C : N ratio may increase the incidence of pathogens such as *Rhizoctonia solani* and *Pythium* spp. and *Fusarium* spp. It has been found that, although fresh crop residues controlled *Pythium ultimum* on lettuce, they increased the incidence of *Fusarium solani* f. sp. *phaseoli* on bean (Palti, 1981). Therefore, dead plant residues are generally safer where infection by fungi such as *Pythium* spp. or *Phytophthora* spp. is expected. Manure applications increased the severity of *Fusarium* wilt and of *Rhizoctonia solani* in the USA (Shipton, 1977). Mulching in banana crops reduced populations of the nematode *R. similis*, but increased the populations of *Pratylenchus goodeyi*, as the latter prefers soils rich in organic matter (Kashaija *et al.*, 2001). A practice that is widespread in Uganda is the use, in banana plantations, of the leaves, the chopped corms and the pseudostems of banana as a mulch. This denies the banana weevil its major breeding sites in whole corms and pseudostems (Karamura and Gold, 2000).

Seasonal incorporation for several years of the quickly decomposing leaves of *T. diversifolia* and *S. sesban* did not reduce the parasitic nematode populations in maize, but greatly increased free-living nematodes (bacterivores and fungivores). By contrast, the slowly decomposing leaves (high in polyphenols) of *Calliandra calothyrsus* resulted in a much lower level of free-living nematodes, a level similar to that achieved through chemical fertilizer applications (Desaeger and Rao, 1999b).

One of the major limitations of the use of organic amendments and mulches is the large amount of organic material necessary, and consequently the high level of human labour required for its application. The use of organic residues for managing certain soilborne pathogens would be feasible in the humid and subhumid tropics where plant growth is rapid and luxurious but of questionable value in the semiarid tropics. The practice may have an advantage in the tropics over temperate regions, as the higher temperatures lead to faster decomposition, greater activity of saprophytes and build-up of potential biocontrol agents.

## 15.6 Conclusions

Many soilborne pathogens account for production losses in tropical agroecosystems. The impact such pests have on the small-scale farms of the tropics is not well recognized, because of a dearth of knowledge about the economic losses caused by them and because of the complexity of those systems practised by farmers in the tropics. One of the main reasons that small-scale farmers practise multiple cropping is to reduce or spread the risks they face, including the risks posed by soilborne pests. Management of soilborne pathogens in these systems requires an integrated approach, which will, in most cases, be based on cultural, physical and biological methods. Chemical means of controlling soilborne pests are expensive, non-remunerative and hazardous in most situations. Four basic approaches for soilborne pest management are suggested here: (i) preventing the introduction of new pests and diseases, which is a basic quarantine procedure; (ii) preventing or avoiding the build-up of diseases to damaging levels, which is the major tactic underlying crop rotations; (iii) reducing pathogen populations by increasing populations of agents antagonistic to the pathogens, for instance through use of organic amendments; (iv) increasing biodiversity, as in, for example, the use of multiple cropping systems and multiple cultivars.[3]

The success of non-chemical approaches in the management of soilborne pests and diseases in the tropics is often hampered by the absence of regional or site-specific information. Although it is true that knowledge about soilborne pests and diseases in the tropics is growing, and is beginning to include subsistence crops such as maize, bean and cassava, it is also true that as more research is done more new problems are disclosed. One of the objectives of this chapter was to show that agricultural diversification, through agroforestry systems, does not by definition exclude soilborne disease problems. A lack of knowledge should not be mistaken for a lack of pests; the potential that newly introduced trees and shrubs have to aggravate such problems is probably equal to their potential to improve the soil pest situation. Many research issues, especially in relation to the tree component, have not been investigated and offer promising opportunities for disease management.

Soilborne pests have a long history of institutional neglect in the IARCs, and corrective measures are required to increase stakeholders' awareness of the existence and potential significance of soil pests (Sharma et al., 1997). With regard to farmers, few of the efforts that have been made to improve farming techniques have focused on the adoption of basic pest management technologies, in spite of the fact that many such practices do exist and could easily be applied. Improving the development and extension of pest management and control tools should, therefore, become a priority, not only in order to increase farmers' awareness, but also to stimulate interest among all the other stakeholders

## Acknowledgements

The authors wish to thank D. Mitchell, S. Sharma and R. Gwynne for their valuable contributions.

---

**Conclusions**

The major principles for reducing the risk of soilborne pathogen problems are:
1. Applying sound sanitary practices.
2. Avoiding pathogen build-up by applying proper rotations (breaking the pathogen cycle).
3. Confronting pathogens – by employing biological agents such as antagonists, parasites and/or predators, by diversifying the cropping systems and by the addition of organic soil amendments.

---

**Future research needs**

1. To determine the economic importance of, and identify major biosecurity concerns for, agroforestry systems in different ecoregions.
2. To understand the ecological relations and interactions that exist between pathogenic and beneficial microorganisms and different components of agroforestry systems.
3. To relate associated microbiological diversity to soil health in different ecosystems.

---

[3] Note that the use of multispecies fallow systems is a 'hybrid' tactic, using the potential of both rotation and diversification.

# 16 Soil Biodiversity and Food Webs

Franciscus X. Susilo, Anje Margriet Neutel, Meine van Noordwijk, Kurniatun Hairiah, Georg Brown and Mike J. Swift

---

**Key questions**

1. How does below-ground biodiversity vary with land-use practices?
2. What is the function of soil biodiversity in ecosystems?
3. Can the food-web concept be used to approach soil biodiversity problems?
4. Under what conditions is managing specific soil biota rather than overall diversity relevant?

---

## 16.1 Introduction

Although it is not apparent to the naked eye, soil is one of the most complex habitats on earth, containing one of the most diverse assemblages of living organisms (Lavelle and Spain, 2002). Over 1000 species of invertebrates were identified in 1 m² of soil in temperate forests in Germany (Schaefer and Schauermann, 1990). The diversity of the microbial component of soil may be even greater than that of the invertebrate component. However, this is only just beginning to be realized, as a result of phylogenetic and ecological studies using molecular methods (Torsvik et al., 1996). A single gram of soil is estimated to contain several thousand species of bacteria (Giller et al., 1997a). Of the 1.5 million species of fungi estimated to exist worldwide, remarkably little is known about soil fungi, apart from common fungal pathogens and mycorrhizal species. Some 100,000 species of protozoa, 500,000 species of nematodes and 7000 species of earthworms are estimated to exist (Reynolds, 1994), not to mention the other invertebrate groups of the mesofauna (e.g. springtails, mites and potworms) and macrofauna (e.g. ants, termites, beetles and spiders) (Brussaard et al., 1997). Few data are available from tropical regions, where it is suspected that the highest levels of diversity may be found. Consequently, although the biological diversity of the community of organisms below the ground is probably higher in most cases than that above ground, it has generally been ignored in surveys of ecosystem biodiversity. The lower visibility of below-ground organisms and an absence of 'charismatic' species (those that attract attention) have previously led to less attention being focused on below-ground biodiversity. Yet, below-ground biodiversity is of direct relevance to the health of crops, trees and other desirable plants, and soil organisms play key roles in the maintenance of soil structure and in closing nutrient cycles by releasing nutrients from dead organic material. But as such studies proceed, driven by the new concerns about biodiversity loss and global change, it becomes more and more apparent that below-ground biodiversity is significantly in excess of earlier predictions (e.g. Eggleton et al., 1995; Giller, 1996).

Soil organisms contribute a wide range of essential services to the sustainable function of all ecosystems. They act as the primary driving agents of nutrient cycling, regulating the dynamics of soil organic matter and soil carbon sequestration (see Chapter 11, this volume). They play key roles in the absorption and emission of greenhouse gases (see Chapter 12, this volume) and modify soil physical structure and water regimes. They can enhance the amount and efficiency of nutrient acquisition by the vegetation through mycorrhiza (see Chapter 14, this volume) and $N_2$-fixing bacteria (see Chapter 13, this volume) and they influence plant health through the interaction of pathogens and pests with their natural predators and parasites (see Chapter 15, this volume). Although the study of soil biology has a long history – including the famous studies by Darwin (1837, 1881) on the role of earthworms in soil formation – the links between the diversity of the soil biota and its functional value are still poorly established (Giller et al., 1997a). The obvious methodological difficulties associated with obtaining species inventories (e.g. taxonomy and sampling) and making assessments of the functional significance of below-ground diversity (in terms of both direct and indirect effects) have hampered such investigations, as have the shortage of conceptual models that might help us answer the 'so what?' question.

Whilst above-ground the conversion of a tropical rainforest, through slash-and-burn land clearing, into food-crop, pasture or tree-crop production systems has an obvious and dramatic effect on all forms of life, the below-ground changes are smaller and take more time (except for the litter layer and top few centimetres of soil directly affected by the burn) to detect. However, due to more limited dispersal means for many soil organisms, changes below-ground may also be more difficult to reverse.

From a purely agronomic perspective, many functions modulated by soil biota can be substituted for by the use of agrochemicals (fertilizer, pesticides) and fossil-fuel energy (soil tillage). However, learning to work through or with, rather than against, the soil biota, is often seen as one of the pillars of a more ecological and sustainable approach to 'healthy agriculture'. As summarized in Fig. 16.1, we can analyse the relationship between land-use practices and below-ground biodiversity in a number of steps or questions:

1. How do land-use change and specific management practices within broad land-use categories impact on soil biota?
2. What are the key functional roles of soil biota in agroecosystems? And which groups play these roles?
3. How do the various soil biota function in below-ground food webs in different ecosystems?
4. How can farmers, as managers of agroecosystems, work through (or with), rather than against soil biota?
5. How does the presence of specific soil biota, and the diversity of the below-ground ecosystem as such, contribute to the overall cost–benefit balance at the farm, landscape and global levels?

In this chapter we will review the first four questions, as the basic data and approaches necessary for answering question 5 are still lacking, except perhaps for some $N_2$-fixing bacteria in annual cropping systems (e.g. Hungria and Campo, 2000). Box 16.1 and Fig. B16.1 give a quick overview of the broad groups of organisms included in this discussion. A conventional indicator of ecosystem diversity is species richness, or the number of species per unit area. This implicitly assumes that all species are of potentially equal value – but how can

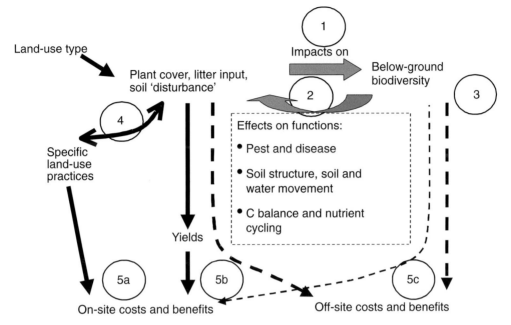

**Fig. 16.1.** Schematic representation of the impacts of land-use practices on below-ground biodiversity (BGBD) and the consequences this may have for internal regulation of BGBD by representatives within below-ground communities and for key functions in the agroecosystem. The numbers refer to questions raised in the text.

we know whether one additional earthworm species is equivalent to 0.5, 1.0, 10 or 100 additional species of mites or fungi? Therefore, in ecological thought this notion is inadequate. As a result, many biologists lost interest in the concept of species richness, after community ecologists showed that one predator species could not be treated as functionally equivalent to one plant species.

Hence, as the overall diversity of organisms is too large and too complex to be fully understood and interpreted, ecologists often use the concept of 'functional groups'. This does not mean that there also are non-functional (or 'redundant') groups, but merely introduces a term for groups of soil organisms that contribute to ecosystem functioning in a similar way (Brussaard et al., 1997). Functional groups can therefore include soil organisms that are not taxonomically related, e.g. termites and earthworms are often included in the functional group of 'ecosystem engineers'.

Numerous problems are associated with the measurement of biodiversity indicators (Paoletti, 1999). Problems in the sampling and extraction of organisms from soil are common to many groups (Swift and Bignell, 2001). Sample size must be determined by a knowledge of both the ecology of the organism in question and of the spatial heterogeneity extant within the habitat under study, and thus cannot be generalized across groups. Larger soil animals (such as termites) can forage over distances of more than 50 m from their nests (Wood, 1988), and can disperse over much larger distances when they fly, whereas smaller animals are relatively sedentary (Giller et al., 1997a). Basidiomycete fungi can forage for several metres (Dowson et al., 1988), and a single individual has been shown to cover an area of more than 15 ha (Smith et al., 1992). By contrast, the habitat range for bacterial colonies is better estimated in microscopic terms of (micro) aggregates.

**Box 16.1.** A quick refresher on 'who is who in the below-ground zoo'.

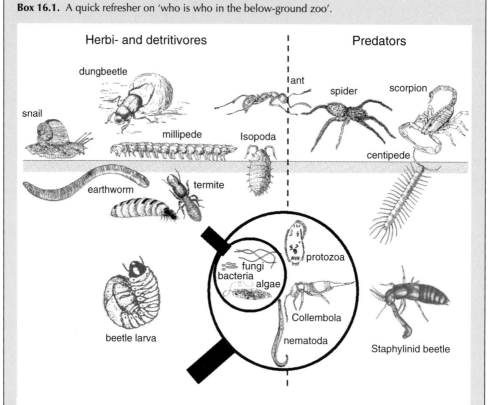

**Fig. B16.1.** Litter layer and soil with some of its inhabitants (see Table B16.1 for details, drawing by Wiyono).

Three groups of 'below-ground' soil biota can be distinguished (Swift and Bignell, 2001).

**1.** Epigeic species are biota that live and feed on the soil surface. These invertebrates effect litter comminution (reduction in litter size) and mineralization (nutrient release), but do not actively redistribute plant materials. They mainly consist of predatory and/or saprophagic arthropods (e.g. ants, beetles, cockroaches, centipedes, millipedes, woodlice and orthopterans (grasshopper-type insects)), as well as gastropods (snails) and small, entirely pigmented (dark-coloured) earthworms. These 'surface-active' macrofauna can be sampled using baits and/or pitfall traps (water-filled containers sunk into the ground into which the animals tumble and are caught).
**2.** Anecic species are biota that take litter from the soil surface and transport it to the deeper soil layers. Through their feeding activities, a considerable amount of topsoil, minerals and organic materials become distributed through the soil profile; this is also accompanied by the formation of channels that alter soil structure, increasing soil porosity. Fauna included in this group are earthworms, non-soil-feeding termites, some beetle larvae and ants (e.g. leaf-cutting *Atta* species).
**3.** Endogeic species are biota that live in the soil and feed on organic matter and dead roots, also ingesting large quantities of mineral materials. Fauna included in this group are non-pigmented earthworms and soil-feeding termites.

**Box 16.1.** *Continued.*

**Table B16.1.** Examples of groups of soil biota, their habitats and food preferences.

| Group | Size: approximate body length | Where do they live? Litter layer? | Soil? | Rhizosphere (around plant roots)? | What do they eat? |
|---|---|---|---|---|---|
| Bacteria | < 1–5 µm, i.e. visible only under microscope (× 1000) | X | X | X | • Organic substrates ('organotrophs') <br> • Inorganic substrates like $CO_2$ from the air, $NH_4^+$, $NO_2^-$, $Fe^{2+}$, $S^-$ and $S_2O_3^{2-}$ ('lithotrophs') |
| Actinomycetes (filamentous bacteria) | < 1–5 µm, i.e. visible only under microscope (× 1000) | X | X | X | Organic substrates; they play an important role in the early stages of decomposition, by mineralizing C and N. |
| Fungi | Some microscopic, but some aboveground parts can reach up to 40 cm! | X | X | x | Carbon and other nutrients from organic matter in the litter and soil. Fungi are the major decomposers of cellulose, hemicellulose, lignin and pectin in plant cell walls. |
| Protozoa (single-celled organisms) | 0.002–0.2 mm | X | X | X | Bacteria, yeasts and algae and sometimes other small protozoa as well. They feed by engulfing the other organisms. |
| Nematodes (roundworms) | 250–5500 µm | X | X | X | Bacteria, fungi (including yeasts) protozoa and other small nematodes. Some nematodes parasitize invertebrates, vertebrates (including humans), and particularly plants, affecting roots and all above-ground parts. |
| Springtails (Collembola) | 1–10 mm | X | X | | Grazing on fungi and other microbes in decaying vegetation. Some feed on live plants or their roots and may be predatory. One family (Onychiuridae) may feed in the rhizosphere and ingest mycorrhizas. Springtails are also effective in biological control of various plant pathogenic fungi. |
| Mites (Acarina) | 0.1–6 mm | X | X | | Fungi, decomposing vegetable matter, or both, nematodes, collembola (predatory). |
| Wood lice (Isopoda) | 5–20 mm | X | Under rocks and in decaying wood | | Fungi and dead organic matter: occasionally roots and foliage of seedlings. |
| Millipedes (Diplopoda) | 2–250 mm | X | X | | Organic debris, but they avoid leaf litter with high polyphenol content and favour litter with a high calcium (Ca) content. |

*Continued*

**Box 16.1.** *Continued.*

**Table B16.1.** *Continued.*

| | | Where do they live? | | | |
|---|---|---|---|---|---|
| Group | Size: approximate body length | Litter layer? | Soil? | Rhizosphere (around plant roots)? | What do they eat? |
| Centipedes (Chilopoda) | 25–280 mm | X | X | | Predators of various fauna (collembola, mites, worms). |
| Scorpions (Scorpionidae) | Average 6 cm, minimum 12 mm, maximum 18 cm | X | Under rocks or logs | | Carnivores. Predators of other arthropods, lizards, mice and birds; they are also cannibalistic. |
| Spiders (Arachnida) | 0.5–90 mm | X | X | | Carnivores. Above-ground predators. |
| Ants (Formicidae) | 1–25 mm | X | X | | Predatory (carnivores), fungi, plant leaves, wood. |
| Termites (Isoptera) | 0.5–20 mm | X | X | | Wood, plants, humus, fungi. |
| Beetles (Coleoptera) | 0.5–200 mm | X | X | | Larvae and adults have very diverse feeding habits: roots, plant litter, animal dung and carcasses, predators (millipedes, various arthropods, and worms). |
| Earthworms (Oligochaeta) | 2–200 cm | X | X | | Organic litter, soil, microorganisms. |

## 16.2 Effects of Land-use (Change) on Soil Biota

Below-ground biota include fungi, bacteria, protists and representatives of the majority of terrestrial invertebrate phyla. No survey can realistically hope to cover all groups, and the degree to which any group can be used as an 'indicator' for all other non-studied groups remains hotly debated (Eggleton *et al.*, 2002; Touyama *et al.*, 2002). Contrary to widely held views, there is remarkably little detailed evidence that agricultural intensification results in a loss of biodiversity in the soil (Giller *et al.*, 1997a). 'Intensification' is here used in a broad sense, and includes an increase in the time that land is cropped, the use of fertilizer, pesticides, mechanization and/or control of soil water content by irrigation and drainage. Even less is known about the thresholds in biodiversity change that are likely to affect soil functions with little reversibility (Swift *et al.*, 1996). Indeed, a number of hypotheses and questions that need to be answered have been identified by Giller *et al.* (1997a).

Studies on soil macrofauna, particularly earthworms (Fragoso *et al.*, 1997), have shown that significant changes in soil biodiversity do indeed occur with land-use change (Lavelle and Pashanasi, 1989) and that these can have functional consequences (Pashanasi *et al.*, 1996). Chauvel *et al.* (1999) showed that conversion of Amazonian rainforest into pasture led to a major reduction in macrofaunal and earthworm diversity, to the extent that only a single species of earthworm survived, resulting in soil compaction due to its surface-casting activity. That changes in land-use have impacts on termites and nematodes has been shown for the Cameroon rainforest (Eggleton *et al.*, 1996, 2002; Hodda *et al.*, 1997). As a result, significant shifts in system carbon fluxes were anticipated. Swift *et al.* (1998) have summarized a number of other studies across a range of environments. Provided external inputs replace biological functions, many of the soils used for temperate-zone agriculture can, however, be mistreated to a remarkable extent and still continue to support crop yields that are close to the theoretical maximum.

A comprehensive approach to the investigation of the relationship between land-use change and soil biodiversity in Indonesia, Cameroon, Brazil and Peru has been initiated by the Alternatives to Slash-and-Burn (ASB) Program. Methods have been standardized (Swift and Bignell, 2001) and overviews of the initial results are now available (Hairiah et al., 2001; Bignell et al., 2002). The main hypotheses underlying this work relate below-ground biodiversity to above-ground biodiversity and the overall C balance of land-use systems, so the work was carried out as part of an integrated survey. Results of the ASB surveys in Indonesia (Fig. 16.2) showed that dramatic land-use change (from rainforest to *Imperata cylindrica* grassland or cassava fields) had a relatively small effect on the number of broad functional groups (orders) represented in a single sample point for the soil fauna. However, it had a greater effect on the surface fauna and especially on the total number of orders encountered per land-use class in the survey. Where we expected to find a drastic difference between the forest and agroforest land-use category (which maintain a year-round litter layer) and the more open land-use classes (which lack such a layer), the survey found that most groups (including spiders, centipedes, millipedes) were still present (at least at the 'order' level). When we 'zoom in' on specific groups, we see little effect on the number of soil nematode genera, or on the diversity of arbuscular mycorrhizal fungi (as judged from the numbers of

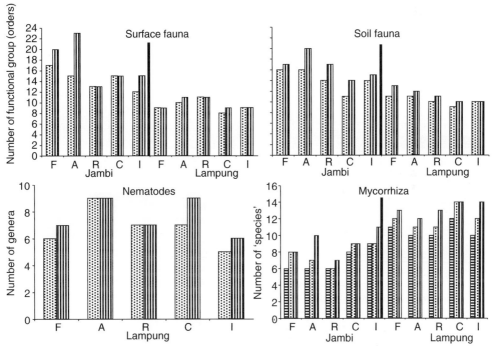

**Fig. 16.2.** Results of below-ground biodiversity surveys by ASB (Alternatives to Slash-and-Burn) of representative land-use systems in Jambi and Lampung in Indonesia. The data were grouped into five classes: F, forest (in Jambi: approximately natural or logged-over, in Lampung: logged-over, degraded); A, agroforest (rubber agroforest or mixed fruit trees); R, regrowing trees, young plantations and agroforests; C, cassava, potentially in rotation with *Imperata* grassland; I, *Imperata* grassland, potentially in rotation with cassava. Data collection included soil fauna of the litter layer and the upper layers of the soil. As a first approximation presence/absence was evaluated of a number of 'functional groups', roughly representing Orders as taxonomic units (e.g. millipedes, centipedes, cockroaches, beetles, spiders ... ). Data are here represented as the maximum number of groups observed for a given location or land cover. Modified from Hairiah et al. (2001).

spore 'morphospecies'). Although such overall diversity indicators change little (and may remain below the threshold of statistical significance as there is considerable variation between replicates), considerable shifts can be observed if we compare the presence/absence of species or genera (Hairiah et al., 2001). Specific nematodes, such as the plant-parasitic *Meloidogyne* spp., were prominent in cassava fields in the ASB surveys, and occurred in smaller numbers in other land-use classes.

Two other cases are worthy of note: the first in the Colombian savannahs (Decäens et al., 2001) and the second in Brazilian Amazonia (Barros, 1999). In these studies, soil macrofauna diversity was assessed at the 'morphospecies' level, in order to evaluate the impact of transformation of the native vegetation to various agricultural uses. In Colombia, between 31 and 57 morphospecies were found in the native savannah, whereas between 42 and 55 were found in various types of pasture. Although the total was similar, only 54% of all morphospecies found occurred in both the natural and the disturbed ecosystems. North of Manaus (Brazil), 151 morphospecies were found in the rainforest, whereas only 48 were found in pasture. Only 15% of the morphospecies were found to occur in both the natural and the disturbed ecosystems.

Although survey data can establish to what degree the presence/absence or relative abundance of various groups has apparently changed, we have very little understanding of how these changes result from changes in reproduction and mortality at the population level. A comparison of the pathways for the impact that land use change has on above-ground and below-ground organisms (Fig. 16.3, Table 16.1) indicates that directly induced mortality (pesticides) and lack of food sources may be the dominant causes of a loss of soil fauna and flora. 'Loss of habitat' (which is a dominant cause of loss of fauna and flora above-ground), may only cause the loss of below-ground organisms that are susceptible to the reduction of soil macropore space that occurs as a result of soil compaction or the loss of all topsoil under serious erosion.

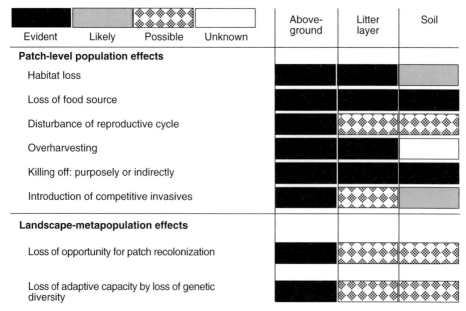

**Fig. 16.3.** Schematic comparison of the various pathways by which land-use change is likely to indirectly affect the above- and below-ground biodiversity, through specific effects on the patch and landscape level populations of soil biota (Table 16.1 gives further details for each of the cells).

**Table 16.1.** Pathways for impacts of land use on biodiversity (see also Fig. 16.3).

| | Agricultural intensification aspects | | |
|---|---|---|---|
| | Above-ground | | 'Below-ground' |
| | Vegetation and fauna | Litter layer microflora and fauna | Soil microflora and fauna |
| **Effects on 'patch-level' populations** | | | |
| Habitat loss | Strong effects on vegetation structure and species composition | Removal of litter layer by slash-and-burn practices, tillage, or overland water-flows; lack of litterfall to reestablish litter layers | Soil compaction reduces macro-pores and habitat availability |
| Loss of food source | Effects on flora and fauna via change in species composition of vegetation | Decrease in quantity and/or diversity or resource qualities, e.g. C/N ratio, content of lignin, poly-phenolics and other secondary metabolites; shifts between 'fungal' and 'bacterial' pathways | Clear, for specific rhizosphere symbionts and 'soilborne pests' |
| Disturbance of reproductive cycle | Loss of pollinators and seed dispersers; loss of nesting and breeding sites | No evidence | No evidence? |
| Overharvesting | Common in 'bush meat' and birds | Discussion (but little evidence) on fungal fruiting bodies (mushrooms) | Large worms for fish bait in Brazil |
| Killing off: purposely or indirectly | Insecticides, rodenticides | Surface-applied insecticides and fungicides | Use of fungicides, nematicides, soil fumigation |
| Introduction of invasive species | Competition with 'weeds', exotic fauna, grazing animals | Leaf-burying worms and dung beetles | E.g. flatworm effects on earthworms |
| **Landscape-level effects** | | | |
| Loss of opportunity for patch recolonization | Coarsening of land-use mosaic pattern can substantially change dispersal opportunities | Coarsening of land-use mosaic pattern | Dispersal distances either small ('creepy crawlies') or large (airborne spores or adults) |
| Loss of adaptive capacity by loss of genetic diversity | Potentially serious, evidence mainly for 'domesticated' plants and animals | Potentially serious, little hard evidence | Potentially serious, little hard evidence |

## 16.3 Functions of Soil Biota in Ecosystems

### 16.3.1 General ecological roles

The terms 'epigeic', 'anecic', and 'endogeic' (see Box 16.1) indicate the overall habitat of soil organisms. The 'anecics', by definition, play a role in functionally connecting the litter layer and mineral soil. Further indications of 'functional roles' are needed, however. A different way of classifying organisms into functional groups distinguishes rhizosphere biota ('rhizospherics'), decomposers and litter transformers, ecosystem engineers, grazers, micropredators and mid-top predators.

*Rhizosphere biota* are organisms that live close to, or in symbiotic relationship with, plant roots, and can thus directly influence plant performance in a positive or negative way. From a plant or an agronomic perspective, a conventional subdivision distinguishes: (i) organisms that increase nutrient amounts (such as the symbiotic $N_2$-fixing bacteria discussed in Chapter 13, this volume); (ii) organisms that increase nutrient availability for the plant (such as the mycorrhizal fungi discussed in Chapter 14, this volume); and (iii) organisms that interfere with root functions, such as plant-pathogenic fungi, plant-parasitic nematodes (Chapter 15, this volume) and rhizovorous ('root herbivorous') insects. From the perspective of all rhizosphere biota, growing roots release an appreciable amount of organic C and N into the rhizosphere (Chapter 5, this volume). The three major sources of this organic C and N are:

- free exudates (substances exuded from roots, i.e. low-molecular-weight organic compounds);
- mucilage (high-molecular-weight gelatinous ('slimy') materials, which are produced by root tips);
- sloughed-off cells and tissues and their decomposition products ('lysates').

The true symbionts and invasive parasites have access to the resources of the living root cells as well.

*Ecosystem engineers* have a major influence on the structure of a soil, creating a network of pores and contributing to aggregation (the way elementary soil particles, clay, silt and/or sand, stick together). Earthworms, termites and some ants can create macropores by pushing their bodies into the soil (thus causing compaction of a zone of soil around the channel that can persist for some time), or by eating their way through the soil and removing soil particles. Earthworms and other animals that feed on soil, produce excrement that contains resistant organomineral structures that may persist for long periods of time (from months to years) and which profoundly affect nutrient cycling and the environment for smaller organisms (Lavelle *et al.*, 1997; Brown *et al.*, 2000). Sometimes these are built into elaborate structures (e.g. termite nests). Earthworms and termites can do this because they have a 'gut flora' of bacteria. These organisms condition the soil as a habitat for other organisms by 'bioturbation' and aggregate formation. Bioturbation is the moving of soil particles, from one horizon to another, by soil biota, in such a way as to affect and determine physical structure and the distribution of organic material in the soil profile. Examples of ecosystem engineers are earthworms and the larvae of some species of scarab beetle of the anecic group (species that live in the soil and feed on surface litter) and earthworms, ants and termites of the endogeic group (species that live and feed in the subsoil).

*Grazers and micropredators* are small invertebrates, mainly protozoa and nematodes, which feed on microorganisms and fungi. They live free in the soil and do not develop mutualistic relationships with microflora. Predation of microorganisms, particularly by nematodes and protozoa, plays an important role in regulating the biomass of microorganisms and is likely to assist in the maintenance of diversity, by preventing the dominance of particular groups. This is arguably more important for bacteria, which tend to be strongly regulated by predation, than for fungi, which are less susceptible to grazing as they are more complex both chemically and structurally (Wardle and Lavelle, 1997). Current models of belowground food-webs (see below) are reasonably successful in predicting the time pattern of N mineralization for a given structure of the foodweb and abundance of functional groups (de Ruiter *et al.*, 1995). Application of such models to tropical ecosystems is, however, still very fragmentary.

*Mid-top predators* are the ants, centipedes, beetles (e.g. carabids), spiders and other arachnids (scorpions, pseudoscorpions, harvestmen (opilionids) and predatory mites) that prey on decomposers and micropredators. As they are more common in the litter layer than in the soil as such, they may cross the below-ground versus above-ground divide. Where they also climb into plants and prey on herbivores they may play an important ecological role relative to their biomass.

Finally, the most important functional group is probably that of the *decomposers and litter transformers*, as all nutrient cycling would come to a halt in the absence of decomposition. The major contributors to decomposition are fungi and bacteria that have the enzymatic capacity to breakdown complex organic molecules and release the nitrogen, phosphorus and sulphur compounds they contain. Soil fauna can improve the access of bacteria and fungi to these molecules by their 'comminutive' action, breaking the material into smaller particles with a larger surface area. Important 'detritivorous' fauna include:

- some species of earthworms (epigeics living in the litter layer on top of the mineral soil);
- wood-eating ('xylophagous') termites;
- pot worms (Enchytraeidae, in the class of Oligochaeta).

### 16.3.2 Relationships between above-ground and below-ground biota

Above-ground diversity consists of plants and nearly all animal groups; but plants play the dominant role by providing both the 'infrastructure' of the vegetation, and the basis of the foodweb – by capturing energy from sunlight and sequestering $CO_2$ into energy-rich carbohydrates, proteins and other organic substrates. Most plants, however, live only partly above-ground – their below-ground organs (roots) are essential for their survival and functioning. We may expect a strong linkage between above-ground and below-ground diversity, primarily because plants and plant diversity determine the functioning of the below-ground ecosystem via factors (van Noordwijk and Swift, 1999) such as:

- plant litter quality, quantity and timing;
- the soil water balance and microclimate in the surface layer;
- root activity, which changes the rhizosphere.

Plant diversity can lead to a wider array and/or a more continuous supply of substrates for the below-ground system. In return, the below-ground community provides a number of 'environmental services' to the plants. However, the functions involved in mineralization and decomposition are broad-based, and there is little evidence to suggest either that specific groups are needed, or that more diverse systems function better from a plant's perspective. Specific relationships with plants do occur in the symbionts, diseases and their antagonists, and it is here that below-ground diversity may facilitate above-ground diversity.

Functional relations between above-ground and below-ground biodiversity, mediated by roots, are likely to involve time lags and may be poorly reversible. Soil organisms tend to have less effective means of dispersal than most above-ground organisms and may thus become a rate-limiting step for ecosystem adjustment in as far as they are critical to the functioning of above-ground vegetation. This is most likely to be the case for specialized obligate symbionts such as mycorrhizal fungi and specific rhizosphere organisms. The impact of soil biota on vegetation change may have been underestimated. The rate of establishment of plant-parasitic nematodes (and not changes in nutrient availability as previously assumed) was shown to be a major determinant of primary succession in sand dunes in The Netherlands (van der Putten et al., 1993).

As herbivores and their below-ground counterparts (rhizovores; van Noordwijk et al., 1998c) exert a considerable selection pressure, it is understandable that plants devote a considerable part of their energy and nutrient resources to the making of 'secondary metabolites', which play a primary role in making them less attractive as food (Brown and Gange, 1991). Several antinutritional factors, such as silica needles and polyphenols, continue to inhibit animal activity after the death of the plant organ, restricting comminution and decomposition. Such relations have been poorly quantified so far, but recent observations (Min Ha Fagerstrom, 2001, Hanoi, personal communication) of limited earthworm activity under fallows of *Tephrosia candida* (a species with a high rotenone content) may provide an explanation for the surface accumulation of its litter. Crop domestication has often led to a reduction in such substances, in order to increase the crop's harvestable yield and consumption value. Interestingly, where the labour efforts required to guard crops without chemical defences from herbivores exceed the labour required for removing the toxins in food processing (as is true in the case of the 'bitter cassava' preferred by African farmers for out-fields) plant chemical defence properties may be retained during domestication. However, due to a general loss of antinutritional factors in today's crops, decomposition may be accelerated in agroecosystems, as compared with natural ecosystems. In addition, the quantity and diversity of organic inputs are lower in agroecosystems. Thus, there may be less need to maintain an assembly of specialists in order to secure decomposition processes.

## 16.4 Food-web Theory and Below-ground Biodiversity

Important aspects of the structure, functioning and stability of the below-ground part of agroecosystems can be captured in the study of the dynamics of the soil food web. A food web is a network of feeding interactions between species or, more generally, groups of organisms. It is a web of interacting food chains. The food-chain concept is basic to ecology and represents an ecosystem in a simple way, as a chain of species where one species is preyed upon by the next. A generic form can be written as: plant, plant-eater, eater of plant-eater, and eater of eater of plant-eater. Well-known examples are the terrestrial (above-ground) plant–herbivore–carnivore chain and the aquatic algae–zooplankton–fish chain. The chains also depict how energy (fixed by plants as primary producers) is transferred over trophic levels to the top consumers. The transfer of energy through the food chain is necessarily accompanied by loss when organisms convert their food into new biomass. This is why we speak of 'trophic pyramids', a concept referring to the decrease in energy consumption that occurs with an increase in trophic levels, often accompanied by a decrease in biomass (Odum, 1971).

Like most natural communities, soil communities are not simple chains of species. They are diverse, complex systems, in which species often feed from more than one trophic level. Ecologists use food-web models to try to capture the interconnected dynamics of populations in such complex communities. By studying soil food webs, ecologists have gained insight into such system processes as overall energy flow and nutrient cycling, as well as into the relationships between the shapes of trophic pyramids and the stability of communities and their environments.

### 16.4.1 Food-web models

When organisms decompose organic matter they transform a proportion of the energy and nutrients into new biomass, and release the rest in mineral or gaseous form. Hunt *et al.* (1987) described the soil food web as a community of 'functional groups', with detritus (i.e. plant litter or other forms of dead organic matter), and live plant roots as the base. A simple representation of a below-ground food web for tropical agroecosystems is shown in Fig. 16.4.

The functional groups in food-web diagrams are not taxonomic categories, but groups of species having a similar choice of prey and similar population dynamics. The detail with which groups are distinguished depends on practical limitations imposed by field measurements and on the level of interest of the researcher. When constructing a food web, the practical decision to distinguish just two groups of microbes, fungi and bacteria (as in Fig. 16.4, for example), leads logically to the distinction of two microbivorous groups, fungivores and bacteriovores. Using this functional-groups approach, de Ruiter *et al.* (1994) evaluated overall annual nitrogen mineralization and respiration rates in a series of agricultural and natural soils. Using observations on population sizes (biomass per unit land area), and data on species' physiologies (in particular the species' natural turnover rates and food-conversion efficiencies), they calculated annual equilibrium feeding rates in

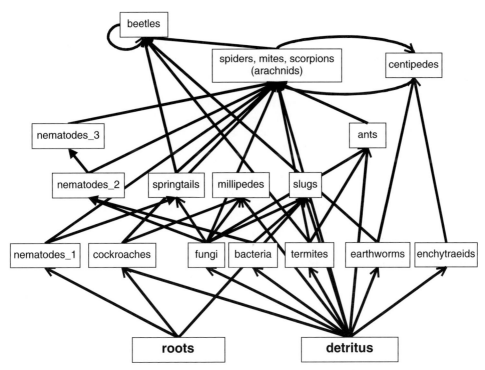

**Fig. 16.4.** Generic food web of the below-ground community in a tropical soil.

the food webs, which determine mineralization and respiration rates. Equilibrium is a state where for each species the population size is constant (i.e. growth rates are equal to loss rates). De Ruiter and colleagues also showed how these functional processes are linked to system stability, based on the stability of all populations in the community. Stability calculations require estimates of the 'interaction strengths' (May, 1973), the effects of species on each other's dynamics near equilibrium, that can be derived from the equilibrium feeding rates (de Ruiter *et al.*, 1995).

### 16.4.2 Stability

When talking about system stability, we may think of many aspects of a system and different types of disturbance. Stability is concerned with the ability to deal with disturbances. It may be associated with concepts such as *'constancy'*, which simply states that a system stays the same over time. However, this does not tell us anything of the way the system handles disturbances.

First we should ask what is a 'disturbance'? Let us say that a disturbance starts with a change in the environment of the system, and can 'propagate' within the system through the response of (local) subcommunities or those populations that make up the biological community. We may distinguish three different levels of system stability: the first is 'resistance' (resisting change), which states that the system and its components withstand change altogether. The second is 'resilience' (coping with change), which states that the components may change, but the system and its components are drawn back, at a certain speed, either to their original state, or to the same domain of attraction. The third is 'adapting' to change: the system and its components end up in another state or domain; but, at a higher level of abstraction, the system stays essentially the same, i.e. it functions in a similar way. Think of the first form (resistance) as a pile of bricks or a brick building, the second (resilience) as a wooden house or a flexible rope bridge, and the third (adapting) as a tree, e.g. it adapts its growth direction, loses some leaves or even branches, but remains the same tree. Our current mathematical concept of the stability of food-web models comes closest to the second level (resilience). But, instead of picturing that level as a more or less static object, it would be better to think of it as a group of people all pulling on the same rope but in different directions. When the rope does not move, all forces are in equilibrium. When an outside disturbance leads to one of the people pulling harder, or letting go, the change in force may affect the other people, leading to all kinds of other changes. When, finally, everyone holding the rope exerts equal force once more, and the centre of the rope is in the same area, we would call this stable behaviour. Instability would be the probability of not returning to this area. The degree of stability could be measured by, for example, determining the maximum disturbance from which a recovery would be possible.

The debate on the relationship between biodiversity and ecosystem stability has a long history. A classical idea in ecology holds that complex and diverse communities or ecosystems are more 'stable', in an unspecified sense, than simple ones (Elton, 1927; MacArthur, 1955; Odum, 1971). This is often illustrated by the example of a well-developed tropical rainforest and a monoculture in an arable field. The rainforest, a highly diverse system with (supposedly) small fluctuations in population abundances and steady nutrient cycles, is regarded as a stable system. The relatively simple agricultural system does not have that many feedback mechanisms and is very susceptible to, for example, pest outbreaks or adverse weather conditions. However, it has been difficult to get operational definitions of stability in order to apply a direct experimental approach to these issues. With the introduction of the use of mathematical models in ecology, it became possible to define and distinguish more explicitly the various aspects of stability. The study of the mathematical stability of a community (i.e. of the ability of the community, when perturbed from a state of equilibrium, to return to equilibrium) has led to the idea that complexity in communities gives instability (May, 1972). The larger the number of species or the higher the density of interactions ('connectance') in a system, the less likely the system is to return to a state of equilibrium following a small perturbation from said equilibrium. This difference between experience from the field and the results of mathematical models, in ideas on the relationship between complexity and stability, is one of the central controversies in ecology. But, the two perspectives can be reconciled (May, 1973), if we reverse the presumed cause–effect chain: diversity itself does not lead to a stable system, but a stable environment is required for diversity to develop. Relative to temperate and subarctic systems, tropical ecosystems have developed under relatively constant conditions. Moreover, when developing, they have also created their own environment (their own microclimate), which in turn acts as a buffer against outside disturbance.

### 16.4.3 Food-web theory on diversity, stability and energetic organization

Recently, the discussion on whether biodiversity leads to stability or instability (or vice versa) has been given new impetus, through studies that point to both the stabilizing effects of (weak) interactions (McCann et al., 1998) and the importance of community organization, indicating that more diversity as such does not make a community more (or less) stable. Rather, it is the organization of diversity (the pattern of strong and weak interactions) that determines stability (McCann et al., 1998; Neutel et al., 2002). Observations on soil food webs give some insight into what constitutes this stabilizing organization. In general, ecosystems are biomass pyramids, i.e. by far the most biomass is at the base of the food chains, and the higher up the chains we look, the less biomass we find (Odum, 1971). Neutel *et al.* (2002) found that this phenomenon is an important stabilizing property of food webs. In two series of soil food webs occurring along gradients of primary vegetation succession, below-ground complexity and food-chain length increase with developmental (successional) age. It was not this complexity that determined the stability of the system, however. Rather, it was the shape of the biomass pyramid of the food webs that turned out to be a good indicator of food-web stability. The stronger the biomass decrease over trophic levels in the bacterial 'chain', the more stable the food web (Neutel, 2001; Fig. 16.5).

### 16.4.4 Perspectives on soil ecology in tropical agroecosystems

For soil ecologists in the tropics, this food-web approach may offer the possibility to compare different soil communities in the field, and analyse their diversity and stability at a system level. There are, however, a number of methodological and conceptual issues to take into account.

The first issue pertains to the complexity of many tropical systems, and the difficulty of collecting complete and detailed data sets. Because of this, simple indicators of community structure are required. Instead of trying to accomplish the almost impossible task of identifying and quantify-

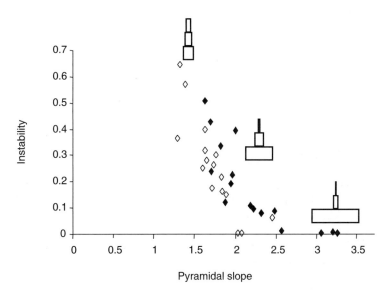

**Fig. 16.5.** Relationship between the slope of the biomass pyramid and an index of instability of the foodweb, in two series of below-ground food webs in Dutch dune soils. Filled diamonds represent food webs from coastal dunes on the island of Schiermonnikoog, unfilled diamonds represent food webs from sand dunes on Hulshorsterzand (part of a nature reserve on the Veluwe), both in The Netherlands. Each symbol represents a food-web replication in one of the series. The pyramidal slope is the decrease over trophic levels of the species' biomass (densities) in the bacterial 'chain'. This decrease was expressed as a power of 10, i.e. pyramidal slopes of 1 and 2 mean a tenfold and a 100-fold decrease in biomass over trophic levels, respectively. In the bacterial chain, trophic level 1 consisted of bacteria, trophic level 2 consisted of bacterivorous nematodes, flagellates and amoebae, and bacterivorous mites, trophic level 2.5 consisted of predatory nematodes, and trophic level 3 consisted of nematophagous mites and predatory collembola. Instability is defined here as the level of intraspecific interaction the populations need for the food web to be (mathematically) stable. Intraspecific interaction, or self-limitation, counteracts destabilizing effects of interaction between the populations. A system that needs a high level of intraspecific interaction is relatively unstable. Based on A.M. Neutel, J. van de Koppel, F. Berendse, P.C. de Ruiter, unpublished.

ing all the feeding relations in a soil food web (Fig. 16.4), we may start by first characterizing systems according to the shape of their associated trophic pyramids. Probably the simplest indicator is the ratio (in orders of magnitude or powers of 10) between the top and the base of the pyramid. In a study of the impacts of above-ground land-use intensification on below-ground biodiversity (Fig. 16.2), we explored this approach. The comparison of five land-use types, in two locations, showed some differences in the top-to-base ratios, i.e. top trophic level biomass and (base) annual organic matter input (Table 16.2) in below-ground pyramids (Fig. 16.6a,b). The first thing to note, however, is the remarkable similarity of the amount of top predator biomass per unit organic input: approximately $5 \times 10^{-6}$ kg top predator biomass per kilogram of annual organic input per year (Fig. 16.6c). The values for Jambi province are higher than those for Lampung, which might suggest that these pyramids would be less stable (and associated with a more stable climate and an environment with less disturbance). Compared with Fig. 16.5, however, we may conclude that extremes such as the pure sands that form the start of the successional series in those data are missing, and all points may be in the range 1.5–2.0 of that figure (assuming three trophic levels and a biomass turnover of 1 year for the top predators as a first guess).

As expected, the (base) organic matter inputs to the soil food web showed a decreasing trend over the land-use gradients (Fig. 16.6a,b). This trend reflects degradation of above-ground vegetation

**Table 16.2.** Assumptions underlying the estimate of annual organic matter input to the soil in five different land-cover types (Fig. 16.2), sampled in Indonesia as part of the Alternatives to Slash-and-Burn Program.

|  | Forest | Agroforestry | Regrowth | Crops | Imperata grassland |
|---|---|---|---|---|---|
| Standing biomass stocks (Mg/ha) | | | | | |
| Woody above-ground | 350 | 120 | 20 | 0 | 0 |
| Leaf and twig | 10 | 10 | 8 | 4 | 6 |
| Coarse roots | 50 | 17 | 3 | 1 | 3 |
| Fine roots | 3 | 3 | 3 | 2 | 3 |
| Coarse roots from previous vegetation | 5 | 10 | 10 | 10 | 10 |
| Estimated turnover rates (per year) | | | | | |
| Woody above-ground | 0.02 | 0.02 | 0.02 | 0.02 | 0.02 |
| Leaf and twig | 1.0 | 1.0 | 1.0 | 1.0 | 1.0 |
| Coarse roots | 0.02 | 0.02 | 0.02 | 0.02 | 0.02 |
| Fine roots | 1.0 | 1.0 | 1.0 | 1.0 | 1.0 |
| Above-ground inputs lost by fire | 0.0 | 0.0 | 0.0 | 0.0 | 0.5 |
| Organic matter input to the soil (Mg/ha/year) | | | | | |
| Above-ground coarse litterfall | 7.0 | 2.4 | 0.4 | 0.0 | 0.0 |
| Leaf and twig litterfall | 10.0 | 10.0 | 8.0 | 4.0 | 3.0 |
| Coarse root turnover | 1.1 | 0.5 | 0.3 | 0.2 | 0.3 |
| Fine root turnover | 2.5 | 2.5 | 2.5 | 1.5 | 2.5 |
| Exudation | 1.3 | 1.3 | 1.3 | 0.8 | 1.3 |
| Annual change in Corg/Cref[a] – Jambi | 0 | 0 | −0.05 | −0.01 | 0 |
| Annual change in Corg/Cref – Lampung | 0 | 0 | 0 | −0.01 | 0 |
| Net change in Corg – Jambi | 0.0 | 0.0 | −3.0 | −0.6 | 0.0 |
| Net change in Corg – Jambi | 0.0 | 0.0 | 0.0 | -0.6 | 0.0 |
| Total (Mg/ha/year) – Jambi | 21.9 | 16.7 | 15.4 | 7.1 | 7.0 |
| Total (Mg/ha/year) – Lampung | 21.9 | 16.7 | 12.4 | 7.1 | 7.0 |

[a]Corg/Cref, soil organic carbon content relative to that for forest soils of the same texture and pH.

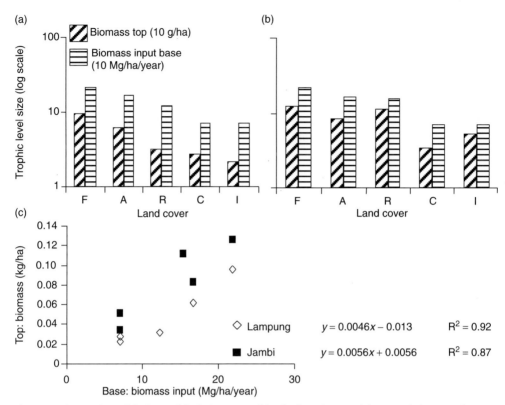

**Fig. 16.6.** Comparison of the base of the below-ground foodweb (estimates of the annual above- and below-ground organic inputs) and the top predators (biomass of arachnids, ants plus centipedes); land cover codes (F, forest; A, agroforest; R, plantation; C, cassava; I, *Imperata* grassland), data from the ASB-Indonesia consortium. (a) Data for Lampung; (b) data for Jambi; (c) correlation graph.

and suggests that below-ground community structure and functioning followed changes in above-ground vegetation. The correspondence between the above-ground and below-ground community is not self-evident; when single trophic groups were compared with organic matter input the relations were much less clear (data not shown).

The second issue to be considered is that of spatial structure. How do spatial compartments affect stability? Where does 'below-ground' end and 'above-ground' start? This is a general issue of course, but is all the more eye-catching in tropical systems. Should we distinguish a litter and a soil food-web compartment? Above-ground vegetation, particularly as regards the presence or absence of trees, could have important consequences for below-ground spatial structure and, consequently, for the dynamics of the litter–soil system (Fig. 16.7). Theory to deal with this aspect is lacking. Investigation of spatial heterogeneity could make an important contribution to our understanding of plant–soil interactions, as well as to food-web theory in general.

The third and final issue pertains to the fact that, although the energy flow constraints placed on food-web structure do not directly allow us to make predictions about below-ground biodiversity, there may be opportunities to link them with the theory developed by Hubbell (2001). In this 'neutral' theory of biodiversity and biogeography, Hubbell proposed that species richness within functional groups can be understood

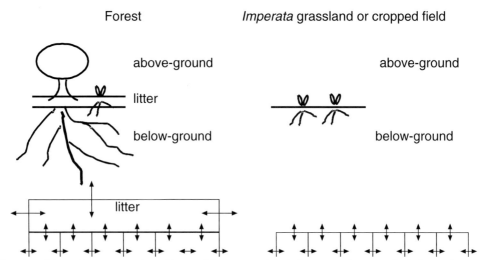

**Fig. 16.7.** Schematic representation of the spatial structure of the below-ground community in forests (or other vegetation with a permanent litter layer) and grasslands of cropped fields without such a layer. The litter layer allows organisms to move over much larger distances than most organisms living in the soil. These organisms transport organic material vertically to and from the soil, and/or horizontally over the surface rather than via the soil. The lateral flows in the litter layer could have important consequences for the dynamics of the soil community and allow for a stronger presence of 'top of the pyramid' organisms.

by assuming that interspecific interactions are approximately equal to intraspecific ones. Increases and decreases in the abundance of individual species can then be seen as 'random walks', with the probability of local disappearance depending on total energy flow to the group and the probability of recolonization from neighbouring sites within the same 'metapopulation' depending on the accessibility and the dispersal means of the species. Combining this theory with food-web theory may prove to be productive.

## 16.5 Farmers and Below-ground Biodiversity: Many Questions Remain

Is below-ground biodiversity a concept that has relevance for farmers? Or should they focus on specific soil biota, especially those that are harmful to their crops (pests, plant pathogens, weeds and their symbionts) or that favour their crops (including symbionts, natural enemies and competitors of the pestiferous biota)? What are the options for 'managing' populations of organisms in the context of the existing food webs and pathways for energy flow? Is 'feeding the soil food web' through organic matter inputs a generally safe way of securing 'healthy soil life'? Is there a need for more specific management? In general, questions such as these cannot yet be answered, despite all the research efforts made thus far; however, we can see that, for specific situations, the balance between 'beneficial' and 'harmful' is a delicate one (Swift and Anderson, 1993; Susilo et al., 1999; Bekunda, 2000).

It is no surprise that changes in land management may change below-ground diversity. But does it matter if one species disappears? Can other species replace its function? Is the function consistently predictable for a wide array of agroecosystems? We all believe that under low-input agricultural systems, earthworms play very important (beneficial) roles in terms of maintaining soil fertility and crop productivity (Box 16.2). But an example exists of earthworms becoming destructive pests in irrigated rice fields in the Philippines (Box 16.3).

> **Box 16.2.** Case study: decline in species richness of earthworms as a result of land management (Hairiah et al., 2001).
>
> Changes in land management may change soil organic matter (SOM) status and subsequently affect the abundance and diversity of 'soil engineers'. Most soil biota respond to litter quality (e.g. termites respond more to low-quality material, ants respond to high-quality, and earthworms appear not respond to litter quality). Brown et al. (1998) showed that when forest was converted into agricultural land in Kenya, Tanzania, Zambia and Zimbabwe, faunal diversity and density were reduced from an average of > 16 orders and 9 g/m² to < 7 orders and 5 g/m². However in some pastures and crop fields, biomass was higher than that found in forests, reaching > 20 g/m². This is primarily a result of the stimulation of earthworm, Coleoptera (beetle) or termite populations.
>
> Organic matter management practices (such as hedgerow intercropping systems) can have a great impact on decomposition, nutrient mineralization and microbial activity. Work on secondary forest in North Lampung (Indonesia) showed that this area had a higher microbial biomass ($10^6$ mg/kg), a higher total number of microbes ($224 \times 10^4$ colony-forming units (cfu)) and higher microbial activity (7 mg/kg/day of $CO_2$) than 8-year-old plots in hedgerow intercropping systems, which benefited from inputs from the pruning of *Peltophorum*, *Gliricidia*, *Calliandra*, *Leucaena* or *Flemingia* hedgerows (Priyanto, 1996). In the same plots, Wibowo (1999) found that the seven species of earthworm found under secondary forest was reduced to six species in hedgerow intercropping systems (*Peltophorum*, *Gliricidia* and mixed *Peltophorum* + *Gliricidia*) and five species in the control plot used (without hedgerows) (Table B16.2).
>
> **Table B16.2.** Species richness of earthworm under secondary forest and agricultural land in North Lampung in dry (D) and rainy (R) seasons (Wibowo, 1999).
>
> | | | | Hedgerow intercropping | | | | | | | |
> |---|---|---|---|---|---|---|---|---|---|---|
> | | Forest | | Pelto | | Gliri | | Pelto+Gliri | | Control | Ecological |
> | Species | D | R | D | R | D | R | D | R | D | R | group |
> | *Megascolex filiciseta* | V | V | V | V | V | V | V | V | V | V | Anecic |
> | *Glyphidrilus papillatus* | V | V | V | V | V | V | V | V | V | V | Endogeic |
> | *Drawida burchardi* | V | V | V | V | V | V | V | V | V | V | Anecic |
> | *Dichogaster affinis* | V | V | V | V | V | V | V | V | V | V | Endogeic |
> | *Dichogaster crawi* | V | V | V | V | V | V | V | V | - | - | Endogeic |
> | *Pontoscolex corethrurus* | V | V | V | V | V | V | V | V | V | V | Endogeic |
> | *Metapheretima carolinensis* | V | V | - | - | - | - | - | - | - | - | Epigeic |
> | Total number of species | 7 | 7 | 6 | 6 | 6 | 6 | 6 | 6 | 5 | 5 | |
>
> Pelto, *Peltophorum dasyrrachis*; Gliri, *Gliricidia sepium*; V, present.

In a simplified causal diagram of the intermediary role that soil biota play in the relationship between 'crop management' and 'yield', a number of intervention points can be identified. Box 16.4 gives an example of how ants, as epigeic soil biota, can play a role in the control of 'above-ground' pests (intervention point 7 in Fig. 16.8). This indicates that the split between below-ground and above-ground biota is artificial: the litter layer fauna in particular can play an intermediary role.

Similar to the view expressed in Chapter 14, this volume, with regard to the management of mycorrhizal fungi, the dominant paradigm in applied soil ecology is based on maintaining the energy resource base for biota through the regular supply of organic inputs to the soil (which could be summed up by the saying 'no litter, no money'). However, where the relationship between the litter layer and the soil has been disturbed, strategic deep placement of organic inputs in soil trenches, along with the stimulation of worm activity, can have spectacular effects on crop production (Box 16.5).

**Box 16.3.** Case study: earthworms in the Ifugao rice terraces (IRTs), the Philippines.

'Soil engineers' that make macropores in the soil (such as earthworms) are not always welcome. This is particularly true in bunded rice fields. Farmers who own such fields work to destroy soil structure, and so reduce its porosity, by puddling, and build dykes to contain the water in the field. 'Soil engineers' counteract all this hard work.

Surveys were conducted across three municipalities containing IRTs (Banaue, Hungduan and Mayoyao). A total of 150 farmers were randomly selected to be respondents (Joshi et al., 1999). The survey was intended to help researchers learn about farmers' knowledge, attitudes and practices concerning both the extent and nature of the problem caused by earthworms in such irrigated rice fields. Of 150 farmer-respondents interviewed, 125 farmers ranked earthworms as the most important pest of terraced rice fields. The farmers described the problem as follows:

*The earthworms seem to cause damage to the rice fields by making tunnels along the terrace walls, causing leaks, resulting in undesired water drainage from the fields.*

The species of earthworm in the area can be divided into two groups:

1. Terrace-dwelling species
- *Polypheretima elongata* (the dominant species)
- Large worm species belonging to either of the genera *Pheretima* or *Metaphire*
- *Pontoscolex corethrurus* (Müller)
- *Pithemera bicinta* (Perrier)
- *Amynthas diffringens* (Baird)
2. Non-terrace-dwelling species
- *Polypheretima* sp. – a hitherto undescribed species.
- *Pheretima* sp. – a hitherto undescribed species
- *Pleinogaster* sp.

Of all the terrace-dwelling species, only *Pontoscolex corethrurus* was found to be present in the neighbouring forest area. The others may be native, new to the area or exotic species. Such invasions occur most often in locations affected by human activity, and rarely in natural vegetation with a resident earthworm fauna. In general, native earthworms are vulnerable to habitat disturbance and invasion by exotic species (Fragoso et al., 1997).

Besides mechanical control (e.g. tillage), farmers used indigenous methods of control, e.g. the use of ground wild sunflower (*Tithonia diversifolia*) or ground seeds of the neem tree (*Azadirachta indica*) mixed with water and poured evenly over the plot, in order to kill the worms.

---

**Box 16.4.** Case study: the changing roles of ants in agroecosystems in Lampung Province (Indonesia).

From a farmers' perspective, the fire ant *Solenopsis* sp. can be both beneficial and destructive. The ants function as a very effective predator of the eggs of the noxious armyworms (*Spodoptera litura*), which affect soybean (Waraspati, 1997; Table B16.3), and may in fact be the key factor limiting outbreaks of armyworms in Lampung soybean fields (Sudarsono et al., 1995).

However, the same ants are seen as pests themselves in pineapple plantations (Rusmiati, 2001). The ants have a symbiotic relationship with the mealy bug (*Dysmicoccus brevipes*), which lives on and sucks the sap of the pineapple leaves, and indirectly functions as the vector of pineapple wilt virus (PWV). The mealy bug picks up the virus from diseased plants and transmits it into new, healthy plants upon feeding. The direct and indirect effects of the mealy bug are economically significant. However, the bug is quite difficult to control, because it is resistant to various insecticides. The plantation companies have considered the possibility that managing the ants (e.g. intervention point 9 on Fig. 16.8) would be a more promising way of controlling the PWV than directly controlling its vector (mealy bug).

**Box 16.4.** *Continued.*

Ants can also be pestiferous when associated with the banana aphid (*Pentalonia nigronervosa*). In this case the ants play a role in the transmission of the bunchy-top virus (BTV) disease in banana plants. In Indonesia the disease is one of the three most important diseases of banana, though in Sumatra it has only been found in Lampung. The aphid pierces the diseased plant and sucks the virus-containing sap for 1–2 h: it then transmits the viruses to a healthy banana plant at a later feeding. The aphid remains capable of transmitting the virus for about 13 days. The ants are surface dwellers (epigeic) involved in a symbiotic relationship with the aphids, which live at, and feed on, the base of the plant. The ants protect the aphid from predators (seraphic flies, coccinellids, chrysopids) and in return the aphids produce honeydew, a sweet tasting, nutritious substance harvested by the ants. More ants mean more aphids, and more aphids result in more damage to the plantation.

**Table B16.3.** Populations of the pestiferous armyworm (*Spodoptera litura*) and the predatory fire ant (*Solenopsis* sp.), and soybean yield in Seputih Mataram-Central Lampung, 1997 (Waraspati, 1997).

| Variables | Time (weeks) | Without exclusion of ants | With exclusion of ants |
|---|---|---|---|
| *Spodoptera litura* (armyworm) | 3 | 46.3 a | 177.5 b |
| population (individuals/hill) | 4 | 19.8 a | 120.8 b |
|  | 5 | 1.3 a | 17.3 b |
| *Solenopsis* sp. (ant) population | 3 | 9.8 a | 16.8 a |
| (individuals/pitfall trap)* | 4 | 9.3 a | 7.5 a |
|  | 5 | 18.8 a | 6.8 a |
| Grain yields (g of soybean seeds/plant) |  | 17.2 a | 11.0 b |

*Pit diameter = 13.5 cm; average values in the same row followed by the same letters are not significantly different at $P < 0.05$.

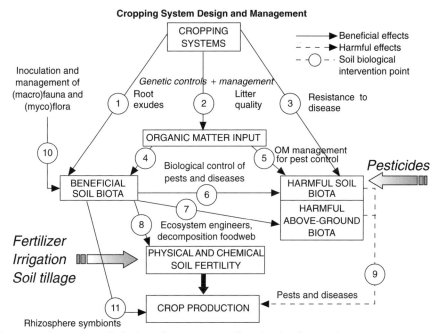

**Fig. 16.8.** The potential entry points for biological management of beneficial and pest soil organisms, cropping systems, organic matter (OM) inputs and soil fertility, affecting crop production. Adapted from Bekunda (2000) and Swift (1999).

**Box 16.5.** Restoring soil fertility and enhancing productivity in tea plantations through the use of earthworms and organic fertilizers: a case study from Tamil Nadu, India (Senapati et al., 2002).

Between the 1950s and the 1980s, tea production in India increased from 1000 to 1800 kg/ha/year, whilst leguminous shade trees disappeared from the tea plantations. This was the result of fertilizer and pesticide use and a shift towards new cultivars that grow better in full sunlight. Currently, national yields have stagnated as decades of intensive cultivation have left soil fertility greatly depleted. On some tea plantations, not even the use of external inputs and plant growth hormones has overcome 100 or more years of intensive exploitation.

Soil degradation on tea plantations is seen in the following: (i) the loss of soil biota (losses can be as high as 70%); (ii) a decrease in organic matter content; (iii) acidification (pH levels can be as low as 3.8); (iv) a decrease in cation exchange capacity; (v) aluminium toxicity; (vi) reduced water retention; (vii) soil compaction; (viii) soil erosion; (ix) nutrient leaching; and (x) the accumulation of toxins (polyphenols) from tea leaves.

In an effort to restore soil fertility and improve tea production using organic matter and earthworms, researchers worked in close collaboration with plantation managers to develop alternative management practices based on locally available organic resources. Tea prunings, high-quality organic matter, and vermicultured earthworms were applied in trenches between tea rows, in order to evaluate the effects had on tea yields. Improvements in the structural and biological properties of soils were expected to produce higher tea yields – and tea yields at one of the estates did indeed increase by a factor of 2.4 in the first year, whilst profits rose by a factor of 3. Trenching is an old practice, and has been mostly abandoned on plantations because it involves high human labour costs. However, researchers in this study thought that trenches would minimize soil loss and improve moisture and aeration conditions, so enhancing nutrient-cycling processes. Lower responses at the other plantations were due to site-specific conditions, including delays in soil recovery that were proportional to the degree of soil degradation.

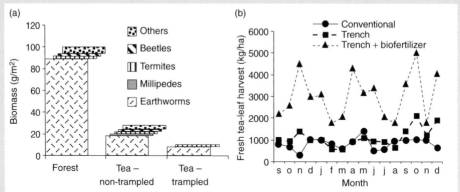

**Fig. B16.2.** (a) Biomass of main groups of soil biota in forests and tea gardens in Tamil Nadu, India. (Redrawn from Senapati et al., 1994.) (b) Effects on harvested tea-leaf yields (fresh weight, kg/ha) of two soil amelioration treatments that involved deep trenches with or without a (patented) biofertilizer formulation, which included cultured earthworms. Conventional = no amelioration treatment (Senapati et al., 1999; Lavelle et al., 1998).

**Conclusions**

**1.** Initial changes in below-ground biodiversity, caused by changes in land-use practices, are less pronounced than those above ground. However, as the dispersal abilities of many soil organisms are more limited than those of organisms above ground, these changes may be less reversible.
**2.** Soil organisms play important roles in agroecosystems by maintaining and restoring soil structure, modifying pest and disease pressures and securing decomposition and nutrient cycles; specific information on the relevance of diversity within 'functional groups' is still scarce.
**3.** The 'food-web' and 'energy pyramid' concepts offer a basic understanding of the abundance of various trophic layers, related to the total amount of organic inputs, and may help to quantify system stability and resilience.
**4.** The scope for managing specific soil biota is limited to the (re-)introduction of species and to the use of specific chemicals. Overall the provision of regular, substantial and diverse organic inputs may be the simplest way to maintain below-ground biodiversity in agroecosystems.

**Future research needs**

**1.** How biodiversity at the 'functional group' level varies with land-use practices should be systematically evaluated, as should the biodiversity of specific target groups at the levels of taxonomic and/or genetic diversity. Such an evaluation should span as wide a range of land uses as possible, including natural, degraded and intensively used lands.
**2.** In order to better understand the function of soil biodiversity in agroecosystems, new models should be developed. These should include soil, litter-layer and above-ground dynamics, a water balance, nutrient and carbon cycles, and pest and disease relationships.
**3.** In order to refine our understanding and improve our models of the food web, the essential biology and life history of the various groups involved should be further investigated. Such investigations should include dispersal, feeding preferences, longevity, seasonality, and energy-use efficiency.
**4.** Research undertaken to address a specific challenge to an agroecosystem (e.g. research linked to the control of pests and diseases) may have to balance the prospects of managing specific soil biota with the opportunity of maintaining the 'below-ground zoo' by feeding it.

# 17 Managing Below-ground Interactions in Agroecosystems

Meka R. Rao, Goetz Schroth, Sandy E. Williams, Sara Namirembe, Michaela Schaller and Julia Wilson

---

**Key questions:**

**1.** Can plant species with complementary root characteristics be used to optimize the exploitation of below-ground resources and limit competition?
**2.** Under what conditions will farmers invest labour and resources in the management of below-ground interactions?
**3.** What practical options are available for manipulating below-ground interactions in agroforestry systems?

---

## 17.1 Introduction

Below-ground interactions (BGI) among component species in mixed agroecosystems encompass the temporal and spatial exploitation of growth resources (water and nutrients), soilborne pests and diseases, and modification of the rhizosphere environment. Interactions can be either direct or indirect, depending largely on whether the system is simultaneous or rotational. Previous chapters of this book (Chapters 1 and 4) have provided clear evidence of the importance of BGI for the functioning of agroforestry and other land-use systems. Of the different BGI, interactions involving growth resources are by far the most important, as they have the greatest effect on productivity (which is where farmers' primary interests lie). Whereas competition for water is the dominant interaction in semiarid environments and during dry seasons or dry spells in humid climates, competition for nutrients is important in many soils across a wide range of environments. Mixed or simultaneous systems in which two or more species are grown together on the same piece of land are most common in the tropics, and traditional agroforestry systems often involve many species. The BGI in such systems are much more complex than in mixed annual systems due to the combination of perennial and short-lived crops, which are of very different sizes, and which occupy overlapping niches below ground. The interactions among these components change over both time and space as trees grow larger and crops are planted, harvested and replanted. A certain degree of competition is inevitable among species in mixed systems when sharing the often fixed and limited below-ground resources (BGR).

To some extent, BGI can be manipulated through choice of plant species, and through soil and plant management. Such manipulation aims to minimize the negative of BGI whilst maximizing their positive effects. For this, a thorough understanding of what happens at the interfaces of the component species is necessary. Unlike above-ground interactions, BGI are difficult to manipulate as they take place unobserved and interventions can be laborious and expensive. Recently available models incorporate soil–plant processes and provide powerful tools for understanding and predicting cause–effect relationships in simple tree–crop agroforestry but not in traditional complex systems. Because of the importance of mixed systems for tropical farmers (in terms of greater yields, the minimization of risks associated with climatic variability and pests and diseases, and better protection of the environment compared with monocrops) ways and means have to be found for improving the productivity of both the traditional and new systems. Although the importance of BGI in agroforestry associations has long been recognized, progress in our understanding of them has been hampered by methodological difficulties. Nevertheless, considerable information is now available from the empirical research of the past decade which can help both with our understanding of BGI and with the development of practical tools for their management. In this chapter we discuss the scope and limitations of different practices for managing BGI in agroforestry land-use systems. Although we confine this discussion to the practices relevant at the plot or field scale, readers are referred to Chapter 18 for those appropriate at the landscape scale.

## 17.2 When and Where are BGI Important?

Use of below-ground resources can be optimized, and BGI minimized, through the combination of species that exploit different niches. Most annual crops do not efficiently utilize below-ground resources, because of their shallow root systems and short growth season. This is particularly so in fertile soils under irrigation or high rainfall, and in areas where acidic or compacted subsoils limit rooting depth. Similarly, perennial crops grown on their own do not fully exploit the inter-row spaces during the early years, because of slow growth and wide spacing. In these situations, total root activity over depth and/or space and time may be increased by: (i) integrating deeper rooting trees of economic value into plots of annual crops to exploit resources at depth and over a longer period of time; and (ii) adding herbaceous cover crops or intercrops between rows of perennial crops for greater exploitation of below-ground resources that otherwise remain unused or are lost to the system (e.g. through leaching or erosion). Some examples of the introduction of trees to annual crops are: planting of trees in rice fields in South India and Bangladesh, poplars in wheat in northern India and *Grevillea robusta* and *Markhamia lutea* in East Africa. In the case of sole systems of perennial tree crops, such as oil palm (*Elaeis guineensis*) and rubber (*Hevea brasiliensis*), patchy occupation of the soil space by the tree root systems and nutrient leaching in the inter-tree spaces may still occur in mature plantations under certain conditions, indicating that associations with shade-tolerant understorey species (intercrops and cover crops) could increase the efficiency of water and nutrient use within the system and increase per-area yields (Schroth *et al.*, 2000a). The exploitation of resources by the perennial component in agroforestry systems can be further enhanced by associating several species of varying growth cycles, so that some species produce early and are then thinned out as others grow and occupy more soil volume.

It is important to consider the conditions under which farmers may, or may not, take below-ground processes into account when designing and managing their land-use systems. Four types of situations may be distinguished, which depend on the abundance of soil resources, the value of the crop (and the relative value of different components), labour availability, and the objectives of the land-use system.

**1.** One situation in which farmers apparently ignore BGI is when soil resources are abundant, i.e. under conditions of high rainfall and soil fertility or the application of sufficient fertilizers. For example, on sites with sufficient water and with adequate fertilization in Costa Rica, coffee growers preferred to use the very fast growing *Eucalyptus deglupta* as a shade tree for coffee, because of its light and homogeneous shade and low pruning requirements (Tavares *et al.*, 1999). Recent research has confirmed that below-ground competition is in fact no problem under these conditions (Schaller *et al.*, 2003). Similarly, coffee growers in eastern Java, where soils are fertile and rainfall is sufficient, prefer to use the very fast-growing *Paraserianthes falcataria*, planted at a high density, as a shade tree for coffee rather than the slower-growing *Pinus merkusii*.

**2.** In other cases farmers may deliberately choose to ignore negative BGI even if they are biophysically relevant, because of socio-economic considerations. Then, the decision whether to invest in manipulating BGI depends on the opportunity costs of the labour necessary for the purpose and the present or future value of the affected crop. An example is the Sumatran 'jungle rubber' system, where rubber planters allow the secondary vegetation to regrow between young rubber trees that have not yet reached the size required for tapping, instead of weeding the trees and establishing a leguminous cover crop as is the recommended practice. Delayed weeding in the early stages can actually protect the rubber seedlings from damage by wild animals, so reducing risk. Of course, tree development is reduced by the competition from the secondary vegetation, so that the planters have to wait 10 years until the first tapping as compared with 5 years in an 'optimally' managed system. Similarly, central Amazonian farmers often abandon their young tree-crop plantations after an initial phase of 1–2 years of intercropping with annual crops. The management of the plantation recommences only when the trees enter the productive phase, which is of course also considerably delayed by such 'poor' early management (Sousa *et al.*, 1999). Still, the decision not to manage competition processes may be rational as long as farmers are more constrained by labour than by the availability of land, and if farmers plant trees as part of a strategy to acquire land, as is the case in Indonesia.

**3.** In rotational systems, and for the rehabilitation of problem sites, maximization of tree root functions can be the objective of management. For the amelioration of compacted, waterlogged or saline soils, the suppression of weeds and the recycling of subsoil nutrients during fallow phases, farmers may use trees with large and competitive root systems at a high planting density and may not be concerned about interactions with crops (Schroth *et al.*, 1996; Mekonnen *et al.*, 1997).

**4.** The management of BGI is most needed and most complex where trees and crops are grown in close association, with the objective of producing multiple products, but where soil resources are limiting, at least during some part of the year. Examples are windbreaks or shade trees in seasonally dry climates, and combinations of trees and crops in the semiarid tropics and on infertile soils. The objective of maximizing the exploitation of soil resources whilst minimizing below-ground competition has to be achieved through complementarity of the associated species in terms of root distribution, phenology and function (e.g. use of different nutrient pools). Such complementarity can be achieved through the selection of species combinations, in conjunction with the management of the component species and the soils. Under these conditions, farmers may invest considerable amounts of time and resources in the management of BGI (e.g. through weeding, shoot and root pruning). The discussion in the following section mainly concentrates on this situation, in which optimization of the use of below-ground resources is the objective. It is assumed that farmers are mainly constrained by land availability and that their primary objective is to increase yields of crops and trees per unit area. This situation is representative of increasingly large areas in the tropics.

## 17.3 Scope and Options for Managing Below-ground Processes

In the surface soil horizon, water availability can fluctuate widely and nutrient availability may be high, nutrients being supplied through leaf litter, mulching or inorganic fertilizers. Plants tend to preferentially exploit these surface layers and shift to deeper ones when the surface resources become limiting. In simultaneous systems with mixtures of perennials and short-lived crops, perennials occupy both surface and deeper layers, whilst annuals often occupy only the former. Therefore, if short-lived crops are not to be outcompeted, plant and soil management that enhances the nutrient and water resources for annual crop roots in the surface horizon, and selection that encourages niche differentiation between species – particularly selection for deep rooting in trees – is desirable. In the horizontal dimension, management that restricts competitive tree roots to areas close to the tree is often preferred. Management should also aim at staggering root occupancy and activity in strategic locations for resource acquisition in the soil profile. Other agroforestry systems require different approaches to tree root architecture. For example, trees with extensive root development in terms of both vertical and lateral spread are preferable for sequential systems as they enhance nutrient capture and transfer to subsequent crops via organic pools. Trees with high root mass tend to suppress understorey weeds, which is an important function of planted tree fallows but is not desirable on sloping lands where soil protection is an issue (Schroth et al., 1996). As another example, trees with high fine-root density at shallower depths are suitable for conservation hedges and filter strips.

The functions of tree roots and their interactions with the soil and the roots of associated plants can be influenced through the selection of species, their arrangement in time and space (system design), and the management of the plant–soil system through practices such as weeding, fertilizing, tillage and pruning (Table 17.1). Given the large differences in root characteristics between plant species, species selection is an important way of influencing below-ground processes, although root properties will normally only be one of several criteria used when selecting tree species and will rarely influence the decision about crop species. System design determines the coarse patterns of below-ground processes at the establishment stage of the system, whereas management can be used for fine-tuning them on a more continuous basis as the system matures. Some of the management techniques routinely applied by farmers (e.g. weeding and soil tillage) have a direct influence on roots, whereas others (e.g. shoot pruning) have indirect effects. Importantly, the more successful these measures are in manipulating below-ground processes, the more flexibility there is for choosing tree species with less-than-optimal root properties, including fast-growing, competitive species.

### 17.3.1 Choice of species/provenances

Species and provenance selection is a suitable way of manipulating BGI in cases where neither yield nor product quality is sacrificed. However, uncompetitive species are often also slow growing, and selecting these for use may defeat the objective of maintaining the productivity of the system. The use of species that demand less soil resources and/or are slow growing is especially important for water- and nutrient-limited sites. In the West African Sahel savannah, a wide array of useful tree species can be used in windbreaks at sites where ground water is accessible and competition with crops for soil water therefore unlikely (Smith et al., 1998). By contrast, at sites where the water table is not within the reach of tree roots, the selection of trees with low water requirements and a limited lateral root spread is crucial in order to avoid competition with crops; furthermore, management measures to reduce the water consumption of the trees, such as pruning, may be required.

Table 17.1. Practices for managing below-ground competition in multispecies systems in different situations and the practicality of these practices.

| Management method | | Aim/effect | Where and when | Practicality |
|---|---|---|---|---|
| Selection of species or provenances | | To maximize complementary use of BGR temporally and spatially | Anywhere, at the establishment stage and when functional niches are available | Feasible, if rooting patterns of species are known and the species possess desirable above-ground characteristics, in order to meet farmers' needs |
| Spacing/design | Boundary plantings | To confine negative effects of BGI to a small area | In drier climates where tree/crop competition for BGR is high | Feasible, but other management practices need to be integrated within the system |
| | Scattered trees or tree clusters | To localize BGI | Relevant for croplands and pastures in all climates but especially in dry areas | Feasible |
| | Row (tree or hedge) intercropping | To maximize positive effects of BGI | In favourable soil and climatic conditions | Feasible |
| | Wider spacing of tree rows | To reduce BGI and force deeper rooting of tree roots | Appropriate for drier areas | Feasible |
| | Thinning of trees over time | To reduce negative effects of BGI | Where the negative effects of trees at a given density increase, and trees gain in value over the years | Feasible |
| | Segregation over space | To avoid BGI | Where tree/crop competition for BGR is intense and trees have to be planted at a high density. Trees and crops planted in separate blocks mostly in semiarid tropics (e.g. woodlots and crops) | Feasible. Below-ground competition still exists at the interface of tree and crop blocks |
| | Segregation over time | To maximize positive effects of tree–soil interactions | For reclamation of compacted soils, saline and alkali soils, nutrient replenishment and lowering the water table. Trees rotated with crops in all climates | Feasible, if land and labour (especially at the tree establishment and clearing stages) are not limited |
| Root pruning | Trenching | To prevent the presence of tree roots in the CRZ | Along tree lines in boundary plantings and around individual trees in croplands | Unfeasible if constrained by labour. May be relevant if combined with other interventions (e.g. fertilizer placement) |

*Continued*

**Table 17.1.** *Continued.*

| Management method | | Aim/effect | Where and when | Practicality |
|---|---|---|---|---|
| Root pruning (continued) | Tillage | To reduce superficial tree roots at 0–15 cm depth | Applied to whole plot at the start of the crop season | Feasible. Depth of root pruning depends on degree of mechanization |
| | Severing of superficial structural roots | To prevent lateral extension of tree roots into CRZ, to avoid conflict with neighbours, and to train roots of young trees | Boundary plantings and individual trees, preferably executed in the dry season to older and younger trees depending on severity of competition | Feasible but constrained by labour and is relevant for only high-value crops |
| Shoot pruning | Pruning side branches; pollarding; lopping branches; pruning to low height | To reduce demand on BGR and root growth | Applicable in all climates depending on the system. Hedges are repeatedly pruned within a year. Shade trees in coffee are pruned at the beginning of dry season and trees in cropland are pruned before rains set in | Feasible. Primary purpose of these practices is to reduce shading of crops by trees but they simultaneously affect tree root growth |
| Mulching | | To increase plant-available water by increasing water infiltration into soil and reducing soil evaporation. To control weeds | Mulches are relevant for water conservation in dry areas and for controlling weeds in wet and dry areas | Feasible where enough organic materials are available. Certain mulches may increase termite activity |
| Nutrient supply through fertilizers or organics (quantity and method) | Broadcast Localized placement (~ 5 cm depth) | To decrease below-ground competition for limiting nutrients | Whenever nutrients are limiting. Localization when nutrients are for valuable species and in the case of less mobile nutrients | Feasible. However, smallholders may be constrained by lack of cash and labour. Organic residues are available in limited quantities |
| Barriers | Physical Chemical Biological | To reduce BGI by preventing the intermingling of tree and crop roots | Installed between tree and crop rows and at the junction of tree and crop blocks, before or together with planting of trees | Physical and chemical barriers involve prohibitive costs. Grass strips are easy to establish but may only have a temporary effect |
| Weed management | Manual Mechanical Chemical Biological (cover crops) | To reduce below-ground competition for water and nutrients from weeds. Cover crops also add N to soil and protect soil | Executed as part of land preparation and whenever weed competition exceeds economic thresholds | Feasible. Constraints are timely availability of labour, cover crop seed and cash for herbicides |

BGI, below-ground interactions; BGR, below-ground resources; CRZ, crop root zone.

Farmers can substitute one tree species for another based on tree root competitiveness, if the trees are grown for low-value products such as green manure, firewood, soil conservation, etc. However, if a tree species is grown for specific, valuable products (such as fruits, nuts, resins or timber) the choice has to be made from among provenances of the particular species. Considerable variability exists among tree species in root system architecture, but the extent of variability among provenances of a species is not known for many agroforestry trees. Selection must not be based solely on root architecture, as root function is also important, and the basis of comparison between species or provenances must be clear (for instance, comparing trees of the same age or size). Lack of a simple and reliable method to evaluate species or provenances for differences in root morphology and function is a major constraint. Recently, the use of competition indices has been explored as a short-cut method for evaluating the competitiveness of the root architecture of different species, with mixed results (Ong et al., 1999; Mulatya, 2000). Substantial differences among provenances in terms of above-ground growth are often reported (e.g. for *Gliricidia sepium*) (Dundson and Simons, 1996), which may be indicative of differences in root system growth and architecture, but there is little evidence available to support this proposition. However, significant differences were observed in the rooting characteristics of *Faidherbia albida* from different seed sources: at Niamey (Niger), material originating from East and southern Africa performed poorly, compared with that from West Africa, due to poor root system development (Vandenbeldt, 1991). The type of planting stock is also important. In Kenya, *Melia volkensii* plants raised from cuttings were more shallow-rooted than those raised from seed of the same provenance (Mulatya et al., 2002).

Contrary to the general belief, most trees have a substantial proportion of their fine roots confined to the same soil horizon as crops (see Chapter 4, this volume), which inevitably leads to competition for below-ground growth resources between trees and adjacent crops. Distribution of the fine root mass of 2-year-old *Senna siamea*, *Eucalyptus tereticornis*, *Prosopis chilensis* and *Leucaena leucocephala* trees was similar to that of maize in the 0–100 cm profile. Only *Eucalyptus camaldulensis* had its roots evenly distributed up to 100 cm (Jonsson et al., 1988). Similarly, roots of 3-year-old *Grevillea robusta* and *G. sepium* possessed a very similar distribution to those of maize in the 0–120 cm profile (Odhiambo et al., 1999), although these trees also possessed roots that penetrated more deeply. As trees age, their root densities increase and their roots spread over ever-increasing distances. Tree root densities often exceed crop root densities and, concomitantly, competition with crops increases. Although the absorption centres of tree roots may tend to become increasingly distant from the tree trunk with increasing tree age (Morales and Beer, 1998), tree–crop competition is often characterized by diminished crop yield close to the tree, correlated with high tree root length densities and reduced soil water (Odhiambo et al., 1999, 2001).

Notwithstanding these features of root distribution, zones of high or low root density are not necessarily indicative of levels of root activity. Tree roots at different depths can adjust their function according to water availability. In the dry season, water uptake by *G. robusta* at a semiarid site in Kenya was predominantly through deep tap roots; but, after rewetting of the topsoil layers with the start of the rains, existing lateral roots became immediately active, competing with the associated crop (Ong et al., 1999). Thus, even if a tree species with desirable root architecture (fewer roots in surface layers) is selected, competition will ultimately depend on the activity of the roots in the surface layers during the cropping season, and not simply on their abundance.

To meet their demands for resources, fast-growing trees tend to have more aggressive roots than slow-growing trees (Jama et al., 1998a), although exceptions to this rule have been reported (Schroth et al., 1996). In western Himalayan valleys, the fast-growing exotic species *E. tereticornis* and *L. leucocephala* were found to have greater root biomass and fine root length density than the slower-growing indigenous trees *Grewia optiva* and *Bauhinia purpurea* (Singh et al., 2000). Cherry (*Prunus cerasoides*) and mandarin (*Citrus reticulata*) extended their fine roots up to 1.5 m from the trunk and had a large number of woody roots close to the surface, which both hindered cultivation under agroforestry, and made the trees more liable to be attacked by pests and diseases when intercropped. In contrast, *Albizia* (*Paraserianthes falcataria*) and alder (*Alnus nepalensis*), which were also classed as faster growing, had the most desirable roots for tree–crop intercrop systems, as their fine roots were confined to within 1 m of the trunk, and they had fewer woody roots (Dhyani and Tripathi, 2000). In Uganda, *Ficus natalensis* is preferred to *Eucalyptus deglupta* as shade for coffee, because below-ground competition with the coffee crop is less (B.L. Oriikiriza, personal communication). In the semiarid northeast of Nigeria, per unit root length, *Acacia nilotica* had a greater negative effect on sorghum above and below ground than did *Prosopis juliflora*, a finding correlated with *A. nilotica*'s higher rates of water extraction from soil layers shared with crop roots (Jones et al., 1998). Species selection for root architecture is also possible for systems in which trees are managed intensively, such as hedgerow intercropping (HI) and conservation hedges. Under a regular pruning regime, *Acioa barteri* and *Peltophorum dasyrrachis* had their fine roots distributed deeply (van Noordwijk et al., 1991b; Ruhigwa et al., 1992) compared with *L. leucocephala*, *Alchornea cordifolia* and *Gmelina arborea*, the fine roots of which were found at a shallow depth (Ruhigwa et al., 1992) and *Erythrina poeppigiana*, the fine roots of which were found at intermediate depths (Nygren and Campos, 1995). Of the 13 woody species screened for HI in subhumid south-western Nigeria, *Lonchocarpus sericeus* had the most desirable root architecture, with only 21% of its fine roots colonizing the 0–30 cm soil layer (as compared with 84% in the case of *Tetrapleura tetraptera*). Although *Enterolobium cyclocarpum* and *Nauclea latifolia* had superior tap root systems and fine root form, they also exhibited extensive root distributions and very large lateral woody root volumes, which may interfere with tillage (Akinnifesi et al., 1999b). As already indicated, there can be seasonal variation in tree root activity in different soil layers, according to the availability of soil water. Root activity may also vary according to species, a fact that could be exploited when selecting tree species for certain agroforestry applications (Broadhead et al., 2003). Some variations are on a short temporal scale, e.g. *Faidherbia albida* has a 'reversed' above-ground phenology (with leaf fall during the rainy season), which implies that the seasonal pattern of root activity in this tree is different from that of other tree species. *P. juliflora* is a conservative water user and does not greatly vary its rate of water uptake in dry and wet conditions (Jones et al., 1998). Use of this species in agroforestry systems may be less risky than that of other species with more variable resource demands, because of its greater predictability. Changes in competitiveness also occur on longer timescales, which must be taken into account when selecting species. For example, *Grevillea robusta* is least competitive as a young tree but depresses crop yields at the pole stage (Lott et al., 2000, 2003). Conversely, trees that are initially competitive may become less competitive when they become older in certain situations. For example, certain parkland trees, despite developing very deep and laterally extensive root systems, have little effect on crop growth and are therefore tolerated by farmers in their fields because the improved soil fertility and microclimate in their vicinity outweigh their negative effects (Rao et al., 1998).

Overall, spatial and temporal separation of tree and crop fine roots and their functions is not easily obtainable unless alternative sources of resources (such as subsoil water and nutrients at depth) are available to trees. Even where there is a certain amount of separation, it is likely that species choice will have to be supplemented with tree management, to improve complementary use of below-ground resources. Plus, management strategies to minimize competition will have to be changed over time, as the tree component ages.

### 17.3.2 Tree spacing and planting arrangement

Tree spacing and planting arrangements are among the most powerful means of managing root (and shoot) interactions in agroforestry systems. Biophysically, the optimum spacing and arrangement of trees in a crop field depends on a somewhat elusive balance between two conflicting objectives: the maximization of favourable tree root effects on soils and nutrient cycles on one hand and the minimization of competition with the crops for soil resources (and light) on the other. If trees are planted at a high density in the whole plot, their root systems intensively exploit the soil, add organic matter, improve the soil structure and reduce nutrient leaching. High tree planting densities also lead to deep tree root systems, as a consequence of competition in the topsoil, and this increases the recycling of subsoil nutrients. Such effects are successfully exploited in planted fallows (Jama et al., 1998a) and also in high-density plantings of coffee (Barros et al., 1995). However, in tree–crop associations the potential for increasing the planting density of the trees to maximize these beneficial effects is limited by simultaneously increasing competition with the crops for soil resources and light.

Increasing the spacing of trees in a crop field or pasture reduces the influence the trees have on the soil and on associated plants simply because there are fewer trees per unit area. It also increases the 'patchiness' of the trees' influence. Another way of reducing interactions between trees and crops (or soil) is to concentrate the trees in groups or rows in some part of the plot, such as the plot boundary (i.e. changing planting arrangement). Ultimately, decisions regarding spacing and arrangement will often be made on economic grounds, depending on the relative value and role of different components of the system. Low-value shade trees need to be distributed throughout coffee and cocoa plantations, and their interactions will be regulated via spacing (i.e. trees will be planted or removed as required). However, if the trees produce valuable products, then it is desirable to keep the tree density high and interactions with the crops (and soil) may be regulated via planting arrangement, with trees often being planted on plot boundaries or in contour rows on slopes. Growing trees and crops in rotation is a third form of reducing direct tree–crop interactions. Root interactions between trees and crops and the effects tree roots have on the soil depend on lateral tree root spread, which may be extensive. The root systems of savannah trees may extend several tens of metres from the trunk (Stone and Kalisz, 1991), and cropped alleys of a few metres width between contour hedgerows can be entirely permeated by tree roots (Schroth, 1995; Rowe et al., 2001). Where trees are planted in rows, with a narrow within-row spacing, the lateral root spread perpendicular to the row is likely to increase further due to competition between neighbouring trees. The decrease of tree root density with increasing distance from a tree row may be roughly logarithmic, but may also show pronounced effects of local soil conditions, e.g. soil tillage, nutrient-rich patches, etc. (Schroth et al., 1995). Pronounced crop yield depressions have been observed in the proximity of boundary plantings, especially on shallow soils and in regions with a pronounced dry season and a deep water table (Malik and Sharma, 1990). In such cases, other management options (such as the selection of less competitive and often slower-growing tree species) may be necessary, assisted by management measures as discussed below.

Some perennial crops including coffee and cocoa are commonly planted under the shade of larger trees, which provide microclimatic protection, assist in nutrient cycling and soil protection and reduce the incidence of certain pests and diseases. As the shade trees are scattered over the whole plot area, the root systems of trees and (tree) crops necessarily interact. Tree spacing, thinning of surplus trees over time and regular shoot pruning are the tools for regulating tree–crop–soil interactions in these systems. Although these measures focus mainly on above-ground interactions, root processes are clearly affected (see Section 17.3.4 'Shoot pruning'). Since the root systems of shade trees and perennial crops necessarily intermingle, one would expect that desirable shade trees would be non-competitive below ground. However, as mentioned above, farmers use surprisingly competitive tree species under conditions of adequate soil moisture and fertilization, apparently without negative effects on the crop. In a commercial coffee plantation near Turrialba in Costa Rica, with 4- to 5-year-old *Eucalyptus deglupta* shade trees planted at a spacing of 8 × 8 m, coffee yields were adequate with no indications of reduced growth in coffee plants in the proximity of trees, despite vigorous growth on the part of the trees (Schaller et al., 2003). Beside the high resource availability in the soil, a further reason for the compatibility of coffee with this aggressive tree species was probably a pronounced small-scale partitioning of the soil space, with coffee roots concentrated near the coffee rows and the tree roots in the inter-row spaces. However, the compatibility of coffee with such fast-growing shade trees is confined to sites with adequate water, as in another region with 6 months of dry season, coffee clearly suffered from competition with *Eucalyptus* shade trees (Jiménez and Alfaro, 1999). For very dry coffee-producing sites in East Africa, it has been recommended that trees be planted on the plot boundary instead of spaced regularly within the plantation, in order to regulate root interactions via planting arrangement (Foster and Wood, 1963). This strategy may become increasingly relevant in regions producing perennial crops in the future, if the climate becomes drier due to climate change. When trees are scattered throughout fields, below-ground competition may not be recognized, as overlapping tree root systems result in competition and reduce yield throughout the cropped area, with no areas being free of tree roots for a comparison to be made.

### 17.3.3 Tillage and root pruning

Tillage is a standard method used by farmers to manage BGI, especially between crops and weeds. Under agroforestry conditions, it can also temporarily reduce tree root length density in the crop rooting zone at the beginning of the cropping season and stimulate tree root turnover, with a corresponding release of nutrients into the soil from decomposing roots. Zero-tillage may therefore not be a suitable practice for agroforestry systems. In dry lands, tillage also exercises a favourable effect on soil water storage. A variety of different tillage methods are practised in the tropics; in West African savannahs the soil is mostly tilled with a hand hoe, forming ridges on which the crops are sown, and the weeds covered within these ridges; where animal traction is available, the soil may also be ploughed to form ridges or a level surface, depending on the equipment.

Tillage destroys most of the tree roots in the top 10–15 cm of soil. This should give a temporary advantage to the crops, which are usually sown shortly after tillage. However, tree roots recolonize the ploughed layer within the cropping season, although the speed with which this happens is not well known. In an experiment in central Togo on a very shallow sandy soil, ridging did not alleviate competition between the crop and *Senna siamea* (as compared with that on land that was tilled to give a level surface). This was because the tree roots invaded the ridges (Schroth et al., 1995).

Additional control of tree roots can be achieved through root pruning, either as part of the tillage process, or separately, which can be achieved by deep tillage and subsoiling along tree rows. This technique is practicable and relevant, especially when it is combined with and incorporates other interventions, such as deep placement of fertilizer for trees, water conservation in dry areas and the improvement of drainage. Korwar and Radder (1994) obtained positive results in south India by ploughing several times per year between hedgerows and adjacent crops, thereby removing tree roots. Soil water contents under the crops and yields were increased, suggesting that tree root competition was reduced. Another option is to cut the superficial lateral coarse roots of trees close to the trunk with an axe, which eliminates large quantities of subtended coarse and fine tree roots in the crop rooting zone. This is often practised when competition from roots of trees in boundary plantings causes conflicts with a neighbours' crops. Farmers in Bangladesh were observed to prune tree roots in the first year after planting; root pruning to plough depth became a routine during cultivation. Intensive and deliberate pruning of tree roots as a separate operation tended to be neglected after the second and third year (Hocking and Islam, 1998). Root pruning combined with top pruning may reduce the overall growth of trees depending on the intensity of pruning. In the case of trees planted in rice fields, combined root and top pruning reduced stem diameter at breast height (dbh) and total volume growth of trees by up to 19% and 41%, respectively (Hocking and Islam, 1998). Root pruning is also being tested as a management option in eastern Africa. At the start of the rainy season, lateral roots were severed with a machete or axe to a depth of 30 cm, about 50 cm away from the trunk. The yields of beans and maize within 5 m of the trees increased in the first season by between 0% and 300%, depending on site and tree species (Raussen and Wilson, 2001). Rapid regrowth of roots indicates that root pruning should probably be repeated every 1–2 years. Two years after the pruning treatment, the dbh of root-pruned trees was 12% less than that of trees that had not been pruned. Although the first root pruning of mature trees was hard work, farmers have found that repeated pruning became much easier, and that it could be easily done at the time of site preparation (J. Wilson, personal communication). Digging trenches along the tree rows or around the trees, severing roots and refilling the spaces with soil is a sure way of avoiding tree root competition for a period of time. In semiarid India, root pruning to 0.5 m depth virtually eliminated below-ground competition between trees and crops (Singh et al., 1989). However, severing tree roots up to such depth could, in many situations, be laborious and uneconomical.

### 17.3.4 Shoot pruning

Pruning the shoots of trees offers a convenient way of managing below-ground competition in simultaneous agroforestry systems, provided that the trees are not being grown for their fruits, in which case further considerations concerning the impacts of management on the development of flowering shoots are important. In addition to reducing competition with crops, farmers also benefit from the products of pruning (fuelwood, poles, etc.), and the process also provides opportunities to improve timber quality. Shoot pruning: (i) controls the water demand by reducing leaf area; (ii) reduces fine roots by changing the functional equilibrium between above- and below-ground components; and (iii) alters fine root distribution within the soil profile. Shoot pruning also affects the timing of root growth and tree demand for below-ground resources. The young leaves formed after shoot pruning may also be more susceptible to drought than the old leaves of unpruned trees, and a resulting midday depression in their transpiration may further reduce competition with crops for water (Namirembe, 1999). Smith et al. (1998) recommended strategic shoot pruning of windbreak trees in the Sahel savannah, in order to reduce tree water use and competition with crops under dry conditions.

The severity of tree pruning varies considerably with systems, from the side pruning of trees in boundary tree plantings, to the lopping of branches in the case of dispersed trees, to severe and frequent pruning (to 0.3–1.0 m in height) in hedgerow intercropping. Few studies have been made of the extent of changes in root morphology and function as a consequence of above-ground pruning. However, studies of several species in Indonesia showed that a low pruning height induced a shallow root system with more fine and adventitious roots, probably due to reduced carbohydrate reserves in the short stems and/or a hormonal imbalance (van Noordwijk and Purnomosidhi, 1995). This suggests, first, that it may be appropriate to initiate some types of pruning only after a deep taproot has developed and, secondly, that trees should not be pruned too low; early and severe shoot pruning may induce excessive root branching in the topsoil and consequently increase competition in the topsoil and reduce tree root interception of nutrients in the subsoil.

Moderate pruning of tree branches may not make much difference to the tree's water demand, and hence may affect competition between trees and crops only slightly. Only the removal of a substantial amount of tree canopy reduces the water demand of trees and allows recharge of the soil profile for use by the associated crop (Jackson *et al.*, 2000). Severe pruning or pollarding of trees to a height of 1.5 m caused a decline in the fine root mass of *Erythrina poeppigiana* (Nygren and Campos, 1995). The effects of shoot pruning on fine roots depend on the soil water conditions. Under water stress, pruning caused an immediate increase in the fine roots of *L. leucocephala* and *S. siamea* in the 20–40 cm soil layer (which was followed by the death of those roots). However, when water was not limiting, pruning caused a significant reduction in fine root density and root biomass (Govindarajan *et al.*, 1996; Namirembe, 1999). Natural senescence and leaf fall in deciduous trees may have an effect on fine roots similar to that had by pruning. Intensive shoot pruning of *Gliricidia sepium* hedges in alley cropping during the rainy season also displaced the root maximum of the trees into the dry season, indicating increased temporal complementarity with the crops in the exploitation of soil resources (Schroth and Zech, 1995a).

The impact of pruning on competition may vary with species. In semiarid Nigeria, crown pruning substantially reduced the competitive effect that *P. juliflora* had on crop yield, but it did not reduce the competition of *A. nilotica* on intercropped sorghum (Jones *et al.*, 1998). In Kenya and Uganda, pollarding was found to be an effective means of reducing competition by five tree species. In the first season after pruning, competition was virtually removed, but the effects diminished as crowns regrew, so that pruning needed to be repeated every 2–3 seasons for the beneficial effects on crop yield to be reliably maintained; also, the magnitude of the interactions was sensitive to rainfall. Pruning of crown and root, separately and in combination, were beneficial and had different effects over time (Fig. 17.1). Although farmers benefited from an improved crop yield and tree products, there were trade-offs in terms of the long-term impacts on tree growth, which pollarding reduced by about 15% in terms of dbh. Acceptability of this to farmers will depend on their short- and long-term objectives and on the relative value of different farm products. In many instances, farmers may not resort to above-ground pruning for the sole purpose of reducing below-ground competition. But, provided it is severe enough, above-ground pruning done for other purposes (such as to remove shade, harvest firewood, remove pest- and disease-affected parts, etc.) simultaneously benefits the associated crop by reducing below-ground competition.

### 17.3.5 Mulching

Mulching is a common practice in multistrata, perennial tree-crop and banana-based agroforestry systems and may influence BGI in various ways. Its effects on root processes have, however, been little studied. Mulching can reduce the formation of surface crusts and thereby increase water infiltration into the soil, especially when the mulch is applied before the onset of the rainy season. It also reduces soil water evaporation. The consequent increase in soil water availability should reduce competition for water between the associated species in dry areas. On the other hand, a mulch layer is known to promote the formation of superficial fine roots because of increased water status in the topsoil layers, potentially leading to increased competition (as there are more roots in the superficial soil layers) after the mulch has decomposed. It may therefore be important to provide mulch on a continuous basis, in order to avoid increasing root interactions. However, experimental evidence for

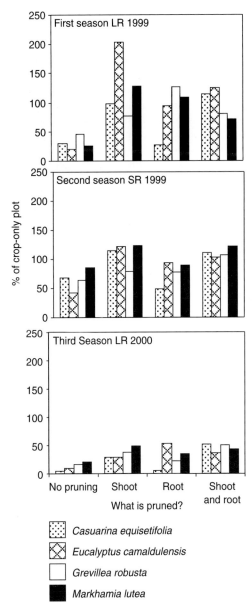

**Fig. 17.1.** Effects of different types of pruning on maize yield (% of no-tree plot) with different tree species. Data are for the first 3 years after pruning at Siaya, western Kenya. LR and SR are long and short rains, respectively. A. Tefera et al. (unpublished results).

had no effect on competition (Schroth et al., 1995). Organic materials used as mulches add nutrients to soil in the course of their decomposition and increase soil biological activity, and thus favourably affect BGI (see Chapter 15, this volume).

### 17.3.6 Fertilizer use and placement

The addition of fertilizer is an intuitive solution to the problem of below-ground competition for nutrients between different components of diverse tropical agroecosystems. Fertilization of intercrops in a system often results in increased growth of the associated tree crops (see Williams, 2000, for a review regarding rubber) and Schroth et al. (2001) for other tree crops. However, fertilization may have unexpected effects, or even no effect (Schroth, 1998). Also, many management-related questions arise: Where should farmers place fertilizer in a mixed species system? When would be the best time/season to apply fertilizer? How should the fertilizer be applied (broadcast uniformly or with localized point placement or injected at depth or spread on the surface)? Which species respond to patches or pulses of added nutrients? In a multispecies system, will all components benefit equally, or will some species take a disproportionate share? What other effects will fertilization have on the system?

Fertilizer should be placed in the zones where there is greatest demand for nutrients by the target component species of the system. These locations can be identified either by systematically measuring nutrient and water distributions in the soil within the agroecosystem (in order to identify areas of depletion) or by studying the root distribution patterns in the system. Knowledge of the location of fine roots and of the occurrence of active root uptake for different species will help target fertilizer application, especially for relatively immobile nutrients such as P. For example, in the coffee–*Eucalyptus* system (Section 17.3.2) coffee plants would benefit most from fertilizer applied around their bases, as is common farming practice.

increased root competition in (discontinuously) mulched systems is not available. In a study in Togo, biomass application either as mulch or as green manure (i.e. ploughed in)

If roots in a tree–crop system extend across the area between tree rows (inter-row area), then it may be best to fertilize within this area of associated crops/vegetation (where interactions are likely to occur at the level of individual roots). Wycherley and Chandapillai (1969) found, when studying 5-year-old rubber in Malaysia, that tree girths were significantly greater when P fertilizer was placed in the inter-row area dominated by secondary vegetation than when the P was applied to the clean-weeded rubber tree rows. In a study in Jambi, Indonesia, growth of rubber trees in three different situations was compared: (i) 'low weeding, no N'; (ii) 'low weeding, plus N' wherein N fertilizer was placed around rubber trees, directly within a weedy inter-row area at 3-month intervals; and (iii) 'high weeding, no N' treatments (Williams, 2000). At 21 months after planting, the mean rubber tree height and trunk volume in the low weeding plus N treatment was significantly greater than in the low weeding, no N treatment ($LSD_{0.05}$), but not significantly different from the high weeding treatment (Fig. 17.2). Thus, addition of nitrogen appeared to partly compensate for the higher belowground competition in the low-weeding plots, so that tree growth reached levels comparable with high-weeded trees. However, addition of fertilizer to plots of early-successional vegetation in Costa Rica decreased the dominance of woody shrubs and trees and increased the dominance of herbaceous species, relative to unfertilized plots. Therefore, increasing the belowground resource of mineral nutrients gave a competitive advantage to the herbs, over the first year of colonization (Harcombe, 1977). This may have important implications for the fertilization of tropical agroecosystems – the desirable tree-crops/woody species may be outcompeted by herbs or aggressive grasses if the system is fertilized too early or too intensively.

The degree of competition for added nutrients exerted by different species in an agroecosystem may also change with season, and this could be exploited by careful timing of fertilization. Seasonal change in the uptake of $^{15}N$ was observed in a mixed fruit tree plantation of *Theobroma grandiflorum* (cupuaçu) and *Bactris gasipaes* (peach palm) with a legume cover crop (*Pueraria phaseoloides*) (Lehmann et al., 2000). In the dry season, the highest N uptake by all three components occurred within the area underneath their own canopies. Yet in the wet season, *Pueraria* took up a greater proportion of N from under the trees, and the trees increased their N uptake from the area under *Pueraria* (although to a lesser extent). Seasonal differences in uptake by different components of mixed systems, related to periods of active root growth, could thus be exploited by fertilizing the specific components at strategic times. For

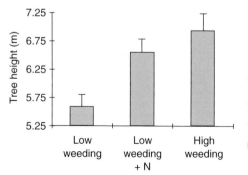
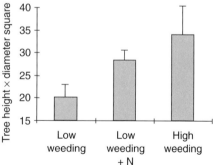

**Fig. 17.2.** Size of rubber trees 21 months after planting in response to weeding and fertilization treatments in Jambi, Indonesia. 'Low weeding', strip-weeding 1 m either side of the rubber trees at 3 and 6 months after planting then no subsequent weeding; 'Low weeding + N', as above, but with 50 g urea per tree applied in a circle every 3 months (equivalent to 55 kg N/ha/year); 'High weeding', clean weeding of the entire plot, nine times per year. Error bars represent one standard error of the mean. Data from Williams (2000).

example, Munoz and Beer (2001) found that fine root productivity of shade trees was greatest at the end of the rainy season, whereas that of underplanted cacao was greatest at the start of the rains, so they suggested early fertilization during the beginning of the rains immediately after pruning the shade trees.

Fertilizer intended for one species within an agroecosystem may actually be taken up by the other associated species. For example, Woods *et al.* (1992) found that weeds took up 68% of the N applied in one experimental treatment in an Australian *Pinus radiata* plantation. Increased biomass production by associated species in response to fertilization may in turn lead to increased competition with the target species. In an association of hazel trees (*Corylus avellana* L.) and a grass (*Dactylis glomerata* L.) in France, for instance, surface application of N mainly benefited the shallow-rooted grass, which increased in biomass and thus competitive strength. This caused severe competition with the trees for water as well as for N (de Montard *et al.*, 1999). However, applying N locally to the area around the tree stem (in addition to surface N application) alleviated the effect of competition for N by the grass, in terms of tree girth increment. In this case, placement of N fertilizer in deep soil horizons, close to the tree stems, was recommended.

If fertilizer is added to a system in order to alleviate below-ground competition, and roots proliferate in response to it, then once the nutrient patch is depleted, intra- or interspecific competition is likely to be even greater than before, so fertilizer application should be repeated regularly (Schroth, 1998). However, if other nutrients or water then become limiting, the situation becomes more complex. Furthermore, if roots proliferate in surface soil layers in response to surface application of fertilizers, then it is possible that during dry periods these plants may become more susceptible to drought.

The physiological characteristics that dictate the response to added nutrients of many of the tree species used in tropical agroecosystems are, at present, incompletely understood and the species' response to this in diverse systems is even less so.

### 17.3.7 Root barriers

As mentioned above, in most agroforestry situations it is desirable that tree roots have access to the soil under the associated crops (because these roots are expected to have soil-improving and nutrient-conserving effects). However, there are situations in which partitioning the soil into tree and crop root compartments can be expected to improve the performance of the system. For example, when trees are planted as windbreaks or shelterbelts in crop fields in dry areas, water uptake from the cropped area by lateral tree roots may counteract the positive microclimatic effects of the trees on the crops. In such a situation, reduced root interactions between trees and crops would lead to higher crop yields. Another example is offered by the invasion of crop fields by lateral tree roots from adjacent tree-fallow plots. It has been shown that fast-growing fallow trees, such as *Sesbania sesban*, can extend their lateral roots to several metres within a few months (Torquebiau and Kwesiga, 1997). Through these roots, the fallow trees may redistribute nutrients from the cropped plot into the tree-fallow plot, instead of recycling nutrients from the subsoil of the fallow plot itself (van Noordwijk, 1999). A certain separation of tree and crop root zones in the crop/planted-fallow interface may be beneficial in two ways: (i) by increasing crop yields by reducing root competition between trees and crops; and (ii) by allowing tree roots to penetrate more deeply into the soil through lateral restriction of the available soil volume, thereby increasing the potential for nutrient recycling and physical subsoil improvement.

Barriers to the lateral development of tree root systems can be chemical, physical or biological. Chemical root barriers can be created inadvertently when trees are planted in very acid and infertile soil and are only locally supplied with fertilizer and lime. The infertile soil surrounding the fertilized planting hole can then impede lateral tree root development. This configuration can be observed in tree–crop plantations on acid soils (Schroth *et al.*, 2000a), but is not a feasible option for managing tree roots in tree–crop associations as the soil under the crops is generally fertilized to a greater degree than that under the trees. The potential for using chemicals other than fertilizers to restrict lateral extension of tree roots at the field scale without causing any detrimental effects is not known.

Physical root barriers such as polyethylene or galvanized iron sheets have often been used in experimental studies to separate the rooting zones of tree rows and adjacent crop rows. Open trenches can be dug along the tree line or around the trees. They are effective only for a short period as, after some time, tree roots tend to pass under the barriers and then grow upwards again, so that the barrier's effect is decreased.

Biological root barriers consist of narrow strips of vegetation with competitive root systems planted alongside tree rows to impede the lateral spread of tree roots. Research into the potential of biological barriers for manipulating tree root distribution is based on reports that the roots of certain trees could be laterally confined and forced to go deeper if herbaceous intercrops or cover crops with competitive root systems are planted at a short distance from trees. Schaller *et al.* (1999) hypothesized that a similar effect could be achieved by planting perennial grasses, with their typically dense root systems, in narrow strips, in order to manipulate the root systems of recently planted trees. The success of this technique depends on the degree of competitiveness of the barrier strips with respect to trees and crops: it could theoretically be used to minimize the competition exerted on adjacent crops by trees in boundary plantings, contour strips or planted fallows.

In a series of experiments in Costa Rica, it was found that the effect of grass barriers depends on both the tree and the grass species. Whereas grass strips induced drastic alterations in the root architecture of *Cordia alliodora* seedlings (Fig. 17.3), the roots of the faster growing, more aggressive *Eucalyptus deglupta* trees were much less affected and generally passed through the barriers. Guinea grass (*Panicum maximum*) and *Brachiaria brizantha* formed more effective barriers than sugarcane (*Saccharum* sp.), vetiver (*Vetiveria zizanioides*) and lemon grass (*Citronella* sp.). Increasing the barrier width from one to three grass rows did not increase the barrier effect of the most competitive species (guinea grass) but tended to increase the effectiveness of the *Brachiaria* barriers (Schaller, 2001). The sugarcane barriers were ineffective against the aggressive *Eucalyptus* roots at all the tested widths. Contrary to expectation, and inexplicably, the grass barriers led to shallower and not deeper root systems in the case of *Eucalyptus* trees.

These early results indicate that the technique may have most potential when it is used with tree species such as *Cordia alliodora*, whose root architecture can be strongly modified at an early stage of development by the presence of grasses. To what extent changes in the root architecture of tree seedlings translate into a more desirable root distribution in older trees remains to be seen.

With regard to physical root barriers, farmers may not be expected to plant grass strips solely for the purpose of tree root management. However, in sloping areas, grass strips also aid soil and water conservation, and the fodder value of the grasses may provide additional benefits. Thus, the root management effect is only one of several functions biological barriers have. However, many more long-term experiments are necessary before it is possible to draw a final conclusion as to their potential.

### 17.3.8 Managing interactions with weeds

Weeds affect BGI by appropriating resources that would otherwise be utilized by the main crop(s) in an agroecosystem, and so limit their growth. For example, in Sumatra (Indonesia), the stem diameter and trunk volume of rubber associated with a mixture of woody and non-woody weeds at 21 months after planting were 17% and 37% lower, respectively, than those of clean-weeded rubber (Williams, 2000). This significant retardation of rubber tree growth was mediated entirely by below-ground interference, as the weeds were low-growing and did not shade the rubber. This was borne out

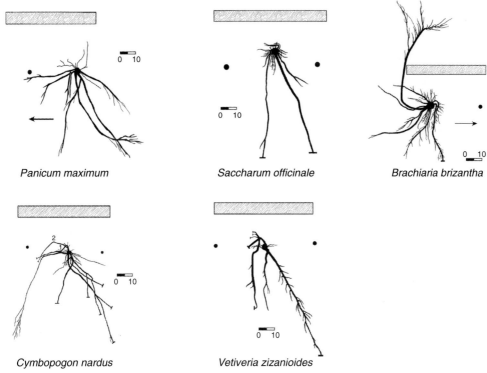

**Fig. 17.3.** Avoidance reaction of the root systems of 8-month-old *Cordia alliodora* saplings in response to strips of different grass species, seen from above, at Turrialba, Costa Rica. The grey bars symbolize the grass strips that were planted at 30 cm from the trees; dots indicate the position of trees whose root systems are not depicted. The scale shown in each figure corresponds to 10 cm. In the case of *Brachiaria brizantha*, a border tree is shown whose roots grew around the end of the grass strip. Modified from Schaller (2001).

by the fact that soil nitrate-nitrogen in the unweeded rubber was 2.83 mg/kg compared with 7.37 mg/kg under weeded rubber.

Biomass allocation within trees may change in response to below-ground competition. For example, rubber trees that experienced competition, either from the noxious weed *Imperata cylindrica* or from a pineapple intercrop, were found to be significantly smaller above ground (in terms of squared stem diameter, Dsq) than were clean-weeded trees (Table 17.2). They were also found to have allocated a greater proportion of biomass to their roots than clean-weeded trees. Competition, especially from the weed *Imperata*, also led to a shift from horizontally to vertically oriented root cross-sectional area (Table 17.2). Implications for management are that regular weeding will favour above-ground tree growth relative to below-ground growth and may also result in a greater concentration of roots in the upper soil layers. This in turn may decrease the severity of future weed infestations due to increased shading and the presence of already well-established tree roots in the surface soil.

Parasitic weeds should be considered as a specific case in below-ground interactions. *Striga hermonthica* and *Striga asiatica* are two major biological constraints to the production of staple cereals (maize, sorghum and millets) in sub-Saharan Africa. The *Striga* problem in smallholders' farms is exacerbated by severe nutrient-depletion as a result of continuous cropping and limited or no use of inorganic inputs. *Striga* remains a pernicious problem as it produces millions of

**Table 17.2.** Root and stem characteristics of 39-month-old rubber trees grown under three inter-row management regimes: no competition (A), competition from an intercrop (B) and competition from a noxious weed (C), at Sembawa Rubber Research Station, South Sumatra, Indonesia. (Source: Williams, 2000.)

| Management of inter-row area | Stem Dsq (cm$^2$) | Shoot : root ratio (Dsqs)[a] | % Horizontally oriented roots (Dsqs)[b] |
|---|---|---|---|
| A. Clean weeded | 74.8 | 0.46 | 60.7 |
| B. Intercrop (pineapple) | 38.4 | 0.23 | 34.1 |
| C. Weed (*Imperata cylindrica*) | 13.9 | 0.28 | 23.7 |
| F-probability | < 0.001 | 0.022 | 0.008 |
| SED[c] | 8.1 | 0.074 | 10.1 |

[a]Shoot : root ratio (diameter squares) = $\Sigma D_{stem}^2 / (\Sigma D_{hor}^2 + \Sigma D_{ver}^2)$. Shoot : root ratios were calculated on the basis of the cross-sectional areas of tree stems and 'proximal' roots (the roots originating from the stem collar or tap root), as the latter can be used as a surrogate for total root system size when applying a fractal branching method (see Chapter 4).
[b]Percent horizontal root diameter squares = $100 \times \Sigma D_{hor}^2 / (\Sigma D_{hor}^2 + \Sigma D_{ver}^2)$.
[c]SED = Standard error of differences between means.

tiny seeds each season, which remain viable for many years in the soil. As the *Striga*–host interaction starts with the establishment of haustorial connections soon after germination of the *Striga* seed, it causes considerable damage to the host crop before it emerges from the soil. Management practices appropriate for small-scale farmers should be based on the principle of depleting the *Striga* seed reserve in the soil. Therefore, rotation of *Striga*-susceptible crops with trap-crops that stimulate *Striga* seeds to germinate without being parasitized (to deplete the soil seed bank) and repeated hand removal of the weed before it sets seed (to avoid additions of seed to the soil) are recommended. However, these practices are not widely adopted for economic reasons. Obviously, an integrated approach with a suite of practices that deplete the soil seed reserve and replenish soil fertility is required to overcome the *Striga* problem and increase crop production (Parker and Riches, 1993).

Agroforestry systems that replenish soil fertility, such as biomass transfer (synonymous with green-leaf manuring) and short-rotation planted fallows, have been examined for their potential to reduce *Striga*. Of the biomass of a number of trees and shrubs tested, only the high-quality biomass of *Tithonia diversifolia* and *Sesbania sesban* (with a low C : N ratio and low concentrations of lignins and polyphenols), reduced the amount of *Striga* present on continuous application to the soil at 5 t/ha (dry weight) over four years (Gacheru and Rao, 2001). The biomass of these species was rapidly decomposed and mineralized to maintain a high level of inorganic N in the topsoil, which has a negative effect on *Striga*. There was no evidence to indicate that *in situ* decomposition of organic residues stimulated *Striga* germination due to the production of *Striga* seed stimulant. None of the organic materials reduced *Striga* as much as inorganic N fertilizers, so use of organic materials should only be considered to be complementary to other methods.

A number of leguminous tree/shrub species have been found to stimulate *Striga* germination in laboratory conditions (Oswald *et al.*, 1996). Of the promising species tested under field conditions, *Senna* spp., *Sesbania* spp. and *Desmodium distortum* depleted *Striga* seeds in the soil after being grown for 12 months, and decreased *Striga* infestation in the subsequent maize crop. However, only the planted fallows of *S. sesban* and *Desmodium* increased the yield of the following maize crop in comparison with that of monocropped maize. This is because both these fallow species produced large amounts of high-quality foliar biomass, which has a direct bearing in terms of increasing soil fertility (Gacheru et al., 1999). Although *Tithonia* and *Tephrosia* also improved maize yields, the decreases they caused in *Striga* infestation were primarily due to increased soil fertility. For use in fallows, farmers will be interested in those species that fix atmospheric nitrogen, produce a high biomass that has multiple uses, and substantially improve maize yields. In this respect, 1- to 2-year-old *S. sesban* fallows are more attractive than others, as *S. sesban* produces firewood and its foliar biomass has fodder value. In soils with moderate *Striga* infestation, repeated cycles of *Sesbania* fallow–crop rotations may overcome *Striga*. However, under conditions of high infestation, fallows alone may not greatly reduce *Striga* infestation. In P-deficient soils (as, for example, in western Kenya) use of phosphorus fertilizers is essential to exploit the benefits of *Striga* reduction gained by the use of the planted-fallow and green manuring technologies.

## 17.4 Conclusions

Optimum use of soil resources requires that below-ground niches (vertical, horizontal, temporal and functional) be exploited by species and life forms with complementary root properties (functional diversity). Exploitation of niches is maximized, for example, by adding deep-rooted trees to shallow-rooted crops or pastures, associating annual crops with perennial trees, or adding temporary intercrops to systems with young tree crops. Whether, and to what extent, farmers attempt to control negative BGI depends on site factors and socioeconomic conditions. Where the availability of soil resources is high, farmers may associate rather fast-growing and competitive tree species with their crops without negative consequences for crop yields. Where labour is more limiting than land and tree-crops are not yet in the productive phase, farmers may also decide not to manage BGI that are having adverse effects on trees, even though technically it would be advantageous (e.g. in the case of young jungle rubber). However, when BGI are a limiting factor in the functioning of land-use systems (e.g. in dry areas) farmers need to consider BGI in their decisions about tree (and crop) species, planting designs and management.

Options for managing BGI include germplasm selection, spatiotemporal arrangement of species, planting density, tillage/root pruning, shoot pruning, fertilizer use and placement, weeding and possibly (as an added benefit of anti-erosion strips) biological root barriers (Table 17.1). The more successful planting design and management are in terms of manipulating BGI, the more flexibility farmers will have to choose tree species with less-than-optimal root characteristics for their systems.

BGI cannot be managed without affecting above-ground interactions and the growth of species, implying the need for a holistic approach to the management of interactions among species in complex systems. For example, delayed weeding and pruning of trees may promote deeper penetration of tree roots, but both operations are likely to reduce the growth of young trees and associated crops. While system design in terms of the spatiotemporal arrangement of trees and tree density radically changes BGI, tillage, weeding, mulching and light shoot pruning have relatively small and/or temporary effects on the root systems of trees and BGI. Severe crown pruning, however, can substantially reduce competition, and the benefits of harvesting tree products can be attractive, but how pruning affects overall growth needs to be considered. Although root pruning is a safe, effective and direct way of reducing below-ground competition, it may be unattractive to farmers because it involves additional work, without the benefit of an immediate tree product (unlike shoot pruning). Farmers' needs and resources and market forces dictate the design of systems, and management of BGI within the context of a given system often demands that a combination of practices be applied. There is much still to be learned about optimizing agroforestry systems: we must improve our understanding of how to optimize resource use as well as our understanding of the short- and long-term effects of such optimization; we must increase our understanding of individual species; we must improve our capacity to predict these interactions, through modelling; and, we must understand these systems within the context of the socioeconomic drivers that dictate what systems will be adopted and how they will be managed. Field experimentation with trees is time consuming, but as models have not been developed to the stage whereby they can be employed for this type of decision making in agroforestry, long-term field experiments as well as the use of indigenous knowledge are essential in order to improve our understanding.

**Conclusions**

**1.** Optimum use of BGR requires the selection of species that exploit different soil resources, or the same resource over different timeframes. Although enhanced interactions between soil and tree roots may have positive effects on subsequent annual crops in rotational systems, increased BGI among component species beyond a certain degree or stage would have negative effects in mixed systems.
**2.** A holistic approach is needed for managing BGI in mixed systems, as most practices will have concomitant and often conflicting effects on below- and above-ground processes, including plant growth.
**3.** Of the different options for managing BGI, germplasm selection, spatiotemporal arrangement of species, planting density (especially of the tree component), and fertilizer use and its placement have greater effects compared with tillage/root pruning, shoot pruning, mulching, weeding, and biological root barriers. In any given system a combination of practices may be desirable to manage BGI, as none would alone minimize the negative effects of BGI.
**4.** The choice of whether or not to manage BGI depends on both site factors and socioeconomic conditions. The need to manage BGI is greater in sites characterized by low rainfall and poor soils than in sites characterized by high rainfall and deep and fertile soils.

**Future research needs**

**1.** How much functional root diversity is needed for a given agroforestry system?
**2.** How many and which types of species are needed to provide this diversity?
**3.** What long-term implications does optimizing the use of below-ground resources have on the resource base and on productivity?
**4.** How do different tree species respond to management practices in the short and long term; and, what differences are there between species and provenances in terms of their flexibility/ability to respond to root management measures?
**5.** What are the costs and benefits of different strategies for managing roots and tree crowns (in terms of tree growth, yields, effects on soil properties, etc.)?
**6.** Can simple predictors and models for the prediction of tree root responses to environmental and management factors be developed?

# 18 Managing Movements of Water, Solutes and Soil: from Plot to Landscape Scale

Simone B.L. Ranieri, Richard Stirzaker, Didik Suprayogo, Edi Purwanto, Peter de Willigen and Meine van Noordwijk

---

**Key questions**

1. How do trees, crops, soil cover and soil properties affect surface and subsurface water movement?
2. What implications do vegetative filters have for soil erosion, nutrient transport and salt movement in the landscape?
3. How effective can vegetative filters be at different scales, and what does this imply for landscape 'design'?

---

## 18.1 Introduction

The following watershed functions have the potential to be modified by changes in land use: (i) the amount of water that flows out of a catchment area; (ii) the timing and regularity of the flow; and (iii) the quality of the water. The latter depends on the concentrations of soil particles, nutrients, salt, agrochemicals, organic material and biota carried by water flowing over or below the surface. In this chapter we will consider how the 'plot-level' understanding of below-ground interactions discussed in the preceding chapters can be used to predict such landscape-level interactions.

Unlike water in unsaturated soil (which mainly moves vertically), runoff and groundwater mainly move laterally. Thus, any change in land use that occurs at a plot scale and that affects infiltration or recharge is likely to have effects at the landscape scale (beyond the plot) via runoff and groundwater movement. Standard representations of the water balance at the plot scale include connections to three types of lateral flow: (i) lateral flows over the surface; (ii) flows through the upper layers of the soil profile; and (iii) 'groundwater' flows. These lateral flows hydrologically connect any 'plot' to its landscape context.

Movement of water leads to the lateral movement of soil, nutrients and other solutes (such as salt), which can cause a range of generally negative environmental effects downhill/downstream (although under some circumstances inflows of soil and nutrients are perceived as being positive). The three lateral flows mentioned above in fact represent a continuum of flow pathways with very different residence times. Surface flows of water 'runoff' and 'run-on' are directly visible, can lead to substantial redistribution of soil and light-fraction organic residues and are generally considered under the headings 'erosion' and 'sedimentation'. Surface flow responds on a second-to-minutes timescale to current rainfall intensity, and its pathway can be easily modified by surface roughness and through the management of surface litter. By contrast, groundwater movement is measured in days, months, years or decades, and responds to the cumulative balance of rainfall and evapotranspiration, rather than to extreme events. The pathways of groundwater movement can be influenced much less easily than those of surface flows, and there is generally a considerable time lag between any management intervention and its effects. This means: (i) that problems are not at first directly apparent; and (ii) that there is little one can do about such problems in the short term once they do become directly apparent. These characteristics of groundwater problems at the landscape scale have consequences for both the degree to which natural resources can be managed and the way in which they can be managed (Lovell et al., 2002). In between the extremes of surface and deep subsoil movement of water, issues of subsurface flows of water and solutes have received relatively little attention. The spatial and timescales at which these flows operate makes them more amenable to management interventions than groundwater flows, yet they are less obvious than surface movements.

In this chapter we will focus on the biophysical aspects of lateral water movement, and its consequences for the movement of solutes and soil. We will also consider how different types and arrangements of land use can affect these types of lateral flow. As indicated by the numbers and letters in Fig. 18.1, we can distinguish four ways in which land cover at the plot level can cause environmental effects outside the plot.

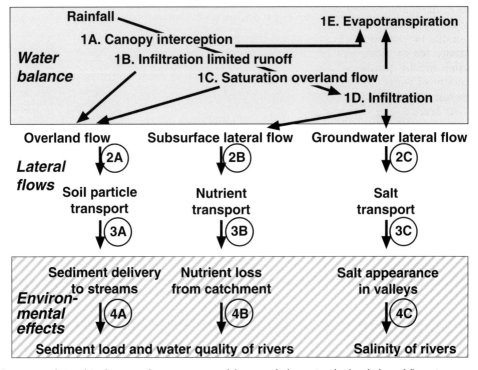

**Fig. 18.1.** Relationships between the components of the water balance (at plot level), lateral flows (at landscape level) and environmental effects. The numbers given refer to types of interventions in the causation of these environmental effects, as discussed in the text.

1. Influences via the interrelated terms of the water balance that determine the total amount of water leaving a plot (rainfall + lateral inflows – evapotranspiration – changes in storage), and its partitioning over surface, subsurface and deep pathways.
2. Partial decoupling of the flow of water and that of soil, nutrients or salt through forms of 'bypass flow'.
3. Filters or interception of the lateral flows of soil, nutrients or salt through changes in the rate of flow of the carrier (water flow) or concentration by processes such as sedimentation, uptake, sorption and precipitation.
4. Interventions that mitigate the environmental effects of the subsurface and deep subsoil lateral flows at their point of re-emergence at the soil surface.

At level 1 (see Fig. 18.1 and list above), land cover influences the pathway of the 'excess' water (rainfall minus water used in evapotranspiration), and thus its partitioning between the flows 1B, 1C and 1D. Infiltration (1D) depends on the characteristics of the soil surface and topsoil, and hence on the balance between soil structure formation due to root turnover and soil biological activity fed by litter inputs, as well as on water use by plants (which increases the amount of water that can infiltrate to refill the soil to field capacity).

At level 2, the dynamic aspects of soil structure also influence the degree of 'bypass flow' that decouples nutrient transport from the mass flow of water (2B). For surface flows such decoupling may occur if water is channelled through channels with a firm bed (2A). Bypass flow for groundwater may occur once all salt in preferential flow pathways is washed out, and will last as long as the amount of groundwater flow remains unchanged.

Level 3 involves filters of various types. The term 'filter' is used here in a generic sense of anything that can intercept a vertical or lateral resource flow (van Noordwijk et al., 2001). Typically, filters occupy a small fraction of the total area and have a large impact per unit area occupied, so they can be seen as 'keystone' elements of a landscape. Important questions on the way filters function in natural resource management are:

- How effective are different types of filters in terms of intercepting the flows of nutrients and soil particles that can be expected in different rainfall regimes?
- To what extent does filter or safety-net efficiency depend on nutrient sorption to the soil and on the 'mesh size' of the safety net, as determined by root length density and the thickness of the soil layer involved?
- How quickly will filters saturate under high inflows?
- How fast can the filters regenerate between events?
- Do filters have a direct value and can they be treated as a separate 'land-use practice'?

Level 4 will not be discussed in detail here as it strongly depends on the 'downstream' situation. 'Mitigation' of negative environmental impacts downstream may be easier to implement if the stakeholders suffering from the negative impacts can see the immediate effects of their actions, whereas addressing lateral flow issues at the 'root cause' may involve considerable time delays and 'transaction costs'.

## 18.2 Understanding the Water Balance as the Basis for Lateral Flows

The water balance at the plot scale (see also Chapter 9, this volume, and Fig. 18.2) can be represented by Equation 18.1:

$$\Delta S = P - (I + R + L + E + T + D) \qquad (18.1)$$

where $\Delta S$ = change in water storage in the soil (mm/day), $P$ = precipitation (mm/day), $I$ = interception by plant canopies followed by evaporation (mm/day), $R$ = runoff – run-on (mm/day), $L$ = subsurface lateral flows (out – in) (mm/day), $E$ = evaporation from the soil surface (mm/day), $T$ = transpiration by plants (mm/day) and $D$ = drainage below the root zone (mm/day).

The terms $R$ and $L$ above represent lateral flows at plot scale and can modify the $T$ and $E$ terms (and hence plant production). At

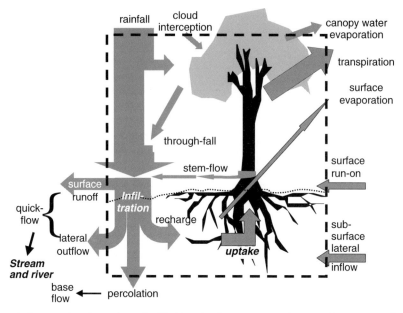

**Fig. 18.2.** Water balance at the plot scale, embedded in a landscape context that provides run-on and subsurface lateral inflow.

the landscape scale, the drainage term $D$ eventually generates a lateral flow. Water moves vertically through the unsaturated soil layers at a rate determined by the soil's hydraulic conductivity. When water reaches an impermeable or low-conductivity layer, the soil becomes saturated above that layer. At this point the water moves laterally downslope in or below rooted soil layers of adjacent (downslope) vegetation and may emerge at the soil surface in valleys in the form of springs.

Forests and partial tree cover in agricultural landscapes have important implications for the water balance of a catchment (Fig. 18.3). Trees, on average, use more water than any other form of land cover (1E in Fig. 18.1) and intercept more rainfall on their canopies than shorter plants (1A). Many studies have shown a strong, often linear, relationship between the clearing of trees and an increase in total river flow and recharge to groundwater. For example, the clearing of native woodland in Australia for cereal production has resulted in water tables rising, over millions of hectares, at rates of 0.1–2.5 m/year (George et al., 1997). The reverse is seen during reforestation, where total river flow and groundwater recharge are generally reduced as water consumption increases, an effect that is generally proportional to the growth rate of the trees. Australian woody plants have become weeds in South Africa and are the subject of massive eradication campaigns because of the effect they have on river flows. The planting of *Eucalyptus* species has also been implicated in the drying-up of drinking wells in India (Calder et al., 1997). Differences between plants in water use per unit growth have been largely linked to the photosynthetic pathway (C3 versus C4 and CAM plants), but differences in leaf phenology, the ageing of leaves and the time of year at which canopies are most active are also potentially important modifiers of the rate of dry matter production achieved per unit of water consumed. In this regard there is nothing special about eucalypts: any tree with a similar growth rate will consume a similar amount of water.

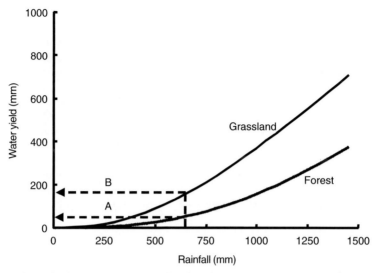

**Fig. 18.3.** The relationship between annual rainfall and catchment water yield under grassland and forest. In this case, water yield includes both runoff and drainage, with no attempt being made to separate them. From Holmes and Sinclair (1986) and re-examined by Zhang et al. (1999). Although the absolute difference in water use by grasslands and forests increases with mean annual rainfall (to a maximum of about 300 mm/year), the relative difference is highest at low rainfall, with a doubling of water yield indicated by lines A and B.

Rising or falling groundwater, peak flow and seasonality of streams have major implications for the supply and quality of water for use in the domestic and irrigation industrial sectors. Such whole-catchment responses depend largely on the proportion, location and arrangement of trees and/or other crops across the area.

Agroforestry differs from forestry in that trees are often mixed with crops or grown in short rotations with them. Trees may be planted in particular locations, such as on hillsides (to capture lateral flow), in areas with a high water table or on thin or stony soils where recharge to groundwater is highest. Therefore, in comparison with conventional forestry, trees in agroforestry designs may capture proportionally more of the rainfall, runoff or recharge for the area of the landscape they cover.

## 18.3 Trees, Groundwater and Salt Movement

The water balance at the catchment scale in terms of groundwater content can be represented by:

$$\Delta S_{gw} = R - G \qquad (18.2)$$

where $\Delta S_{gw}$ = change in the groundwater storage (mm/day), $R$ = recharge to groundwater (mm/day) and $G$ = amount of groundwater that leaves the catchment (mm/day).

The term $R$ in Equation 18.2 may be less than $D$ in Equation 18.1, since not all drainage from the root zone becomes recharge to groundwater. This is because shallow lateral flows may be intercepted by deep-rooted vegetation, or may intersect the soil surface lower in the catchment and produce springs or seeps. The amount of water that leaves the catchment – $G$ (m³/day) – is determined by the transmissivity, the hydraulic gradient and the width of the aquifer through which water is discharged, and can be represented by:

$$G = \Delta h \, K_{sat} \, A \qquad (18.3)$$

where $\Delta h$ = is the hydraulic gradient or slope of the water table (or the pressure gradient in the case of a confined aquifer) [−], $K_{sat}$ = the saturated conductivity of the aquifer (m/day) and $A$ = the cross-sectional area of the aquifer (m²).

A useful concept is the 'discharge capacity' of an aquifer. Discharge capacity represents the maximum amount of water that can leave a groundwater system without the groundwater reaching the surface. Discharge capacity is set at the point in the aquifer where the product of $\Delta h$, $K_{sat}$ and $A$ in Equation 18.3 is lowest.

The drainage term ($D$) in Equation 18.1 represents unsaturated flow below the root zone and becomes the major determinant of $R$, recharge to the groundwater. Unlike water in unsaturated soil (which moves vertically), groundwater moves laterally, so any change in land use at a plot scale that affects drainage is likely to have effects at the landscape scale (beyond the plot).

There are five ways in which tree crops affect the recharge term ($R$) in Equation 18.2 and hence contribute to falling or rising groundwater levels (intervention 2C, Fig. 18.1).

### 18.3.1 Spatial variability

A catchment may contain several soil types with varying soil physical properties. The depth of 'rootable' soil (measured to the first layer that constricts root growth) varies as well. The drainage term increases as soils become lighter in texture (more sandy) and shallower. Furthermore, there may be a rainfall gradient within a catchment, with the highest rainfall often coinciding with land on steeper slopes and with shallower soils. Ringrose-Voase and Cresswell (2000) examined what would happen if the positions of existing land-use practices (native vegetation, crop rotations and continuous cropping) were rearranged in the catchment. The study showed that matching current land use to catchment position had a major effect on runoff and recharge.

### 18.3.2 Belts of trees

A plantation of trees has a small edge effect relative to the whole stand, and thus the productivity in fertile soils is limited by the amount of rainfall received per unit area, with little opportunity for 'lateral resource capture'. Trees planted in widely spaced belts or alleys also have access to water beyond their canopies, if their roots penetrate laterally into the cropped zone between the belts. Thus, trees in alleys are likely to grow faster than their counterparts in a plantation. The benefits of alley cropping, from the perspective of productivity, have been hotly debated in the literature on this subject (Ong, 1995; Chapter 1, this volume); however, widely spaced trees (with many opportunities for lateral resource capture) represent the most powerful means (per unit of tree planted) for reducing the field-level recharge term (Stirzaker et al., 1999). For catchments with a low discharge capacity, where it is essential to reduce recharge but where farming must remain viable, there exists a trade-off between productivity and drainage. Stirzaker et al. (2002) have provided a methodology for evaluating this trade-off using the leaf area of trees in alleys relative to that in plantations or native stands, and crop yields obtained at different distances from the trees.

### 18.3.3 Short rotations

Intensive competition between trees and crops often means that it is better to opt for temporal rather than spatial separation. For example, a short rotation of leguminous trees or shrubs and crops may prove to be a better option than alley cropping. The tree phase is likely to dry out the subsoil and create a buffer for water that would be refilled during the subsequent cropping phase. Thus, the rotation would reduce drainage during both the tree and crop phases.

### 18.3.4 Direct use of groundwater

Plants can use groundwater directly or they can use water from the capillary fringe above the water table (the latter process being more common). In Australia, root densities as high as 0.7 cm/cm$^3$ have been measured at a depth of 14 m for native vegetation such as jarrah (*Eucalyptus* sp.) above a water table 15 m below the soil surface. It appears that groundwater is used predominately for survival, with trees switching to groundwater use after the soil water store has been depleted. This point is illustrated in Fig. 18.4, which gives data regarding tagasaste (*Chamaecytisus proliferus*) grown over fresh groundwater at a depth of 5 m. Tree water use was similar throughout the year, despite the large difference in potential evaporation between summer and winter. Trees used soil water during the wet winter, when evapotranspiration was limited by the atmosphere, and switched to groundwater during the summer. The only time during the summer that evapotranspiration approached potential rates was after a cyclone, when the soil store was replenished, confirming tagasaste's preference for soil water.

When soil water is saline, even only slightly so, salt accumulates in the capillary fringe (Thorburn, 1996; Stirzaker *et al.*, 1999). This occurs because trees exclude most of the salt at the root surface. In such a situation it is virtually impossible to lower the level of the water table by planting trees, unless there is some way by which salt can be flushed out of the root zone.

### 18.3.5 Tree belts on hillsides

Belts of trees on hillsides can be a powerful agroforestry design in both the control of runoff and the recharge of groundwater. Hillsides often have shallow soils (so the saturated zone will be within the reach of the tree roots), and sufficient slope to allow water to flow to the belts. Silberstein *et al.* (2001a) calculated the rate of water supply (per unit length of tree belt, $q$, in dm$^3$/m/day) to a belt of trees as:

$$q = 10^3 \, \Delta l \, K_{sat} \, z \qquad (18.4)$$

where $\Delta l$ = slope [−], $K_{sat}$ = saturated hydraulic conductivity (m/day) and $z$ = depth of the saturated layer (m).

Figure 18.5 indicates the combination of slope and conductivity likely to generate significant lateral flow. The analysis assumes that the zones between the tree belts generate drainage water and that the saturated depth ($z$) is not so deep that trees become waterlogged.

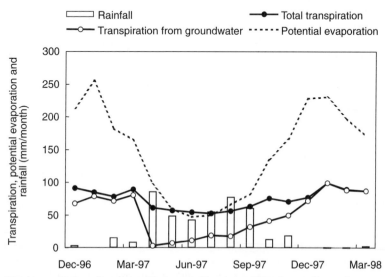

**Fig. 18.4.** Total transpiration of tagasaste (*Cytisus proliferus*), partitioned according to the source of water between soil and groundwater, using a combination of neutron probe and isotope methods. The trees used mainly groundwater during periods of high evaporation in summer, but switched to soil water after the autumn/winter rains. Redrawn from Lefroy *et al.* (2001).

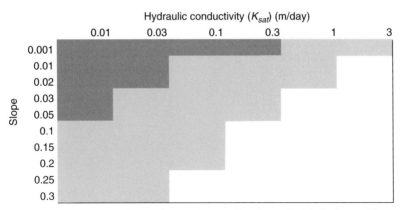

**Fig. 18.5.** Combinations of slope and hydraulic conductivity that show which hillsides could generate lateral flows to belts of trees. Black, insignificant lateral movement; grey, some lateral movement; white, significant lateral movement.

The optimum design is one that ensures that the amount of groundwater consumed by the tree belt is equal to the amount of groundwater recharge generated between the tree belts:

$$L \times R = W \times G \quad (18.5)$$

where $L$ = the distance between the tree belts (m), $R$ = the recharge between the tree belts (mm/day), $W$ = the width of the tree belt (m) and $G$ = groundwater consumption by the tree belt (mm/day).

Real hillsides are more complicated than the steady-state analysis above allows for, and may be convergent or divergent (i.e. there is a decrease or increase in the length of contour lines when going downhill), concave or convex (i.e. there is a decrease or increase in slope when going downhill). Recharge is also likely to be seasonal or episodic. Silberstein et al. (2001b) demonstrated the importance of waterlogging in concave slopes, by using a more detailed model that can take spatial and temporal variation into account.

The five strategies noted above all hinge on the correct siting or arrangement of the trees involved, so that the proportion of land covered by trees has a greater impact on $R$ than an equivalent area under a plantation. Moreover, the strategies above (except the first) have implications for productivity as well, since the trees receive more water than is provided by incident rainfall per unit area and can thus be expected to grow faster than if they were grown under plantation conditions.

## 18.4 Consequences of Subsurface Flows for Nutrient Transport

Nutrient transport is conventionally described as a one-dimensional process (vertical). This conceptualization at the plot scale may be accurate for land that is perfectly flat or in soils of high hydraulic conductivity. At the landscape scale, even on relatively shallow slopes, a reduction in saturated hydraulic conductivity with depth may be enough to make water flow laterally in the soil profile.

Usually such lateral flow in the soil profile is referred to as 'throughflow' or 'subsurface flow'. Throughflow generally travels relatively slowly through the soil matrix, causing near-saturated sections around stream channels and in topographic depressions, thereby maintaining the baseflow of the stream (Hewlett and Hibbert, 1963).

Although throughflow is slow, if natural pipes exist (such as decayed root channels, animal burrows and other 'macropores') lateral flow may be faster and may cause rapid subsurface flow during or immediately after storms. However, though fast lateral flow may occur in certain cases (e.g. where subsurface pipes have developed), rates of throughflow through the soil are generally far too slow to enable 'new' rainfall to reach a stream during a storm event (Dunne, 1978). Therefore, Hewlett and Hibbert (1963) advanced the concept of 'translatory flow' or 'piston flow': a 'push-through' mechanism whereby each new volume of water added by rain to a hillside displaces an approximately equivalent amount of 'old' water, thus causing the oldest water to exit from the bottom of the slope into the stream (Bruijnzeel, 1990).

Eshleman et al. (1993) suggested that the relative significance of vertical and lateral flow depends on the intensity of each rainfall event. During high-intensity events, saturation occurs because the vertical flow velocity greatly exceeds horizontal flow velocities and water table 'mounds' can develop. During low-intensity rainfall events, vertical flow approaches the soil hydraulic conductivity and, hence, there may be little lateral water flow. Wenzel et al. (1998), in a study in East Kalimantan (Indonesia), found a reduction in saturated hydraulic conductivity at a depth of 80 cm, and suggested that lateral flow through the permeable cross section (at a depth of 40–60 cm) was limited to between 18.5 and 92.9 m/year. With a rainfall excess of, say, 1 m (for a rainfall of 2.5 m and an evapotranspiration rate of 1.5 m/year), this implies that slopes 10–100 m long can be drained laterally.

Subsurface flow can be divided into steady-state flow and non-steady-state flow. In agrohydrological literature, much emphasis is placed on steady-state water flow under saturated conditions to describe the performance of subsurface flow (van Schilfgaarde, 1974). The Dupuit approach is often used, and assumes: (i) that the flow is horizontal; (ii) that the upper boundary of the flow is the groundwater (phreatic) table (the height of which determines the water potential in the concerned vertical direction); and (iii) that the slope of the phreatic table determines the gradient in water potential. In equation form, the flux density ($q$) for the cross-section PQ (Fig. 18.6) can thus be represented by (van der Molen, 1983):

$$q = -k \frac{\partial H}{\partial x} z \quad (18.6)$$

where $q$ = flux density or discharge per meter of contour line (m²/day), $k$ = hydraulic conductivity (m/day), $H$ = hydraulic head (m) and $z$ = height above the impermeable layer (m).

The water potential in the cross-section PQ is determined by the height above the impermeable layer of the phreatic table $H$ (Fig. 18.6). So:

$$q = -k \frac{\partial H}{\partial x} H \quad (18.7)$$

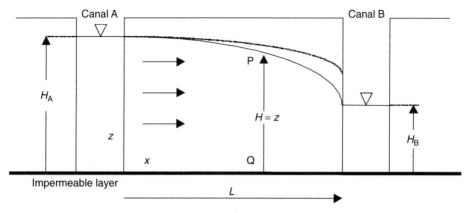

**Fig. 18.6.** Dupuit's assumption applied to subsurface flow between two canals. Continuous line, phreatic table (water table) according to Dupuit; dashed line, real phreatic table. (Redrawn from Van der Molen, 1983.) See text for definitions of terms.

Integration gives:

$$x + c = -\frac{k}{2q} H^2 \quad (18.8)$$

The shape of the (calculated) phreatic table is a parabola. Integration between positions $x = 0$ and $x = L$ gives:

$$q = \frac{k}{2L}\left(H_A^2 - H_B^2\right) \quad (18.9)$$

To calculate the subsurface flow of nutrients, the concentrations of nutrients in cross-section PQ ($C_n$, mg/l), can then be converted from mg/l to g/m² using the equation:

$$Nut_{lateral} = 10^{-3}\, C_n\, q\, H \quad (18.10)$$

where $Nut_{lateral}$ = the amount of subsurface flow of nutrient (g/m²) out of cross-section PQ. This equation allows a first-order estimate of the amounts of nutrients involved in subsurface flows.

Some soils are more susceptible to lateral flow than others. In Ultisols the clay content typically increases with depth, and we can thus expect saturated hydraulic conductivity to decrease with depth. Even on mild slopes, this may result in subsurface lateral water flow according to Fig. 18.5, and a significant impact on the hydraulic behaviour of the soil profile as a whole (Herron and Hairsine, 1998). Suprayogo (2000) tested this in Lampung (Indonesia) and found that, with an increase in clay content with depth, saturated hydraulic conductivity decreased sharply (Fig. 18.7a,b) from the topsoil (0–0.2 m) to the subsoil (0.2–1.0 m). Conductivity dropped to very low values at a plinthic horizon at a depth of 1.2 m. Measurements of the water table during a period with two major storm events provided strong evidence for lateral water flow in the Lampung experiment (Fig. 18.7c). Before, during and after heavy rains, the lateral discharge of water varied from 0.32 to 0.35, from 0.84 to 0.98, and from 0.55 to 0.73 cm³/cm²/day, respectively. These values are considerably lower than those measured on a layered silt loam soil in The Netherlands, where de Vos (1997) estimated the soil to have a maximum lateral discharge rate of 3.5 cm³/cm²/day.

Such lateral movement of water, and of nutrients carried along in mass flow, has important consequences for the possible location of 'safety-net' tree roots. The 'safety net' concept described in Chapter 6 (this volume) is usually considered to act in a vertical direction, where tree roots intercept nutrients that would otherwise be lost to the deep soil zones. It occurs at the plot scale. At the landscape scale, however, even on mild slopes, the safety-net concept can be extended to the lateral effect of tree roots, since they can intercept lateral subsurface flow (intervention 2B, Fig. 18.1) and nutrients (interventions 3B and 4B, Fig. 18.1). Besides tree roots, the charges on soil particles can also affect lateral movement of nutrients, so retaining ions and preventing the pollution of groundwater and rivers (see Chapters 6 and 10, this volume). Identifying which process (root capture or soil retention) is the dominant one is a task that future research should undertake, since it will affect land-management decisions.

## 18.5 Soil Cover, Runoff and Its Consequences for Sediment Transport

If rainfall intensities exceed the infiltration capacity of the soil, the unabsorbed excess runs off to areas downslope where it re-enters the soil as 'run-on' and may either infiltrate or continue as 'runoff' until it reaches a stream channel. Two constraints to infiltration capacity are normally distinguished: (i) situations where rainfall exceeds the saturated hydraulic conductivity of the surface layer ('Hortonian' or 'infiltration excess' overland flow; Horton, 1933); (ii) situations where the transmissivity of the surface layer is a constraint, as lower soil layers are saturated and water cannot enter the profile at the top any faster than it can leave it at the bottom or at the side ('saturation overland flow' or SOF).

Soil erosion can be defined as a process of soil detachment and movement by mass flows of air or water. In the latter case, raindrop impacts that overcome the coherence of aggregates at the soil surface are the main cause of detachment. Surface water flows transport the particles detached by splash impacts, by shallow sheet flow (sheet ero-

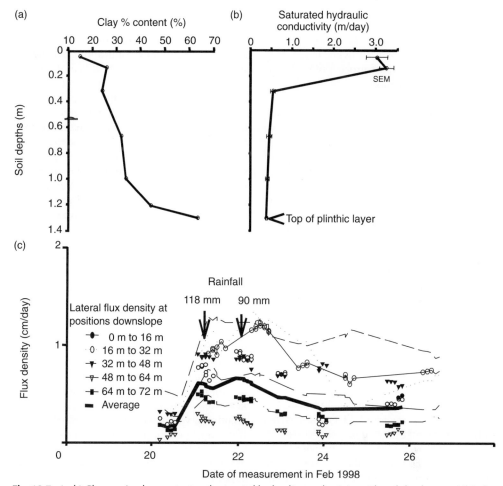

**Fig. 18.7.** (a, b) Changes in clay content and saturated hydraulic conductivity with soil depth on an Ultisol in Lampung, Indonesia; horizontal bar, standard error of mean (SEM); (c) observed lateral subsurface flux density during two storm events (Suprayogo, 2000).

sion), or by concentrated flow in rills; splash transport is a few centimetres at most, sheet flows may end in depressions in the field or may reach rills, while rill flows tend to enter streams (Flanagan and Nearing, 1995). The erosion process stops when surface flow stops (as current rainfall intensity plus run-on become less than the infiltration capacity of the soil) or when the amount of sediments in the runoff exceed the soil particle transport capacity of the flow, leading to net sedimentation.

The soil loss process can be described using empirical models such as the Universal Soil Loss Equation (USLE; Wischmeier and Smith, 1978), or using physical equations such as that found in the Griffith University Erosion System Template (GUEST) model (Misra and Rose, 1990, 1996).

USLE is described as:

$$A = R K L_S C P \qquad (18.11)$$

where $A$ = average annual soil loss (Mg/ha/year), $R$ = rainfall erosivity factor (MJ/ha mm/h), $K$ = soil erodibility factor ((Mg/ha/(MJ/ha mm/h))/year), $L_S$ = slope length and steepness factor (non-dimensional), $C$ = cover-management factor (non-dimensional) and $P$ = land-use practice factor (0–1).

The model has primarily been used to summarize data for soil loss from standardized 'Wischmeier' plots of 22 m slope length. The equation predicts 'universal soil loss' because the counterpart process of sedimentation ('negative erosion') is absent from the equation. The results are zero or positive, never negative (as erosion plots that exclude run-on can, by definition, not yield negative results).

In the GUEST model, the concentration of soil particles in overland flow is multiplied by the total volume of water involved in each runoff event. Rose and Yu (1998) show how this model estimates soil loss at the plot scale, considering a situation without rill erosion:

$$M = k^\beta Q_e^{0.4\beta} \sum Q \qquad (18.12)$$

where $M$ = total mass of soil lost during an erosion event (Mg/ha), $k$ = approximately constant in any given context (slope, soil type), $\beta$ = soil erodibility parameter, $Q_e$ = effective runoff flow rate and $\Sigma Q$ = total runoff amount.

The basis of the GUEST model is thus surface runoff (usually derived from a water balance model) rather than total rainfall.

Susceptibility of a soil to erosion is not only determined by average soil texture, but also by the distribution of soil particles in the profile. As explained above, soils with abrupt textural changes (e.g. sandy at the surface and with a high clay content in subsurface layers) are more susceptible to lateral flow and erosion, promoted by the strong difference in the infiltration velocity. Methods for measuring this in the context of agroforestry were recently reviewed by McDonald et al. (2003).

The role land use plays in reducing soil erosion can be seen in Equation 18.12. In order to reduce soil loss, it is necessary to reduce the total amount of runoff ($\Sigma Q$) and the rate of runoff per unit area ($Q$) or the velocity of runoff flow, which is related to $Q_e$. Increasing soil coverage using litter or live biomass can be effective in achieving this objective.

Some general effects specifically attributed to tree crops in terms of soil erosion control are frequently referred to as 'filter effects'. These include protecting the soil against raindrop impact; decreasing runoff velocity by increasing the soil's surface roughness and water infiltration; decreasing soil particle transport downhill and, consequently, reducing the pollution of stream water (Lowrance et al., 1997; Trimble, 1999). These filter effects require the presence of a litter layer and of tree roots, which create channels in the soil; they are not related to the above-ground parts of trees. Unlike the situation for groundwater and subsurface water movements where we saw earlier that trees can play a specific role, erosion control does not require a forest: good soil coverage (by live biomass or dead biomass from cropland areas) can reduce erosion just as well as a forest can. For example, soil erosion rates are small in traditional cropping systems in South Brazil (mainly soybeans and maize, in rotation with legumes), which maintain soil cover throughout the year. Erosion effects of logging are largely due to the loss of a protective litter layer (Haranto et al., 2003).

Tree filters are more efficient during low- or medium-intensity rainfall events than during heavy storms. High amounts of rainfall often saturate the soil profile and any additional water will become surface runoff. However, the amount of soil particles carried by runoff from tree filter areas is, normally, much less than that carried from other crop systems.

Soil cover plays a key role in controlling erosion. When we consider 'plot', 'hillside' and 'landscape' scales, we can see an increasing number of processes that jointly determine the overall effect had (Table 18.1). The main role of soil cover is to promote infiltration, reduce the velocity of runoff (situation 2A in Fig. 18.1) and, as a consequence, reduce soil particle transport. On the other hand, the role of soil coverage in situation 3A is to retain the soil particles transported by runoff, by promoting sediment deposition on areas with high surface roughness. Since sediment comes from upslope areas, filter strips can promote sediment deposition (case 4A); this is one of the roles of riparian forests (in addition to the role they play in controlling stream bank erosion, see Box 18.1). Riparian vegetation may be

**Table 18.1.** Effects of soil cover on runoff and erosion/mass-movement, at different spatial scales.

|  | Scale | | |
| --- | --- | --- | --- |
| Effects of soil cover | Plot | Hillside | Landscape |
| Reduces splash erosion – due to raindrop interception | X | X | X |
| Reduces runoff velocity | X | X | X |
| Reduces rill erosion – due to decreased runoff and soil particle transport capacity | X | X | X |
| Increases deposition | X | X | X |
| Increases infiltration due to increased soil porosity and permeability promoted by biological actors (roots and earthworms), and improved soil structure caused by organic matter | X | X | X |
| Controls soil particle transport – due to increased surface roughness | X | X | X |
| Controls gully erosion |  | X | X |
| Controls landslide |  | X | X |
| Controls soil creep |  | X | X |
| Controls soil particle discharge to river |  |  | X |
| Controls stream bank erosion via the stabilization effect of roots |  |  | X |

most effective if it is in a rapid growth phase, after disturbance (Dignan and Bren, 2003; Giese et al., 2003).

In Boxes 18.1 and 18.2 we present two case studies of how land-use patterns control runoff and sediment yield. It is clearly illustrated that sediment yield measurements differ substantially at different scales; therefore, simply multiplying average (plot-level) sediment yields by the total area of land in question is unlikely to produce realistic results.

Van Noordwijk et al. (1998e) applied a physical erosion model to a number of hypothetical agroforestry arrangements and showed that a 50% tree cover using the most favourable spacing had, effectively, the same effect in terms of reducing the sediment load of streams as full forest cover. It was also found that a tree cover of 25% could reduce the negative impacts crops have by 80% in the case of sediment loss and by 70% in the case of storm flow. According to the model, the largest reduc-

---

**Box 18.1.** Case study: Cikumutuk catchment, West Java, Indonesia.

Some of the issues that arise when plot-level assessments of erosion are compared with soil losses at the catchment level can be seen in Purwanto's (1999) study of the 125 ha Cikumutuk catchment in the volcanic uplands of West Java (Indonesia). The research was carried out in a small catchment on the slopes at the foot of the inactive Cakrabuana volcano, near Malangbong, some 60 km east of the city of Bandung. The catchment has been almost entirely converted into agricultural uses, with some agroforestry practices.

*Sediment yield at multiple scales*
Starting in October 1994, runoff and sediment output were measured at five successive levels of scale in a 'nested arrangement', which involved: (i) individual terrace risers or beds, using small 'artificial boundary erosion plots' (ABEPs); (ii) single backsloping bench terraces comprising the cultivated bed plus the adjacent toe drain and upslope terrace riser, using so-called 'non-imposed (natural) boundary erosion plots'; (iii) groups of multiple terraces comprising a part of the hillside (containing 10–25 individual terraces); (iv) two 4–5 ha subcatchments, each drained by a zero-order gully with ephemeral flow (containing up to 100 terraces); and (v) the entire 125 ha catchment. In addition, observations were made in a settlement area and on irrigated rice.

*Continued*

**Box 18.1.** *Continued.*

Catchment sediment yield proved high (e.g. 70 Mg/ha in the 1995/96 wet season), although surface runoff volumes were not very great (in the order of 15% in the 1995/96 season). However, sediment production by the terraced rainfed agricultural fields was very high indeed (in the order of 100–250 Mg/ha in the 1994/95 and 1995/96 wet seasons). This was mainly because of the high erosion rate found on the bare terrace risers. The data listed in Table B18.1 were mainly collected during the rainy season of October 1995–April 1996, when rainfall was 7% above average. Therefore, the quoted figures are probably slightly above average.

**Table B18.1.** Sediment output (Mg/ha) from rainfed bench-terraced areas measured from plot to landscape scale, in Cikumutuk catchment, West Java, Indonesia.

|  | 1994/1995 | 1995/1996 |
|---|---|---|
| Precipitation (mm) | 2422 | 2345 |
| Terrace risers |  |  |
| • Gentle slope plots | –[a] | 325 |
| • Steep slope plots | – | 280 |
| Individual terrace units |  |  |
| • Gentle slope plots | 100–137 | 97–112 |
| • Steep slope plots | 209–242 | 140–175 |
| Multiple-terrace (hill-slope) system |  |  |
| • Concave slope | 27 | 220 |
| • Convex slope | – | 35 |
| Micro-catchment unit (4 ha) | – | 53 |
| Landscape unit (125 ha) | 49 | 63 |

[a]Not measured.

*Runoff and sediment delivery*
Of the roughly 7000 Mg of sediment leaving the catchment during the 1995/96 rainy season, the bulk was supplied by rainfed agricultural fields, with only modest volumes being supplied by settlement areas and trails or agroforestry and grasslands, or being associated with an expansion in the area of irrigated rice, with river bank erosion or mass wasting.

The research showed that runoff from rainfed terraces typically amounts to 15–35% of rainfall. This result depended on rainfall characteristics, the dimensions and gradients of terrace risers, beds and toe drains (running along the foot of the riser) and the presence of vegetation cover. However, it was found that runoff could exceed 50% for individual heavy storms. Most of this runoff was generated on the compacted terrace drains and, to a lesser extent, on the steep, bare terrace risers. On the other hand, terrace risers with a well-established protective plant cover were found to produce hardly any surface runoff even during the largest storms. In contrast, the runoff produced by a settlement area varied from around 40% to around 70% of the rainfall, depending on the fraction of the land area occupied by impervious surfaces such as roofs and compacted yards. Irrigated rice fields also showed a very high runoff coefficient (close to 100%), but their cascade-like design effectively slowed down the arrival of the peak runoff at the stream by several hours. Only after more than 50–60 mm of intense rain did the bund around the terraces overflow occasionally, resulting in a much quicker response to rainfall.

Overall, opportunities to store eroded material on its way to the nearest gully or stream proved quite limited (on average only about 4 Mg/ha on the unirrigated hillsides). This is related to the fact that the preferred pathways of the runoff carrying the sediment followed trails, gullies and the main stream, all of which were incised into massive, not readily erodible substrates. As such, sediment contributions made by stream-bank erosion or gully-wall collapse were rather minor. Likewise, erosion rates for trails and the settlement were distinctly lower than those for rainfed agricultural terraces, despite their much higher runoff coefficients (50% and 20% of annual rainfall, respectively). Soil losses from agroforestry (young tree plantations in combination with maize and rice) and from fallow land, as measured in later years, were an order of magnitude lower than for settlement areas and terraced fields.

**Box 18.2.** Case study: optimal riparian forest (RF) width to control sediment yield in southeastern Brazil.

Riparian forests are recognized as land-use units essential in protecting streams against pollution from sediments carried by runoff. This role is related to a RF's ability to retain sediments, preserve floodplain channels, filter and decompose nutrients and pollutants (a result of its high biological activity), and improve water infiltration. Despite recognition of RF's essential role, no agreement exists between ecologists and farmers on the desirable width of RF strips. This inability to reach an agreement reflects not only the desire of farmers to occupy riparian land (as such land is very fertile), but also the different scales that must be considered when addressing the issues of water quality and supply (large scale) and conservation or reclamation actions (small and local scale). Finally, the lack of agreement also reflects a lack of quantitative data proving the efficiency of RF, with regard to improving water quality.

Sparovek et al. (2001) developed a quantitative method with which to check the efficiency of RF with regard to controlling net sediment loss from a catchment of 77 ha under sugarcane cultivation in southeastern Brazil. The method used the WEPP erosion prediction model (Flanagan and Nearing, 1995). The researchers hypothesized that it was possible to determine an optimum RF width based on certain variables, some based on physics and biology, others reflecting farmer decisions. They calculated the minimum width of RF that would be needed to reduce the sediment yield below a target level. They also defined the width that maximized sediment capture (Fig. B18.1); for RF widths below this width much sediment would still pass through the riparian zone, for RF widths above this value the landscape-level gross erosion would start to decrease.

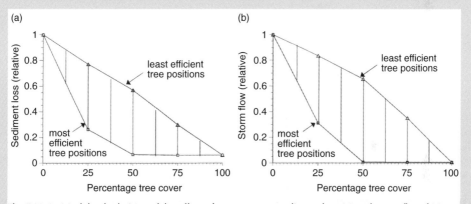

**Fig. B18.1.** Model calculations of the effect of tree cover on sediment loss (a) and storm flow (b) in the spatially distributed model and a set of parameters for Machakos experimental station (Kenya). Source: van Noordwijk et al. (1998e).

The researchers found an RF width of 52 m maximized sediment capture for that particular situation, where the RF trapped 54% of the sediment flows in the landscape. This width is substantially greater than the 30 m prescribed by Brazilian Federal Law. The study illustrates that, on a case-by-case basis, quantitative methods can be combined with local targets for maximum acceptable sediment loads of rivers to achieve effective results in terms of both water quality improvement and the provision of data to support land-use change recommendations.

tion in net sediment loss was achieved when trees were placed at the bottom of hillsides (as riparian forests) or on well-spaced contour lines. The first arrangement was able to intercept sediment from the hillside, and the second arrangement worked to prevent gully erosion. Figure 18.8 shows the efficiency of various tree arrangements with regard to reducing sediment loss and storm flow.

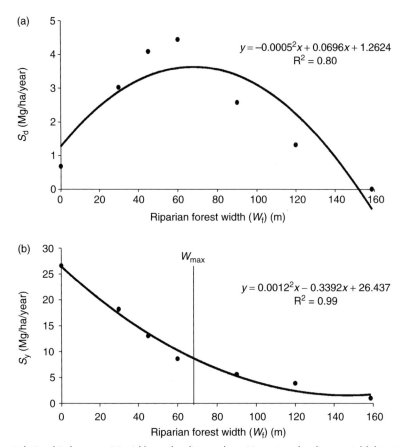

**Fig. 18.8.** Relationship between RF widths and sediment deposition (a) and sediment yield (b). $W_{max}$ represents the maximum efficiency width of riparian forests.

## 18.6 Discussion: Scaling up the Effects of Land-use Change on River Flow

Water provided by surface, subsurface and/or groundwater flows will feed streams at any time. The discussion presented above considered the roles soil properties, soil coverage and tree roots play, at the plot and landscape scales, with regard to these three types of water flows. As considered above, land-use changes usually affect water movements; but, land management action can be taken in order to avoid the negative effects of the vertical and lateral movement of water and nutrients (the pollution of rivers for example).

The 'mental model' of a forest as a sponge that receives rainfall and gradually feeds it to a stream is a familiar concept. Although the model is easily communicated, it has been controversial among forest hydrologists ever since it was formulated (see Box 18.3 for some of the debate in the 1930s in Indonesia for example), as the vegetation only controls the access of water to the subsoil, which gradually releases water to streams at a rate that essentially depends on the geology of the landscape. In the humid tropics the 'sponge' may be continuously wet and not able to absorb much of the incoming rainfall.

Though it has received much less attention than the 'sponge' model, there is an alternative explanation for even river-flow patterns: spatial heterogeneity of rainfall. Put simply, if today it rains here and tomorrow it rains there, the river that receives water from both areas may have a fairly steady flow, despite poor buffering in either area (see Fig. 18.9).

If this second model is dominant, changes in river flow may be due to a change in the spatial distribution of rainfall, and not to changes in land use in any of the subcatchments *per se*. A distinction between the above two types of explanation for patterns in river flow is thus essential both to evaluate the likely impact of current land-use change in forested areas and to assess which types of interventions may be effective.

---

**Box 18.3.** Debate on forests and hydrological functions in Indonesia.

Kartasubrata (1981) summarized the development of ideas about forest and water in Indonesia, as they were reflected in the debate on the issue during the colonial era. This debate still resonates today, so it may be interesting to see the arguments as phrased at that time.

The debate started with a statement by Heringa (1938) who pleaded for a substantial increase of forest cover on Java, both for the production of timber, resin, turpentine and tannin, as well as for the hydrological significance of forests. On the island of Java, with its high volcanoes, the rivers have such a steep gradient that, in the wet season, rain water flows rapidly into the sea, transporting, as a result of the force of its flow, much fertile soil and mud from the fields and from the river beds. This is then deposited into the sea. Heringa formulated a theory, which stirred up much of the debate, when he said (in a translation by Kartasubrata, 1981):

> The forest works as a sponge; it sucks up water from the soil in the wet season, and then releases it gradually in the dry monsoon at the time when there is a shortage of irrigation water. A decrease in forest cover therefore will bring about a decrease in discharge during the East monsoon ('dry season') and cause a shortage of the needed irrigation water. Therefore, a certain balance is needed between the condition of the forest and the output of agricultural lands (rice fields). Consequently one has to determine a minimum forest cover for every catchment area.

Roessel (1938) applauded the idea of expanding the industrial forests; however, he criticized the other motivation for reforestation (i.e. the hydrological aspects). In contrast to the 'forest as a sponge' theory, Roessel adhered to the 'infiltration theory', which emphasized that percolation of water through the subsoil produces spring water, not the forests as such. Coster (1938), working at the Forest Research Institute in Bogor, provided some quantitative data and suggested a synthesis of the sponge and infiltration theories: vegetation determines the recharge to the 'sponge', but water is held in the subsoil, not in the forest as such (Table B18.2).

**Table B18.2.** Three different viewpoints on forests and hydrological functions in the 1930s in Indonesia. (After Kartasubrata, 1981.)

| Aspect | 'Forest as a sponge' theory (Heringa, 1937) | 'Infiltration' theory (Roessel, 1933) | Synthesis and quantification (Coster, 1938) |
|---|---|---|---|
| Dry season river flow | Depends on afforestation | Depends on geological formations | Vegetation determines soil permeability |
| Required forest area for hydrological functions | A minimum required fraction can be calculated from the area of rice fields to be irrigated with dry season flow | There is no minimum forest cover | Discharge of springs depends on the amount of water that percolates into the soil *minus* the loss of water because of evaporation |
| What to do if forest target is not met? | Farmland owned by farmers and agricultural estates has to be purchased and reforested | Reforestation is only carried out if certain soil types are susceptible to erosion if exposed, but only after other measures, such as terracing, use of 'catching holes' and soil cover have proved insufficient | Depends on *elevation*. Lysimeter measurements indicated that evaporation from a bare soil surface is 1200, 900 and 600 mm/year at locations with elevations of 250, 1500 and 1750 m a.s.l., respectively |

*Continued*

**Box 18.3.** *Continued*

**Table B18.2.** *Continued.*

| Aspect | 'Forest as a sponge' theory (Heringa, 1937) | 'Infiltration' theory (Roessel, 1933) | Synthesis and quantification (Coster, 1938) |
|---|---|---|---|
| Forests or ground cover? | All soil types are equal; afforestation with industrial timber species has the same hydrological effect as natural forest and is (always) better than agricultural estates | An agricultural estate succeeds in stopping surface runoff by terracing etc. or use of soil cover is hydrologically more valuable than an industrial forest, where, for example, because of steep slopes, poor undergrowth or poor humus formation, superficial runoff still takes place | Measurements by the Forest Research Institute showed that well-maintained tea, coffee, rubber and *Cinchona* plantations are, from the hydrological point of view, nearly the same as forests (planted or natural) but superior to agricultural fields. Fires in the grass wilderness in the mountains stimulate water outflow and erosion |
| Scope of reforestation | All problems related to 'watershed functions' can be cured by reforestation | Recovery by reforestation can only be expected in cases where surface runoff and erosion can be controlled with 'good' forests. Forests without undergrowth and without good humus formation are usually not sufficient. However, a soil cover consisting of grass, or dense herbaceous or shrubby vegetation, would do. | It is probable that afforestation in the lowlands may decrease discharge (including that in the dry season), because of the high evaporation rate from the forest; in the mountains the increased infiltration of abundant rain into the soil more than offsets the increased water use by trees. |

In much of the current debate the more 'synthetic' viewpoints of Coster (1938), which consider both the positive and negative impacts of trees on river flow, have not yet been understood, and existing public perceptions and policies are based on Heringa's point of view.
A final quote:

> Formerly the view was generally accepted, that forests had the tendency to increase rainfall to a large extent. Nowadays this view is combated by many investigators, who deny any appreciable influence; others support the view that the *distribution* is changed by the forest, and not the total amount of rainfall ... (Braak, 1929).

The relative importance of the two explanations clearly depends on scale, i.e. the size of the area being considered. In small subcatchments there is hardly space for the second explanation: the first must dominate. In areas of several hundreds of square kilometres or at a subcontinental scale, the second explanation is likely to dominate. So, at some point at the intermediate scale the two may break even. But, can we assess where this occurs? Unfortunately, most previous research was undertaken in small plots and, when 'scaling up', the possible impact of the second explanation was not recognized. Chapter 19 further confronts our perceptions of watershed functions, farmer knowledge and what current models can tell us.

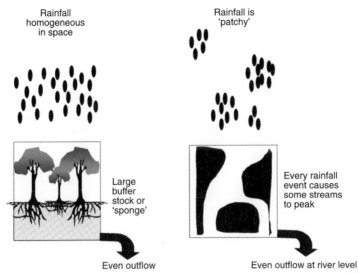

**Fig. 18.9.** Two alternative explanations for relatively even flow of a river: left, all rainfall passes through a sponge that only gradually releases water; right, rainfall is spatially heterogeneous and the river integrates over peaks in flow from different streams that occur on different days.

---

**Conclusions**

1. The ways water flows in landscapes via surface, subsurface and deep groundwater pathways depends on permanent features of the landscape (such as slope and basic soil properties) and climate (duration, intensity and distribution of rainfall). It also depends on the spatial distribution of land-cover types that modify: (i) total flows, via the amounts of water intercepted and used by vegetation; (ii) the pathway, via the relative distribution of roots with depth and effects on soil macroporosity; and (iii) surface infiltration.

2. Vertical and lateral transport of soil particles, nutrients and salt at the landscape scale can be strongly affected by the spatial distribution of land-cover types, via total water use by the vegetation, and via the degree to which water flow is coupled with the transport of soil particles, nutrients and salt.

3. Both the quantity and arrangement (spatial and temporal) of trees have different impacts on the movement of surface water, subsurface water and groundwater. Trees located on lower slopes (riparian forests) play an important role in trapping sediments from incoming overland flows (filter effect). Trees on the middle slopes (belts) are able to trap sediments, decrease runoff velocity and reduce groundwater recharge and the subsurface lateral flow of water and nutrients.

---

**Future research needs**

1. Models should be improved so that they better simulate the effect of agroforestry systems and mosaics of land-cover types, which have channel and filter effects, on surface and subsurface water flows at the landscape scale.

2. Attempts should be made to better define the spatial scale at which land-use change becomes of secondary importance in determining the regularity and quality of river flows.

3. Attempts should be made to better understand the decoupling mechanisms ('preferential flow') for solutes in lateral flows and the way they depend on soil structure and hence on the balance of soil structural decay and creation of macropores.

4. Attempts should be made to better quantify the way filter effects, for surface and subsurface lateral flows, can change with time (in terms of saturation and recharge of filter capacity) under different land-use scenarios.

# 19 Soil and Water Movement: Combining Local Ecological Knowledge with that of Modellers when Scaling up from Plot to Landscape Level

Laxman Joshi, Wim Schalenbourg, Linda Johansson, Ni'matul Khasanah, Endy Stefanus, Minh Ha Fagerström and Meine van Noordwijk

---

**Key questions**

1. How do generally held perceptions of the relationship between 'forest' and 'watershed functions' compare with available data and hydrological theory?
2. How, in practical situations, can we compare and combine local ecological knowledge with that of modellers?
3. How can such an analysis be used to reduce conflict and negotiate landscape-level land-use patterns?

---

## 19.1 Introduction

Chapter 2 introduced a specific method for documenting, representing and analysing local ecological knowledge of below-ground interactions in agroecosystems. Most of the examples given focused on 'plot-level' soil fertility issues. The succeeding chapters described various aspects of the 'scientific' exploration of the contributing processes, leading to a discussion of management options for farmers in Chapter 17. The management options that farmers actually use, however, depend on the specific constraints that those farmers face (e.g. with regard to labour and capital), and the degree to which the farmers wish to maximize profitability. In Chapter 18, emphasis was placed on landscape-level impacts of local land use, via the lateral flows of water. In this chapter we will 'pick up' the discussion from Chapter 2, and discuss examples of how local ecological knowledge can be contrasted with the ecological knowledge of modellers as well as examples of how the two can be combined to gain a better understanding of the way landscape-level resources can be managed. Both types of knowledge may contrast with the apparent logic underlying existing policies and with the perceptions generally held by people living in urban areas. Almost by definition, the management of resources at this scale involves multiple stakeholders: thus, conflicts can easily emerge. We therefore have a number of reasons to carefully explore local ecological knowledge and perceptions.

- In as far as local ecological knowledge is based largely on site-specific local experience and observations, it provides a rich testing ground for the supposedly generic concepts reflected in current models.
- Local ecological knowledge tends to be 'operational' or 'functional' and linked to the implications of various human interventions, and therefore provides a practical perspective, which is directly relevant to (and thus very useful in) 'system-oriented' science.
- In landscape-level interventions there is little scope for the 'recipe'-type outcomes sometimes associated with traditional agronomic research; so, a blending of 'hard' and 'soft' science that tries to extract and adapt 'rule-based knowledge' is needed.
- Communication with and the provision of extension services to farmers requires a common 'language'.
- Negotiations between multiple stakeholders are probably easier if they are based on a shared understanding of underlying ecological phenomena.

This chapter will begin by analysing the current confusion surrounding watershed functions in relation to land-use change (specifically deforestation), and discussing an analytical framework that can be used to predict relationships between plot-level interventions and landscape-level effects. Finally, the chapter will consider two case studies, documenting and comparing both local concepts and those of modellers, before addressing associated issues.

## 19.2 Myths, Misunderstandings and Analytical Frameworks

Usually the general public attributes watershed functions directly to standing trees, the entity that is seen as that promoting and maintaining most forest functions (Calder, 2002). Removal of trees is most usually blamed for events such as floods, landslides and reductions in the baseflow of downstream rivers (which subsequently cause siltation). Media reports claiming calamities due to the loss of trees are accepted as 'the truth' by the public at large. As explored by Grove (1995), perceptions of the relationships between deforestation, subsequent changes in rainfall, land degradation and siltation of rivers date back to experiences in the Mediterranean region, with Theophrastus ($c.$ 372–287 BC) being one of the earliest writers to document such perceptions.

Experience gained as a result of European colonial expansion into the tropics, particularly that gained on small islands such as Mauritius (which have a dry, non-forested side and a wet, forested side), strengthened the perception that forests generate rainfall. Yet, hard evidence of a change in documented rainfall as a consequence of deforestation is still lacking. The causal relationship between forests and rainfall ('rainfall leads to forest') is generally actually the reverse of what is perceived to be true ('forest leads to rainfall'). For example, a re-analysis of rainfall patterns in Indonesia (Kaimuddin, 2000) indicated shifts in isohyets (zones of equal rainfall) that are not obviously related to local land-cover change: some areas that lost forest cover became wetter whereas others became drier. For Indonesia as a whole, average rainfall did not change, despite a considerable loss of forest cover, though there may have been a change in the overall circulation pattern that affects local rainfall. Although at a local scale real changes in rainfall may have coincided with real changes in forest cover, there is no convincing evidence to support the hypotheses concerning causal relationships. However, the way a landscape 'processes' the incoming rainfall depends directly on land cover – the total amount of water in streams, the regularity of the flow and the quality of the stream water can be directly affected by changes in cover (see Box 18.3, this volume, for a debate concerning the link between forests and hydrological functions).

Globally, the 'community' practising soil and water conservation and integrated watershed management is in a state of confusion. On the one hand therefore, multi-billion dollar efforts are being made to rehabilitate degraded watersheds based on the 'scaling up' of results obtained from erosion plot experiments and the expectation that conserving forest and planting trees are, respectively, the best and next-best methods by which to guarantee dry-season flows of water and secure land productivity. However, on the other hand, there is a remarkable absence of documented evidence regarding the impact of such methods, whilst the rules regarding the 'scaling up' of results have been seriously questioned (as the results of reforestation efforts made to restore watershed functions are generally disappointing). On the basis of an e-conference and a search of published literature, Kiersch and Tognetti (2002) could not find any convincing evidence that land use affects the major 'watershed functions' related to flow rates and sediment loads for areas larger than $10 \times 10$ km$^2$. We should therefore ask is 'watershed management' a fiction and a waste of public resources? Or, has research not yet addressed the right questions?

Of course, as Kiersch and Tognetti (2002) state, 'lack of evidence of effect' is no 'evidence for lack of effect'. Those authors discuss a number of reasons why a measurable impact may be lacking – given that rainfall variability occurs in short-term studies and climate change affects extrapolations at larger scales in longer-term studies. It should also be remembered that intersite comparisons are complex, and attributing measured changes in water to specific factors requires a full understanding of both internal and external feedbacks in the system. However, evidence given in Table 19.1 (regarding the effects that land-use change has on salinity, pesticides and heavy metals) shows that there is no lack of studies at the 10$^5$ km$^2$ scale. The fact that 'sediment delivery ratios' (the ratio of erosive losses from uplands and the sediment load of streams) tend to decrease continuously with an increase in the size of the area under consideration (van Noordwijk et al., 1998d) indicates that landscape-scale sedimentation processes have been overlooked, or underestimated, in most attempts to scale up erosion studies on small plots. Calder (2002) has called attention to the many 'myths' surrounding forests, tree planting and water resources, while leading tropical forest hydrologists (Bruijnzeel, 2003) have made valiant attempts to summarize the available empirical data. Their data show that increases occur in total river flow on the conversion of

Table 19.1. Documented impacts of land-use change on 'watershed functions', by basin size (Kiersch and Tognetti, 2002).

| Impact type | Basin size (km$^2$) | | | | | | |
|---|---|---|---|---|---|---|---|
| | 0.1 | 1 | 10 | 10$^2$ | 10$^3$ | 10$^4$ | 10$^5$ |
| Thermal regime | x | x | – | – | – | – | – |
| Pathogens | x | x | x | – | – | – | – |
| Average flow | x | x | x | x | – | – | – |
| Peak flow | x | x | x | x | – | – | – |
| Base flow | x | x | x | x | – | – | – |
| Groundwater recharge | x | x | x | x | – | – | – |
| Organic matter | x | x | x | x | – | – | – |
| Sediment load | x | x | x | x | – | – | – |
| Nutrients | x | x | x | x | x | – | – |
| Salinity | x | x | x | x | x | x | x |
| Pesticides | x | x | x | x | x | x | x |
| Heavy metal | x | x | x | x | x | x | x |

x, Measured impact; –, no well-documented impact.

forest into agriculture, that variable impacts are made over time on the baseflow/peakflow ratio and that there is a marked lack of evidence to indicate the return of baseflow after reforestation.

Watershed management projects have evolved away from the largely technical focus of the past towards one governed by participatory practices and the need for consultation with local stakeholders. However, rigid project frameworks hardly ever allow for a critical questioning of the basic premises of these projects, and the gap between 'science' and the 'community of practice' may be widening. In the following sections, we look at local farmers' mental models of watershed ecosystems and compare them with models developed through a scientific approach. We take two examples – one from Indonesia, the other from Vietnam.

## 19.3 Case Study 1: Sumberjaya, West Lampung, Sumatra (Indonesia)

The island of Sumatra is composed of a chain of (inactive) volcanoes and mountains (running parallel to its west coast) and a vast lowland peneplain with generally acid sedimentary soils on its eastern side (van Noordwijk et al., 1998e). The richer soils are found in the mountains and foothills (piedmont). Many of the valleys in the mountains have been used for agriculture for thousands of years, with pottery and other archaeological remains providing evidence of long-term external trade links via the rivers. Sumberjaya is one of these valleys, having an elevation of between 500 and 800 m a.s.l. and rainfall averaging 2614 mm/year (Agus et al., 2002). Until the middle of the 20th century, the valley remained relatively inaccessible by road and was sparsely populated. Population densities have now reached 147 per km$^2$ (BPS, 1999), as a result of immigrants flowing into the area either from traditional coffee-growing areas to the north, or from the island of Java. Coffee (*Coffea robusta*) is the main component of the majority of gardens. A considerable part of the area has been designated 'protection forest', and hundreds of households have been evicted from the area in the name of 'watershed-protection functions'. Only after the political changes of the late 1990s have farmers resettled the area, and they are currently negotiating tenurial rights in the context of 'community forest management' arrangements. Perceptions of watershed functions thus have a direct, political relevance in this area.

Coffee cultivation methods and garden typology vary widely across the district (Verbist et al., 2002). Gardens range from young monocultures of coffee, through simple shaded coffee to complex multistrata agroforests. Increasing land scarcity has resulted in the cultivation of steeper land and the conversion of most primary and secondary forest into agriculture, except in the case of some of the steepest slopes and the top of a ridge which formally held the status 'protection forest'. Soil conservation in these erosion-susceptible areas is a priority, in order to sustain coffee yields in the short term and prevent a longer-term decline in productivity. Consequently, various soil management strategies and garden typologies have developed to suit different locations. A variety of soil conservation measures are applied in coffee gardens – from physical barriers such as terraces, trenches, ridges and pits, to the choice, positioning and manipulation of the plant components within the garden. These measures are often practised in conjunction with soil improvement through cultivation, and fertilizer and compost application. The effects of companion tree species in a mixed coffee system are well understood by farmers in Sumberjaya, where trees are classified based on their 'friendliness' to coffee (Box 19.1).

A study of the local ecological knowledge held by farmers was carried out using the knowledge-based-systems (KBS) approach (Sinclair and Walker, 1999), the same method used in the investigation of local ecological knowledge discussed in Chapter 2. Over 30 farmers were interviewed and asked to articulate their knowledge and understanding of the ecological processes occurring in their fields and in the surrounding landscape.

> **Box 19.1.** Grouping of trees in coffee gardens based on their influence on coffee plants (source: farmer interviews in Sumberjaya in 2000/2001; Chapman, 2002).
>
> **'Coffee-friendly' trees**
> Trees considered 'friendly' to coffee demonstrate the following:
>
> - non-competitive roots variously described as 'cold', 'deep' or 'water holding';
> - a light, airy crown with small leaves (allowing penetration by sunlight);
> - regular leaf shedding;
> - leaves that decompose readily;
> - leaves with a good compost value (e.g. improving soil fertility);
> - leaf retention during dry season.
>
> Examples of such trees are kayu hujan (*Gliricidia sepium*), lamtoro (*Leucaena leucocephala*), sengon (*Paraserianthes falcataria*) and dadap (*Erythrina orientalis*).
>
> **'Coffee-neutral' trees**
> Trees considered neutral in terms of their interaction with coffee provide some shade and help in soil conservation, although they do compete with coffee to some extent. This category includes fruit and spice trees (grown for household consumption and for sale), which are mostly maintained around homesteads. Examples of such trees are nangka (*Artocarpus heterophyllus*), rambutan (*Nephelium lappaceum*) and jambu air (*Syzigium aqueum*).
>
> **'Coffee-harming' trees**
> Trees considered harmful to coffee are usually productive, being grown for timber, spices or fruit. The economic gains of such trees outweigh the negative effects they have on coffee production. However, the negative effects they have on coffee are acknowledged, and are mitigated by planting position (boundary) and spacing (wide) used. Such trees have:
>
> - strong, 'hot', expansive root systems;
> - high nutrient requirements;
> - leaves with a poor composting value – such leaves are described as keras ('hard', 'difficult to decompose').
>
> It is preferable not to have such trees in one's field, so they are mostly maintained on the garden boundary. Examples of such trees are kemiri (*Aleurites moluccana*), jati (*Tectona grandis*), pohon afrika (*Maesopsis eminii*) and mahogani (*Swietenia macrophylla*).

### 19.3.1 Erosion and water quality and flow

Farmers in Sumberjaya hold the view that a decline in forest cover affects uniformity of water flow in rivers, resulting in an increase in river flooding in the rainy season and greatly reducing the amount of water in rivers in the dry season. They also believe that water turbidity increases with the destruction of forest cover (Fig. 19.1).

Cultivation methods strongly influence the efficiency with which coffee gardens maintain watershed functions. Earthen constructions (such as terraces, furrows and composting holes) can help reduce erosion. On the other hand, weeds and weeding techniques also affect soil erosion, as intensive weeding increases erosion whereas the presence of weeds can be used to reduce erosion, as can weed strips, ring weeding and mulching.

### 19.3.2 Riparian vegetation

Riverside vegetation is believed to be crucial to watershed function at a landscape level, significantly influencing flooding, landslides, bank erosion and changes in the courses of rivers. There was no consistency among the farmers with regard to how wide this vegetation should be: estimates ranged from 50 to 500 m. Trees along river banks, even if they occur only in thin strips a few metres wide, are considered to be effective filters by

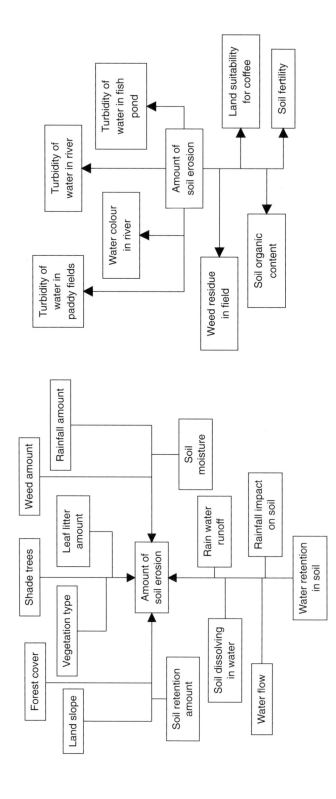

**Fig. 19.1.** Sumberjaya farmers' understanding of the causes and consequences of soil erosion in coffee gardens and the surrounding landscape. Change in values (such as an increase or decrease) of source nodes determines values in associated target nodes.

farmers. Additionally, the root systems of vegetation are believed to hold soil, thereby reducing the occurrence of landslides and soil loss. Shrubs and bushes along riverbanks are also believed to have similar functions. Bamboo, which has many fine and intricate roots, is considered a very efficient plant for planting along riverbanks.

Farmers see turbid water flowing down from upslope coffee gardens and forests as something that contributes to soil fertility in paddy fields (represented in the second diagram in Fig. 19.1), even though excessive water flow and sedimentation are physically detrimental to paddy plants. By carefully monitoring and regulating water flow in and out of paddy fields, farmers control both water speed and the duration for which that water remains in paddy fields, and hence the deposition of soil particles. It is common knowledge among farmers that, if water flow is properly regulated, such sedimentation leads to a reduction in the turbidity of the water flowing out of the fields. Cultivation practices that disturb soil (installing paddy fields, building terraces, hoeing and even planting rice), however, increase water turbidity.

## 19.4 Case Study 2: Dong Cao Catchment, Vietnam

The Dong Cao catchment lies 60 km south of Hanoi (20°57′N, 105°29′E) in the Luong Son district in Hoa Binh province, northern Vietnam. It is inhabited by 40 households, from the Muong and Kinh ethnic groups. The area receives a mean annual rainfall of 1500 mm, which falls mainly between April and September (Fagerström et al., 2002). Ferralsols and Acrisols, classified as 'clay' and 'clay loam' soils, dominate the area. There are patches of secondary forest, mainly at higher altitudes. Cassava, maize, arrowroot and soybean are the major annual crops grown on hill slopes, whereas paddy is the major crop grown at lower altitudes. The gradients of the slopes in the catchment range from 15% to 60% (Toan et al., 2001). On the gentle slopes and on the foothills, legume-based cropping systems are common.

An investigation of soil–plant interactions in the Dong Cao catchment was carried out using a suite of methods that included a Participatory Landscape Analysis (PaLA) survey, and biophysical data gathering, as well as the use of Participatory Rural Appraisal tools. The KBS methodology (Sinclair and Walker, 1999) was adopted to explore farmers' ecological knowledge, with ten purposively selected farmers being interviewed in order to gain an insight into their knowledge and understanding. Farmers in both the upper and lower parts of the catchment were consulted and an electronic knowledge base developed and tested, as recommended by Dixon et al. (2001). Farmer knowledge was analysed in terms of the farmers' understanding of erosion and filter functions in the landscape. Particular attention was focused on the filter efficiency of species such as *Acacia mangium*, *Vernicia montana* and bamboos. Farmer knowledge was then compared with scientific knowledge, as represented in the WaNuLCAS (Water, Nutrient and Light Capture in Agroforestry Systems) model (van Noordwijk and Lusiana, 1999).

### 19.4.1 Trees, Soil and Water

In the interviews, farmers articulated their knowledge about and perceptions of soil movement and the processes of terrace formation, and also the influences that earthworms and organic and inorganic fertilizers have on soil fertility. Using examples of *A. mangium* and *V. montana*, the two most common tree species in the catchment, farmers explained the mechanisms by which, according to their understanding, both the leaves (in terms of size, colour and density) and the rooting behaviour of different plants influence soil erosion and soil fertility (Fig. 19.2). Furthermore, they also stated that tree roots can hold soil and absorb and retain moisture when it rains and that this moisture is later slowly released into the surrounding soil. The farmers also stated their belief that trees retain water during the day and so resist heat from the sun.

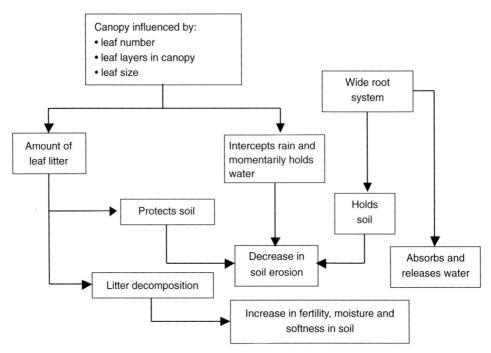

**Fig. 19.2.** Dong Cao farmers' understanding of tree–soil interactions. Change in values (such as an increase or decrease) of source nodes determines values in associated target nodes.

Farmers in the Dong Cao catchment identified factors, due to physical processes, plant growth and human activities, that they believed influence soil erosion. Rainfall intensity and duration, slope of land, weeding, soil cultivation and tillage all increase soil erosion in fields. Farmers said that short duration rainfall (< 20 min) merely translocates soil within a field (from the upper slope to lower down the slope); long duration rainfall (> 1 h) can permanently wash away soil from the field. Conversely, the presence of bamboo hedgerows slows down the downward movement of soil, as its fine, widespread roots hold soil and prevent it being washed away. By the same token, leaf litter covers the soil and also absorbs rainwater, and tree crowns reduce splash erosion by intercepting raindrops before they hit the soil. Farmers also stated that ditches on the lower slopes also accumulate soil, preventing it from being permanently lost. The overall impact of soil erosion means that good soil is gradually lost from the field, and the water in the streams becomes turbid. Some of the soil lost may also be deposited in paddy fields below. Where live fences (hedgerows) exist along lower borders of a field, good soil can be retained.

The Dong Cao farmers believed that tree roots release water into the soil, leading to a higher and continuous water flow in the streams. Farmers also believed that the more trees there are in the catchment, the higher the uniformity of water flow and the higher the volume of water in the streams.

Farmers regard bamboo as a very good hedgerow plant for use along field boundaries. In addition to preventing animals moving into the field, bamboos trap and retain soil because they grow in dense clumps (locally called *boi*) and have far-reaching, dense roots. Bamboo stems also reduce water runoff. However, farmers said that the extensive, fine roots of bamboo also absorb or 'eat' soil fertility, significantly affecting annual crops in their vicinity. Similar observations are reported by hill farmers in Nepal, who only maintain bamboo along a field's boundary, and never in the middle of a field (Thapa *et al.*, 1995).

## 19.5 Science-based Models of Watershed Functions

The behaviour of rivers and the relation of such behaviour to land use and land cover can be studied using either a 'spatial pattern' approach (a common starting point in geographical studies) or a 'process' perspective (an approach commonly used in physical hydrology). When the two approaches are applied to one particular situation (e.g. the evergreen forests found at higher elevations in northern Thailand), apparently contradictory statements may arise (Table 19.2).

The contradiction apparent between the two statements given in Table 19.2 can be resolved by realizing that, as in Thailand, evergreen forest tends to occur in locations where rainfall is highest. The real question, then, is whether this higher level of rainfall is the cause or the effect of the presence of evergreen forest. If either model is used to predict the impacts of land-use change on watershed functions, uncertainty with regard to the causes and effects of rainfall plays a key role.

Remnants of the 'spatial pattern' approach still exist in public perceptions; however, the theory of river flow that dominates current scientific thinking is based on our understanding of 'hydrological processes'. The validity of many of the hydrological-process models appears to be constrained, however, by incomplete data on rainfall, due to spatially inadequate sampling schemes resulting from, for example, too few rainfall gauges and a bias towards easily accessible locations.

Relevant to the construction of science-based models of watershed functions are the four types of controls (see Fig. 19.3), which can normally be distinguished in the infiltration process. Of these, the following three can be influenced by land cover:

- the rate of water use between rainfall events (relative to the potential evapotranspiration dominated by the energy balance);
- soil surface structure and macroporosity (which influence the potential rate of infiltration);
- the difference between field capacity and saturated soil water content.

Nearly all models, even those applied at a global or river-basin scale (Vörösmarty *et al.*, 2000), include the first control listed above in their predictions of the impact that land-use change will have on river behaviour. The effects of land use on the second and third controls listed above are only included in models such as DHSVM (http://www.hydro.washington.edu/Lettenmaier/Models/DHSVM/index.htm; Wigmosta *et al.*, 1994) and WaNuLCAS (van Noordwijk and Lusiana, 1999), which were developed for high-resolution applications.

**Table 19.2.** Some characteristics of two 'modelling approaches' applied to the relationships between land cover and watershed functions.

| Starting point | Spatial patterns | Hydrologic (water balance, processes) |
|---|---|---|
| General characteristics | • Approach starts with existing land cover and river flow properties, as they vary across space<br>• Correlations are analysed and used for extrapolation<br>• Models can be based on data obtained at different scales, and can apply to various map resolutions | • Approach starts with rainfall and traces water, through various pathways, to evapotranspiration or delivery to oceans<br>• Land-use change is taken into account, as it can affect interception, infiltration and evapotranspiration (seasonality)<br>• Models can be strongly spatially disaggregated, 'lumped' or 'parsimonious' |
| Typical statement | 'Evergreen forest is associated with highest water yields …' | 'Evergreen forest uses more water and allows less rainfall to reach associated streams than other land-use types …' |

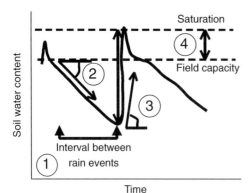

**Fig. 19.3.** Schematic time course of soil water content and soil physical understanding of the determinants of the infiltration process: (1) time interval between rainfall events; (2) rate of soil water depletion between rainfall events, creating soil storage space; (3) potential rate of infiltration into the soil, in relation to the intensity of rainfall and (slope-dependent) opportunities for temporary water storage at the soil surface; and (4) difference between 'field capacity' (= soil water content 24 h after a heavy fall of rain, when the rate of water seepage to deeper layers tends to reach a small value) and 'saturated' soil water content, when all soil pores are water-filled.

We will now look more closely at the WaNuLCAS model, which has been used to compare modellers' understanding of the erosion process with that of farmers. The WaNuLCAS model was developed to simulate a range of tree–soil–crop interactions in agroforestry systems, for a wide range of soil, climate and slope conditions (see also Chapter 10, this volume). Basic ecological principles and processes, as understood from a scientific perspective, are incorporated into the model using modules such as climate, soil erosion, sedimentation, water and nutrient balance, tree growth and uptake, competition for water and nutrients, root growth, and soil organic matter and light capture (Fig. 19.4; Khasanah et al., 2002).

In WaNuLCAS, physical soil properties (i.e. texture, bulk density and organic matter content) and soil structure dynamics (i.e. biological activity, dependent on nutrition provided by plants through litterfall and root decay) determine saturated hydraulic conductivity ($K_{sat}$), and condition the processes of lateral flow and vertical infiltration. Rain intensity, plant growth (through the interception of rain) and lateral flow (over the surface and as subsurface flows) influence infiltration, which determines the amount of runoff water. Soil erosion is influenced by the amount of runoff water, the flow velocity (which determines the maximum transport capacity for particulate matter) and the actual concentrations of sediment (which depends on the particles' 'entrainment' or 'propensity to join the flow'). Actual sediment concentrations in overland flow thus depend on the steepness of the slope (determining the runoff velocity), the soil's surface cover (canopy of trees, shrubs, weeds, and litter: all of which reduce flow velocity at the surface and thus cause the sedimentation of particulate matter) and the coefficient of entrainment (which mainly depends on aggregate stability at the soil's surface).

## 19.6 Soil Erosion – Farmer Perception Versus Simulation Modelling

In this section we will compare farmers' mental models of surface runoff and erosion with the way such processes are represented in current scientific simulation models. The overall concept of a water balance, in which all losses and gains can be accounted for (in = out ± change in storage), appears to be absent from farmers' perceptions and interpretations of events. This is probably because water use by plants (evapotranspiration) is invisible, whereas rainfall, stream flow and changes in soil water content are observable. However, with regard to the phenomena of overland flow and erosion, the observational basis of local ecological knowledge differs little from the basis of 'scientific' models, and agreement is stronger with regard to the underlying concepts.

The major components of farmers' understanding of the erosion process include:

1. Rain – duration and intensity.
2. Standing trees, bamboos and shrubs – crown morphology and root system.
3. Ground cover – leaf litter and live ground vegetation.
4. Soil – e.g. physical properties and nutrient content.

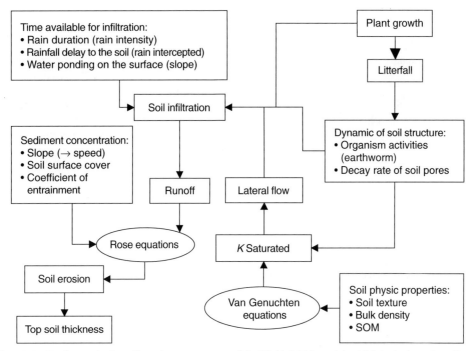

**Fig. 19.4.** Key factors in the soil erosion component of the WaNuLCAS model, which include such well-established process descriptions as the 'van Genuchten' functions for soil water conductivity under saturated ($K_{sat}$) and unsaturated conditions, and the 'Rose equations' for overland flow of soil particles (see also Chapter 18). Source: Khasanah *et al.* (2002).

A generalized representation of the process of soil erosion as understood by farmers is shown in Fig. 19.5. Farmers believe that rain intensity and duration play an important role in determining the intensity of soil erosion. Farmers also believe that trees and other tall vegetation have multiple functions: their crowns intercept raindrops, and so reduce splash erosion, and also cause shading (a positive or negative effect depending on density and crop type); their roots (depending on the spread and type) hold soil in place; their stems (especially in the case of clumps of bamboo) slow water runoff, whilst the leaf litter they produce prevents soil being washed away by rain and also reduces excessive evaporation in dry periods. Decayed leaf litter is also an important source of the organic matter and plant nutrients soil contains. The presence of earthworms is considered to be an indicator of a good soil, as they are known to contribute to increasing soil fertility.

Farmers' conception of soil compactness, and the influence it has on surface runoff, is similar to the representation given in WaNuLCAS (Fig. 19.4), which represents it as the coefficient of entrainment, determined by aggregate stability. Farmers understand well that loose soil will erode more quickly, and that soil compaction can therefore increase overland flow of water and reduce soil movement.

In the WaNuLCAS model, soil structure dynamics are caused by biological activity in the soil (mainly represented by earthworm activity), which results from inputs of plant material such as leaf litter, prunings and decaying roots. Farmers relate the presence of leaf litter and roots directly to infiltration; thus, both their understanding and that of researchers are comparable in this matter. They also see and value the role earthworms play in improving soil fertility. Farmers directly link soil physical properties, usually linked to organic content (loose and sandy

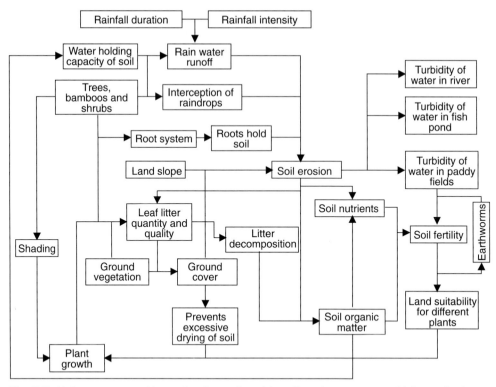

**Fig. 19.5.** Main components of farmers' understanding of the soil erosion process, combining studies in Indonesia, Vietnam and Nepal. Arrows show cause–effect relationship between linked 'nodes'.

versus compact and clayey soils) and soil structure dynamics (biological activity: leaf litter and roots) to infiltration. Although they do not allude to the processes that appear in the model (such as saturated hydraulic conductivity or lateral flow), both farmers' and researchers' conclusions are similar. Farmers, however, do not mention macropores or microbial-related aspects of the soil.

On the other hand, farmers do emphasize the important role trees and their leaf litter play in breaking the fall of the raindrops, thereby reducing the strength of their impact on the soil and so decreasing soil erosion. Interception was incorporated into the WaNuLCAS model as a factor that diminishes the amount of rainfall reaching the ground. However, the impact rainfall has in terms of directly increasing erosion, caused by splash erosion, has not been incorporated in WaNuLCAS (although it is indirectly represented in the effect soil cover has on entrainment).

Some differences do exist between local and scientific understandings of soil processes, e.g. in the way information is represented (qualitative in the case of the former and quantitative in the case of the latter). As with most simulation models, WaNuLCAS uses a mathematical approach. By contrast, farmers' understanding is largely qualitative in nature and simple cause–effect relations are a norm in farmers' mental models. So, whereas WaNuLCAS can predict a final output quantitatively, farmers' models reflect the direction of change in the form of an increase or decrease. Furthermore, due to the site-specific nature of local knowledge, the ranges that occur in terms of local variation will condition farmer knowledge. Simulation models, on the other hand, are supposed to be more generic and able to cover a wider range of situations, thus requiring more parameters to specify components' behaviour on any given site.

Both farmer knowledge and science-based models allow for a much more nuanced approach to soil conservation than that implied by the simple 'forest' versus 'non-forest' dichotomy that dominates public discussions on soil conservation and land classification systems. By considering the underlying processes of soil and water movement the category 'forest' can be split, so making clear the dominant role played by the surface litter layer. Within the 'non-forest' category we are able to see the presence or absence of a litter layer emerge as a distinguishing element in predicted soil and water movement. We may note that a surface litter layer is both an 'indicator' of the absence of surface flows with enough energy to move the litter particles and a link in the causal chain to soil biological activity, which maintains soil structure and increases infiltration rates. Wiersum's (1984) observation of the occurrence of high erosion rates under tropical plantation forests with uniformly high canopies (and thus with high-energy drips falling from such canopies) and an absence of surface litter (due to farmers harvesting it) is easy to understand using both local and scientific paradigms.

## 19.7 The Gap Between Knowledge and Practice

Knowledge is conceptually different from practice or action, as discussed in Chapter 2. What farmers know may not always be reflected by what they do and vice versa. Farmers cultivating cassava in the Dong Cao catchment (Vietnam) are well aware of the heightened soil erosion problem that results when cassava is interplanted with *Acacia* in its establishment phase. They know that cassava cultivation can significantly affect *Acacia* growth in its early stages. They also know that if the *Acacia* or *Vernicia* being grown are taller than the cassava (a state normally reached after 2 years) the cassava will be severely affected. WaNuLCAS simulations of an *Acacia* and cassava intercropping system have also shown that erosion is a major problem during the first year of the cycle. Despite this knowledge, farmers still continue to cultivate cassava with *Acacia* and *Vernicia*, because doing so provides them with much-needed food and income and also reduces animal damage to trees and crops, due to the presence of people in the field (PaLA survey).

In Sumberjaya (Indonesia) all farmers interviewed were aware of deforestation, erosion and water problems: their knowledge was detailed and commonly shared. Farmers also know about the processes and reasons behind these problems and possess a substantial range of possible technical solutions. However, in reality, not all farmers practice soil and water conservation measures when cultivating steep slopes. What constrains farmers from translating this sophisticated knowledge into practice? In the Indonesian case study of farmer knowledge and practice in Sumberjaya, Schalenbourg (2002) identified the following common constraints as being those that make it difficult for farmers to translate their knowledge into practice.

**1.** Lack of capital investment (money, labour and time). Most soil conservation practices require time, money and labour, and often involve construction work and maintenance. Farmers, particularly the poorer members of the community, simply do not have the resources necessary to invest in soil conservation innovations.

**2.** Lack of enthusiasm ('laziness') or lack of the necessary incentives. Many farmers reported that they are too *malas* or 'lazy' (note that the Indonesian word does not have the same negative connotations as the English word 'lazy'). The farmers probably meant to imply that soil conservation is not their priority or that implementing soil conservation practices does not yield sufficient benefits to make it worthwhile.

**3.** Uncertain land tenure. Many farmers cultivate coffee on government-designated 'forest land', and the region has seen numerous evictions (by the government). Land tenure largely remains uncertain, and this has been an important factor with regard to influencing farmers' decisions not to spend their resources on long-term soil conservation methods.

**4.** Low returns to labour, or a low price for coffee, result in emphasis being placed on short-term cash gains (including alternative annual cash crops) rather than on long-term productivity and sustainability. Like that of many commodity crops, the price of coffee has 'nose-dived' in recent years. Many farmers have converted their fields to the production of other cash crops, and thus are involved in vegetable production and fish farming. Again, farmers are not prepared to invest in any soil conservation activity that requires additional resources, especially if that activity only facilitates long-term coffee production.

**5.** Isolated efforts with regard to soil conservation are ineffective. Only a concerted effort can yield tangible results, which perhaps to a great extent explains why farmers do not practice soil conservation practices. On-going land disputes (both between settlers and the government and settlers who arrived in the area at different times) mean that there is little possibility that farmers' groups will be organized.

Biophysical, social, economic and market environments are likely to vary between sites. Therefore, methodological guidelines need to be developed both in order to provide a more holistic view of the constraints farmers face and to develop strategies to address them.

## 19.8 Discussion

Numerous studies concerning local ecological knowledge (including the two reported in this chapter and those reported in Chapter 2) provide convincing evidence that farmers have detailed plot-level knowledge, which they may use when managing their resources. Farmers' knowledge at the landscape level seems: (i) to consist of logical explanations for various natural processes; and (ii) to be based on their plot-level knowledge. In comparison with plot-level processes, farmers have a less intimate knowledge of landscape-level interactions, which perhaps explains why landscape-level local ecological knowledge is less developed at the 'process' or 'explanatory' level. There is, of course, a fairly detailed understanding of the variations that occur in topography, vegetation and microclimate at the landscape scale: the language spoken by farmers contains many words to describe such variation. In the Indonesia case study at least, although plot-level knowledge varied to some degree between farming communities (Chapman, 2002) landscape-level knowledge varied less between farmers and between farming communities (Schalenbourg, 2002).

It would be unrealistic to think that we can 'quantify' farmers' models as simulation models. However, it may be possible to gather data in order to quantify certain key components in farmers' models (such as the fact that live and dead vegetation reduce splash erosion and that plant roots have the ability to 'hold' soil). Likewise, the reasoning that farmers apply when 'running' their models can be tested and represented in scientific models. The understanding that farmers have developed can complement scientific understanding, thereby enriching scientific models. At the same time farmers will be better able to comprehend, accept and benefit from such synthesized models if their knowledge is represented. A combined model will, therefore, be richer than either the stand-alone 'local' or the scientific model. Better understanding, appreciation and representation of local knowledge, terminology, and perceptions can likewise contribute towards improved communication and negotiation between farmers, professionals and decision makers.

From the perspective of translating knowledge into practice, there is increasing evidence (the Sumberjaya study, for example) that farmers not only rely on their ecological knowledge to make management decisions about their resources, but also take into consideration available resources (land, labour, capital), markets, and social relationships. An additional factor that is becoming clearer from our work (particularly that undertaken in Indonesia) is the important role social capital plays when scaling up plot-level actions to the landscape-level management of natural resources. In other words, no individual effort will accrue benefits on a landscape scale in those cases where neighbours' practices are detrimental to soil conservation. Under such circumstances, no farmer will expend personal resources to seriously practice soil conservation measures. The need for collective action in soil conservation is obvious. In the case of the study made in Sumberjaya, perhaps the main limitation to control erosion is not a lack of knowledge of conservation practices, but the constraints associated with farmers functioning as an effective unit and the fact that they do not have secure land tenure. The 'scientific understanding' that we have discussed so far answers only one of the five issues in any natural resource management issue (Box 19.2).

As we have seen, local and science-based perceptions of landscape-level watershed issues contrast with public perceptions of the same issues. This is a matter that requires some form of 'negotiation support' if landscape-level watershed issues are to be resolved (van Noordwijk et al., 2001; Verbist et al., 2002). Such support should involve a shared vision of the likely consequences of various land-use alternatives in combination with a social process of stakeholder negotiation, whilst retaining respect for the various positions held by the stakeholders involved.

---

**Box 19.2.** Knowledge and natural resource management.

Improved natural resource management, for example watershed management, may not be limited by a lack of scientific understanding of the issue. There are five important issues that are for 'problem solving', answering different question in natural resource management:
1. Emotional links: The first question we should ask, if put bluntly, is 'why should I (or anybody) care?'
2. Scientific understanding: 'how does the system work?' Which elements, patterns, processes and system dynamics are associated with the natural resource in question.
3. The current problem: 'what or who is causing the current problem or perceived problem?', 'what are the consequences', 'what are possible remedies?'
4. Stakeholders: 'who benefits from causing the problem?', 'who suffers the consequences?', 'who will pay for remedies and solutions?'
5. Governance opportunities: 'how can a working solution be achieved?', 'is it better to spatially segregate activities or go for an 'integrated' multifunctionality solution?', 'how can the different stakeholders and actors negotiate solutions that meet their various sets of objectives?'

---

**Conclusions**

1. Farmers' knowledge of water and soil movement at the landscape level provides logical explanations for several 'observable' phenomena and is closely linked to farmers' plot-level knowledge. However, their landscape-level knowledge may be less clearly articulated.
2. Farmers generally have a fairly detailed understanding of the variations that occur in topography, vegetation and microclimate at the landscape scale.
3. Because of the constraints imposed by the observational capacity of farmers, less visible processes (such as evapotranspiration) are not included in the farmers' mental models and, hence, for them 'water balance' is a qualitative, rather than a quantitative concept.
4. Farmer knowledge, like scientific knowledge, is cumulative; it evolves as farmers adopt, adapt and formulate new ideas and innovations, try them out in different settings, evaluate and assess the results and make decisions about their potential value for continuously improving their farming methods.
5. Collective action and social capital are important with regard to the scaling up of plot-level actions to the landscape-level management of natural resources.

**Future research needs**

**1.** There is a need to further explore the obvious contrast between the 'process-based' conceptual models held by farmers and researchers and the 'black-or-white' public perceptions of the landscape-level watershed issues (which underlie current policies and regulatory frameworks).

**2.** Research should be conducted to find replicable ways to provide 'negotiation support', by combining a shared vision of the likely consequences of various land-use alternatives with a social process that involves stakeholder negotiation and respect for the various positions.

# 20 Challenges for the Next Decade of Research on Below-ground Interactions in Tropical Agroecosystems: Client-driven Solutions at Landscape Scale

Meine van Noordwijk, Georg Cadisch and Chin K. Ong
(General Editors)

---

**Key questions**

1. So what? After we have answered all the questions previously asked in this book, how can farmers and agroecosystems benefit?
2. How can different types of knowledge and understanding be integrated to allow an evaluation of 'complex agroecosystems'?
3. Which research topics emerge as particularly relevant to this subject?

## 20.1 Introduction

Over the past five decades, great progress has been made in terms of agricultural productivity, both globally and in a number of tropical regions – although per capita food production in sub-Saharan Africa did decline. This increase in the production of main staple food crops is generally referred to as the 'Green Revolution'. In the early stages of the Green Revolution, research was directed mainly at plant breeding, fertilizer use and plant protection. However, the pace at which advances are being made is slowing, and there is increased awareness of the 'downside' associated with the loss of crop diversity.

The annual increase in cereal yields in developing countries has fallen to just under 1%, compared with 2.9% from 1967 to 1982. As a consequence, more attention has recently been directed at increasing the productivity of land and water resources (Kijne et al., 2003). Further incentive for the agricultural research and development community to take this direction has come from considerations of sustainability.

When applied to the use of land and water, 'sustainability' means meeting the production needs of present land users whilst conserving, for future generations, the resources on which production depends. Such an approach is known as 'integrated natural resource management' (INRM), and requires a major shift in both research and management approaches. It differs from the Green Revolution approach in several ways. First, it often focuses on the needs of the poor farmers, integrating the interests of community-level land users and managers as well as national and international policy makers. Secondly, it focuses on environments with a high diversity, in which Green Revolution solutions are not readily acceptable or applicable. Thirdly, it focuses on the functions of natural capital in agriculture, in order to increase productivity whilst ensuring the sustainability and stability of any increases.

The majority of this book (Chapters 1–17) was devoted to helping the reader gain a better understanding of below-ground interactions at the plot level. The last part of the book (Chapters 18 and 19) made steps towards a landscape-level understanding of natural resource management issues (Fig. 20.1).

In this chapter, using examples from Africa, Asia and Latin America, we will illustrate how a better understanding of below-ground interactions is vital if some of the major INRM problems are to be solved. In each example, we begin by stating the problems that should be addressed by research, followed by the approaches currently being taken and the major challenges for the future. We hope that this will help the reader 'put the pieces of the puzzle together' and see how a better understanding of below-ground interactions can help in solving real-world problems and rural livelihood issues.

## 20.2 Example 1. Lake Victoria Basin

Lake Victoria, surrounded by Kenya, Uganda and Tanzania, is the world's second largest freshwater lake, having a surface area of approximately 68,000 km² and an adjoining

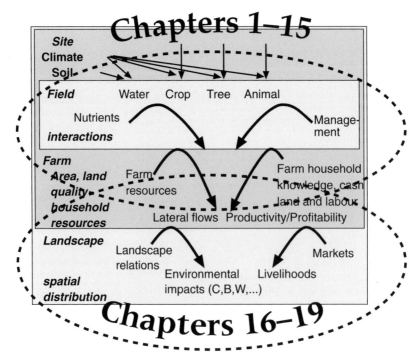

**Fig. 20.1.** Nesting of relations between local conditions (such as site, climate and soil), the field-level interactions between water, crops, trees and animals, farm-level use of resources and their impact on lateral flows, productivity and profitability and the landscape-level issues of environmental impacts (C, carbon stocks; B, biodiversity; W, watershed functions) and livelihoods.

catchment area of 155,000 km², and comprising 12 major river basins. Lake Victoria is a world-renowned site of vertebrate diversity, containing an estimated 500 species of fish (mostly endemic members of the family Cichlidae). It is ranked second in the world in terms of fish species richness, and is probably the site of the world's fastest proliferation of vertebrate diversity (because hundreds of fish species evolved in less than 12,400 years). However, it is also well known as the site of the world's largest contemporary species extinction event, linked to the introduction of one of the largest freshwater fish (Nile perch, *Lates niloticus*). This fish now supports the world's largest freshwater fishery, in Lake Victoria.

The Lake Victoria basin supports an estimated 40 million people, who produce an annual gross economic product in the order of US$3–4 billion (Sida, 1999). With the exception of the city of Kampala (Uganda), the lake catchment economy is principally agricultural, involving a number of cash crops, fisheries and types of subsistence agriculture. The quality of the physical environment is crucial if the living standards of the growing population are to be maintained or increased. It is estimated that a 5% reduction in the productivity of the region would lead to a loss of US$150 million annually. The lake basin is a major source of food, energy, shelter, transport and drinking and irrigation water. It is also a repository for human, agricultural and industrial waste.

Recently, the environment of Lake Victoria has attracted the attention of policy makers, following its colonization by water hyacinth (*Eichhornia crassipes*), which blocked water transport and fishing activities. The communities living around the lake, who lived, predominantly, by fishing, were the most affected, because they could no longer go out to fish. For a few years, the problem was compounded by the fact that any fish that were caught were unfit for either export or local consumption (due to contamination with high levels of *Escherichia coli*, derived from human waste that had been disposed of in the lake). Colonization of the lake by the water hyacinth is largely attributed to: (i) increased levels of nutrients (particularly phosphorus and nitrogen) entering the lake from urban, agricultural and industrial sources; (ii) sediment deposits, originating from soil erosion due to poor upland management practices; and (iii) damage to the natural filter function of wetlands, especially around the mouths of rivers.

### 20.2.1 Root causes of the lake's ecological problems

The ecosystem around the Lake Victoria basin has undergone substantial changes during the last three decades as a result of two major human interventions in the basin. The first was the introduction of the Nile perch, which altered the structure of the food web. The second was an increase in nutrient flows into the lake from the surrounding catchments. The most urgent problems in Lake Victoria are the presence of water hyacinth, decreasing fish numbers, and poor water quality; however, the lake is not the *source* of these problems. The root cause of eutrophication (besides the disposal of sewage) is continuing land degradation, resulting from deforestation, settlement, farming and accelerated soil erosion. Sediment and nutrient loads in the lake are high and will further accelerate the process of eutrophication. Low levels of soil P are an agricultural problem in the area, and are partially linked to the excess of P in the lake. Urban runoff is also an important component in the pollution of the lake, and will worsen in the next few years. Industrial discharges and sewage will exacerbate this problem, especially as there are a large number of agroindustries in the Nyando River basin.

The Lake Victoria Environmental Management Program (LVEMP), a basin-wide project that has been funded by the World Bank and the European Union since 1995, recognizes the fundamental importance that wetlands play in the lake's ecology. The programme is responsible for implementing policy concerned with the sustainable development of such wetlands. This policy pays due regard to both their economic value and their ecological importance, including their value as a filter surrounding the lake.

Consultants working for LVEMP (Bullock et al., 1995) have made recommendations that highlight the importance of monitoring the buffering capacity of the lake basin wetlands and integrating the socioeconomic concerns of people using the wetlands. However, wetlands continue to be drained for the purposes of agriculture, despite growing evidence that this practice is unsustainable and economically unsound (Ong and Orego, 2003).

Although it is widely agreed that nutrient levels in the lake have increased in recent decades, we do not have reliable information concerning the major nutrient sources and sinks. Nutrients enter the lake from different sources (i.e. agricultural, atmospheric, urban and industrial sources), but there is still debate as to the relative importance of each. Quantifying the relative contribution made by the various sources, as a first step towards abatement, is still a major challenge. Most methods are flawed or inadequate, because they are based on values extrapolated from either North America (Bullock et al., 1995), small catchments in Tanzania (Scheren et al., 2000) or from minor streams in Uganda, in which case the values were extrapolated from data covering only a single year (Lindenschmidt et al., 1998). The three countries bordering Lake Victoria (Kenya, Tanzania and Uganda) have very different agroecosystems, topography and discharges into the lake. Such extrapolations are therefore grossly unreliable when applied to the whole basin, though the latter two studies might reflect differences in land use between the two countries. Nevertheless, these rough estimates indicate that the contribution made by agricultural lands accounts for about half the nutrient load entering the lake. Of the total amount of water entering the lake, 85% comes from rain falling directly on to the lake, whilst the remainder is carried to the lake by its 12 major rivers. There is an urgent need for more reliable data on the relative importance of each source, especially for the major rivers (such as the Kagera in the west). Recent analysis of these major river basins (made using satellite images) indicates that the Nyando and Kagera river basins stand out in terms of sediment transport capacity and average slope.

## 20.2.2 How will an understanding of below-ground processes help?

Farmers in the Lake Victoria basin have realized that the decline in soil fertility that has occurred over the last five decades is a major concern for them. They have also realized that this decline is largely attributable to depletion caused by continuous cultivation, because only 10 kg of nutrients per hectare are applied annually (Smaling et al., 1997).

Because P limitation is common in the area, crops, fallow species and trees with a high P use efficiency and specific P mobilizing properties are preferred (Chapter 7, this volume). Although agricultural plot or farm-level studies of erosion have consistently suggested the existence of high erosion rates in western Kenya, it is now commonly understood that much of the sediment is probably deposited elsewhere in the landscape, as opposed to directly into the lake (Chapter 18, this volume). Downward leaching and substantial accumulation of mobile nutrients, especially nitrate, have been well-documented in western Kenya (Chapter 6, this volume). Much progress has been made in terms of understanding and quantifying the processes involved in the retrieval of nitrate by perennials in agroecosystems (Chapters 6 and 10, this volume). Lateral movement of mobile nutrients has, however, not been considered in these studies. It was only recently that the magnitude of the erosion problem has been appreciated, through a combination of remote sensing, laboratory analysis of field samples (Shepherd and Walsh, 2002) and sediment coring. Sediment coring is currently being undertaken at the mouths of the Nyando, Sondu–Miriu and Yala rivers, in order to reconstruct the recent history of sedimentation in the lake. Preliminary analysis of the cores taken indicates that the sedimentation rate has been strongly influenced by El Niño events, which have a long-lasting effect. Dramatic increases in erosion were detected only in the last two decades: these coincided with the clearing of forests on a massive scale, for agriculture. Reflectance spectroscopy offers a potentially powerful 'pedo transfer' method with which to rapidly assess land degradation as well as soil fertility status (Shepherd and Walsh, 2002). The major challenge is to determine whether this approach can be used to detect changes in soil physical properties and pinpoint the 'erosion hotspots' associated with land degradation across the whole lake basin.

### 20.2.3 Challenges and opportunities for the management of the lake basin

Political leaders have recognized that there is a need to act to address the problems faced in the lake basin.

> The problems, challenges and opportunities posed by the Lake are today of such magnitude that they cannot be faced by the riparian states acting separately. Our challenge is to contribute to the development of a collective approach to roll back the environmental threat that hangs over the Lake and unlock the vast potential of the Lake for the benefit of the people in the region.
>
> (Minister of Foreign Affairs and International Cooperation, Tanzania, quoted in Sida, 1999)

Political recognition of the lake's problems and potential has resulted in the East African States making a concerted effort to set up a secretariat to coordinate and address cross-sectoral and transboundary issues (such as harmonizing laws and regulations, mapping pollution sources, and promoting sustainable development).

Information and knowledge are often the most limiting factors in watershed and basin management (El-Swaify, 2000). Information brokers (research organizations and universities) can assist in efforts to provide all stakeholders with a good information base, which will help them make decisions that affect their lives, their farm enterprises and their communities. Better information skills may also assist negotiations to manage or solve conflicts among stakeholders with competing interests (Chapters 18 and 19, this volume). Swallow et al. (2001) argue that research organizations can play key roles in the provision of both information and training, especially to governmental and non-governmental organizations.

A major challenge faced by policy makers is their limited understanding of the filter functions of vegetation in the landscape. For example, wetlands play a vital role in the lake basin, both as a major source of income for local communities and as a filter for sediments. However, they are traditionally undervalued by governments. Therefore, it is common for policy makers to declare state ownership of wetlands and riparian forests. Consequently, plans are still being formulated to convert critical wetlands to agricultural uses. Another challenge policy makers need to overcome is the misconception that there is a strong link between the ecological concept of lateral flows and deforestation. For example, the Mara river basin, which straddles Kenya and Tanzania, is critical to the survival of the pastoralists, farmers and fishermen in the area. The basin also controls the world's largest migration of wildlife, which takes place in the Mara–Serengeti ecosystem. The loss of a third of the forest at the river's headwaters in Kenya was believed to have decreased water flow by 40%. However, in actual fact, the reverse has been found to be true (Mungai, 2003).

In western Kenya, research has focused on the use of perennials to retrieve nitrate from depths and the need for the addition of phosphorus (Buresh et al., 1997; Chapters 6 and 10, this volume). These technologies involving perennials (known as improved fallows) have proved to be tremendously successful in providing short-term benefits to small-scale farmers, as well as offering reasonable protection in terms of runoff and erosion control (Boye and Albrecht, 2002). On the steep slopes of southwestern Uganda, tree fallows have been developed that are more profitable and productive than traditional bush fallow or continuous maize (Siriri and Raussen, 2003). Although these simplified agroecosystems are not as stable and sustainable as the original multistrata rainforests, and although they are beginning to suffer from pests and diseases (Chapter 15, this volume), they provide some improved ecosystem and economic functions. They do, however, need to be further developed and integrated into farmers livelihood needs. An alternative is the development of multistrata agroforestry systems (such as those in Indonesia), which mimic the major ecological functions of the original rainforests (Chapter 1, this volume), but provide greater productivity and a high income (from fruits), in addition to carbon sequestration (Chapter 12, this volume) and sustainability benefits.

## 20.3 Example 2. Sumberjaya Benchmark for Watershed Function Conflicts

Sumberjaya (literally 'source of wealth') is a subdistrict in the mountain range (Bukit Barisan), which spans the island of Sumatra (Indonesia) and forms the upper watersheds of all major rivers on the island. Of the island's provinces, the southernmost (Lampung) has the highest population density (8 million people living on 3.5 million hectares), as well as the highest level of poverty. A considerable number of people in the lowlands depend on irrigation water and hydroelectric power provided by these upper watersheds. Thus, the relationship between the lower and upper watersheds is considered to be critical with regard to provincial natural resource management plans.

Although 32% of the province is officially considered to be 'state forest land', only 20% of this state forest land actually had intact forest cover in the mid 1990s; and, 'forest encroachment' has not stopped since then. At least half a million recent settlers live in state forest lands. The majority of people in Lampung province, for example, either consider themselves to be of Javanese or Sundanese descent, even though they were born in Lampung, or are recent Javanese or Sundanese immigrants. However, many settlers also come from neighbouring provinces in Sumatra. The ethnic diversity apparent among the 'Lampungese' is therefore considerable, reflecting multiple waves of immigration for which the rivers were the main conduit.

Economically, coffee production in the uplands is one of the major income earners in the province, with Sumberjaya district being one of the important production areas to have attracted large numbers of immigrants over the past few decades. Much of the land settled is, however, classified as 'protection forest', and thousands of families have been evicted and forcefully removed to 'transmigration' sites on the much poorer soils of the lowland peneplain, greatly increasing their poverty. During these evictions, villages were destroyed and the coffee plants of the pioneer farmers uprooted. During the 'Tanggamus' operation of 1990–1991, more than 3000 people were evicted from Purawiwitan, which was at that time a 'village' of 7000 people. During that campaign alone, more then 8500 people were evicted – almost 10% of the population of the Sumberjaya subdistrict at that time. In 1994, in a second campaign, more than 3000 people were evicted from the neighbouring village of Purajaya, and a further 1200 were evicted from Purawiwitan: almost 500 people were evicted from the smaller village of Muarajaya. Newspaper reports of the time reflect the violence associated with these evictions, including incidents in which people were burned inside their houses. These evictions were undertaken to safeguard 'watershed protection forest'. However, the areas designated 'protection forests' have not returned to natural forest, and the livelihoods of the settlers were seriously affected.

The Sumberjaya district coincides, approximately, with the catchment area of the Way Besai, one of the tributaries of the Tulang Bawang River (one of three major rivers in Lampung, the other two being the Way Sekampung and Way Seputih). The Way Besai is used as the water source for a hydroelectric power scheme, which involves a runoff dam. Electricity generation started in 2001, but was interrupted in 2002 by a landslide near the turbines. Though the landslide was actually the result of the construction of a poorly designed road, the slip has generally been blamed on 'deforestation'.

Recently, ICRAF initiated a project, in conjunction with national and international partners, to assess the relationship between land-use change and the efficiency of watershed functions. A key hypothesis for the Sumberjaya benchmark states that:

some farmer-developed agroforestry mosaics are as effective as the original forest cover in protecting watershed functions related to water yield and water quality, and hence a substantial share of current conflicts between state forest managers and local population can be resolved to mutual benefit.

### 20.3.1 How will an understanding of below-ground processes help?

Current research (some of which is considered in Chapters 10, 18 and 19, this volume) has made clear that the evictions in Sumberjaya were based on an incomplete understanding of the consequences that coffee production systems would have for the watershed functions in that area (Fig. 20.2). On the other hand, transforming open, monocultural coffee systems into 'shade coffee' systems, where coffee is grown in conjunction with fruit, timber and service trees (e.g. *Gliricidia* and *Erythrina*) would provide a win–win situation, in which rural poverty could be reduced without risk to stream flow or (probably) water quality.

In an innovative form of 'negotiation support', a multi-institutional team is facilitating dialogue between local farm communities, local government (which is currently 'in charge' of natural resource management), forestry officials and representatives of the hydroelectric scheme. The initial outcome of these negotiations was the signing of a series of community forest management agreements in April 2002. These provide at least temporary security of tenure for coffee farms within 'protection forest' domain, provided that the groups concerned actively protect the remaining forest and maintain watershed functions (e.g. by introducing trees into the coffee gardens). Although

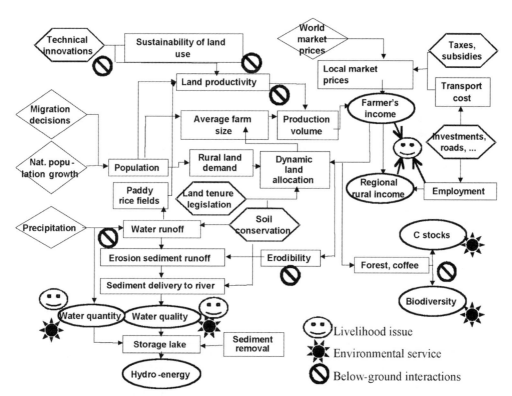

**Fig. 20.2.** Natural resource management system analysis of main relations in the Sumberjaya area, with special symbols for the key livelihood and environmental service outcomes, as well as the main places where 'belowground interactions' take place (nat., national).

these initial agreements are seen as 'policy experiments' (they have so far benefited only a few hundred families) other groups are anxious to follow the example being set. So far, however, no transparent mechanisms have been agreed upon that will allow either the monitoring of compliance with the agreements made or the evaluation of the environmental services provided. However, as part of the ongoing biophysical and ecological research in the area, progress is being made in testing candidate criteria that can be used to evaluate these environmental services. Furthermore, researchers have identified participatory methods of water quality monitoring, which have been used elsewhere in South-East Asia, as options that should be pursued. Clarifying these compliance and evaluation issues is important if we are to take this policy experiment beyond its current limits. For the farmers in the area, however, further economic benefits (beyond a reduced chance of being evicted) are important, especially as, globally, current coffee prices are low. Mechanisms by which farmers could benefit financially from the environmental services (i.e. the increase in terrestrial carbon stocks (van Noordwijk et al., 2002) and maintenance of bird and other diversity) that their shade-coffee systems provide would be very welcome. Use of these mechanisms could further solidify the farmers' choice for environmentally friendly land-use practices.

Although the Sumberjaya subdistrict has experienced some of the most violent conflict, the conflicts over watershed function experienced by farmers in the upper watersheds have been repeated elsewhere, both in Indonesia and in other densely populated parts of Asia.

### 20.3.2 Challenges and opportunities for negotiation support informed by 'science'

The main opportunity Sumberjaya currently offers is that of learning from the process of negotiating land-use rights within the 'protection forest' domain. The primary rewards in such a situation are 'recognition' and 'loss of fear of eviction'. As the economic opportunities for productive land use are considerable, such rewards may be sufficient inducement at this stage for a switch to more environmentally friendly land-use practices. However, this still requires a degree of community coherence and collaborative action, which can be a challenge in areas where migrations have occurred recently. Research aimed at verifying and improving simple criteria could facilitate the process of switching to more environmentally friendly land-use practices, e.g. by blending our understanding of below-ground interactions, surface phenomena, agricultural productivity and landscape-level land-use planning. Such simple criteria could, in turn, be useful both in negotiations and in the subsequent monitoring of effects.

## 20.4 Example 3. Alternatives to Slash-and-burn in the Western Amazon Basin

The Amazon basin is the largest remaining area of intact tropical rainforest, although forest conversion is 'nibbling' at its outer edges and along all major new access routes. Much of the converted forest is used for pasture, which rapidly degrades in terms of productivity and which will only support low stocking rates per hectare and low human population densities. This contrasts with the 'slash-and-burn' conversion of forests in central Africa and South-East Asia, where the predominance of tree crops in land-use systems after forest conversion supports much higher population densities and where logging is a more important 'trigger' factor, creating and paying for access. Although the rate at which the Amazon is being 'colonized' is largely driven by macro-economic policies (increasing road access, subsidizing *colonistas*, and the pricing of main agricultural outputs), the wastefulness of converting forest into low-intensity land-use systems has stimulated efforts to develop more sustainable forms of land use in the form of 'improved pasture' or 'tree crops'. The widespread expectation that agricultural intensification can help relieve the pressure being placed on natural forests is based on an oversimplification. Agricultural intensification is certainly not in itself *sufficient* for the protection of old-growth forests (Tomich *et al.*, 1998; Angelsen and Kaimowitz, 2001; Lee and Barrett, 2001). Such intensification can, however, help in the context of specific forest protection, spatial land-use planning linked to the selective stimulation of road access, and a policy framework that balances economic growth and resource protection (Wunder, 2003).

### 20.4.1 Ecological problems associated with land-use intensification

The productivity of pastures derived from forest declines rapidly, unless specific efforts are made to introduce improved pasture grasses and adequate management practices (Toledo *et al.*, 1985). Nematode infestation (see Chapter 15, this volume), nutrient (N and P) deficiencies, overgrazing – which results in soil compaction and the predominance of a single species of (introduced) earthworm (*Pontoscolex corethrurus*) (Chapter 16, this volume) – can all be associated with pasture degradation. The recent use of spatial analysis in the state of Acre (Brazil) has indicated that, if planted with the main forage grass (*Brachiaria brizantha*), 72% of the area has soils that demonstrate a high risk of pasture degradation, whilst 9% demonstrate a medium risk and 20% a low risk (Valentim *et al.*, 2000). A better understanding of below-ground interactions is necessary for more sustainable pasture management.

An initial approach to the problem of pasture degradation is the development of mixed-species swards through the integration of pasture legumes. Such legume-based pastures have been shown to be more sustainable in relation to N cycling (Cadisch *et al.*, 1994b) and able to support a higher diversity of fauna (Decaëns *et al.*, 1994), but legume persistence is still a major challenge. One of the main agricultural alternatives to pasture development is the development of sustainable tree-crop production systems. Rather than the more intensive monoculture system of coffee production, agroforestry forms are receiving increasing attention. Of the 44,000 ha of agroforestry systems in the Amazonian state of Rondônia, 45% have coffee as one of their components. However, in this area, associations of trees and coffee (*Coffea canephora*) have only recently received research attention. Farmers quote the fluctuation of coffee prices, the forecast of persistently low prices in the coming years and the increase in production costs associated with using inputs (fertilizers, herbicides, fungicides, etc.) as the main reasons for the inclusion of trees among the coffee. Cordia wood (*Cordia alliodora*), bandarra (*Schizolobium amazonicum*), rubber (*Hevea brasiliensis*), pinho cuiabano (*Parkia multijuga*) and peach palm (*Bactris gasipaes*) are among the trees preferred.

### 20.4.2 How will an understanding of below-ground processes help?

The main limitations that farmers in this area mention as being those that affect the establishment of forest species in conjunction with coffee are: (i) a lack of information on the growth, planting density and the spacing of trees associated with coffee; (ii) a lack of seed; and (iii) the mishandling of seed (seeds of many species quickly lose their viability). A range of forest species exists among the 'forest coffee' systems. The decision to associate trees with coffee has been found, in this area, to depend more on socioeconomic factors than on biophysical ones, even though most of the producers interviewed (70%) were aware that the presence of trees benefits the environment. The choice of species used depends on the seeds and seedlings that happen to be available: farmers do not take into account the root type, crown form and size of the tree, even though their objective is, always, to produce wood for the market.

Many practical aspects of the interactions between trees and coffee remain obscure (Rodrigues *et al.*, 2002). Research is now beginning to consider the matter at a more fundamental level. For example, as part of a broader ecological comparison, researchers compared mycorrhizal spore diversity between coffee agroforestry and coffee monocultures in Rondônia (Brazil). This research found that, whereas spore numbers were lowest in coffee monoculture, spore diversity was lowest in the combination of pinho cuiabano (*Parkia multijuga*) and coffee (compare Chapter 14, this volume). Neither the expectation that complex systems will directly solve the problems associated with low nutrient availability and disease pressure, nor the thesis that they will only lead to competition, is particularly useful. Finding suitable combinations requires a careful search, especially in the case of migrant communities, which do not have access to traditional ecological knowledge of tree–tree interactions.

## 20.5 In Praise of Complexity?

### 20.5.1 Ecological versus agronomic functions

In the first chapter of this book, Ong *et al.* started, using a relatively simple method, to unravel the positive and negative aspects of plant–plant interactions, in so far as they determine the overall production of agroecosystems that have more than one component. Most of that analysis targeted tree + annual food crop systems; however, the basic concepts used are also valid for combinations of perennials, such as those found in multistrata coffee systems. Empirical separation of positive and negative effects has been found to be feasible, though the results are site-specific. Extrapolation of such 'interactions' to other conditions is difficult as, with relatively small shifts in overall growth-limiting factors, the interactions can change in character. The classical paradigm, of developing technology in pilot sites and then extrapolating ('scaling out') to similar places may be less feasible for more complex systems than it is for 'simple' agroecosystems.

Rather than trying all $2^n$ possible interactions in a system with $n$ components (which is difficult even if $n$ is only 3 or 4) we may benefit from using summary characteristics, such as system-level productivity and system-level environmental impact. Expanding on an analysis by Sitompul (2002), we can take the classical trade-off analysis between two components of an intercropping system and apply it to the 'relative ecological function' (REF) versus the 'relative agronomic function' (RAF) of agroecosystems. In schematic form, in the case of agroecosystems in dynamic landscapes that are undergoing intensification, a number of stages (A to E in Fig. 20.3) can be distinguished. During a trajectory such as that shown here, the trade-off between the REF and RAF alternates between being negative (lose–win), positive (lose–lose, win–win) and neutral (no change, win).

The following attributes of an agroecosystem can be included in the 'ecological function'.

- Local nutrient cycling and capture of nutrients before they leave the system via vertical or lateral flows of water or in a gaseous form (in which case they escape into the atmosphere). In Chapters 6 and 10, we saw that systems that include deeply rooted components with an undersaturated 'demand' for nutrients can indeed play a significant role in this regard.
- Full use of incoming water, as long as annual rainfall is less than evaporative demand. If nearly all incoming water is used by an agroecosystem in approximately the same way as the natural vegetation to which the landscape has adjusted, outgoing lateral flows of water will be small. This is important, because an increase in such flows can bring salt into circulation and have other negative effects downstream (Chapters 9 and 18, this volume).
- A level of organic inputs that allows the 'ecosystem engineers' among the soil fauna to play their role in the creation and restoration of soil structure, complementing plant roots that create channels in the soil and improve its structure by causing wetting/drying cycles (Chapters 11 and 16, this volume).
- Permanence of a litter layer on the surface of the soil, which protects the mineral soil layers from erosion through wind or water (Chapters 12, 18 and 19, this volume).

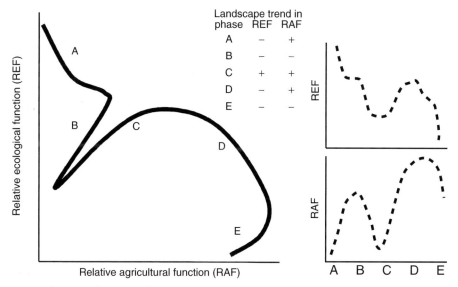

**Fig. 20.3.** Schematic relationships between REF (relative ecological function) and RAF (relative agronomic function) during a trajectory of 'agricultural intensification' (A–E); the diagram suggests that the overall trade-off is a negative one and usually the two move in opposite directions, but that phases where both REF and RAF decline (letters B, E) or increase (letter C) can defy this overall trend.

- A food web and ecological pyramid that: (i) is close to the potential dictated by energy flow in organic inputs to the below-ground system; and (ii) maintains internal controls that make pest population explosions (pest outbreaks) less likely (Chapters 15 and 16, this volume).

### 20.5.2 How much diversity is enough?

Although considering the broad trends of ecological and agronomic functions can both help us understand the big picture and help us identify those conditions in which a (temporary) 'win–win' situation is possible (such as phase C in Fig. 20.3), we may need to be more specific when considering the opportunities for beneficial interactions. Generally 'diversity' is associated with a 'cost' to agronomic functions, because of a quantity–quality trade-off that affects the use that farmers, as managers of agroecosystems, can make of their knowledge, and because of 'economies of scale' in mechanization, which favour more uniform crops and farm animals. So, how much diversity is enough?

Diversity of natural ecosystems tends to have a non-linear relationship with resource availability and levels of environmental 'stress' factors. The most diverse systems generally do not have the highest standing biomass or biological productivity (Fig. 20.4). Van Noordwijk and Ong (1999) tried to answer the question of how much diversity is actually needed to reduce the risk of production failure given the degree of variability that the environment imposes, which creates uncertainty for the farmer. As uncertainty increases the relevance of diversity, one can speculate that we will need more biodiversity in future, as global climate predictions for many tropical countries forecast more 'extreme' (less certain or predictable) weather conditions and hence increased stress and risk.

The results presented in this book show that below-ground diversity is, in fact, much greater than above-ground diversity. At the same time, however, there appear to be large 'redundancies', or at least aspects of diversity for which no functional value can be quantitatively assigned: we still know little about how much of this below-ground biodiversity

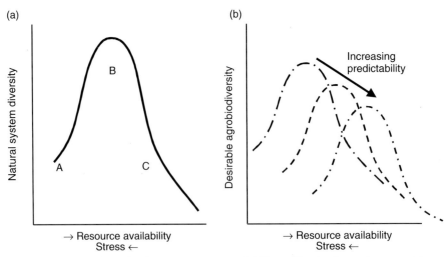

**Fig. 20.4.** (a) Schematic relationship between resource availability and levels of stress factors and the diversity of natural ecosystems (A, poor sites with only a few specialist species; B, intermediate sites with high diversity; C, rich sites with only a few dominant species). (b) Transfer of this concept to agroecosystems suggests that the optimum amount of diversity depends on the resource supply and stress factors, and on the predictability of the environment and the way products of agroecosystems are valued in human society.

we need for the system to be resilient. Some of the recent research presented here has shown that, despite drastic changes in land use, a large proportion of the organisms either remains present (in an active or dormant state), or is replaced by other organisms in the same 'functional' group.

We do, however, know that there exist a few essential specialists, which have a fundamental, unique impact (e.g. rhizobia, mycorrhizas, nitrifiers, and some pests, such as nematodes) and which today probably represent the most promising target group for interventions (e.g. improved management) by farmers. Examples of pest incidence still give the most compelling case for adopting a diverse system. Such a system need not necessarily be 'vertically' diverse (e.g. multistrata); the need could be fulfilled equally well by 'lateral' biodiversity (e.g. sequential systems) – the latter being easier for farmers to manage and mechanize. Even though 'vertical biodiversity' can lead to intense competition events that are often difficult to manage, or whose management is labour intensive, new approaches are emerging. In intercropping systems, there is a trend away from 1:1 species row schemes towards a system that reduces the interface between species by employing alternative options (such as 2 : 2, 4 : 2 or similar strip designs). These systems appear to be less risky and more profitable for farmers, and so merit further investigation. Mechanization of farm operations may also play a role in changing planting patterns (e.g. where bullock ploughing replaces manual tillage).

### 20.5.3 Knowledge

In the public debate on the 'desirability' of changes in agroecosystems, we may have to acknowledge important differences between the various interacting knowledge systems (Fig. 20.5). Such differences are related to whether a knowledge system is primarily based on categories and definitions or on observable phenomena and processes, plus inferences made regarding underlying processes and 'balance sheets'.

Farmer management of complex agroecosystems has to be based on monitoring, diagnosis, remediation, mitigation and adaptation, rather than on a blueprint predictability of the behaviour of the

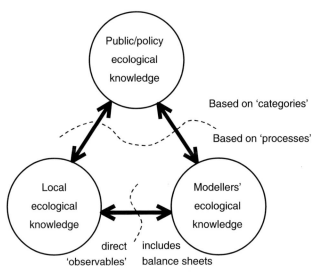

**Fig. 20.5.** Three types of ecological knowledge that interact in the public debate on the 'desirability' of changes in agroecosystems.

agroecosystem. The agricultural research approach, which involves clear-cut hypotheses and empirical tests of interventions in the form of technologies, has had a poor track record in this regard.

Increased and more quantitative understanding of below-ground interactions can help farmers improve their agroecosystems as long as it is:

- Able to provide explanatory knowledge, augmenting existing farmer knowledge of mechanisms.
- Articulated in ways farmers can relate to their own understanding of trade-offs between current productivity, lateral flows and long-term resource conservation.
- Phrased as generic principles that have a value that goes beyond a specific site.
- Embedded in diagnostic tools that can identify what we learn from and/or extrapolate to other places if we know a certain 'system' works at a given place.

As explored by Matthews *et al.* (Chapter 3, this volume) current simulation models still have some way to go before they will fully meet the above requirements. However, they have still become an essential part of the overall knowledge exchange. Models are not only a powerful tool with which to assess multiple interactions that would otherwise be too complex to evaluate but, crucially, they also allow us to quantify the importance (or unimportance) of potential interactions and improved resource uses. In future, this may help us better prioritize resources so that a significant impact can be achieved, rather than chasing scientifically interesting interactions that are of little consequence to the sustainability of the system or the livelihood of farmers. In this respect, recognition of who our clients are, and thus for whom such tools are being developed, is already pointing towards a more focused approach. But, we still lack the data necessary to better validate and calibrate our models.

For relatively simple systems, one can envisage simulation models that evaluate all feasible combinations of management interventions and search for the best 'package' of available options. In more complex systems, however, the number of feedback effects related to any farmer management decision is substantial, and we need to include the farmer's decision rules in the model. Models of such systems need to include equations that represent how farmers are likely to take decisions, based on information derived from the actual performance of the system and external variables. Simulations can then focus on the parameters of these decision rules rather than on a calendar listing of all the individual decisions. For example, rather than stating the desired frequency of pruning, a model user specifies at what size of tree and intensity of shading an intervention is desirable. This will then increase the actual pruning frequency if the growth rate of the tree increases. Pruning the tree has various effects on the other components of the system, via light intensity at ground (or crop) level, as well as the regrowth pattern of the tree, the 'leakiness' of the system in terms of water and nutrients (Chapter 10, this volume), the expected time till the next pruning intervention, and the use of labour, etc. Based on all such consequences, the model can evaluate the overall effect of the decision rules, and help to optimize management decisions.

## 20.6 Challenges for the future

The preceding chapters have yielded some conclusions that may have surprised the reader. Some of these deserve to be highlighted here.

- Legume-based systems are more 'leaky', in terms of leaching and greenhouse gas emissions, than low/no input conventional systems. However, when we consider that we need to intensify production in order to increase food supply, it is likely that legume-based systems will fare equally as well, or better, than traditional, fertilizer-based systems.
- Although we have in the past given a lot of attention to the use of spatial complementarity to capture resources (water, nutrients) located at depth, we have so far put little effort into understanding how these resources are recharged. This is a challenge, and points to the need to look at lateral flows and catchment events. Here we need to see erosion and leaching in a wider context (e.g. what is one farmer's loss may be a gain for the farmer further downstream).
- Management of water is critical (e.g. with respect to erosion, leaching, competition, P mobilization, rooting pattern and lateral flows). Given the predicted changes in global warming, water management will become even more important.
- What is or is not acceptable with regard to competition effects depends on the value of the components. However, the challenge associated with trees is that an originally attributed value (e.g. market value) may change during the lifetime of the tree and hence change the balance that governs what is viable and what enhances livelihoods of farmers. Thus, crucially, all components of a mixed-species system should have a considerable 'value' of their own. Then, not only does biodiversity provide a 'safety net' against risk, it also provides an economic buffer against human-induced changes.

In the preceding chapters, a large number of recommendations for further research have been made. Overall, we think that static concepts and tests (such as the 10–14 hypotheses formulated about agroforestry a decade ago, see Young, 1987) have played their role in stimulating research, and are no longer of much use. Instead, new experimental and modelling approaches are required, both at the level of system dynamics and at the level at which interactive management decisions are made by farmers.

Even though we have become increasingly aware of the potential that diverse systems have in terms of reducing external effects and increasing resource use efficiency, we still face the problem of actually measuring such events (particularly leaching and nutrient capture in 'filter' zones). To effectively manage a complex agroecosystem, a farmer needs ways to monitor or measure the real-time performance of the system. Currently, the way dynamic elements of below-ground growth and interactions are measured at the research level is very time and labour consuming. New methods need to be developed that will allow researchers to make real-time measurements with automated analytical software, allowing large datasets to be processed efficiently and immediately and so meaning those data would not take months or years to analyse. Examples include continuous non-destructive monitoring of root dynamics, soil solution nutrient concentrations, and water availability. New approaches are currently being developed, which allow, at least partially, more real-time observations such as these to be made (e.g. spectrometric approaches or non-destructive observations via tomography). However, all of these new 'gadgets' and tools may widen the gap between what a farmer understands and appreciates about a system and what a researcher can see. Therefore, efforts are needed to 'calibrate' such detailed methods against the simpler concepts that farmers use.

Ultimately, decades of research on managed tropical systems has given way to the recognition that the relevance of multispecies systems is largely determined by the preferences and priorities of farmers or communities. Thus a feasible beneficial species combination is not necessarily in itself sufficient argument for adoption. The various facets of a farmer's livelihood are thus crucial elements, and need to be considered when evaluating mixed-species systems.

But it is not only farmers who wish to capitalize on the benefits of mixed-species systems. In most societies, recognition is growing that natural and managed ecosystems have value beyond the farm products they provide. In this context, mixed-species systems may play an important role in providing biodiversity for recreational purposes, as well as being more efficient with regard to reducing negative externalities (e.g. pollution and greenhouse gases), which will help to provide a more stable world ecosystem. These increasingly and equally important considerations will drive research further away from the purely plot-oriented approach, towards approaches that consider the matter both at a landscape level and, finally, from a global perspective.

**Conclusions**

**1.** Multispecies agroecosystems, with their potential for synergy in terms of below-ground interactions, can offer improved farmer livelihoods and sustainability and basic ecosystem functions, at levels of complexity far below those of natural ecosystems.
**2.** Farmers increase or decrease the complexity of agroecosystems depending on their livelihood strategies. This has consequences for other stakeholders in the landscape, in terms of below-ground and above-ground lateral flows of water, nutrients, soil and organisms.
**3.** Understanding the root causes of land-use problems and the different ways they are perceived by different farmers, as well as by other local stakeholders, scientists and policy shapers, is essential for the development and introduction of sustainable solutions.

**Future research needs**

**1.** Identification of a wider portfolio of simple, manageable and profitable multispecies options, which exploit synergies in below-ground interactions, and which serve farmers' livelihoods and provide beneficial environmental service functions.
**2.** Development of methods that researchers can use to support farmers as managers of 'complex' agroecosystems. Rather than being based on 'packaged technology' options, such methods should be based on an understanding of basic principles, and should utilize tools for monitoring and evaluating the current status of the system.
**3.** Development of improved tools to monitor the impacts, at the landscape scale, of land-use decisions made at the plot scale.
**4.** Development of better ways to communicate results between different stakeholders.

# References

Abao, E.B. Jr, Bronson, K.F., Wassmann, R. and Singh, U. (2000) Simultaneous records of methane and nitrous oxide emissions in rice-based cropping systems under rain fed conditions. *Nutrient Cycling in Agroecosystems* 58, 131–139.

Abawi, G.S. and Chen, J. (1998) Concomitant pathogen and pest interactions. In: Barker, K.R., Pederson, G.A. and Windham, G.L. (eds) *Plant and Nematode Interactions*. Agronomy Monograph no. 36. ASA, CSSA and SSSA, Madison, Wisconsin, pp. 135–158.

Achouak, W., Christen, R., Barakat, M., Martel, M.H. and Heulin, T. (1999) *Burhoilderia caribensis* sp. Nov. exopolysaccharide-producing bacterium isolated from Vertisol microaggregates in Martinique. *International Journal of Systematic Bacteriology* 49, 787–794.

Adegbidi, H.G., Volk, T.A., White, E.H., Abrahamson, L.P., Briggs, R.D. and Bickelhaupt, D.H. (2001) Biomass and nutrient removal by willow clones in experimental bioenergy plantations in New York State. *Biomass and Bioenergy* 20, 399–411.

Adhya, T.K., Bharati, K., Mohanty, S.R., Mishra, S.R., Ramakrishnan, B., Rao, V.R., Sethunathan, N. and Wassmann, R. (2000) Methane emission from rice fields at Cuttack, India. *Nutrient Cycling in Agroecosystems* 58, 95–105.

Ae, N. and Otani, T. (1997) The role of cell wall components from groundnut roots in solubilizing sparingly soluble phosphorus in low fertility soils. *Plant and Soil* 196, 265–270.

Ae, N., Arihara, J., Okada, K., Yoshihara, T., Otani, T. and Johansen, C. (1993) The role of piscidic acid secreted by pigeon pea roots grown on an Alfisol with low-P fertility. In: Randall, P.H., Delhaize, E., Richards, A.R. and Munns, E. (eds) *Genetic Aspects of Plant Nutrition*. Kluwer Academic Publishers, Dordrecht, The Netherlands, pp. 279–288.

Agrawal, A. (1995) Dismantling the divide between indigenous and scientific knowledge. *Development and Change* 26, 413–439.

Agrios, G.N. (1997) *Plant Pathology*, 4th edn. Academic Press, San Diego, California, 635 pp.

Agus, F., Gintings, A.N. and van Noordwijk, M. (2002) *Pilihan Teknologi Agroforestri/Konservasi Tanah untuk Areal Pertanian Berbasis Kopi di Sumberjaya, Lampung Barat* [in Indonesian]. ICRAF-SE Asia, Bogor, Indonesia.

Aitken, R.L., Moody, P.W. and McKinley, P.G. (1990) Lime requirement of acidic Queensland soils. 1. Relationships between soil properties and pH buffer capacity. *Australian Journal of Soil Research* 28, 695–701.

Akinnifesi, F.K., Kang, B.T. and Tijani-Eniola, H. (1995) Root size distribution of a *Leucaena leucocephala* hedgerow as affected by pruning and alley cropping. *Nitrogen Fixing Tree Research Reports* 13, 65–69.

Akinnifesi, F.K., Kang, B.T. and Tijani-Eniola, H. (1996) Root–soil interface in agroforestry systems. In: Agboola, A.A., Sobulo, O. and Obatolu, C.R. (eds) *Proceedings of the Third African Soil Science Society Conference*, Lagos, Nigeria, pp. 201–210.

Akinnifesi, F.K., Smucker, A.J.M. and Kang, B.T. (1999a) Belowground dynamics in agroforestry systems. *Annals of Arid Zone* 38, 239–273.

Akinnifesi, F.K., Kang, B.T. and Ladipo, D.O. (1999b) Structural root form and fine root distribution of some woody species evaluated for agroforestry systems. *Agroforestry Systems* 42, 121–138.

Akinnifesi, F.K., Tijani-Eniola, H. and Adesanya, O.O. (1999c) Soil impedance on early root growth and nodulation in *Enterolobium cyclocarpum* and *Leucaena leucocephala*. *Forest, Farm and Community Trees Research* 4, 28–32.

Akinnifesi, F.K., de Araujo, M.A. and Moura, E.G. (1999d) Root distribution of *Cajanus cajan* and alley-cropped maize in response to inter-hedgerow spacing. *Forest, Farm and Community Trees Research* 4, 64–67.

Alexander, I., Ahmad, N. and See, L.S. (1992) The role of mycorrhizas in the regeneration of some Malaysian forest trees. *Philosophical Transactions of the Royal Society of London, Series B* 335, 379–388.

Allen, M.F. (ed.) (1992) *Mycorrhizal Functioning*. Chapman and Hall, New York, 534 pp.

Allen, O.N. and Allen, E.K. (1981) *The Leguminosae. A Source Book of Characteristics, Uses and Nodulation*. University of Wisconsin Press, Wisconsin, 812 pp.

Altieri, M.A. (1991) How best can we use biodiversity in agroecosystems? *Outlook in Agriculture* 20, 15–23.

Altieri, M.A. and Liebman, M.Z. (1986) Insect, weed and plant disease management in multiple cropping systems. In: Francis, C.A. (ed.) *Multiple Cropping Systems*. Macmillan, New York, pp. 183–218.

Anderson, G. (1980) Assessing organic phosphorus in soils. In: Khasawneh, F.E., Sample, E.C. and Kamprath, E.J. (eds) *The Role of Phosphorus in Agriculture*. American Society of Agronomy, Madison, Wisconsin, pp. 411–431.

Anderson, L.S. and Sinclair, F.L. (1993) Ecological interactions in agroforestry systems. *Agroforestry Abstracts* 6(2), 57–91.

Anderson, L.S., Muetzelfeldt, R.I. and Sinclair, F.L. (1993) An integrated research strategy for modelling and experimentation in agroforestry. *Commonwealth Forestry Review* 72(3), 166–174.

Andrade, D.S., Murphy, P.J. and Giller, K.E. (2002) Effects of liming and legume/cereal cropping on populations of indigenous rhizobia in an acid Brazilian Oxisol. *Soil Biology and Biochemistry* 34, 477–485.

Andreae, M.O. and Merlet, P. (2001) Emission of trace gases and aerosols from biomass burning. *Global Biogeochemical Cycles* 15, 955–966.

Angelsen, A. and Kaimowitz, D. (eds) (2001) *Agricultural Technologies and Tropical Deforestation*. CAB International, Wallingford, UK, 422 pp.

Angers, D.A. and Caron, J. (1998) Plant-induced changes in soil structure: processes and feedbacks. *Biogeochemistry* 42, 55–72.

Araujo, A.P., Teixeira, M.A. and de Almeida, D.L. (1998) Variability of traits associated with phosphorus efficiency in wild and cultivated genotypes of common bean. *Plant and Soil* 203, 173–182.

Archer, E., Swanepoel, J.J. and Straus, H.C. (1988) Effect of plant spacing and trellising systems on grapevine root distribution. In: van Zyl, J.L. (ed.) *The Grapevine Root and its Environment*. Department of Agriculture and Water Supply Technical Communication No. 215, Pretoria, South Africa, pp. 74–87.

Arihara, J., Ae, N. and Okada, K. (1991) Root development of pigeonpea and chickpea and its significance in different cropping systems. In: Johansen, C., Lee, K.K. and Sahrawar, K.L. (eds) *Phosphorus Nutrition of Grain Legumes in the Semiarid Tropics*. ICRISAT, India, pp. 183–194.

Ashley, C. and Carney, D. (1999) *Sustainable Livelihoods: Lessons from Early Experience*. Department for International Development, London, UK, 55 pp.

Asturias, M.A. (1949) *Men of Maize*. Editorial Losada, SA, Buenos Aires, Argentina, 335 pp.

Azam, F., Malik, K.A. and Sajjad, M.I. (1985) Transformations in soil and availability to plants of $^{15}N$ applied as inorganic fertilizer and legume residues. *Plant and Soil* 86, 3–13.

Azam-Ali, S.N., Crout, N.M.J. and Bradley, R.G. (1994) Perspectives in modelling resource capture by crops. In: Monteith, J.L., Scott, R.K. and Unsworth, M.H. (eds) *Resource Capture by Crops*. Nottingham University Press, Nottingham, UK, pp. 125–148.

Baanante, C.A. (1997) Economic evaluation of the use of phosphate fertilizers as a capital investment. In: Johnston, A.E. and Syers, J.K. (eds) *Nutrient Management for Sustainable Food Production in Asia*. CAB International, Wallingford, UK, pp. 109–120.

Baggs, E.M., Rees, R.M., Smith, K.A. and Vinten, A.J.A. (2000) Nitrous oxide emission from soils after incorporating crop residues. *Soil Use and Management* 16, 82–87.

Baggs, E.M., Millar, N., Ndufa, J.K. and Cadisch, G. (2001) Effect of residue quality on $N_2O$ emissions from tropical soils. In: Rees, R.M., Ball, B.C., Campbell, C. and Watson, C.A. (eds) *Sustainable Management of Soil Organic Matter*. CAB International, Wallingford, UK, pp. 120–125.

Baggs, E.M., Cadisch, G., Stevenson, M., Pihlatie, M., Regar, A. and Cook, H. (2003) Nitrous oxide emissions resulting from interactions between cultivation technique, residue quality and fertiliser application. *Plant and Soil* 254, 361–370.

Bago, B. (2000) Putative sites of nutrient uptake in arbuscular mycorrhizal fungi. *Plant and Soil* 226, 263–274.

Bago, B., Azcón-Aguilar, C., Goulet, A. and Piché, Y. (1998) Branched absorbing structures (BAS): a feature of the extraradical mycelium of symbiotic arbuscular mycorrhizal fungi. *New Phytologist* 139, 375–388.

Baker, K.F. and Cook, R.J. (1974) *Biological Control of Plant Pathogens*. W.H. Freeman, San Francisco, California, 433 pp.

Bala, A. (1999) Biodiversity of rhizobia which nodulate fast growing tree legumes in tropical soils. PhD thesis, Imperial College London, Wye Campus, UK, 216 pp.

Bala, A. and Giller, K.E. (2001) Symbiotic specificity of tropical tree rhizobia for host legumes. *New Phytologist* 149, 495–507.

Barber, S.A. (1984) *Soil Nutrient Bioavailability*. John Wiley & Sons, New York, 353 pp.

Barber, S.A. (1995) *Soil Nutrient Bioavailability, a Mechanistic Approach*, 2nd edn. John Wiley & Sons, New York, 414 pp.

Barraclough, P.B. and Tinker, P.B. (1981) The determination of ionic-diffusion coefficients in field soils. 1. Diffusion-coefficients in sieved soils in relation to water-content and bulk-density. *Journal of Soil Science* 32, 225–236.

Barrios, E., Kwesiga, F., Buresh, R.J. and Sprent, J.I. (1997) Light fraction soil organic matter and available nitrogen following trees and maize. *Soil Science Society of America Journal* 61, 826–831.

Barros, M.E. (1999) Effet de la Macrofaune sur la Structure et les Processus Physiques du Sol de Pâturages Dégradés d'Amazonie. PhD Thesis, Université de Paris VI, Paris, France.

Barros, R.S., Maestri, M. and Rena, A.B. (1995) Coffee crop ecology. *Tropical Ecology* 36, 1–19.

Barrow, N.J. (1980) Evaluation and utilisation of residual phosphorus in soils. In: Khasawneh, F.E., Sample, E.C. and Kamprath, E.J. (eds) *The Role of Phosphorus in Agriculture*. American Society of Agronomy, Madison, Wisconsin, pp. 333–359.

Basso, B., Ritchie, J.T., Gallant, J.C. and Baer, B.D. (2000) Digital terrain modelling to predict spatial variability of soil water balance. In: *2000 ASAE Annual International Meeting, Milwaukee, Wisconsin, USA, 9–12 July 2000*. ASAE Paper No. 003041. American Society of Agricultural Engineers, St Joseph, USA, pp. 1–15.

Batjes, N.H. (1996) Total carbon and nitrogen in the soils of the world. *European Journal of Soil Science* 47, 151–163.

Baujard, P. and Martiny, B. (1994) Transport of nematodes by wind in the peanut cropping area of Senegal, West Africa. *Fundamental and Applied Nematology* 17, 543–550.

Beck, M.A. and Sanchez, P.A. (1994) Soil phosphorus movement and budget during 18 years of cultivation on a Typic Paleudult. *Soil Science Society of America Journal* 58, 1424–1431.

Begon, M., Harper, J.L. and Townsend, C.R. (1996) *Ecology: Individuals, Populations and Communities*, 3rd edn. Blackwell Science, Oxford, UK, 1068 pp.

Beinroth, F.H., Jones, J.W., Knapp, E.B., Papajorgji, P. and Luyten, J. (1998) Evaluation of land resources using crop models and a GIS. In: Tsuji, G.Y., Hoogenboom, G. and Thornton, P.K. (eds) *Understanding Options for Agricultural Production. Systems Approaches for Sustainable Agricultural Development*. Kluwer Academic Publishers, Dordrecht, The Netherlands, pp. 293–311.

Bekunda, M. (2000) TSBF action to integrate management of soil fertility and pests for better crop production. *TSBF Comminutor* 5(3), 10–13.

Bengough, A.G., Castrignano, A., Pages, L. and van Noordwijk, M. (2000) Sampling strategies, scaling and statistics. In: Smit, A.L., Bengough, A.G., Engels, E., van Noordwijk, M., Pellerin, S. and van de Geijn, S.C. (eds) *Root Methods: a Handbook*. Springer-Verlag, Berlin, Germany, pp. 147–173.

Bentley D., Grierson, P.F., Bennett, L.T. and Adams, M.A. (1999) Anion exchange membranes to estimate bioavailable P in native grasslands of semi-arid northwestern Australia. *Communications in Soil Science and Plant Analysis* 30 (15, 16), 2231–2244.

Berkes, F., Colding, J. and Folke, C. (2000) Rediscovery of traditional ecological knowledge as adaptive management. *Ecological Applications* 10(5), 1251–1262.

Bethlenfalvay, G.J. and Linderman, R.G. (eds) (1992) *Mycorrhizae in Sustainable Agriculture*. American Society of Agronomy Special Publication 54, Madison, Wisconsin, 124 pp.

Beven, K. and Germann, P. (1982) Macropores and water flow in soils. *Water Resources Research* 18, 1311–1325.

Bever, J.D., Schultz, P.A., Pringle, A. and Morton, J.B. (2001) Arbuscular mycorrhizal fungi: more diverse than meets the eye, and the ecological tale of why. *BioScience* 51, 923–931.

Bezdicek, D.F. and Granatstein, D. (1989) Crop rotation efficiencies and biological diversity in farming systems. *American Journal of Alternative Agriculture* 4, 111–119.

Bhogal, A., Murphy, D.V., Fortune, S., Shepherd, M.A., Hatch, D.J., Jarvis, S.C., Gaunt, J.L. and Goulding, K.W.T. (2000) Distribution of nitrogen pools in the soil profile of undisturbed and reseeded grasslands. *Biology and Fertility of Soils* 30, 356–362.

Bidartondo, M.I., Redecker, D., Hijri, I., Wiemken, A., Bruns, T.D., Domínguez, L., Sérsic, A., Leake, J.R. and Read, D.J. (2002) Epiparasitic plants specialized on arbuscular mycorrhizal fungi. *Nature* 419, 389–392.

Bignell, D.E., Tondoh, J., Dibog, L., Huang, S.P., Moreira, F., Nwaga, D., Pashanasi, B., Susilo, F.-X. and Swift, M. (2002) Belowground biodiversity assessment: the ASB functional group approach. In: Ericksen, P.J., Sanchez, P.A. and Juo, A. (eds) *Alternatives to Slash-and-Burn: a Global Synthesis*. American Society of Agronomy Special Publication, Madison, Wisconsin.

Birch, H.F. (1958) The effects of soil drying on humus decomposition and nitrogen availability. *Plant and Soil* 10, 9–31.

Black, A.S. and Waring, S.A. (1979) Adsorption of nitrate, chloride and sulfate by some highly weathered soils from south-east Queensland. *Australian Journal of Soils Research* 17, 271–282.

Black, C.R. and Ong, C.K. (2000) Utilisation of light and water in tropical agriculture. *Agricultural and Forest Meteorology* 104, 25–47.

Blake, C.D. (1966) The histological changes in banana roots caused by *Radopholus similis* and *Helicotylenchus multicinctus*. *Nematologica* 12, 129–137.

Blanchart, E. (1992) Role of earthworms in the restoration of the macro-aggregate structure of a destructured savanna soil under field conditions. *Soil Biology and Biochemistry* 24, 1587–1594.

Blanchart, E., Lavelle, P., Braudeau, E., Le Bissonnais, Y. and Valentin, C. (1997) Regulation of soil structure by geophagous earthworm activities in humid savannas of Côte d'Ivoire. *Soil Biology and Biochemistry* 29, 431–439.

Blanchart, E., Chevallier, T., Albrecht, A., Cabidoche, Y.M., Hartmann, C., Lavelle, P. and Feller, C. (2002) Consequences of earthworm (*Polypheretima elongata*) activities in the restoration of the properties of a Vertisol under pasture (Martinique). 7th International Symposium on Earthworm Ecology, Cardiff, UK, *Agriculture, Ecosystems and Environment*.

Blokker, K.J. (1986) A view of information technology as an aid to decision-making by farmers. *Research and Development Agriculture* 3, 79–81.

Blumenthal, J.M. and Russelle, M.P. (1996) Subsoil nitrate uptake and symbiotic dinitrogen fixation by alfalfa. *Agronomy Journal* 88, 909–915.

Boddington, C.L. and Dodd, J.C. (1999) Evidence that differences in phosphate metabolism in mycorrhizas formed by species of *Glomus* and *Gigaspora* may be related to their life-cycle strategies. *New Phytologist* 142, 531–538.

Bohm, W. (1979) *Methods of Studying Root Systems*. Ecological Studies, Springer-Verlag, Berlin, Germany, 188 pp.

Bolan, N.S., Hedley, M.J. and White, R.E. (1991) Processes of soil acidification during nitrogen cycling with emphasis on legume based pastures. *Plant and Soil* 134, 53–63.

Bowers, J.H. and Mitchell, D.J. (1990) Effect of soil–water matric potential and periodic flooding on mortality of pepper caused by *Phytophthora capsici*. *Phytopathology* 80, 1447–1450.

Boye, A. (2000) Soil erodibility under improved fallows practices. MSc thesis, University of Copenhagen, Copenhagen, Denmark, 112 pp.

Boye, A. and Albrecht, A. (2002) Soil erodibility control by short-term tree fallow under simulated rainfall conditions in western Kenya. Poster presented at 17th International Soil Science Congress, Bangkok, Thailand.

BPS (1999) Sumberjaya Dalam Angka 1998. Badan Pusat Statistik, Kabupaten Lampung Barat, 58 pp.

Braak, C. (1929) *The Climate of the Netherlands Indies*. Koninklijk Magnetisch en Meteorologisch Observatorium te Batavia, Verhandelingen No. 8.

Bradley, R.G. and Crout, N.M.J. (1994) *The PARCH Model for Predicting Arable Resource Capture in Hostile*

*Environments: Users' Guide.* Tropical Crops Research Unit, University of Nottingham, Sutton Bonington, Leicestershire, UK, 122 pp.

Brandon, N.J. and Shelton, H.M. (1997a) Factors affecting the early growth of *Leucaena leucocephala* 1. Effects of nitrogen, phosphorus, lime and irrigation at three sites in southeast Queensland. *Australian Journal of Experimental Agriculture* 37, 27–34.

Brandon, N.J. and Shelton, H.M. (1997b) Factors affecting the early growth of *Leucaena leucocephala*. Role of indigenous arbuscular mycorrhizal fungi and its importance in determining yield of leucaena in pots and in the field. *Australian Journal of Experimental Agriculture* 37, 45–53.

Bravdo, B.A., Levin, I. and Asaf, R. (1992) Control of root size and root environment of fruit trees for optimal fruit production. *Journal of Plant Nutrition* 15, 699–712.

Breman, H. and Kessler, J.J. (1995) *Woody Plants in Agro-Ecosystems of Semi-arid Regions.* Springer Verlag, Heidelberg/Berlin, Germany.

Brenner, A.J. (1996) Microclimatic modifications in agroforestry. In: Ong, C.K. and Huxley, P. (eds) *Tree–Crop Interactions: a Physiological Approach.* CAB International, Wallingford, UK, pp. 157–187.

Bridge, J. (1987) Control strategies in subsistence agriculture. In: Brown, R.H. and Kerry, B.R. (eds) *Principles and Practice of Nematode Control in Crops,* Academic Press, Marrickville, NSW, Australia, pp. 389–420.

Bridge, J. (1996) Nematode management in sustainable and subsistence agriculture. *Annual Review of Phytopathology* 34, 201–225.

Bridge, J. (1998) Theory and practice of non-chemical management of nematode pests in tropical farming systems. In: *Brighton Crop Protection Conference: Pests and Diseases – 1998 (Volume 3). Proceedings of an International Conference held in Brighton, UK, 16–19 November 1998.* British Crop Protection Council, Farnham, UK, pp. 761–768.

Bridge, J. (2000a) Nematodes of bananas and plantains in Africa: research trends and management strategies relating to the small-scale farmer. In: Karamura, E. and Vuylsteke, D. (eds) *Proceedings of the First International Conference on Banana and Plantain for Africa. Acta Horticulturae* 540, 391–408.

Bridge, J. (2000b) *The Assessment of Major Soil Borne Pests of the Multi-purpose Agroforestry Tree, Sesbania sesban, and the Development and Promotion of IPM Strategies. Final Technical Report.* Department for International Development (DFID) Crop Protection Programme, UK, 24 pp.

Bridge, J., Luc, M. and Plowright, R. (1990) Nematode parasites of rice. In: Luc, M., Sikora, R.A. and Bridge, J. (eds) *Plant Parasitic Nematodes in Tropical and Subtropical Agriculture,* 2nd edn. International Institute of Parasitology, CAB International, Wallingford, UK, pp. 69–108.

Broadhead, J.S., Ong, C.K. and Black, C.R., (2003). Tree phenology and water availability in semi-arid agroforestry systems. *Forest Ecology and Management* 180, 61–73.

Brockwell, J., Bottomley, P.J. and Thies, J.E. (1995) Manipulation of rhizobia microflora for improving legume productivity and soil fertility: a critical assessment. *Plant and Soil* 174, 143–180.

Bronson, K.F., Neue, H.U., Singh, U. and Abao, E.B. Jr (1997a) Automated chamber measurements of methane and nitrous oxide flux in a flooded rice soil: I. Residue, nitrogen and water management. *Soil Science Society of America Journal* 61, 981–987.

Bronson, K.F., Neue, H.U., Singh, U. and Abao, E.B. Jr (1997b) Automated chamber measurements of methane and nitrous oxide flux in a flooded rice soil: II. Fallow period emissions. *Soil Science Society of America Journal* 61, 988–993.

Brouwer, R. (1963) Some aspects of the equilibrium between overground and underground plant parts. *Jaarboek IBS Wageningen* 1963, 31–39.

Brouwer, R. (1983) Functional equilibrium: sense or nonsense? *Netherlands Journal of Agricultural Science* 31, 335–348.

Brown, G.G., Moreno, AG., Moller, J., Martinsen, H.J., Lavelle, P., Bueno, J., Irisson, S., Mwabvu, T., Socotela, S., Goma, H., Msumali, G., Carter, S., Ojiem, J., Chegue, A., Kamwe, M., Mwakalombe, B. and Odour, P. (1998) Soil fertility, soil macrofauna and greenhouse gas emissions under different land use systems in E and SE Africa. In: Carter, S. and Riley, J. (eds) *Biological Management of Soil Fertility in Small-scale Farming Systems in Tropical Africa.* EU Project Number TS3*-CT94-0337, Final Report, Rothamsted International, Harpenden, UK, pp. 89–121.

Brown, G.G., Barois, I. and Lavelle, P. (2000) Regulation of soil organic matter dynamics and microbial activity in the drilosphere and the role of interactions with other edaphic functional domains. *European Journal of Soil Biology* 36, 177–198.

Brown, V.K. and Gange, A.C. (1991) Effects of root herbivory on vegetation dynamics. In: Atkinson, D. (ed.) *Plant Root Growth: an Ecological Perspective.* Blackwell, Oxford, pp. 453–470.

Bruehl, G.W. (1987) *Soilborne Plant Pathogens*, 1st edn. Macmillan, New York, 368 pp.

Bruijnzeel, L.A.S. (1990) Hydrology of moist tropical forests and effects of conversion: a state of knowledge review. UNESCO, Paris.

Brundrett, M., Bougher, N., Dell, B., Grove, T. and Malajczuk, N. (1996) Working with mycorrhizas in forestry and agriculture. ACIAR Monograph 32, Canberra, Australia.

Brundrett, M.C. (2002) Coevolution of roots and mycorrhizas of land plants. *New Phytologist* 154, 275–304.

Brussaard, L., Behan-Pelletier, V.M., Bignell, D.E., Brown, V.K., Didden, W., Folgarait, P., Fragoso, C., Wall-Freckman, D., Gupta, V.V.S.R., Hattori, T., Hawksworth, D.L., Klopatek, C., Lavelle, P., Malloch, D.W., Rusek, J., Söderström, D., Tiedje, J.M. and Virginia, R.A. (1997) Biodiversity and ecosystem functioning in soil. *Ambio* 26, 563–570.

Buckles, D. and Triomphe, B. (1999) Adoption of mucuna in the farming systems of northern Honduras. *Agroforestry Systems* 47, 67–91.

Bullock, A., Keya, S.O., Muthuri, F.M., Baily-Watts, T., Williams, R. and Waughray, D. (1995) Lake Victoria Environmental Management Program Task Force 2. Final report by regional consultants on Tasks 11, 16 and 17 (Water quality, land use and wetlands). Centre for Ecology and Hydrology (CEH), Wallingford, UK and FAO, Rome.

Buresh, J.R. and Niang, I.A. (1997) *Tithonia diversifolia* as a green manure: awareness, expectations and realities. *Agroforestry Forum* 8, 29–30.

Buresh, R.J. and Tian, G. (1997) Soil improvement by trees in sub-Saharan Africa. *Agroforestry Systems* 38, 51–76.

Buresh, R.J., Smithson, P.C. and Hellums, D.T. (1997) Building soil phosphorus capital in Africa. In: Buresh, R.J., Sanchez, P.A. and Calhoun, F. (eds) *Replenishing Soil Fertility in Africa*. SSSA Special Publication 51, Soil Science Society of America, Madison, Wisconsin, pp. 111–149.

Burgess, S.S.O., Adams, M.A., Turner, N.C. and Ong, C.K. (1998) The redistribution of soil water by tree root systems. *Oecologia* 115, 306–311.

Burgess, S.S.O., Adams, M.A., Turner, N.C. and Ward, B. (2000) Characterisation of hydrogen isotope profiles in an agroforestry system: implications for tracing water sources of trees. *Agricultural Water Management* 45, 229–241.

Burgess, S.S.O., Adams, M.A., Turner, N.C., White, D.A. and Ong, C.K. (2001a) Tree roots: conduits for deep recharge of soil water. *Oecologia* 126, 158–165.

Burgess, S.S.O., Adams, M.A., Turner, N.C., Ong, C.K., Khan, A.A.H., Beverly, C.R. and Bleby, T.M. (2001b) An improved heat pulse method to measure low and reverse rates of sap flow in woody plants. *Tree Physiology* 21, 589–598.

Burnham, C.P. (1989) Pedological processes and nutrient supply from parent material in tropical soils. In: Proctor, J. (ed.) *Mineral Nutrients in Tropical Forest and Savanna Ecosystems*. Blackwell Scientific Publications, Oxford, UK, pp. 27–41.

Bushby, H.V.A., Date, R.A. and Butler, K.L. (1983) Rhizobium strain evaluation of *Glycine max* cv. Davis, *Vigna mungo* cv. regur and *V. unguiculata* cv. Caloona for three soils in glasshouse and field experiments. *Australian Journal of Experimental Agriculture and Animal Husbandry* 23, 43–53.

Buwalda, J.G. (1993) The carbon costs of root systems of perennial fruit crops. *Environmental and Experimental Botany* 33, 131–140.

Cadisch, G., and Giller, K.E. (eds) (1997) *Driven by Nature: Plant Litter Quality and Decomposition*. CAB International, Wallingford, UK, 409 pp.

Cadisch, G., Giller, K.E., Urquiaga, S., Miranda, C.H.B., Boddey, R.M. and Schunke, R.M. (1994a) Does phosphorus supply enhance soil-N mineralization in Brazilian pastures? *European Journal of Agronomy* 3, 339–345.

Cadisch, G., Schunke, R.M. and Giller, K.E. (1994b) Nitrogen cycling in a pure grass pasture and a grass-legume mixture on a red Latosol in Brazil. *Tropical Grasslands* 28, 43–52.

Cadisch, G., Imhof, H., Urquiaga, S., Boddey, R.M. and Giller, K.E. (1996) Carbon ($^{13}$C) and nitrogen mineralization potential of particulate light organic matter after rainforest clearing. *Soil Biology and Biochemistry* 28, 1555–1567.

Cadisch, G., Rowe, E. and Van Noordwijk, M. (1997) Nutrient harvesting – the tree-root safety net. *Agroforestry Forum* 8(2), 31–33.

Cadisch, G., Handayanto, E., Malama, C., Seyni, F. and Giller, K. E. (1998) N recovery from legumes prunings and priming effects are governed by the residue quality. *Plant and Soil* 205, 125–134.

Cadisch, G., Gathumbi, S.M., Ndufa, J.K. and Giller, K.E. (2002a) Resource acquisition of mixed species fallows – competition or complementarity? In: Vanlauwe, B., Diels, J., Sanginga, N. and Merckx, R. (eds) *Balanced Nutrient Management Systems for the Moist Savanna and Humid Forest Zones of Africa*. Kluwer Academic Publishers, Dordrecht, The Netherlands, pp. 143–154.

Cadisch, G., Ndufa, J.K., Yasmin, K., Mutuo, P., Baggs, E., Keerthisinghe, G. and Albrecht, A. (2002b) Use of stable isotopes in assessing belowground contributions to N and soil organic matter dynamics. In: International Union of Soil Science, The Soil and Fertilizer Society of Thailand, Ministry of Agriculture and Cooperatives of Thailand (eds) *17th World Soil Science Conference 'Soil Science: Confronting New Realities in the 21st Century'*. International Soil Science Society, Bangkok, Thailand, CD – Paper No. 1165, pp. 1–10.

Cahn, M.D., Bouldin, D.R., Cravo, M.S. and Bowen, W.T. (1993) Cation and nitrate leaching in an oxisol of the Brazilian Amazon. *Agronomy Journal* 85, 334–340.

Cairney, J.W.G. and Chambers, S.M. (eds) (1999) *Ectomycorrhizal Fungi – Key Genera in Profile*. Springer-Verlag, Heidelberg, Germany, 369 pp.

Calder, I.R. (2002) Forests and hydrological services: reconciling public and science perceptions. *Land Use and Water Resources Research* 2, 2.1–2.12 [online] (www.luwrr.com).

Calder, I.R., Rosier, P.T.W. and Prasanna, K.T. (1997) *Eucalyptus* water use greater than rainfall input – a possible explanation from southern India. *Hydrology and Earth Systems Science* 1, 249–256

Caldwell, M.M. and Richards, J.H. (1989) Hydraulic lift – water efflux from upper roots improves effectiveness of water uptake by deep roots. *Oecologia* 79, 1–5.

Caldwell, M.M., Dawson, T.E. and Richards, J.H. (1998) Hydraulic lift – consequences of water efflux from the roots of plants. *Oecologia* 113, 151–161.

Caldwell, R.M. and Fernandez, A.A.J. (1998) A generic model of hierarchy for systems analysis and simulation. *Agricultural Systems* 57(2), 197–225.

Caldwell, R.M. and Hansen, J.W. (1993) Simulation of multiple cropping systems with CropSys. In: Penning de Vries, F.W.T., Teng, P. and Metselaar, K. (eds) *Systems Approaches for Agricultural Development*. Kluwer Academic Publishers, Dordrecht, The Netherlands, pp. 397–412.

Campbell, C.L. and Benson, D.M. (eds) (1994) *Epidemiology and Management of Root Diseases*. Springer-Verlag, New York, 344 pp.

Campos, V.P., Sivapalan, P. and Gnanapragasam, N.C. (1993) Nematode parasites of coffee, cocoa and tea. In: Luc, M., Sikora, R.A. and Bridge, J. (eds) *Plant Parasitic Nematodes in Tropical and Subtropical Agriculture*, 2nd edn. International Institute of Parasitology, CAB International, Wallingford, UK, pp. 387–430.

Cannell, M.G.R., van Noordwijk, M. and Ong, C.K. (1996) The central agroforestry hypothesis: the tree must acquire resources that the crop would not otherwise acquire. *Agroforestry Systems* 34, 27–31.

Cannell, M.G.R., Mobbs, D.C. and Lawson, G.J. (1998) Complementarity of light and water use in tropical agroforests II. Modelled theoretical tree production and potential crop yield in arid to humid climates. *Forest Ecology and Management* 102(2–3), 275–282.

Carberry, P.S., Adiku, S.G.K., McCown, R.L. and Keating, B.A. (1996) Application of the APSIM cropping systems model to intercropping systems. In: Ito, O., Johansen, C., Adu-Gyamfi, J.J., Katayama, K., Kumar Rao, J.V.D.K. and Rego, T.J. (eds) *Dynamics of Roots and Nitrogen in Cropping Systems of the Semi-arid Tropics*. International Agricultural Series No. 3. Japan International Research Center for Agricultural Sciences, pp. 637–648.

Cardoso, I.M. (2002) Phosphorus in agroforestry systems: a contribution to sustainable agriculture in Zona da Mata of Minas Gerais, Brazil. PhD thesis, Wageningen Agricultural University, The Netherlands.

Carney, D. (1998) *Implementing the Sustainable Rural Livelihoods Approach, Sustainable Rural Livelihoods – What Contribution Can We Make?* Department for International Development, London, pp. 3–23.

Carsky, R.J., Nokoe, S. and Lagoke, S.T.O. (1998) Maize yield determinants in farmer-managed trials in the Nigerian northern Guinea savanna. *Experimental Agriculture* 34, 407–422.

Carter, W.E. (1969) *New Lands and Old Traditions. Kekchi Cultivators in the Guatemalan Lowlands*. University of Florida Press, Gainesville, Florida, 153 pp.

Castellanos, J., Jaramillo, V.J., Sanford, R.L., Jr Kauffman, J.B. (2001) Slash-and-burn effects on fine root biomass and productivity in a tropical dry forest ecosystem in Mexico. *Forest Ecology and Management* 148, 41–50.

Castro, M.S., Peterjohn, W.T., Melillo, J.M. and Steudler, P.A. (1994) Effects of nitrogen fertilization on the fluxes of $N_2O$, $CH_4$ and $CO_2$ from soils in a Florida slash pine plantation. *Canada Journal of Forest Research* 24, 9–13.

Cermak, J., Jenik, J., Kucera, J. and Zidek, V. (1984) Xylem water flow in a crack willow tree (*Salix fragilis* L.) in relation to diurnal changes in environment. *Oecologia* 64, 145–151.

Chang, S.C. and Jackson, M.L. (1957) Fractionation of soil phosphorus. *Soil Science* 84, 133–144.

Chapela, I.H., Osher, L.J., Horton, T.R. and Henn, M.R. (2001) Ectomycorrhizal fungi introduced with exotic pine plantations induce soil carbon depletion. *Soil Biology and Biochemistry* 33, 1733–1740.

Chapin, F.S. (1980) The mineral nutrition of wild plants. *Annual Review of Ecology and Systematics* 11, 233–260.

Chapman, M.G. (2002) Local ecological knowledge of soil and water conservation in the coffee gardens of Sumberjaya, Sumatra. BSc Honours thesis, University of Wales, Bangor, UK.

Chaturvedi, O.P. and Das, D.K. (2003) Studies on rooting patterns of 5-year-old important agroforestry tree species grown in north Bihar, India. *Forest, Trees and People*.

Chauvel, A., Grimaldi, M., Barros, E., Blanchart, E., Desjardins, T., Sarrazin, M. and Lavelle, P. (1999) Pasture damage by an Amazonian earthworm. *Nature* 398, 32–33.

Cheng, X. and Bledsoe, C.S. (2002) Contrasting seasonal patterns of fine root production for blue oaks (*Quercus douglasii*) and annual grasses in California oak woodland. *Plant Soil* 240, 263–274.

Chesney, P. and Nygren, P. (2002) Fine root nodule dynamics of *Erythrina poeppigiana* in an alley cropping system in Costa Rica. *Agroforestry Systems* 56, 259–269.

Chevallier, T., Girardin, C., Mariotti, A., Blanchart, E., Albrecht, A. and Feller, C. (2001) The role of biological activity (roots, earthworms) on medium-term C dynamics in Vertisol under a *D. decumbens* pasture. *Applied Soil Ecology* 16, 11–21.

Chidumayo, E.N. (1987) A shifting cultivation land use system under population pressure in Zambia. *Agroforestry Systems* 5, 5–25.

Christensen, B.T. (1987) Decomposability of organic matter in particle size fractions from field soils with straw incorporation. *Soil Biology and Biochemistry* 19, 429–435.

Clarholm, M. (1985) Interactions of bacteria, protozoa and plants leading to mineralization of soil nitrogen. *Soil Biology and Biochemistry* 17, 181–187.

Coleman, M.D., Dickson, R.E. and Isebrands, J.G. (2000) Contrasting fine-root production, survival and soil $CO_2$ efflux in pine and poplar plantations. *Plant and Soil* 225, 129–139.

Collinson, M.P. and Tollens, E. (1994) The impact of the international agricultural centres: measurement, quantification and interpretation. *Experimental Agriculture* 30, 395–419.

Comerford, N.B., Porter, P.S. and Escamilla, J.A. (1994) Use of Theissen areas in models of nutrient uptake in forested ecosystems. *Soil Science Society of America Journal* 58, 210–215.

Conklin, H.C. (1957) *Hanunóo Agriculture: a Report on an Integral System of Shifting Cultivation in the Philippines*. FAO Forestry Development Paper 12. FAO, Rome, 209 pp.

Conradie, W.J. (1988) Effect of soil acidity on grapevine root growth and the role of roots as a source of nutrient reserves. In: van Zyl, J.L. (ed.) *The Grapevine Root and its Environment*. Department of Agriculture and Water Supply Technical Communication No. 215, Pretoria, South Africa, pp.16–29.

Cook, R.J. and Baker, K.F. (1983) *The Nature and Practice of Biological Control of Plant Pathogens*. American Phytopathological Society, St Paul, Minnesota, 539 pp.

Coolman, R.M. (1994) Nitrous oxide emissions from amazonian ecosystems. PhD thesis, North Carolina State University, USA.

Cools, N., de Pauw, E. and Deckers, J. (2003) Towards an integration of conventional land evaluation methods and farmers' soil suitability assessment: a case study in northwestern Syria. *Agriculture, Ecosystems and Environment* 95(1), 327–342.

Copping, L.G. and Menn, J.J. (2000) Biopesticides: a review of their action, applications and efficiency. *Pest Management Science* 56, 651–676.

Corbeels, M., Shiferaw, A. and Haile, M. (2000) Farmers' knowledge of soil fertility and local management strategies in Tigray, Ethiopia. Managing Africa's Soils No. 10, ii and 23.

Corkidi, L., Rowland, D.L., Johnson, N.C. and Allen, E.B. (2002) Nitrogen fertilization alters the functioning of arbuscular mycorrhizas at two semiarid grasslands. *Plant and Soil* 240, 299–310.

Costa, F.J.S.A., Bouldin, D.R. and Suhet, A.R. (1990) Evaluation of N recovery from mucuna placed on the surface or incorporated in a Brazilian Oxisol. *Plant and Soil* 124, 91–96.

Coster, C. (1938) Naschrift: Herbebossching op Java (Postcript: Reafforestation in Java). *Tectona* 31, 602–605.

Coutts, M.P. (1983) Development of the structural root systems of Sitka spruce. *Forestry* 56, 1–6.

Coveney, P. and Highfield, R. (1995) *Frontiers of Complexity: the Search for Order in a Chaotic World*. Faber and Faber Ltd, London, 462 pp.

Cox, F.R., Kamprath, E.J. and McCollum, R.E. (1981) A descriptive model of soil test nutrient levels following fertilization. *Soil Science Society of America Journal* 45, 529–532.

Cox, P.G. (1996) Some issues in the design of agricultural decision support systems. *Agricultural Systems* 52(2–3), 355–381.

Crill, P.M., Keller, M., Weitz, A., Grauel, B. and Veldkamp, E. (2000) Intensive field measurements of nitrous oxide emissions from a tropical agricultural soil. *Global Biogeochemical Cycles* 14, 85–95.

Cuenca, G. and Lovera, M. (1992) Vesicular-arbuscular mycorrhizae in disturbed and revegetated sites from La Gran Sabana, Venezuela. *Canadian Journal of Botany* 70, 73–79.

Cuevas, E. and Medina, E. (1988) Nutrient dynamics within Amazonian forests. I. Fine root growth, nutrient availability and leaf litter decomposition. *Oecologia* 76, 222–235.

Dalal, R.C. (1977) Soil organic phosphorus. *Advances in Agronomy* 29, 83–117.

Daniels, J. and Chamala, S. (1989) Practical men or dreamers? *Australian Journal of Adult Education* 29(3), 25–31.

Darrah, P.R. (1993) The rhizosphere and plant nutrition: a quantitative approach. *Plant and Soil* 155/156, 1–20.

Darwin, C. (1837) On the formation of mould. *Proceedings of the Geological Society* 2, 574–576.

Darwin, C. (1859) *The Origin of Species*. (Harvard facsimile, 1st Edn. 1964)

Darwin, C. (1881) *The Formation of Vegetable Mould through the Action of Worms with Observations on their Habits*. Murray, London.

Date, R.A. (1977) Inoculation of tropical pasture legumes. In: Vincent, J.M., Whitney, A.S. and Bose, J. (eds) *Exploiting the Legume–Rhizobium Symbiosis*. University of Hawaii College of Agriculture, Miscellaneous Publication No. 145, Hawaii, pp. 293–311.

Date, R.A. (1991) Nitrogen fixation in *Desmanthus*: strain specificity of *Rhizobium* and responses to inoculation in acidic and alkaline soil. *Tropical Grasslands* 25, 47–55.

Date, R.A. (2000) Inoculated legumes in cropping systems of the tropics. *Field Crops Research* 65, 123–136.

Davidson, E.A. (1991) Fluxes of nitrous oxide and nitric oxide from terrestrial ecosystems, In: Rogers, J.E. and Whitman, W.B. (eds) *Microbial Production and Consumption of Greenhouse Gases: Methane, Nitrogen Oxides and Halomethanes*. American Society for Microbiology, Washington, DC, pp. 219–236.

Davidson, E.A. and Ackerman, I.L. (1993) Changes in soil carbon inventories following cultivation of previously untilled soils. *Biogeochemistry* 20, 161–193.

Davidson, E.A. and Trumbore, S.E. (1995) Gas diffusivity and production of $CO_2$ in deep soils of the eastern Amazon. *Tellus* 47B, 550–565.

Davies, W.J. and Zhang, J. (1991) Root signals and the regulation of growth and development of plants in dry soil. *Annual Review of Plant Molecular Biology* 42, 55–76.

Dawson, T.E. (1993) Water sources of plants as determined from xylem-water isotopic composition: perspectives on plant competition, distribution and water relations. In: Ehleringer, J.R., Hall, A.E. and Farquhar, G.D. (eds) *Stable Isotopes and Plant Carbon–Water Relations*. Academic Press, New York, pp. 465–496.

Debaeke, P., Caussanel, J.P., Kiniry, J.R., Kafiz, B. and Mondragon, G. (1997) Modelling crop–weed interactions in wheat with ALMANAC. *Weed Research Oxford* 37(5), 325–341.

Decaëns, T.P., Lavelle, P., Jean Jiménez, J.J., Escobar, G. and Rippstein, G. (1994) Impact of land management on soil macrofauna in the Oriental Llanos of Colombia. *European Journal of Soil Biology* 30, 157–168.

Decaëns, T., Galvis, J.H. and Amezquita, E. (2001) Properties of the structures created by ecological engineers at the soil surface of a Colombian savanna. *Comptes Rendus de L'Academie des Sciences Series III Sciences De La Vie Life Sciences*, 324, 465–478.

Deenik, J., Ares, A. and Yost, R.S. (2000) Fertilization response and nutrient diagnosis in peach palm (*Bactris gasipaes*): a review. *Nutrient Cycling in Agroecosystems* 56, 195–207.

de Faria, S.M. and de Lima, H.C. (1998) Additional studies of the nodulation status of legume species in Brazil. *Plant and Soil* 200, 185–192.

de Faria, S.M., Lewis, G.P., Sprent, J.I. and Sutherland, J.M. (1989) Occurrence of nodulation in the Leguminosae. *New Phytologist* 111, 607–619.

de Foresta, H. and Kahn, F. (1984) Un système racinaire adventif dans un tronc creux d'*Eperua falcata*. *Revue d'Ecologie La Terre et Vie* 39, 347–350.

Degens (1997) Macroaggregation in soil by biological bonding and binding mechanisms and factors affecting this: a review. *Australian Journal of Soil Research* 35, 431–459.

de Jong, R. (1982) Assessment of empirical parameters that describe soil water characteristics. *Canadian Journal of Agricultural Engineering* 24, 65–70.

Dekkers, T.B.M. and van der Werff, P.A. (2001) Mutualistic functioning of indigenous arbuscular mycorrhizae in spring barley and winter wheat after cessation of long-term phosphate fertilization. *Mycorrhiza* 10, 195–201.

de Lajudie, P., Willems, A., Pot, B., Dewettinck, D., Maestrojuan, G., Neyra, M., Collins, M.D., Dreyfus, B., Kersters, K. and Gillis, M. (1994) Polyphasic taxonomy of rhizobia: emendation of the genus *Sinorhizobium* and description of *Sinorhizobium meliloti* comb. nov., *Sinorhizobium saheli* sp. nov. and *Sinorhizobium teranga* sp. nov. *Journal of Systematic Bacteriology* 44, 715–733.

de Lajudie, P., Willems, A., Nick, G., Moreira, F., Molouba, F., Torck, U., Neyra, M., Collins, M.D., Lindstrom, K., Dreyfus, B. and Gillis, M. (1998) Characterization of tropical tree rhizobia and description of *Mesorhizobium plurifarium* sp. nov. *International Journal of Systematic Bacteriology* 48, 369–382.

de Montard, F.X., Rapey, H., Delpy, R. and Massey, P. (1999) Competition for light, water and nitrogen in an association of hazel (*Corylus avellana* L.) and cocksfoot (*Dactylis glomerata* L.). *Agroforestry Systems* 43, 135–150.

de Neergaard, A., Porter, J.R. and Gorissen, A. (2002) Distribution of assimilated carbon in plants and rhizosphere soil of basket willow (*Salix viminalis* L.). *Plant and Soil* 245, 307–314.

de Ruiter, P.C., Neutel, A.M. and Moore, J.C. (1994) Modelling food webs and nutrient cycling in agroecosystems. *Trends in Ecology and Evolution* 9, 378–383.

de Ruiter, P.C., Neutel, A.M. and Moore, J.C. (1995) Energetics, patterns of interaction strengths and stability in real ecosystems. *Science* 269, 1257–1260.

Desaeger, J. (2001) Implications of plant-parasitic nematodes for improved fallows in Africa. PhD thesis, University of Leuven, Belgium, 216 pp.

Desaeger, J. and Rao, M.R. (1999a) The root-knot nematode (*Meloidogyne* spp.) problem in sesbania fallows and scope for managing it in western Kenya. *Agroforestry Systems* 47, 273–288.

Desaeger, J. and Rao, M.R. (1999b) Effects of organic and inorganic nutrient sources on nematode communities with maize in western Kenya. *Poster Presentation at International Workshop on Tropical Soil Biology and Fertility: Opportunities and Challenges for African Agriculture. March 16–19, 1999, TSBF, Nairobi, Kenya* (unpublished).

Desaeger, J. and Rao, M.R. (2000) Parasitic nematode populations in natural fallows and improved cover crops and their effects on subsequent crops in Kenya. *Field Crops Research* 65, 41–56.

Desaeger, J. and Rao, M.R. (2001a) Effect of field establishment methods on root-knot nematode (*Meloidogyne* ssp.) infection and growth of *Sesbania sesban* in western Kenya. *Crop Protection* 20, 31–41.

Desaeger, J. and Rao, M.R. (2001b) The potential of mixed cover crops of *Sesbania, Tephrosia* and *Crotalaria* for minimising nematode problems in subsequent crops. *Field Crops Research* 70, 111–125.

de Swart, P.H. and van Diest, A. (1987) The rock-phosphate solubilizing capacity of *Pueraria javanica* as affected by soil pH, superphosphate priming effect and symbiotic $N_2$ fixation. *Plant and Soil* 100, 135–147.

de Vos, J.A. (1997) Water flow and nutrient transport in a layered silt loam soil. PhD thesis, Wageningen Agricultural University, Wageningen, The Netherlands.

de Willigen, P. and van Noordwijk, M. (1987) Roots for plant production and nutrient use efficiency. PhD thesis, Wageningen Agricultural University, Wageningen, The Netherlands, 282 pp.

de Willigen, P. and van Noordwijk, M. (1994) Diffusion and mass flow to a root with constant nutrient demand or behaving as a zero-sink. *Soil Science* 157, 162–175.

de Willigen, P., Nielsen, N.E., Claassen, N. and Castrignano, A.M. (2000) Modelling water and nutrient uptake. In: Smit, A.L., Bengough, A.G., Engels, E., van Noordwijk, M., Pellerin, S. and van de Geijn, S.C. (eds) *Root Methods: a Handbook.* Springer-Verlag, Berlin, Germany, pp. 509–543.

de Willigen, P., Heinen, M., Mollier, A. and van Noordwijk, M. (2002) Two-dimensional growth of a root system modelled as a diffusion process. I. Analytical solutions. *Plant and Soil* 240, 225–234.

de Wit, C.T. (1992) Resource use efficiency in agriculture. *Agricultural Systems* 40, 125–151.

Dhyani, S., Naran, P. and Singh, R.K. (1990) Studies on root distribution of five multipurpose tree species in Doon valley, India. *Agroforestry Systems* 12, 149–161.

Dhyani, S.K. and Tripathi, R.S. (2000) Biomass and production of fine and coarse roots of trees under agrisilvicultural practices in north-east India. *Agroforestry Systems* 50, 107–121.

Dierolf, T.S., Arya, L.M. and Yost, R.S. (1997) Water and cation movement in an Indonesian Ultisol. *Agronomy Journal* 89, 572–579.

Dignan, P. and Bren, L. (2003) Modelling light penetration edge effects for stream buffer design in mountain ash forest in southeastern Australia. *Forest Ecology and Management* 179, 95–106.

Dinkelaker, B., Hengeler, C. and Marschner, H. (1995) Distribution and function of proteoid roots and other root clusters. *Botanica Acta* 108, 183–200.

Dixon, H.J., Doores, J.W., Joshi, L. and Sinclair, F.L. (2001) *Agroecological Knowledge Toolkit for Windows: Methodological Guidelines, Computer Software and Manual for AKT5*. School of Agricultural and Forest Sciences, University of Wales, Bangor, UK, 171 pp.

Dodd, J.C. and Thompson, B.D. (1994) The screening and selection of inoculant arbuscular-mycorrhizal and ectomycorrhizal fungi. *Plant and Soil* 159, 149–158.

Dommergues, Y. (1990) *Casuarina equisetifolia*: an old timer with a new future. A publication of the Forest, Farm, and Community Tree Network (FACT Net). *NFT Highlights* (NFTA 90–02). Winrock International, Morrilton, Arkansas, www.winrock.org/forestry/factpub/FACTSH/C_equisetifolia.htm

Doran, J.W. (1980) Soil microbial and biochemical changes associated with reduced tillage. *Soil Science Society of America Journal* 44, 765–771.

Doussan, C., Vercambre, G. and Pagès, L. (1999) Water uptake by two contrasting root systems (maize, peach tree): results from a model of hydraulic architecture. *Agronomie* 19, 255–263.

Dove, M.R. (1985) *Swidden Agriculture in Indonesia: the Subsistence Strategies of the Kalimantan Kantu*. Mouton Press, Berlin, Germany, 515 pp.

Dove, M.R. (2000) The life-cycle of indigenous knowledge, and the case of natural rubber production. In: Ellen, R., Parkes, P. and Bicker, A. (eds) *Indigenous Environmental Knowledge and its Transformations: Critical Anthropological Perspectives*. Harwood Academic Publishers, Amsterdam, The Netherlands, pp. 213–251.

Dowson, C.G., Rayner, A.D.M. and Boddy, L. (1988) The form and outcome of mycelial interactions involving cord-forming decomposer basidiomycetes in homogeneous and heterogeneous environments. *New Phytologist* 109, 423–432.

Drechsel, P. and Zech, W. (1991) Foliar nutrient levels of broad-leaved tropical trees: a tabular review. *Plant and Soil* 131, 29–46.

Dreyfus, B.L. and Dommergues, Y.R. (1981) Nodulation of Acacia species by fast- and slow-growing tropical strains of *Rhizobium*. *Applied and Environmental Microbiology* 41, 97–99.

Duguma, B., Kang, B.T. and Okali, D.U.U. (1988) Effect of liming and phosphorus application on performance of *Leucaena leucocephala* in acid soils. *Plant and Soil* 110, 57–61.

Dunbabin, V.M., Diggle, A.J., Rengel, Z. and van Hugten, R. (2002a) Modelling the interactions between water and nutrient uptake and root growth. *Plant and Soil* 239, 19–38.

Dunbabin, V.M., Diggle, A.J. and Rengel, Z. (2002b) Simulation of field data by a basic three-dimensional model of interactive root growth. *Plant and Soil* 239, 39–54.

Dundson, A.J. and Simons, A.J. (1996) Provenance and progeny trials. In: Stewart, J.L., Allison, G.E. and Simons, A.J. (eds) *Gliricidia sepium: Genetic Resources for Farmers*. Tropical Forestry Papers No. 33, Oxford Forestry Research Institute, Oxford, UK, pp. 93–118.

Dunne, T. (1978) Field studies of hillslope flow processes. In: Kirkby, M.J. (ed.) *Hillslope Hydrology*. John Wiley & Sons, New York, pp. 227–293.

Duponnois, R., Cadet, P., Senghor, K. and Sougoufara, B. (1997) Sensibilité de plusieurs acacias australiens au nématode à galles *Meloidogyne javanica*. *Annales des Sciences Forestieres* 54, 181–190.

Eastham, J. and Rose, C.W. (1990) Tree/pasture interactions at a range of tree densities in an agroforestry experiment. I. Rooting patterns. *Australian Journal of Agricultural Research* 41, 683–695.

Edmeades, D.C. and Ridley, A.M. (2003) Using lime to ameliorate topsoil and subsoil acidity. In: Rengel, Z. (ed.) *Handbook of Soil Acidity*. Marcel Dekker, New York, pp. 297–336.

Eggleton, P., Bignell, D.E., Sands, W.A., Waite, B., Wood, T.G. and Lawton, J.H. (1995) The species richness of termites (Isoptera) under differing levels of forest disturbance in the Mbalmayo Forest Reserve, southern Cameroon. *Journal of Tropical Ecology* 11, 85–98.

Eggleton, P., Bignell, D.E., Sands, W.A., Mawdsley, N.A., Lawton, J.H., Wood, T.G. and Bignell, N.C. (1996) The diversity, abundance and biomass of termites under differing levels of disturbance in the Mbalmayo Forest Reserve, southern Cameroon. *Philosophical Transactions of the Royal Society of London, Series B* 351, 51–68.

Eggleton, P., Bignell, D.E., Hauser, S., Dibog, L., Norgrove, L. and Madong, B. (2002) Termite diversity across an anthropogenic disturbance gradient in the humid forest zone of West Africa. *Agriculture, Ecosystems and Environment* 90, 189–202.

Egunjobi, O.A. (1985). The International *Meloidogyne* Project (IMP) in Region VI: current status, progress and future outlook. In: Sasser, J.N. and Carter, C.C. (eds) *An Advanced Treatise on Meloidogyne. Volume I. Biology and Control.* North Carolina State University Graphics, Raleigh, North Carolina, pp. 353–360.

Eisenback, J.D. and Griffin, G.D. (1987) Interactions with other nematodes. In: Veech, J.A. and Dickson, D.W. (eds) *Vistas on Nematology: a Commemoration of the 25th Anniversary of the Society of Nematologists.* Society of Nematologists, Inc., Hyattsville, Maryland, pp. 313–320.

Eissenstat, D.M. (1992) Costs and benefits of constructing roots of small diameter. *Journal of Plant Nutrition* 15, 763–782.

Eissenstat, D.M., Wells, C.E., Yanai, R.D. and Whitbeck, J.L. (2000) Building roots in a changing environment: implications for root longevity. *New Phytologist* 147, 33–42.

Ellen, R. (1998) Comment [on Sillitoe, 1998]. *Current Anthropology* 39, 238–239.

El-Swaify, S.A. (2000) Operative processes for sediment-based watershed degradation in small, tropical volcanic island ecosystems. In: Lal, R. (ed.) *Integrated Watershed Management in the Global Ecosystem.* CRC Press, Boca Raton, Florida, pp. 35–49.

Elton, C. (1927) *Animal Ecology.* Macmillan, New York.

Emerman, S.H. and Dawson, T.E. (1996) Hydraulic lift and its influence on the water content of the rhizosphere: an example from sugar maple, *Acer saccharum. Oecologia* 108, 273–278.

Erickson, H., Keller, M. and Davidson, E.A. (2000) Nitrogen oxide fluxes and nitrogen cycling during postagricultural succession and forest fertilization in the humid tropics. *Ecosystems* 4, 67–84.

Eshel, A. (1998) On fractal dimensions of a root system. *Plant Cell Environment* 21, 247–251.

Eshleman, K.N., Pollard, J.S. and O'Brien, A.K. (1993) Determination of contributing areas for saturation overland flow from chemical hydrograph separations. *Water Resources Research* 29, 3577–3587.

Estores, R.A. and Chen, T.A. (1972) Interactions of *Pratylenchus penetrans* and *Meloidogyne incognita* as coinhabitants in tomato. *Journal of Nematology* 4, 170–174.

Evans, D.O. and Rotar, P.P. (1987) Productivity of *Sesbania sesban. Tropical Agriculture* 64, 193–200.

Evans, K. and Haydock, P.P.J. (1993) Interactions of nematodes with root-rot fungi. In: Khan, M.W. (ed.) *Nematode Interactions.* Chapman and Hall, London, pp. 104–133.

Ewel, J.J. (1999) The ecosystem mimic concept. In: Lefroy, E.C., Hobbs, R.J., O'Connor, M.H. and Pate, J.S. (eds) *Agriculture as a Mimic of Natural Ecosystems.* Current Plant Science and Biotechnology in Agriculture. Vol. 37, Kluwer Academic Publishers, Dordrecht, The Netherlands, pp. 57–97.

Ewel, J.J., Mazzarino M.J. and Berish, C.W. (1991) Tropical soil fertility changes under monocultures and successional communities of different structure. *Ecological Applications* 1, 289–302.

Fagerström, M.H., Toan, T.D., Iwald, J., Nhuyen, L., Hai, T.S., Schwan, K., Olsson, D., Phai, D.D., Hien, N.T. and Anh, L.X. (2002) A Participatory Landscape Analysis of Dong Cao catchment, Tien Xuan commune, Luong Son district, Hoa Binh province, Vietnam. Fieldwork report of 'Sustainable land use in the uplands of Vietnam and Laos – Science and local knowledge for food security' (LUSLOF) project, 30 pp.

Fairhurst, T., Lefroy, R., Mutert, E. and Batjes, N. (1999) The importance, distribution and causes of phosphorus deficiency as a constraint to crop production in the tropics. *Agroforestry Forum* 9, 2–8.

Fearnside, P.M. and Barbosa, R.I. (1998) Soil carbon changes from conversion of forest to pasture in Brazilian Amazonia. *Forest Ecology and Management* 108, 147–166.

Feller, C. and Beare, M.H. (1997) Physical control of soil organic matter dynamics in the tropics. *Geoderma* 79, 69–116.

Feller, C., Albrecht, A. and Tessier, D. (1996) Aggregation and organic matter storage in Kaolinitic and smectitic tropical soils. In: Carter, M.R. and Stewart, B.A. (eds) *Structure and Organic Matter Storage in Agricultural Soils.* Advances in Soil Science. Lewis Publishers, Boca Raton, Florida, pp. 309–360.

Feller, C., Albrecht, A., Blanchart, E., Cabidoche, Y.M., Chevallier, T., Hartmann, C., Eschenbrenner, V., Larre-Larrouy, M.C. and Ndandou, J.F. (2001) Soil carbon sequestration in tropical areas: general considerations and analysis of some edaphic determinants for Lesser Antilles soils. *Nutrient Cycling in Agroecosystems* 61, 19–31.

Finlay, R.D. and Read, D.J. (1986) The structure and function of the vegetative mycelium of ectomycorrhizal plants. 1. Translocation of C14-labeled carbon between plants interconnected by a common mycelium. *New Phytologist* 103 (1), 143–156.

Firsching, B.M. and Claassen, N. (1996) Root phosphatase activity and soil organic phosphorus utilization by Norway spruce [*Picea abies* (L.) Karst.]. *Soil Biology and Biochemistry* 28, 1417–1424.

Fisher, M.J., Rao, I.M., Ayarza, M.A., Lascano, C.E., Sanz, J.I., Thomas, R.J. and Vera, R.R. (1994) Carbon storage by introduced deep-rooted grasses in the South American savannas. *Nature* 371, 236–238.

Fisher, M.J., Rao, I.M., Lascano, C.E., Sanz, J.I., Thomas, R.J., Vera, R.R. and Ayarza, M.A. (1995) Pasture soils as carbon sink. *Nature* 376, 472–473.

Fitter, A.H. (2001) Specificity, links and networks in the control of diversity in plant and microbial communities. In: Press, M.C., Huntly, N.J. and Levin, S. (eds) *Ecology – Achievement and Challenge.* Blackwell Science, Oxford, UK, pp. 95–114.

Fitter, A.H. and Stickland, T.R. (1992) Fractal characterization of root system architecture. *Functional Ecology* 6, 632–635.

Flanagan, D.C. and Nearing, M.A. (1995) USDA-Water Erosion Prediction Project: hillslope profile and watershed model documentation, 1st edn. USDA-ARS-MWA-SWCS, West Lafayette, pp. 1.1–14.27.

Ford, J. and Martinez, D. (2000) Traditional ecological knowledge, ecosystem science, and environmental management. *Ecological Applications* 10(5), 1249–1250.

Fortin, J.A., Bécard, G., Declerck, S., Dalpé, Y., St-Arnaud, M., Coughlan, A.P. and Piché, Y. (2002) Arbuscular mycorrhiza in root-organ cultures. *Canadian Journal of Botany* 80, 1–20.

Foster, L.J. and Wood, R.A. (1963) Observations on the effects of shade and irrigation on soil-moisture utilization under coffee in Nyasaland. *Empire Journal of Experimental Agriculture* 31, 108–115.

Fragoso, C., Brown, G.G., Patron, J.C., Blanchart, E., Lavelle, P., Pashanasi, B., Senapati, B. and Kumar, T. (1997) Agricultural intensification, soil biodiversity and agroecosystem function in the tropics: the role of earthworms. *Applied Soil Ecology* 6, 17–35.

Francl, L.J. and Wheeler, T.A. (1993) Interaction of plant-parasitic nematodes with wilt-inducing fungi. In: Khan, M.W. (ed.) *Nematode Interactions.* Chapman and Hall, London, pp. 79–103.

Franke-Snyder, M., Douds, D.D., Galvez, L., Phillips, J.G., Wagoner, P., Drinkwater, L. and Morton, J.B. (2001) Diversity of communities of arbuscular mycorrhizal (AM) fungi present in conventional versus low-input agricultural sites in eastern Pennsylvania, USA. *Applied Soil Ecology* 16, 35–48.

Fransen, B. and de Kroon, H. (2001) Long-term disadvantages of selective root placement: root proliferation and shoot biomass of two perennial grass species in a 2-year experiment. *Journal of Ecology* 89, 711–722.

Franzel, S. and Scherr, S.J. (eds) (2002) *Trees on the Farm: Assessing the Adoption Potential of Agroforestry Practices in Africa.* CAB International, Wallingford, UK.

Franzel, S., Cooper, P. and Denning, G.L. (2001) Scaling up the benefits of agroforestry research: reason learned and research challenges. *Development in Practice* 11(4), 524–534.

Freckman, D.W. and Caswell, E.P. (1985) The ecology of nematodes in agroecosystems. *Annual Review of Phytopathology* 23, 275–296.

Friend, A.D., Stevens, A.K., Knox, R.G. and Cannell, M.G.R. (1997) A process-based, biogeochemical, terrestrial biosphere model of ecosystem dynamics (Hybrid v3.0). *Ecological Modelling* 95, 249–287.

Friend, M.T. and Birch, H.F. (1960) Phosphate responses in relation to soil tests and organic phosphorus. *Journal of Agricultural Science* 54, 341–347.

Frossard, E.D., Lopez-Hernandez, D. and Brossard, M. (1996) Can isotopic exchange kinetics give valuable information on the rate of mineralization of organic phosphorus in soils? *Soil Biology and Biochemistry* 28, 857–864.

Fujisaka, S. (1997) Research: help or hindrance to good farmers in high risk systems? *Agricultural Systems* 54(2), 137–152.

Gachengo, C.N., Palm, C.A., Jama, B. and Othieno, C. (1999) *Tithonia* and *Senna* green manures and inorganic fertilizers as phosphorus sources for maize in Western Kenya. *Agroforestry Systems* 44, 21–36.

Gacheru, E. and Rao, M.R. (2001) Managing *Striga* infestation on maize using organic and inorganic nutrient sources in western Kenya. *International Journal of Pest Management* 47, 233–239.

Gacheru, E., Rao, M.R., Jama, B. and Niang, A. (1999) The potential of agroforestry to control *Striga* and increase maize yields in sub-Saharan Africa. In: CIMMYT (International Maize and Wheat Improvement Center) and EARO (Ethiopia Agriculture Research Organization) (eds) *Maize Production Technology for the Future: Challenges and Opportunities—Proceedings of the Sixth Eastern and Southern Africa Regional Maize Conference, 21–25 September, 1998, Addis Ababa, Ethiopia.* CIMMYT, Harare, Zimbabwe, pp. 180–184.

Gacheru, E., Rao, M.R., Jama, B. and Niang, A. (2000) The potential of agroforestry to control *Striga* and increase maize yields in sub-Saharan Africa. In: *Maize Production Technology for the Future: Challenges and Opportunities. Proceedings of the 6th Eastern and Southern Africa Regional Maize Conference, 21–25 September 1998, Addis Ababa.* CIMMYT-Zimbabwe, Harare, Zimbabwe.

Gahoonia, T.S., Claassen, N. and Jungk, A. (1992) Mobilization of phosphate in different soils by ryegrass supplied with ammonium or nitrate. *Plant and Soil* 140, 241–248.

Galindo, J.J., Abawi, G.S., Thurston, H.D. and Gálvez, G. (1983) Effect of mulching on web blight of beans in Costa Rica. *Phytopathology* 73, 610–615.

Gallez, A., Juo, A.S.R. and Herbillon, A.J. (1976) Surface and charge characteristics of selected soils in the tropics. *Soil Science Society of America Journal* 40, 601–608.

Gamliel, A. (2000) Soil amendments: a non chemical approach to the management of soilborne pests. In: Gullino, M.L., Katan, J. and Matta, A. (eds) *Proceedings of the Fifth International Symposium on Chemical and Non-chemical Soil and Substrate Disinfestation, Torino, Italy, 11–15 September 2000. Acta Horticulturae* 532, pp. 39–48.

Ganry, F., Guiraud, G. and Dommergues, Y. (1978) Effect of straw incorporation on the yield and nitrogen balance in the sandy soil-pearl millet cropping system of Senegal. *Plant and Soil* 50, 647–662.

Garcia-Espinosa, R. (1998) Important biological components to be included in the search for alternatives to the use of methyl bromide in Mexico. www.epa.gov./docs/ozone/mbr/1998airc/007espinosa.pdf

Gardner, R., Mawdesley, K., Tripathi, B.R., Gaskin, S. and Stuart, A. (2000) *Soil Erosion and Nutrient Loss in the Middle Hills of Nepal (1996–1998)*. ARS, Lumle; Soil Science Division, NARC, Khumaltar; and Queen Mary and Westfield College, University of London, UK, 57 pp.

Gardner, W.K. and Boundy, K.A. (1983) The acquisition of phosphorus by *Lupinus albus* L. IV. The effect of interplanting wheat and white lupin on the growth and mineral composition of the two species. *Plant and Soil* 70, 391–402.

Gardner, W.K., Parbery, D.G. and Barber, D.A. (1982) The acquisition of phosphorus by *Lupinus albus* L. I. Some characteristics of the soil–root interface. *Plant and Soil* 68, 19–32.

Gardner, W.K., Barber, D.D. and Parbery, D.G. (1983) The acquisition of phosphorus by *Lupinus albus* L. III. The probable mechanisms by which phosphorus movement in the soil/root interface are enhanced. *Plant and Soil* 70, 107–124.

Gathumbi, S.M., Cadisch, G. and Giller, K.E. (2002a) $^{15}$N natural abundance as a tool for assessing $N_2$-fixation of herbaceous, shrub and tree legumes in improved fallows. *Soil Biology and Biochemistry* 34, 1059–1071.

Gathumbi, S.M., Ndufa, J.K., Giller, K.E. and Cadisch, G. (2002b) Do mixed species improved fallows increase above- and below-ground resources capture? *Agronomy Journal* 94, 518–526.

Gathumbi, S.M., Cadisch, G., Buresh, R.J. and Giller, K.E. (2003) Subsoil nitrogen capture in mixed legume stands as assessed by deep nitrogen-15 placement. *Soil Science Society of America Journal* 67, 573–582.

Geelhoed, J.S., Van Riemsdjik, W.H. and Findenegg, G.R. (1999) Simulation of the effect of citrate exudation from roots on the plant availability of phosphate adsorbed on goethite. *European Journal of Soil Science* 50, 379–390.

George, R.J., McFarlane, D.J. Nulsen, R.A. (1997) Salinity threat of agriculture and ecosystems in Western Australia. *Hydrogeology Journal* 5(1), 6–21.

George, T.S., Gregory, P.J., Robinson, J.S., Buresh, R.J. and Jama, B.A. (2001) *Tithonia diversifolia*: variations in leaf nutrient concentration and implications for biomass transfer. *Agroforestry Systems* 52, 199–205.

Gerke, J. (1992) Phosphate, aluminium and iron in the soil solution of three different soils in relation to varying concentrations of citric acid. *Zeitschrift für Pflanzenernahrung und Bodenkunde* 155, 339–343.

Gianinazzi, S. and Schüepp, H. (eds) (1994) *Impact of Arbuscular Mycorrhizas on Sustainable Agriculture and Natural Ecosystems*. Birkhäuser Verlag, Basel, Switzerland, 226 pp.

Gianinazzi, S., Schüepp, H., Haselwandter, K. and Barea, J.M. (eds) (2002) *Mycorrhizal Technology in Agriculture – from Genes to Bioproducts*. Birkhäuser, Basel, Switzerland.

Giese, L.A.B., Aust, W.M., Kolka R.K. and Trettin, C.C. (2003) Biomass and carbon pools of disturbed riparian forests. *Forest Ecology and Management* 180, 493–508.

Gijsman, A., Oberson, A., Tiessen, H. and Friesen, D.K. (1996) Limited applicability of the CENTURY model to highly weathered tropical soils. *Agronomy Journal* 88, 894–903.

Gijsman, A.J., Hoogenboom, G., Parton, W.J. and Kerridge, P.C. (2002a) Modifying the DSSAT crop models for low-input agricultural systems, using a SOM /residue module from CENTURY. *Agronomy Journal* 94, 462–474.

Gijsman, A.J., Jagtap, S.S. and Jones, J.W. (2002b) Wading through a swamp of complete confusion: how to choose a method for estimating soil water retention parameters for crop models. *European Journal of Agronomy* 18, 75–105.

Gill, R.A. and Jackson, R.B. (2000) Global patterns of root turnover for terrestrial ecosystems. *New Phytologist* 147, 3–12.

Giller, K. (2000) Translating science into action for agricultural development in the tropics: an example from decomposition studies. *Applied Soil Ecology* 14, 1–3.
Giller, K.E. (2001) *Nitrogen Fixation in Tropical Cropping Systems*, 2nd edn. CAB International, Wallingford, UK, 423 pp.
Giller, K.E. and Cadisch, G. (1995) Future benefits from biological nitrogen fixation: an ecological approach to agriculture. *Plant and Soil* 174, 255–277.
Giller, K.E., Beare, M.H., Lavelle, P., Izac, A.M.N. and Swift, M.J. (1997a) Agricultural intensification, soil biodiversity and agroecosystem function. *Applied Soil Ecology* 6(1), 3–16.
Giller, K.E., Cadisch, G., Ehaliotis, C., Adams, E., Sakala, W.D. and Mafongoya, P.L. (1997b) Building soil nitrogen capital in Africa. In: Buresh, R.J., Sanchez, P.A. and Calhoun, F. (eds) *Replenishing Soil Fertility in Africa*. ASA, CSSA, SSSA, Madison, Wisconsin, pp. 151–192.
Giller, K.E., Cadisch, G. and Palm, C. (2002) The North-South divide! Organic wastes, or resources for nutrient management? *Agronomie* 22, 703–709.
Giller, P.S. (1996) The diversity of soil communities, the 'poor man's tropical rainforest'. *Biodiversity and Conservation* 5, 135–168.
Gindrat, D. (1979) Biological soil disinfestation. In: Mulder, D. (ed.) *Soil Disinfestation*. Elsevier, Amsterdam, The Netherlands, pp. 253–287.
Glasener, K.M. and Palm, C.A. (1995) Ammonium volatilization from tropical legume mulches and green manures and unlimed and limed soil. *Plant and Soil* 177, 33–41.
Glund, K. and Goldstein, A.H. (1993) Regulation, synthesis and excretion of a phosphate starvation inducible RNase by plant cells. In: Verma, D.P.S. (ed.) *Control of Plant Gene Expression*. CRC Press, Boca Raton, Florida, pp. 311–323.
Godwin, D.C. and Jones, C.A. (1991) Nitrogen dynamics in soil–plant systems. In: Hanks, J. and Ritchie, J.T. (eds) *Modeling Soil and Plant Systems*. Agronomy Society of America, Madison, Wisconsin, pp. 287–322.
Good, J.M., Minton, N.A. and Jaworski, C.A. (1965) Relative susceptibility of selected cover crops and coastal Bermuda grass to plant nematodes. *Phytopathology* 55, 1026–1030.
Goreau, T.J., and de Mello, W.Z. (1988) Tropical deforestation: some effects on atmospheric chemistry. *Ambio* 17, 275–281.
Goss, M.J. and de Varennes, A. (2002) Soil disturbance reduces the efficacy of mycorrhizal associations for early soybean growth and $N_2$ fixation. *Soil Biology and Biochemistry* 34, 1167–1173.
Gouyon, A., de Foresta, H. and Levang, P. (1993) Does 'jungle rubber' deserve its name? An analysis of rubber agroforestry systems in southeast Sumatra. *Agroforestry Systems* 22, 181–206.
Govindarajan, M., Rao, M.R., Mathuva, M.N. and Nair, P.K.R. (1996) Soil-water and root dynamics under hedgerow intercropping in semiarid Kenya. *Agronomy Journal* 88, 513–520.
Gowen, S.R. and Quénéhervé, P. (1993) Nematode parasites of bananas, plantains and abaca. In: Luc, M., Sikora, R.A. and Bridge, J. (eds) *Plant Parasitic Nematodes in Tropical and Subtropical Agriculture*, 2nd edn. International Institute of Parasitology, CAB International, Wallingford, UK, pp. 431–460.
Granier, A. (1987) Evaluation of transpiration in a Douglas-fir stand by means of sap flow measurements. *Tree Physiology* 3, 309–320.
Grau, C. (2001) Soybean cyst nematodes and soil pH, December, 2001. www.plantpath.wisc.edu/soyhealth/scnph.htm
Graves, A.R., Hess, T., Matthews, R., Stephens, W. and Middleton, T. (2002) Crop simulation models as tools in computer laboratory and classroom-based education. *Journal of Natural Resources and Life Sciences Education* 31, 48–54.
Green, S.R. and Clothier, B.E. (1988) Water use of kiwifruit vines and apple trees by the heat-pulse technique. *Journal of Experimental Botany* 39, 115–123.
Gressel, N., McColl, J.G., Preston, C.M., Newman, R.H. and Powers, R.F. (1996) Linkages between phosphorus transformations and carbon decomposition in a forest soil. *Biogeochemistry* 33, 97–123.
Grierson, P.F. (1992) Organic acids in the rhizosphere of *Banksia integrifolia* L.f. *Plant and Soil* 144, 259–265.
Grierson, P.F. and Adams, M.A. (2000) Plant species affect acid phosphatase, ergosterol and microbial P in a Jarrah (*Eucalyptus marginata* Donn ex Sm.) forest in south-western Australia. *Soil Biology and Biochemistry* 32, 1817–1827.
Grierson, P.F. and Attiwill, P.M. (1989) Chemical characteristics of the proteoid root mat of *Banksia integrifolia* L.f. *Australian Journal of Botany* 37, 137–143.
Grierson, P.F., Comerford, N.B. and Jokela, E.J. (1999) Phosphorus mineralization and microbial biomass in a Florida Spodosol: effects of water potential, temperature and fertilizer application. *Biology & Fertility of Soils* 28, 244–252.

Griffith, R. and Koshy, P.K. (1993) Nematode parasites of coconut and other palms. In: Luc, M., Sikora, R.A. and Bridge, J. (eds) *Plant Parasitic Nematodes in Tropical and Subtropical Agriculture*, 2nd edn. International Institute of Parasitology, CAB International, Wallingford, UK, pp. 363–386.

Groot, J.J.R. and Soumare, A. (1995) The roots of the matter – soil quality and tree roots in the Sahel. *Agroforestry Today* 7, 9–11.

Grove, R.H. (1995) *Green Imperialism: Colonial Expansion, Tropical Island Edens and the Origins of Environmentalism*, 1600–1860. Cambridge University Press, Cambridge, UK, 540 pp.

Haggar, J.P., Tanner, E.U., Beer, J.W. and Kass, D.C.L. (1993) Nitrogen dynamics of tropical agroforestry and annual cropping systems. *Soil Biology and Biochemistry* 25, 1363–1378.

Hairiah, K., van Noordwijk, M., Santoso, B. and Syekhfani, M.S. (1992) Biomass production and root distribution of eight trees and their potential for hedgerow intercropping on an ultisol in Lampung. *Agrivita* 15, 54–68.

Hairiah, K., van Noordwijk, M., Stulen, I., Meijboom, F.M. and Kuiper, P.J.C. (1993) P nutrition effects on aluminium avoidance of *Mucuna pruriens* var. *utilis. Experimental and Environmental Botany* 33, 75–83.

Hairiah, K., Adawiyah, R. and Widyaningsih, J. (1996) Amelioration of aluminium toxicity with organic matter: selection of organic matter based on its total cation concentration. *Agrivita* 19(4), 158–164.

Hairiah, K., Utami, S.R., Suprayogo, D., Widianto, Sitompul, S.M., Sunaryo, Lusiana, B., Mulia, R., van Noordwijk, M. and Cadish, G. (2000b) *Agroforestry on Acid Soils in the Humid Tropics: Managing Tree–Soil–Crop Interactions*. ICRAF SEA, Bogor, Indonesia, 38 pp.

Hairiah, K., van Noordwijk, M. and Cadisch, G. (2000c) Crop yield, C and N balances of three types of cropping systems on an ultisol in Northern Lampung. *Netherlands Journal of Agricultural Science* 48(1), 3–17.

Hairiah, K., Williams, S.E., Bignell, D., Swift, M.J. and van Noordwijk, M. (2001) Effect of land-use change on belowground biodiversity. In: van Noordwijk, M., Williams, S.E. and Verbist, B. (eds) *Toward Integrated Natural Resource Management in Forest Margins of the Humid Tropics: Local Action and Global Concerns*. ASB Lecture Note 06A, ASB-ICRAF/SEA, 32 pp.

Hall, D.O., Scurlock, J.M.O., Bolhar-Nordenkampf, H.R., Leegood, R.C. and Long, S.P. (eds) (1993) *Photosynthesis and Production in a Changing Environment*. Chapman and Hall, London, 464 pp.

Halliday, J. and Nakao, P.L. (1982) The symbiotic affinities of woody species under consideration as nitrogen-fixing trees. A resource document. NifTAL Project and Mircen, University of Hawaii, Hawaii.

Halliday, J. and Somasegaran, P. (1983) Nodulation, nitrogen fixation and Rhizobium strain affinities in the genus *Leucaena*. In: Anon (ed.) *Leucaena Research in the Asian-Pacific Region*. IDRC, Ottawa, pp. 27–32.

Hamilton, W.D., Woodruff, D.R. and Jamieson, A.M. (1991) Role of computer-based decision aids in farm decision-making and in agricultural extension. In: Muchow, R.C. and Bellamy, J.A. (eds) *Climatic Risk in Crop Production – Models and Management for the Semi-arid Tropics and Subtropics*. CAB International, Wallingford, UK, pp. 411–423.

Handayanto, E., Cadisch, G. and Giller, K.E. (1994) Nitrogen release from prunings of legume hedgerow trees in relation to quality of the prunings and incubation method. *Plant and Soil* 160, 237–248.

Handayanto, E., Giller, K.E. and Cadisch, G. (1997) Regulating N release from legume tree prunings by mixing residues of different quality. *Soil Biology and Biochemistry* 29, 1417–1426.

Harcombe, P.A. (1977) The influence of fertilization on some aspects of succession in a humid tropical forest. *Ecology* 58, 1375–1383.

Hart, M.M., Reader, R.J. and Klironomos, J.N. (2001) Life-history strategies of arbuscular mycorrhizal fungi in relation to their successional dynamics. *Mycologia* 93, 1186–1194.

Hartanto, H., Prabhu, R., Widayat, A.S.E. and Asdak, C. (2003) Factors affecting runoff and soil erosion: plot-level soil loss monitoring for assessing sustainability of forest management. *Forest Ecology and Management* 180, 361–374.

Hartemink, A.E., Buresh, R.J., Jama, B. and Janssen, B.H. (1996) Soil nitrate and water dynamics in sesbania fallows, weed fallows, and maize. *Soil Science Society of America Journal* 60, 568–574.

Hartemink, A.E., Buresh, R.J., van Bodegom, P.M., Braun, A.R., Jama, B. and Janssen, B.H. (2000) Inorganic nitrogen dynamics in fallows and maize on an Oxisol and Alfisol in the highlands of Kenya. *Geoderma* 98, 11–33.

Hassink, J. (1997) The capacity of soils to preserve organic C and N by their association with clay and silt particles. *Plant and Soil* 191, 77–87.

Hauser, S. (1993) Root distribution of *Dactyladenia* (*Acioa*) *barteri* and *Senna* (*Cassia*) *siamea* in alley cropping on an Ultisol. I. Implications for field experimentation. *Agroforestry Systems* 24, 111–121.

Haynes, R.J. (1986) *Mineral-Nitrogen in the Plant–Soil System*. Academic Press, New York, 484 pp.

Hedley, M.J., Stewart, J.W.B. and Chauhan, B.S. (1982) Changes in inorganic and organic soil phosphorus fractions induced by cultivation practices and laboratory incubations. *Soil Science Society of America Journal* 46, 970–976.

Helal, H.M. (1990) Varietal differences in root phosphatase activity as related to the utilization of organic phosphates. In: El Bassam, N., Dambroth, M. and Loughman, B.C. (eds) *Genetic Aspects of Plant Mineral Nutrition*. Kluwer Academic Publishers, Dordrecht, The Netherlands, pp. 103–105.

Helgason, T., Daniel, T.J., Husband, R., Fitter, A.H. and Young, J.P.W. (1998) Ploughing up the wood-wide web? *Nature* 394, 431.

Helyar, K.R. and Porter, W.M. (1989) Soil acidification, its measurement and processes involved. In: Robinson A.D. (ed.) *Soil Acidity and Plant Growth*. Academic Press, Australia, pp. 61–101.

Helyar, K.R., Conyers, M.K. and Munns, D.N. (1993) Soil solution aluminium activity related to theoretical Al mineral solubilities in four Australian soils. *Journal of Soil Science* 44(2), 317–333.

Henrot J., Hamadina, M.K. and van Noordwijk, M. (1996) Decomposition of fine roots: a field method requiring small amounts of root material. In: Neeteson, J.J. and Henrot, J. (eds) *The Role of Plant Residues in Soil Management for Food Production in the Humid Tropics*. AB-DLO, Haren, The Netherlands, Report 65, pp 37–41.

Heringa, P.K. (1938) De stand der reboisatie in verband met de industrialisatie in Oost-Java [Reafforestation condition in connection with industrialisation in East Java]. *Landbouw-kundig Tijdschrift* 50(611), 359–372.

Herrera, I.C. and Marbán-Mendoza, N. (1999) Effects of leguminous cover crops on plant-parasitic nematodes with coffee in Nicaragua. *Nematropica* 29(2), 223–232.

Herron, N.F. and Hairsine, P.B. (1998) A scheme for evaluating the effectiveness of riparian zones in reducing overland flow to streams. *Australian Journal of Soil Restoration* 36, 683–689.

Hertel, D. and Leuschner, C. (2002) A comparison of four different fine root production estimates with ecosystem carbon balance data in a *Fagus–Quercus* mixed forest. *Plant and Soil* 239, 237–251.

Hewlett, J.D. and Hibbert, A.R. (1963) Moisture and energy conditions within a sloping soil mass during drainage. *Journal of Geophysical Research* 68, 1081–1087.

Hibbett, D.S., Gilbert, L.J. and Donoghue, M.J. (2000) Evolutionary instability of ectomycorrhizal symbioses in Basidiomycetes. *Nature* 407, 506–508.

Hilhorst, T. and Muchena, F. (eds) (2000) *Nutrients on the Move: Soil Fertility Dynamics in African Farming Systems*. International Institute for Environment and Development, London, 146 pp.

Hillocks, R.J. (1997) Cotton and tropical fibres. In: Hillocks, R.J. and Waller, J.M. (eds) *Soilborne Diseases of Tropical Crops*, 1st edn. CAB International, Wallingford, UK, pp. 303–329.

Hillocks, R.J. and Waller, J.M. (1997) *Soilborne Diseases of Tropical Crops*, 1st edn. CAB International, Wallingford, UK, 452 pp.

Hinsinger, P. (2001) Bioavailability of soil inorganic P in the rhizosphere as affected by root-induced chemical changes: a review. *Plant and Soil* 237, 173–195.

Hocking, D. (1998) Trees in wetland rice fields: an exciting project in Bangladesh. *Agroforestry Today* 10 (3), 4–6.

Hocking, D. and Islam, K. (1998) Initial growth of agroforestry trees in wetland rice fields, as influenced by top and root pruning to regulate crop impact. *Bangladesh Journal of Forest Science* 27, 90–102.

Hocking, P.J. (2001) Organic acids exuded from roots in phosphorus uptake and aluminium tolerance of plants in acid soils. *Advances in Agronomy* 74, 63–97.

Hocking, P.J., Keerthisinghe, G., Smith, F.W. and Randall, P.J. (1997) Comparison of the ability of different crop species to access poorly-available soil phosphorus. In: Ando, T. (ed.) *Plant Nutrition for Sustainable Food Production and Environment*. Kluwer Academic Publishers, Dordrecht, The Netherlands, pp. 305–308.

Hodda, M., Bloemers, G.F., Lawton, J.H. and Lambshead, P.J.D. (1997) The effects of clearing and subsequent land use on abundance and biomass of soil nematodes in tropical forest. *Pedobiologia* 41, 279–294.

Hoffland, E. (1989) Quantitative evaluation of the role of organic acid exudation in the mobilization of rock phosphate by rape. *Plant and Soil* 140, 279–289.

Holbrook, N.M. and Putz, F.E. (1996) Physiology of tropical vines and hemi-epiphytes: plants that climb up and plants that climb down. In: Mulkey, S.S., Chazdon, R.L. and Smith, A.P. (eds) *Tropical Forest Plant Ecophysiology*. Chapman and Hall, New York, pp. 363–394.

Holland, E.A., Dentener, F.J., Braswell, B.H. and Sulzman, J.M. (1999) Contemporary and pre-industrial reactive nitrogen budgets. *Biogeochemistry* 46, 7–43.

Holmes, J.W. and Sinclair, J.A. (1986) Water yield from some afforested catchments in Victoria. In: *Hydrology and Water Resources Symposium*. Griffith University, Brisbane, pp. 214–218.

Hooker, J.E., Black, K.E., Perry, R.L. and Atkinson, D. (1995) Arbuscular mycorrhizal fungi induced alteration of root longevity in poplar. *Plant and Soil* 172, 327–329.

Hooker, J.E., Hendrick, R. and Atkinson, A. (2000) The measurement and analysis of fine root longevity. In: Smit, A.L., Bengough, A.G., Engels, E., van Noordwijk, M., Pellerin, S. and van de Geijn, S.C. (eds) *Root Methods: a Handbook*. Springer-Verlag, Berlin, Germany, pp. 273–304.

Horst, W.J. and Waschkies, C.H. (1987) Phosphorus nutrition of spring wheat (*Triticum aestivum* L.) in mixed culture with white lupin (*Lupinus albus* L.). *Zeitschrift für Pflanzenernahrung und Bodenkunde* 150, 1–8.

Horton, R.E. (1933) The role of infiltration in the hydrologic cycle. *Transactions American Geophysics* 14, 446–460

Howieson, J.G., O'Hara, G.W. and Loi, A. (2000) The legume–rhizobia relationship in the Mediterranean basin. In: Sulas, L. (ed.) *Legumes For Mediterranean Forage Crops, Pastures and Alternative Uses*. Cahiers Options Mediterranneenes 45, 305–314.

Huang, J. (1987) Interactions of nematodes with Rhizobia. In: Veech, J.A. and Dickson, D.W. (eds) *Vistas on Nematology: a Commemoration of the 25th Anniversary of the Society of Nematologists*. Society of Nematologists, Inc., Hyattsville, Maryland, pp. 301–306.

Hubbell, S.P. (2001) *The Unified Neutral Theory of Biodiversity and Biogeography*. Princeton University Press, Princeton, New Jersey.

Humphries, S. (1996) Milk cows, migrants, and land markets: unravelling the complexities of forest to pasture conversion in northern Honduras. PhD thesis, University of Guelph, Guelph, Ontario, Canada.

Hungria, M. and Campo, R.J. (2000) Interrelações da microbiologia com a fertilidade do solo. In: *Fertbio 2000, Biodinâmica do Solo*. Universidade Federal de Santa Maria-RS, Brazil. CD-ROM.

Hunt, H.W., Coleman, D.C. and Ingham, E.R. (1987) The detrital food web in a shortgrass prairie. *Biology and Fertility of Soils* 3, 57–68.

Hunt, L.A., Jones, J.W., Hoogenboom, G., Godwin, D.C., Singh, U., Pickering, N., Thornton, P.K., Boote, K.J. and Ritchie, J.T. (1994) General input and output file structures for crop simulation models. In: Uhlir, P.F. and Carter, G.C. (eds) *Crop Modelling and Related Environmental Data*. CODATA, Paris, pp. 35–72.

Hussey, R.S. and McGuire, J.M. (1987) Interactions with other organisms. In: Brown, R.H. and Kerry B.R. (eds) *Principles and Practice of Nematode Control in Crops*. Academic Press, Marrickville, NSW, Australia, pp. 294–329.

Huth, N.I., Snow, V.O. and Keating, B.A. (2001) Integrating a forest modelling capability into an agricultural production systems modelling environment – current applications and future possibilities. In: Ghassemi, F., White, D.H., Cuddy, S. and Nakanishi, T. (eds) *MODSIM 2001: International Congress on Modelling and Simulation Proceedings*, the Australian National University, Canberra, Australia, 10–13 December 2001. ANU, Canberra, Australia. Reprint no. 4335, 1895–1900.

Huth, N.I., Carberry, P.S., Poulton, P.L., Brennan, L.E. and Keating, B.A. (2002) A framework for simulation of agroforestry options for the low rainfall areas of Australia using APSIM. *European Journal of Agronomy* 18, 171–185.

Hütsch, B. (1996) Methane oxidation in soils of two long-term fertilization experiments in Germany. *Soil Biology and Biochemistry* 28, 773–782.

Hutton, E.M. (1990) Field selection of acid-soil tolerant leucaena from *L. leucocephala* × *L. diversifolia* crosses in a tropical Oxisol. *Tropical Agriculture (Trinidad)* 67, 2–8.

Huxley, P.A. (1999) *Tropical Agroforestry*. Blackwell Science, Oxford, UK, 371 pp.

Huxley, P.A., Pinney, A., Akunda, E. and Muraya, P. (1994) A tree/crop orientation experiment with a *Grevillea robusta* hedgerow and maize. *Agroforestry Systems* 26, 23–45.

IAEA (1975) *Root Activity Patterns of Some Tree Crops*. Technical Report Series No. 170. International Atomic Energy Agency, Vienna, Austria, 154 pp.

Ibrikci, H., Comerford, N.B., Hanlon, E.A. and Rechcigl, J.E. (1994) Phosphorus uptake by bahiagrass from Spodosols – modeling of uptake. *Soil Science Society of America Journal* 58, 139–143.

ICRAF (1994) International Center for Research in Agroforestry: Annual Report, 1994. ICRAF, Nairobi, Kenya, pp. 82–84.

ICRAF (1996) Annual Report 1995. ICRAF, Nairobi, Kenya, 288 pp.

ICRAF (1998) Annual Report for 1997. International Center for Research in Agroforestry Nairobi, Kenya, 204 pp.

ICRISAT (1984) Annual Report for 1983. International Crops Research Institute for Semi Arid Tropics, Patancheru-502324, AP, India, 350 pp.

Idol, T.W., Pope, P.E. and Ponder, F. Jr (2000) Fine root dynamics across a chronosequence of upland temperate deciduous forests. *Forest Ecology and Management* 127, 153–167.

IDS (1979) Rural development: whose knowledge counts? *IDS Bulletin* 10(2). Institute of Development Studies, University of Sussex, Brighton, UK, 64 pp.

IPCC (2000) *Land Use, Land-use Change, and Forestry Special Report*. Cambridge University Press, Cambridge, UK, 375 pp.

Ishizuka, S., Tsuruta, H. and Murdiyarso, D. (2000) Relationship between the fluxes of greenhouse gases and soil properties in a research site of Jambi, Sumatra. In: Murdiyarso, D. and Tsuruta, H. (eds) *The Impacts of Land-Use/Cover Change on Greenhouse Gas Emissions in Tropical Asia*. Global Change Impacts Centre, Southeast Asia, Bogor, Indonesia and National Institute of Agro-Environmental Sciences, Tsukuba, Japan, pp. 31–34.

Jackson, N.A., Wallace, J.S. and Ong, C.K. (2000) Tree pruning as a means of controlling water use in an agroforestry system in Kenya. *Forest Ecology and Management* 126, 133–148.

Jackson, R.B., Canadell, J., Ehleringer, J.R., Mooney, H.A., Sala, O.E. and Schulze, E.-D. (1996) A global analysis of root distributions for terrestrial biomes. *Oecologia* 108, 389–411.

Jackson, R.B., Mooney, H.A. and Schulze E.D. (1997) A global budget for fine root biomass, surface areas, and nutrient contents. *Proceedings of the National Academy of Sciences, USA* 94, 7362–7366.

Jain, M.C., Kumar, K., Wassmann, R., Mitra, S., Singh, S.D., Singh, J.P., Singh, R., Yadav, A.K. and Gupta, S. (2000) Methane emissions from irrigated rice fields in Northern India (New Delhi). *Nutrient Cycling in Agroecosystems* 58, 75–83.

Jakobsen, I., Gazey, C. and Abbott, L.K. (2001) Phosphate transport by communities of arbuscular mycorrhizal fungi in intact soil cores. *New Phytologist* 149, 95–103.

Jama, B., Swinkels, A.R. and Buresh, J.R. (1997) Agronomic and economic evaluation of organic and inorganic sources of phosphorus in Western Kenya. *Agronomy Journal* 89, 597–604.

Jama, B., Buresh, R.J., Ndufa, J.K. and Shepherd, K.D. (1998a) Vertical distribution of roots and soil nitrate: tree species and phosphorus effects. *Soil Science Society of America Journal* 62, 280–286.

Jama, B., Buresh, R.J., Smithson, P.C., Mbugua, P.N., Smestad, T. and Tiessen, H. (2000a) Maize yields and soil phosphorus fractions following woody leguminous fallows in western Kenya. *Agronomy Abstracts*, Minneapolis, Minnesota, 2000, 59 pp.

Jama, B., Palm, C.A., Buresh, R.J., Niang, A., Gachengo, C., Nziguheba, G. and Amadalo, B. (2000b) *Tithonia diversifolia* as a green manure for soil fertility improvement in western Kenya: a review. *Agroforestry Systems* 49, 201–221.

Jamaludheen, V., Kumar, B.M., Wahid, P.A. and Kamalan, N.V. (1997) Root distribution pattern of wild jack tree (*Artocarpus hirsutus* Lamk.) as studied by $^{32}$P soil injection method. *Agroforestry Systems* 35, 329–336.

Janos, D.P. (1996) Mycorrhizas, succession and the rehabilitation of deforested lands in the humid tropics. In: Frankland, J.C., Magan, N. and Gadd, G.M. (eds) *Fungi and Environmental Change*. Cambridge University Press, Cambridge, UK, pp. 129–161.

Jansa, J., Mozafar, A., Anken, T., Ruh, R., Sanders, I.R. and Frossard, E. (2002) Diversity and structure of AMF communities as affected by tillage in a temperate soil. *Mycorrhiza* 12, 225–234.

Janse, J.M. (1897) Les endophytes radicaux de quelques plantes javanaises. *Annales Jardin Botanique Buitenzorg* 14, 53–212.

Janzen, H.H. and Schaalje, G.B. (1992) Barley response to nitrogen and non-nutritional benefits of legume green manure. *Plant and Soil* 142, 19–30.

Jenkinson, D.S. (1981) The fate of plant and animal residues in soil. In: Greenland, D.J. and Haynes, M.H.B. (eds) *The Chemistry of Soil Processes*. John Wiley & Sons, Chichester–New York, pp. 505–561.

Jenkinson, D.S. and Rayner, J.H. (1977) The turnover of soil organic matter in some of the Rothamsted classical experiments. *Soil Science* 123, 298–305.

Jenkinson, D.S., Hart, P.B.S., Rayner, J.H. and Parry, L.C. (1987) Modelling the turnover of organic matter in long-term experiments at Rothamsted. *INTECOL Bulletin* 15, 1–8.

Jiménez, F. and Alfaro, R. (1999) Available soil water in *Coffea arabica–Erythrina poeppigiana, C. arabica–Eucalyptus deglupta* and *C. arabica* monoculture plantations. In: Jimenez, J. and Beer, J. (eds) *Proceedings of the International Symposium on Multistrata Agroforestry Systems with Perennial Crops, 22–27 February 1999, Extended Abstracts*. CATIE, Turrialba, Costa Rica, pp. 112–115.

Johns, A. (1996) Conserving biodiversity at tropical forest margins. In: Bowen, M.R., Sequeira, V. and Burgess, J. (eds) *Managing the Forest Boundary: Issues at the Forest–Agriculture Interface*. Tropical Forestry Forum, UK, pp. 25–27.

Johnson, N.C. (1993) Can fertilization of soil select less mutualistic mycorrhizae? *Ecological Applications* 3, 749–757.

Johnson, N.C., Copeland, P.J., Cerrookston, R.K. and Pfleger, F.L. (1992) Mycorrhizae: possible explanation for yield decline with continuous corn and soybean. *Agronomy Journal* 84, 387–390.

Johnson, N.C., Graham, J.H. and Smith, F.A. (1997) Functioning of mycorrhizal associations along the mutualism–parasitism continuum. *New Phytologist* 135, 575–585.

Jones, D.L. (1998) Organic acids in the rhizosphere – a critical review. *Plant and Soil* 205, 25–44.

Jones, D.L. and Darrah, P.R. (1994) Role of root derived organic acids in the mobilization of nutrients from the rhizosphere. *Plant and Soil* 166, 247–257.

Jones, J.W., Keating, B.A. and Porter, C.H. (2001) Approaches to modular model development. *Agricultural Systems* 70, 421–443.

Jones, M., Sinclair, F.L. and Grime, V.L. (1998) Effect of tree species and crown pruning on root length and soil water content in semi-arid agroforestry. *Plant and Soil* 201, 97–207.

Jones, P.G. and Thornton, P.K. (2000) MARKSIM: software to generate daily weather data for Latin America and Africa. *Agronomy Journal* 92, 445–453.

Jonsson, K., Fidjeland, L., Maghembe, J.A. and Hogberg, P. (1988) The vertical distribution of fine roots of five tree species and maize in Morogoro, Tanzania. *Agroforestry Systems* 6, 63–69.

Jose, S., Gillespie, A.R., Seifert, J.R. and Biehle, D.J. (2000). Defining competition vectors in a temperate alley cropping system in the Midwestern USA. *Agroforestry Systems* 48, 41–59.

Joshi, K.D., Sthapit, B.R. and Vaidya, A. (1995a) Indigenous methods of maintaining soil fertility and constraints to increasing productivity in the mountain farming systems. In: Joshi, K.D., Vaidya, A., Tripathi, B.P. and Pound, B. (eds) *Formulating a Strategy for Soil Fertility Research in the Hills of Nepal*. Workshop Proceedings, 17–18 August. Lumle Agricultural Research Centre, Pokhara, Nepal, and Natural Resources Institute, Kent, UK, pp. 20–29.

Joshi, K.D., Tuladhar, J.K. and Sthapit, B.R. (1995b) Indigenous soil classification systems and their practical utility: a review. In: Joshi, K.D., Vaidya, A., Tripathi, B.P. and Pound, B. (eds) *Formulating a Strategy for Soil Fertility Research in the Hills of Nepal*. Workshop Proceedings, 17–18 August. Lumle Agricultural Research Centre, Pokhara, Nepal, and Natural Resources Institute, Kent, UK, pp. 36–42.

Joshi, L. (1998) Incorporating farmers' knowledge in the planning of interdisciplinary research and extension. PhD thesis, University of Wales, Bangor, UK, 272 pp.

Joshi, L. and Sinclair, F.S. (1997) Knowledge acquisition from multiple communities. Report submitted to the Department for International Development, UK, 34 pp.

Joshi, L., Wibawa, G., Beukema, H., Williams, S.E. and van Noordwijk, M. (2003) Technological change and biodiversity in the rubber agroecosystem of Sumatra. In: Vandermeer, J. (ed.) *Tropical Agroecosystems*. CRC Press LLC, Boca Raton, Florida, pp. 133–158.

Joshi, R.C., Matchoc, O.R.O., Cabigat, J.C. and James, S.W. (1999) Survey of earthworms in the Ifugao rice terraces, Philippines. *Journal of Environmental Science and Management* 2(2), 1–12.

Joslin, J.D., Wolfe, M.H. and Hanson, P.J. (2000) Effects of altered water regimes on forest root systems. *New Phytologist* 147, 117–129.

Jungk, A.O. (1991) Dynamics of nutrient movement at the soil–root interface. In: Waisel, Y., Eshel, A. and Kafkafi, U. (eds) *Plant Roots: the Hidden Half*. Marcel Dekker, New York, pp. 455–481.

Kadiata, B.D., Mulongoy, K. and Isirimah, N.O. (1997) Influence of pruning frequency of *Albizia lebbeck*, *Gliricidia sepium* and *Leucaena leucocephala* on nodulation and potential nitrogen fixation. *Biology of Soils and Fertilizers* 24, 255–260.

Kaimuddin (2000) Dampak perubahan iklim dan tataguna lahan terhadap keseimbangan air wilayah Sulawesi Selatan. PhD thesis, Institut Pertanian Bogor, Bogor, Indonesia.

Kaiser, E.A., Hohrs, K., Kücke, M., Schnug, E., Heinemeyer, O. and Munch, J.C. (1998) Nitrous oxide release from arable soil: importance of N-fertilization, crops and temporal variation. *Soil Biology and Biochemistry* 30, 1553–1563.

Kandji, S.T., Ogol, C.K.P.O. and Albrecht, A. (2001) Diversity of plant-parasitic nematodes and their relationships with some soil physico-chemical characteristics in improved fallows in western Kenya. *Applied Soil Ecology* 18, 143–157.

Kapp, G.B. and Beer, J. (1995) A comparison of agrosilvicultural systems with plantation forestry in the Atlantic lowlands of Costa Rica. *Agroforestry Systems* 32, 207–223.

Karamura, E.B. and Gold, C.S. (2000) The elusive banana weevil, *Cosmopolites sordidus* Germar. In: Karamura, E. and Vuylsteke, D. (eds) *Proceedings of the First International Conference on Banana and Plantain for Africa. Acta Horticulturae* 540, 471–485.

Kartasubrata, J. (1981) Pre-war concepts concerning land use in Java in particular related to forest conservation. Presented at Symposium on Forest Land Use Planning, Gajah Mada University, Yogyakarta, Indonesia. Reprinted in: Kartasubrata, J. (ed.) (2003) *Social Forestry and Agroforestry in Asia*, Book 2. Faculty of Forestry, Bogor Agricultural University, Bogor, Indonesia, pp. 3–11.

Kasasa, P., Mpepereki, S., Musiyiwa, K., Makonese, F. and Giller, K.E. (1999) Residual nitrogen benefits of promiscuous soybeans to maize under field conditions. *African Crop Science Journal* 7, 375–382.

Kashaija, I.N., Fogain, R. and Speijer, P.R. (2001) Habitat management for control of banana nematodes. In: Frison, E.A., Gold, C.S., Karamura, E.B. and Sikora, R.A. (eds) *Mobilizing IPM for Sustainable Banana Production in Africa. Proceedings of a workshop on banana IPM held in Nelspruit, South Africa, 23–28 November 1998*. International Network for Improvement of Banana and Plantain (INIBAP), Montpellier, France, 356 pp. www.inibap.org/publications/proceedings/mobilizingIPM.pdf

Kasper, T.C. and Bland, W.L. (1992) Soil temperature and root growth. *Soil Science* 154, 290–299.

Kasper, T.C., Wolley, D.G. and Taylor, H.M. (1981) Temperature effect on the inclination of lateral roots of soybeans. *Agronomy Journal* 73, 383–385.

Keller, M. and Reiners, W.A. (1994) Soil-atmosphere exchange of nitrous oxide, nitric oxide and methane under secondary succession of pasture to forest in the Atlantic lowlands of Costa Rica. *Global Biogeochemical Cycles* 8, 399–409.

Keller, M.E., Veldkamp, A.M., Weitz, A.M. and Reiners, W.A. (1993) Effect of pasture age on soil trace-gas emissions from a deforested area of Costa Rica. *Nature* 365, 244–246.

Kelly, R.H., Parton, W.J., Crocker, G.J., Grace, P.R., Klír, J., Körschens, M., Poulton, P.R. and Richter, D.D. (1997) Simulating trends in soil organic carbon in long-term experiments using the CENTURY model. *Geoderma* 81(1–2), 75–90.

Kemper, W.D. (1986) Solute diffusivity. In: Klute, A. (ed.) Methods of Soil Analysis, Part 1, Physical and Mineralogical Methods. *Agronomy* 9, 1007–1024.

Kendon, G., Walker, D.H., Robertson, D., Haggith, M., Sinclair, F.L. and Muetzelfeldt, R.I. (1995) Supporting customised reasoning in the agroforestry domain. *The New Review of Applied Expert Systems* 1, 179–192.

Kenis, M., Sileshi, G. and Bridge, J. (2001) Parasitism of the leaf beetle *Mesoplatys ochroptera* Stal (Coleoptera: Chrysomelidae) in eastern Zambia. *Biocontrol Science and Technology* 11, 611–622.

Kerven, C., Dolva, H. and Renna, R. (1995) Indigenous soil classification systems in Nothern Zambia. In: Warren, D.M., Slikkerveer, I.J. and Brokensha, D. (eds) *The Cultural Dimension of Development. Indigenous Knowledge Systems*. Intermediate Technology Publications, London, pp. 82–87.

Khalil, M.I., Rosenani, A.B., Van Cleemput, O., Fauziah, C.I. and Shamshuddin, J. (2000) Nitrous oxide emissions from a sustainable land management system in the humid tropics. In: *Proceedings of the International Symposium on Sustainable Land Management*, Kuala Lumpur, Malaysia, pp. 71–72.

Khan, M.W. (1993) *Nematode Interactions*, 1st edn. Chapman and Hall, London, UK, 377 pp.

Khanna, P.K. (1998) Nutrient cycling under mixed-species tree systems in Southeast Asia. *Agroforestry Systems* 38, 99–120.

Khasanah, N., Lusiana, B., Farida and van Noordwijk, M. (2002) Simulating run-off and soil erosion using WaNuLCAS: Water, Nutrient and Light Capture in Agroforestry Systems. In: Backgrounds for ACIAR Project Planning Meeting, Sumberjaya, 12–16 October 2002. World Agroforestry Centre (ICRAF), SE Asia, Bogor, Indonesia, pp. 122–127.

Kho, R.M. (2000a) On crop production and the balance of available resources. *Agriculture, Ecosystems and Environment* 80, 71–85.

Kho, R.M. (2000b) A general tree–environment–crop interaction equation for predictive understanding of agroforestry systems. *Agriculture, Ecosystems and Environment* 80, 87–100.

Kho, R.M. (2002) Approaches to tree–environment–crop interactions. *Journal of Crop Production*.

Kho, R.M., Yacouba, B., Yayé, M., Katkoré, B., Moussa, A., Iktam A. and Mayaki, A. (2001) Separating the effects of trees on crops: the case of *Faidherbia albida* and millet in Niger. *Agroforestry Systems* 52, 219–238.

Kidd, C.V. and Pimental, D. (1992) *Integrated Resource Management. Agroforestry for Development*. Academic Press, San Diego.

Kiers, E.T., West, S.A. and Denison, R.F. (2002) Mediating mutualisms: farm management practices and evolutionary changes in symbiont co-operation. *Journal of Applied Ecology* 39, 745–754.

Kiersch, B. and Tognetti, S. (2002) *Land–Water Linkages in Rural Watersheds: Proceedings of an Electronic Workshop.* FAO Land and Water Development Division. FAO Land and Water Bulletin Vol. 9. FAO, Rome, 80 pp.

Kijne, J.W., Barker, R. and Molden, D. (2003) *Water Productivity in Agriculture: Limits and Opportunities for Improvement.* CAB International, Wallingford, UK.

Killham, K. and Yeomans, C. (2001) Rhizosphere carbon flow measurement and implications: from isotopes to reporter genes. *Plant and Soil* 232, 91–96.

Kiniry, J.R. and Williams, J.R. (1993) Simulating intercropping with the ALMANAC model. In: Sinoquet, H. and Cruz, P. (eds) *Ecophysiology of Tropical Intercropping.* Proceedings of an International Meeting held in Guadeloupe on 6–10 December, 1994. Institut National de la Recherche Agronomique (INRA), Paris, pp. 387–396.

Kiniry, J.R., Williams, J.R., Gassman, P.W. and Debaeke, P. (1992) A general process-oriented model for two competing plant species. *Transactions of the American Society of Agricultural Engineers* 35(3), 801–810.

Kinraide, T.B. and Parker, D.R. (1987) Cation amelioration of aluminum toxicity in wheat. *Plant Physiology* 83(3), 546–551.

Kiptot, E. (1996) An investigation of farmers' ecological knowledge about fruit trees grown on farms in South Yatta, Kenya. MPhil thesis, University of Wales, Bangor, UK, 155 pp.

Kirk, G.J.D. (1999) A model of phosphate solubilization by organic anion excretion from plant roots. *European Journal of Soil Science* 50, 369–378.

Kirk, G.J.D. (2002a) Modelling root-induced solubilization of nutrients. *Plant and Soil* 245, 49–57.

Kirk, G.J.D. (2002b) Use of modeling to understand nutrient acquisition by plants. *Plant Soil* 247, 123–130.

Knapp, R. (1973) *Die Vegetation von Afrika.* Gustav Fischer, Stuttgart, Germany.

Knight, J.D. (1997) The role of decision support systems in integrated crop protection. *Agriculture Ecosystems and Environment* 64, 157–163.

Kochain, L.V. and Jones, D.L. (1996) Aluminium toxicity and resistance in plants. In: Yokel, R. and Golub, M.S. (eds) *Research Issues in Aluminium Toxicity.* Francis Publishers, Washington, DC, pp. 69–89.

Korwar, G.R. and Radder, G.D. (1994) Influence of root pruning and cutting interval of *Leucaena* hedgerows on performance of alleycropped *rabi* sorghum. *Agroforestry Systems* 25, 95–109.

Kranz, J., Schmutterer, H. and Koch, W. (1977) *Diseases, Pests and Weeds in Tropical Crops.* Paul Parey, Berlin, Germany, 666 pp.

Kroeze, C., Mosier, A.R. and Bouwman, L. (1999) Closing the global $N_2O$ budget: a retrospective analysis 1500–1994. *Global Biogeochemical Cycles* 13, 1–8.

Kwesiga, F.R., Franzel, S., Place, F., Phiri, D. and Simwanza, C.P. (1999) *Sesbania sesban* improved fallows in eastern Zambia: their inception, development and farmer enthusiasm. *Agroforestry Systems* 47, 49–66.

Lambert, D.H., Baker, D.E. and Cole, H.J.R. (1979) The role of mycorrhizae in the interactions of phosphorus with zinc, copper, and other elements. *Soil Science Society of America Journal* 43, 976–980.

Lamberti, F., Waller, J.M. and Van der Graaff, N.A. (1983) *Durable Resistance in Crops.* Plenum, New York, 454 pp.

Lamers, J.P.A. and Feil, P.R. (1995) Farmers' knowledge and management of spatial soil and crop growth variability in Niger, West Africa. *Netherlands Journal of Agricultural Science* 43, 375–389.

Lamont, B.B. (1986) The significance of proteoid roots in proteas. *Acta Horticultura* 185, 163–170.

Lavelle, P. and Pashanasi, B. (1989) Soil macrofauna and land management in Peruvian Amazonia (Yurimaguas, Loreto). *Pedobiologia* 33, 283–291.

Lavelle, P. and Spain, A.V. (2002) *Soil Ecology.* Kluwer Academic Publishers, Dordrecht, The Netherlands, 653 pp.

Lavelle, P., Spain, A.V., Blanchart, E., Martin, A. and Martin, S. (1992) Impact of soil fauna on the properties of soils in the humid tropics. In: Lal, R. and Sanchez, P.A. (eds) *Myths and Science of Soils of the Tropics.* Soil Science Society of America and American Society of Agronomy, Madison, Wisconsin, USA. SSSA Special Publication Number 29, pp. 157–185.

Lavelle, P., Bignell, D., Lepage, M., Wolters, V., Roger, P., Ineson, P., Heal, O.W. and Ghillion, S. (1997) Soil function in a changing world: the role of invertebrate ecosystem engineers. *European Journal of Soil Biology* 33, 159–193.

Lavelle, P., Barros, I., Blanchart, E., Brown, G.G., Brussaard, L., Decaëns, T., Fragoso, C., Jimenez, J.J., Ka Kajondo, K., Martínez, M.A., Moreno, A.G., Pashanasi, B., Senapati, B.K. and Villenave, C. (1998) Earthworms as a resource in tropical agroecosystems. *Nature and Resources* 34, 28–44.

Le Page, M. (2002) Village-life.com: the web's the way to catch a fish or arrange a marriage. *New Scientist* 2341(4 May, 2002), 44–45.

Lee, D.R. and Barrett, C.B. (eds) (2001) *Tradeoffs or Synergies? Agricultural Intensification, Economic Development and the Environment.* CAB International, Wallingford, UK.

Lefroy, E.C., Hobbs, R.J., O'Connor, M.H. and Pate, J.S. (eds) (1999) *Agriculture as a Mimic of Natural Ecosystems.* Current Plant Science and Biotechnology in Agriculture. Vol. 37, Kluwer Academic Publishers, Dordrecht, The Netherlands.

Leggett M., Gleddie, S. and Holloway, G. (2001) Phosphate-solubilizing microorganisms and their use. In: Ae, N., Arihara, J., Okada, K. and Srinivasan, A. (eds) *Plant Nutrient Acquisition: New Perspectives.* Springer-Verlag, Tokyo, Japan, pp. 299–318.

Lehmann, J. and Muraoka, T. (2001) Tracer methods to assess nutrient uptake distribution in multi-strata agroforestry systems. *Agroforestry Systems* 53, 133–140.

Lehmann, J., da Silva, J.P., Schroth, G., Gebauer, G. and da Silva, L.F. (2000) Nitrogen use in mixed tree–crop plantations with a legume cover crop. *Plant and Soil* 225, 63–72.

Lehmann, J., Günther D., Socorro da Mota, M., Pereira de Almeida, M., Zech, W. and Kaiser, K. (2001a) Inorganic and organic soil phosphorus and sulfur pools in an Amazonia multistrata agroforestry system. *Agroforestry Systems* 53, 113–124.

Lehmann, J., Muraoka, T. and Zech, W. (2001b) Root activity patterns in an Amazonian agroforest with fruit–trees determined by $^{32}$P, $^{33}$P, and $^{15}$N applications. *Agroforestry Systems* 52, 185–197.

Lendzemo, V.W. and Kuyper, T.W. (2001) Effects of arbuscular mycorrhizal fungi on damage by *Striga hermonthica* on two contrasting cultivars of sorghum, *Sorghum bicolor. Agriculture, Ecosystems and Environment* 87, 29–35.

Lenné, J.M. and Boa, E.R. (1994). Diseases of tree legumes. In: Gutteridge, R.C. and Shelton, H.M. (eds) *Forage Tree Legumes in Tropical Agriculture.* CAB International, Wallingford, UK, pp. 292–308.

Lerat, S., Gauci, R., Catford, J.G., Vierheilig, H., Piché, Y. and Lapointe, L. (2002) $^{14}$C transfer between the spring ephemeral *Erythronium americanum* and sugar maple seedlings via arbuscular mycorrhizal fungi in natural stands. *Oecologia* 132, 181–187.

Levi-Strauss, C. (1966) *The Savage Mind.* Weidenfeld and Nicolson, London, UK, 290 pp.

Li, M., Osaki M., Rao, I.M. and Toshiaki, T. (1997) Secretion of phytase from the roots of several plant species under phosphorus-deficient conditions. *Plant and Soil* 195, 161–169.

Liddell, C.M. (1997) Abiotic factors and soilborne diseases. In: Hillocks, R.J. and Waller, J.M. (eds) *Soilborne Diseases of Tropical Crops,* 1st edn. CAB International, Wallingford, UK, pp. 365–376.

Lieffers, M.L., and Rothwell, R.O. (1987) Rooting of peatland black spruce and tamarak in relation to depth of water table. *Canadian Journal of Botany* 65, 817–821.

Lindenschmidt, K.E., Suhr, M., Magumba, M.K., Hecky, R.E. and Bugenyi, F.E.W. (1998) Loading of solutes and suspended solids from rural catchment areas flowing into Lake Victoria in Uganda. *Water Research* 32, 2776–2786.

Lindsay, W.L. (1979) *Chemical Equilibria in Soils.* John Wiley & Sons, New York, 449 pp.

Linford, M.B., Yap, F. and Oliveira, J.M. (1937) *Reduction of Soil Populations of the Root-knot Nematode During Decomposition of Organic Matter.* Technical Paper No. 108, Pineapple Experiment Station, University of Hawaii, Honolulu, pp. 127–141.

Linquist, B.A., Singleton, P.W., Cassman, K.G. and Keane, K. (1996) Residual phosphorus and long-term management strategies for an Ultisol. *Plant and Soil* 184, 47–55.

Linquist, B.A., Singleton, P.W. and Cassman, K.G. (1997) Inorganic and organic phosphorus dynamics during a build-up and decline of available phosphorus in an Ultisol. *Soil Science* 162, 254–264.

Livesley, S.J., Gregory, P.J. and Buresh, R.J. (2002) Competition in tree row agroforestry systems. 2. Distribution, dynamics and uptake of soil inorganic N. *Plant and Soil* 247, 177–187.

Liyanage, A. de S. (1997) Rubber. In: Hillocks, R.J. and Walle, J.M. (eds) *Soilborne Diseases of Tropical Crops,* 1st edn. CAB International, Wallingford, UK, pp. 331–347.

Loos, C.E. (1961) Eradication of the burrowing nematode, *Radopholus similis,* from bananas. *Plant Disease Reporter* 45, 457–461.

Loreau, M. (2000) Biodiversity and ecosystem functioning: recent theoretical advances. *Oikos* 91, 3–17.

Lott, J.E., Howard, S.B., Ong, C.K., Black, C.R. (2000) Long term productivity of a *Grevillea robusta*-based agroforestry system in semi-arid Kenya. 2. Crop growth and system performance. *Forest Ecology and Management* 139, 187–201.

Lott, J.E., Khan, A.A.H., Black, C.R. and Ong, C.K. (2003) Water use in a *Grevillea robusta*–maize overstorey agroforestry system in semi-arid Kenya. *Forest Ecology and Management* 180, 45–59.

Lovell, C., Mandondo, A. and Moriarty, P. (2002) The question of scale in integrated natural resource management. *Conservation Ecology* 5(2), 25. [online] URL: http://www.consecol.org/vol5/iss2/art25

Lowrance, R., Altier, L.S., Newbold, J.D., Schnabel, R.R., Groffman, P.M., Denver, J.M., Correll, D.L., Gilliam, J.W., Robinson, J.L., Brinsfield, R.B., Staver, K.W., Lucas, W. and Todd, A.H. (1997) Water quality functions of riparian forest buffers in Chesapeake Bay watersheds. *Environmental Management* 21, 687–712.

Luc, M., Bridge, J. and Sikora, R.A. (1990) Reflections on nematology in subtropical and tropical agriculture. In: Luc, M., Sikora, R.A. and Bridge, J. (eds) *Plant Parasitic Nematodes in Subtropical and Tropical Agriculture*. International Institute of Parasitology, CAB International, Wallingford, UK, pp. xi–xvii.

Luizão, F., Matson, P., Livingston, G., Luizão, R. and Vitousek, P. (1989) Nitrous oxide flux following tropical land clearing. *Global Biogeochemical Cycles* 3, 281–285.

Lupwayi, N.Z., Arshad, M.A., Rice, W.A. and Clayton, G.W. (2001) Bacterial diversity in water-stable aggregates of soils under conventional and zero tillage management. *Applied Soil Ecology* 16, 251–261.

Lynch, J.M. and Whipps, J.M. (1991) Substrate flow in the rhizosphere. In: Keister, D.L. and Cregan, P.B. (eds) *The Rhizosphere and Plant Growth*. Kluwer Academic Publishers, Boston, Massachusetts, pp. 15–24.

Macadam, R., Britton, I., Russell, D., Potts, W., Baillie, B. and Shaw, A. (1990) The use of soft systems methodology to improve the adoption by Australian cotton growers of the Siratac computer-based crop management system. *Agricultural Systems* 34, 1–14.

MacArthur, R. (1955) Fluctuations of animal populations, and a measure of community stability. *Ecology* 36, 533–536.

Mafongoya, P.L., Nair, P.K. and Dzowela, B.H. (1997) Multipurpose tree prunings as a source of nitrogen to maize under semiarid conditions in Zimbabwe. *Agroforestry Systems* 35, 57–70.

Mafongoya, P.L., Giller, K.E. and Palm, C.A. (1998) Decomposition and nitrogen release patterns of tree prunings and litter. *Agroforestry Systems* 38, 77–97.

Magid, J., Cadisch, G. and Giller, K.E. (2002) Short and medium term plant litter decomposition in a tropical ultisol, as elucidated by physical fractionation in a dual $^{13}C$ and $^{14}C$ isotope study. *Soil Biology and Biochemistry* 34, 1273–1281.

Mahieu, N., Olk, D.C. and Randall, E.W. (2000) Analysis of phosphorus in two humic acid fractions of intensively cropped lowland rice soils by $^{31}P$-NMR. *European Journal of Soil Science* 51, 391–402.

Mai, W.F. (1986) Plant-parasitic nematodes: their threat to agriculture. In: Sasser, J.N. and Carter, C.C. (eds) *An Advanced Treatise on Meloidogyne*. Volume I. *Biology and Control*. North Carolina State University, pp. 11–17.

Makumba, W., Akinnifesi, F.K. and Kwesiga, F.R. (2001) Above and belowground performance of *Gliricidia*/maize mixed cropping in Makoka, Malawi. In: Kwesiga, F., Ayuk, E. and Aggumya, A. (eds) *Proceedings of the 14th Southern Africa Regional Review and Planning Workshop, 3–7 September 2001, ICRAF Regional Office, Harare, Zimbabwe*. ICRAF, Nairobi, Kenya.

Malajczuk, N. and Cromack, K. Jr. (1982) Accumulation of calcium oxalate in the mantle of ectomycorrhizal roots of *Pinus radiata* and *Eucalyptus marginata*. *New Phytologist* 92, 527–531.

Malik, R.S. and Sharma, S.K. (1990) Moisture extraction and crop yield as a function of distance from a row of *Eucalyptus tereticornis*. *Agroforestry Systems* 12, 187–195.

Manlay, R.J., Chotte, J.-L., Masse, D., Laurent, J.-Y. and Feller, C. (2002) Carbon, nitrogen and phosphorus allocation in agro-ecosystems of a West African savanna. III. Plant and soil components under continuous cultivation. *Agriculture, Ecosystems and Environment* 88, 249–269.

Mapfumo, P., Mpepereki, S. and Mafongoya, P. (2000) Pigeonpea rhizobia prevalence and crop response to inoculation in Zimbabwean smallholder-managed soils. *Experimental Agriculture* 36, 423–434.

Maroko, J.B., Buresh, R.J. and Smithson, P.C. (1999) Soil phosphorus fractions in unfertilized fallow-maize systems on two tropical soils. *Soil Science Society of America Journal* 63, 320–326.

Marschner, H. (1995) *Mineral Nutrition of Higher Plants*. Academic Press, London, 889 pp.

Marsden, W.H. (1783) *The History of Sumatra*, 3rd edn (1811). Oxford University Press, Oxford, UK.

Martinez-Romero, E., Segovia, L., Mercante, F.M., Franco, A.A., Graham, P.H. and Pardo, M.A. (1991) *Rhizobium tropici*: a novel species nodulating *Phaseolus vulgaris* L. beans and *Leucaena* sp. trees. *International Journal of Systematic Bacteriology* 41, 417–426.

Marx, D.H. (1975) The role of ectomycorrhizae in the protection of pine from root infection by *Phytophthora cinnamomi*. In: Bruehl, G.W. (ed.) *Biology and Control of Soil-Borne Plant Pathogens*. American Phytopathological Society, St Paul, Minnesota, pp. 112–115.

Matthews, R.B. and Hunt, L.A. (1994) GUMCAS: a model describing the growth of cassava (*Manihot esculenta* L. Crantz). *Field Crops Research* 36, 69–84.

Matthews, R.B. and Lawson, G.J. (1997) Structure and applications of the HyCAS model. *Agroforestry Forum* 8(2), 14–17.

Matthews, R.B. and Stephens, W. (eds) (2002) *Crop–Soil Models: Applications in Developing Countries*. CAB International, Wallingford, UK, 304 pp.

Matthews, R.B., Wassmann, R., Knox, J. and Buendia, L.V. (2000) Using a crop/soil simulation model and GIS techniques to assess methane emissions from rice fields in Asia. IV. Upscaling of crop management scenarios to national levels. *Nutrient Cycling in Agroecosystems* 58, 201–217.

Matthews, R., Stephens, W., Hess, T., Mason, T. and Graves, A. (2002a) Applications of crop/soil simulation models in tropical agricultural systems. *Advances in Agronomy* 76, 31–124.

Matthews, R.B., Stephens, W. and Hess, T.M. (2002b) Impacts of crop/soil models. In: Matthews, R.B. and Stephens, W. (eds) *Crop–Soil Simulation Models: Applications in Developing Countries*. CAB International, Wallingford, UK, pp. 195–205.

May, R.M. (1972) Will a large complex system be stable? *Nature* 238, 413–416.

May, R.M. (1973) *Stability and Complexity in Model Ecosystems*. Princeton University Press, Princeton, New Jersey.

Mayers, P.E. and Hutton, D.G. (1987) Bacterial wilt, a new disease of custard apple: symptoms and etiology. *Annals of Applied Biology* 111(1), 135–141.

Mayus, M., Keulen, H.V. and Stroosnijder, L. (1998a) A model of tree–crop competition for windbreak systems in the Sahel: description and evaluation. *Agroforestry Systems* 43(1–3), 183–201.

Mayus, M., van Keulen, H. and Stroosnijder, L. (1998b) Analysis for dry and wet years with the WIMISA model of tree–crop competition for windbreak systems in the Sahel. *Agroforestry Systems* 43(1–3), 203–215.

McCann, K., Hastings, A.G. and Huxel, R. (1998) Weak trophic interactions and the balance of nature. *Nature* 395, 794–798.

McClellan, G.H. and Gremillion, L.R. (1980) Evaluation of phosphatic raw materials. In: Khasawneh, F.E., Sample, E.C. and Kamprath, E.J. (eds) *The Role of Phosphorus in Agriculture*. American Society of Agronomy, Madison, Wisconsin, pp. 43–80.

McCown, R.L., Hammer, G.L., Hargreaves, J.N.G., Holzworth, D.P. and Freebairn, D.M. (1996) APSIM: a novel software system for model development, model testing and simulation in agricultural systems research. *Agricultural Systems* 50, 255–271.

McDonald, J.D. (1994) The soil environment. In: Campbell, C.L. and Benson, D.M. (eds) *Epidemiology and Management of Root Diseases*. Springer-Verlag, Berlin, Germany, pp. 82–116.

McDonald, M.A., Lawrence, A. and Shresta, P.K. (2003) Soil erosion. In: Schroth, G. and Sinclair, F.L. (eds) *Trees, Crops and Soil Fertility: Concepts and Research Methods*. CAB International, Wallingford, UK, pp. 325–343.

McIntyre, B.D., Riha, S.J. and Ong, C.K. (1997) Competition for water in a hedge-intercrop system. *Field Crops Research* 52, 151–160.

McLaughlin, M.J., Alston, A.M. and Martin, J.K. (1988) Phosphorus cycling in wheat-pasture rotations. III. Organic phosphorus turnover and phosphorus cycling. *Australian Journal of Soil Research* 26, 343–353.

McLaughlin, M.J., Malik, K.A., Memon, K.S. and Idris, M. (1990) The role of phosphorus in nitrogen fixation in upland crops. In: *Phosphorus Requirements for Sustainable Agriculture in Asia and Oceania*. International Rice Research Institute, Los Banos, the Philippines, pp. 295–305.

McNeill, A.M., Chunya, Z. and Fillery, I.R.P. (1997) Use of in situ 15-N labelling to estimate the total below-ground nitrogen of pasture legumes in intact soil–plant systems. *Australian Journal of Agricultural Research* 8, 295–304.

McSorley, R. (1981) Plant-parasitic nematodes associated with tropical and subtropical fruits. *IFAS Technical Bulletin 823*, University of Florida, Gainesville, Florida, 49 pp.

McSorley, R. (2001) Multiple cropping systems for nematode management: a review. *Soil and Crop Science Society of Florida* 60, 132–142.

McSorley, R., Dickson, D.W. and de Brito, J.A. (1994) Host status of selected tropical rotation crops to four populations of root-knot nematodes. *Nematropica* 24, 45–53.

Meinke, H., Baethgen, W.E., Carberry, P.S., Donatelli, M., Hammer, G.L., Selvaraju, R. and Stockle, C.O. (2001) Increasing profits and reducing risks in crop production using participatory systems simulation approaches. *Agricultural Systems* 70, 493–513.

Mekonnen, K., Buresh, R.J. and Jama, B. (1997) Root and inorganic nitrogen distributions in sesbania fallow, natural fallow and maize fields. *Plant and Soil* 188, 319–327.

Mekonnen, K., Buresh, R.J., Coe, R. and Kipleting, K. (1999) Root length and nitrate under *Sesbania sesban*: vertical and horizontal distribution and variability. *Agroforestry Systems* 42, 265–282.

Melendez, P.L. and Powell, N.T. (1969) The influence of *Meloidogyne* on root decay in tobacco caused by *Pythium* and *Trichoderma*. *Phytopathology* 59, 1348.

Melillo, J.M., Steudler, P., Feigl, B.J., Neill, C., Garcia, D., Piccolo, M.C., Cerre, C.C. and Tian, H. (2001) Nitrous oxide emissions from forests and pastures of various ages in the Brazilian Amazon. *Journal of Geophysical Research* 106, 34179–34188.

Meroto, A. Jr and Mundstock, C.M. (1999) Wheat root growth as affected by soil strength. *Revista Brasileira de Ciencia de Solo* 23, 197–202.

Michori, P.K. (1993) Nitrogen budget under coffee. PhD thesis, University of Reading, UK (British Thesis Service DX175716), 334 pp.

Millar, N. (2002) The effect of improved fallow residue quality on nitrous oxide emissions from tropical soils. PhD thesis, University of London.

Miller, M.H. (2000) Arbuscular mycorrhizae and the phosphorus nutrition of maize: a review of Guelph studies. *Canadian Journal of Plant Science* 80, 47–52.

Miller, R.M. and Jastrow, J.D. (1990) Hierarchy of root and mycorrhizal fungal interactions with soil aggregation capacities. *Soil Biology and Biochemistry* 22, 579–584.

Minasny, B., McBratney, A.B. and Bristow, K.L. (1999) Comparison of different approaches to the development of pedotransfer functions for water-retention curves. *Geoderma* 93, 224–253.

Misra, R.K. and Rose, C.W. (1990) Manual for use of program GUEST. Monograph, Division of Australian Environmental Studies, Griffith University, Brisbane, Australia.

Misra, R.K. and Rose, C.W. (1996) Application and sensitivity analysis of process-based erosion model GUEST. *European Journal of Soil Science* 47, 593–604.

Mobbs, D.C., Cannell, M.G.R., Crout, N.M.J., Lawson, G.J., Friend, A.D. and Arah, J. (1998) Complementarity of light and water use in tropical agroforests I. Theoretical model outline, performance and sensitivity. *Forest Ecology and Management* 102(2–3), 259–274.

Mobbs, D.C., Lawson, G.J., Friend, A.D., Crout, N.M.J., Arah, J. and Hodnett, M. (2001) *HyPAR v4.5 Technical Manual*. Centre for Ecology and Hydrology, Edinburgh, UK, 11 pp.

Mokwunye, A.U., Chien, S.H. and Rhodes, E. (1986) Phosphorus reaction with tropical African soils. In: Mokwunye, A.U. and Vlek, P.L.G. (eds) *Management of Nitrogen and Phosphorus Fertilizers in Sub-Saharan Africa*. Martinus Nijhoff, Dordrecht, The Netherlands, pp. 253–281.

Monteith, J.L. and Unsworth, M.H. (1990) *Principles of Environmental Physics*, 2nd edn. Edward Arnold, London, 291 pp.

Monteith, J.L., Scott, R.K. and Unsworth, M.H. (eds) (1994) *Resource Capture by Crops*. Proceedings of Easter Schools in Agricultural Science. Nottingham University Press, Nottingham, UK, 469 pp.

Morales, E. and Beer, J. (1998) Distribución de raíces finas de *Coffea arabica* y *Eucalyptus deglupta* en cafetales del Valle Central de Costa Rica. *Agroforesteria en las Americas* 5(17–18), 44–48.

Morton, J.B. and Redecker, D. (2001) Two new families of *Glomales*, Archaeosporaceae and Paraglomaceae, with two new genera *Archaeospora* and *Paraglomus*, based on concordant molecular and morphological characters. *Mycologia* 93, 181–195.

Mosier, A., Schimel, D., Valentine, D., Bronson K. and Parton, W. (1991) Methane and nitrous oxide fluxes in native, fertilized and cultivated grasslands. *Nature* 350, 330–332.

Moss, C., Frost, F., Obiri-Darko, B., Jatango, J., Dixon, H. and Sinclair, F.L. (2001) *Local Knowledge and Livelihoods: Tools for Soils Research and Dissemination in Ghana*. School of Agricultural and Forest Sciences, University of Wales, Bangor, UK.

Mpepereki, S., Javaheri, F., Davis, P. and Giller, K.E. (2000) Soyabeans and sustainable agriculture: 'promiscuous' soyabeans in southern Africa. *Field Crops Research* 65, 137–149.

Mulatya, J.M. (2000) Tree root development and interactions in drylands: focusing on *Melia volkensii* with socio-economic evaluations. PhD thesis, University of Dundee, UK.

Mulatya, J.M., Wilson, J., Ong, C.K., Deans, J.D. and Sprent, J.I. (2002) Root architecture of provenances, seedlings and cuttings of *Melia volkensii*: implications for crop yield in dryland agroforestry. *Agroforestry Systems* 56, 65–72.

Mulligan, D.R. (1988) Phosphorus concentrations and chemical fractions in *Eucalyptus* seedlings grown for a long period under nutrient-deficient conditions. *New Phytologist* 110, 479–486.

Mulligan, D.R. and Sands, R. (1988) Dry matter, phosphorus and nitrogen partitioning in three *Eucalyptus* species grown under a nutrient deficit. *New Phytologist* 109, 21–28.

Mungai, D.N. (2003) An assessment of hydrological impact of land use change in the Lake Victoria basin. A case study of the Mara and South Ewaso-Ngiro watersheds.

Munoz, F. and Beer, J. (2001) Fine root dynamics of shaded cacao plantations in Costa Rica. *Agroforestry Systems* 51, 119–130.

Munyanziza, E., Kehri, H.K. and Bagyaraj, D.J. (1997) Agricultural intensification, soil biodiversity and agro-ecosystem function in the tropics: the role of mycorrhiza in crops and trees. *Applied Soil Ecology* 6, 77–85.

Murniati (2002) From *Imperata cylindrica* grasslands to productive agroforestry. PhD thesis, Wageningen Agricultural University, The Netherlands, 170 pp.

Murphy, D.V., Bhogal, A., Shepherd, M., Goulding, K.W.T., Jarvis, S.C., Barraclough, D. and Gaunt, J.L. (1999a) Comparison of N-15 labelling methods to measure gross nitrogen mineralization. *Soil Biology and Biochemistry* 31, 2015–2024.

Murphy, D.V., Fortune, S., Wakefield, J.A., Stockdale, E.A., Poulton, P.R., Webster, C.P., Wilmer, W.S., Goulding, K.W.T. and Gaunt, J.L. (1999b) Assessing the importance of soluble organic nitrogen in agricultural soils. In: Wilson, W. (ed.) *Managing Risks of Nitrates to Humans and the Environment*. Royal Society of Chemistry, Cambridge, Special Publication No. 237, pp. 65–86.

Mutsaers, H.J.W. and Wang, Z. (1999) Are simulation models ready for agricultural research in developing countries? *Agronomy Journal* 91, 1–4.

Nadkarni, N.M. (1981) Canopy roots: convergent evolution in rainforest nutrient cycles. *Science* 214, 1023–1024.

Nair, P.K.R. (1993) *An Introduction to Agroforestry*. Kluwer Academic Publishers, Dordrecht, The Netherlands, 499 pp.

Nair, P.K.R., Kang, B.T. and Kass, D.C.C. (1995) Nutrient cycling and soil erosion control in agroforestry systems. In: Juo, A.S.R. and Freed, R.D. (eds) *Agriculture and the Environment: Bridging Food Production and Environmental Protection in Developing Countries*. American Society Of Agronomy, Madison, Wisconsin, pp. 115–136.

Namirembe, S. (1999) Tree shoot pruning to control competition for below-ground resources in agroforestry. PhD thesis, University of Wales, Bangor, UK.

Nannipieri, P., Sastre, I., Landi, L., Lobo, M.C. and Pietramellara, G. (1996) Determination of extracellular neutral phosphomonoesterase activity in soil. *Soil Biology and Biochemistry* 28, 107–112.

Narwal, S.S. (2000) Allelopathy in ecological agriculture. In: Narwal, S.S., Hoagland, R.E., Dilday, R.H. and Reigosa, M.J. (eds) *Allelopathy In Ecological Agriculture and Forestry. Proceedings of the III International Congress on Allelopathy in Ecological Agriculture and Forestry*, Dharwad, India, 18–21 August, 1998. Kluwer Academic, Dordrecht, The Netherlands, pp. 11–32.

Natscher, L. and Schwertmann, U. (1991) Proton buffering in organic horizons of acid forest soils. *Geoderma* 48, 93–106.

Ndufa, J.K. (2001) Nitrogen and soil organic matter benefits to maize by fast-growing pure and mixed species legume fallows in western Kenya. PhD thesis, Imperial College London, Wye Campus, UK, 323 pp.

Ndufa, J.K., Sheperd, K.D., Buresh, R.J. and Jama, B. (1999) Nutrient uptake and growth of young trees in a P-deficient soil: tree species and phosphorus effects. *Forest Ecology and Management* 122, 231–241.

Nelson, R., Grist, P., Menz, K., Paningbatan, E. and Mamicpic, M. (1997) A cost-benefit analysis of hedgerow intercropping in the Philippine uplands using the SCUAF model. *Agroforestry Systems* 35, 203–220.

Nepstad, D.C., de Carvalho, C.R., Davidson, E.A., Jipp, P.H., Lefebvre, P.A., de Negreiros, G.H., da Silva, E.D., Stone, T.A., Trumbore, S.E. and Vieira, S. (1994) The role of deep roots in the hydrological and carbon cycles of Amazonian forests and pastures. *Nature* 372, 666–669.

Neutel, A.M. (2001) Stability of complex food webs: pyramids of biomass, interaction strengths and the weight of trophic loops. PhD thesis, University of Utrecht, The Netherlands, 128 pp.

Neutel, A.M., Heesterbeek, J.A.P. and de Ruiter, P.C. (2002) Stability in real food webs: weak links in long loops. *Science* 296, 1120–1123.

Newman, E.I. (1969) Resistance to water flow in soil and plant. I. Soil resistance in relation to amounts of root: theoretical estimates. *Journal of Applied Ecology* 6, 1–12.

Newsham, K.K., Fitter, A.H. and Watkinson, A.R. (1995) Multifunctionality and biodiversity in arbuscular mycorrhizas. *Trends in Ecology and Evolution* 10, 407–411.

Niang, A., Amadalo, B., Gathumbi, S. and Obonyo, C. (1996a) Maize yield response to green manure application from selected shrubs and tree species in western Kenya: a preliminary assessment. In: *Proceedings of the First Kenya National Agroforestry Conference on People and Institutional Participation in Agroforestry for Sustainable Development*. KEFRI, Nairobi, Kenya, pp. 350–358.

Niang, A., Gathumbi, S. and Amadalo, B. (1996b) The potential of short duration improved fallow for crop productivity enhancement in the highlands of western Kenya. *East Africa Agricultural and Forestry Journal* 62, 103–114.

Nick, G., de Lajudie, P., Eardly, B.D., Soumalainen, S., Paulin, L., Zhang, X., Gillis, M. and Lindstrom, K. (1999) *Sinorhizobium arboris* sp. nov. and *Sinorhizobium kostiense* sp. nov., two new species isolated from leguminous trees in Sudan and Kenya. *International Journal of Systematic Bacteriology* 49, 1359–1368.

Nickel, J.L. (1972) Pest situation in changing agricultural systems – a review. In: *Symposium on Agricultural Entomology in Developing Countries, 14th International Congress of Entomology*, Canberra, Australia, pp. 136–142.

Nickle, W.R. (ed.) (1984) *Plant and Insect Nematodes*. Marcel Dekker, New York, 925 pp.

Nielsen, K.L., Miller, C.R., Beck, D. and Lynch, J.P. (1998) Fractal geometry of root systems: field observations of contrasting genotypes of common bean (*Phaseolus vulgaris* L.) grown under different phosphorus regimes. *Plant and Soil* 206, 181–190.

Niemeijera, D. and Mazzucato, V. (2003) Moving beyond indigenous soil taxonomies: local theories of soils for sustainable development. *Geoderma* 111, 403–424.

Noble, A.D., Zenneck, I. and Randall, P.J. (1996) Leaf litter ash alkalinity and neutralisation of soil acidity. *Plant and Soil* 179(2), 293–302.

Noda, H., Bueno, C.R. and Silva Filho, D.F. (1994) Guinea arrowroot. In: Hernando Bermejo, J.E. and Leon, J. (eds) *Neglected Crops: 1492 from a Different Perspective*. Plant Production and Protection Series No. 26. FAO, Rome, pp. 239–244.

Noe, J.P. (1998) Crop and nematode-management systems. In: Barker, K.R., Pederson, G.A. and Windham, G.L. (eds) *Plant and Nematode Interactions*. Agronomy Monograph no. 36. ASA, CSSA and SSSA, Madison, Wisconsin, pp. 159–172.

Norby, R.J., and Jackson, R.B. (2000) Root dynamics and global change: seeking an ecosystem perspective. *New Phytologist* 147, 13–31.

Norris, J.R., Read, D.J. and Varma, A.K. (eds) (1994) *Techniques for Mycorrhizal Research. Methods in Microbiology*. Academic Press, San Diego, California, 928 pp.

Norton, J.B., Pawluk, R.R. and Sandor, J.A. (1998) Observation and experience linking science and indigenous knowledge at Zuni, New Mexico. *Journal of Arid Environments* 39, 331–340.

Nyberg, G. (2001) Carbon and nitrogen dynamics in agroforestry systems. Temporal patterns of some important soil processes. PhD thesis, Acta Universitatis Agriculturae Sueciae Silvestria no. 181, SLU, Umea, Sweden, 145 pp.

Nye, P.H. (1986) Acid–base changes in the rhizosphere. In: Tinker, B. and Läuchli, A. (eds) *Advances in Plant Nutrition*, Vol. 2. Praeger Scientific, New York, pp. 129–153.

Nye, P.H. and Tinker, P.B. (1977) *Solute Movement in the Soil–Root System*. Blackwell, Oxford, UK, 342 pp.

Nygren, P. (1995) Above-ground nitrogen dynamics following the complete pruning of a nodulated woody legume in humid tropical field conditions. *Plant, Cell and Environment* 18, 977–988.

Nygren, P. and Campos, A. (1995) Effect of foliage pruning on fine root biomass of *Erythrina poeppigiana* (Fabaceae). In: Sinoquet, H. and Cruz, P. (eds) *Ecophysiology of Tropical Intercropping*. Institute Nationale de la Recherche Agronomique (INRA), Paris, pp. 295–304.

Nygren, P. and Ramirez, C. (1995) Production and turnover of $N_2$ fixing nodules in relation to foliage development in periodically pruned *Erythrina poeppigiana* (Leguminosae) trees. *Forest Ecology and Management* 73, 59–73.

Nziguheba, G., Merckx, R. and Palm, C.A. (2000) Organic residues affect phosphorus availability and maize yields in a Nitisol of western Kenya. *Biology and Fertility of Soils* 32, 328–339.

Nziguheba, G., Merckx, R. and Palm, C.A. and Mutuo, P. (2002) Combining *Tithonia diversifolia* and fertilizers for maize production in phosphorus deficient soil in Kenya. *Agroforestry Systems* 55, 165–174.

Oades, J.M., Gillman, G.P., Uehara, G., Hue, N.V., van Noordwijk, M., Robertson, G.P. and Wada, K. (1989) Interactions of soil organic matter and variable-charge clays. In: Coleman, D.C., Oades, J.M. and Uehara, G. (eds) *Dynamics of Soil Organic Matter in Tropical Ecosystems*. University of Hawaii Press, Honolulu, Hawaii, pp. 69–95.

Oberson, A., Friesen, D.K., Rao, I.M., Bühler, S. and Frossard, E. (2001) Phosphorus transformations in an Oxisol under contrasting land-use systems: the role of the soil microbial biomass. *Plant and Soil* 237, 197–210.

Odee, D.W., Sutherland, J.M., Kimiti, J.M. and Sprent, J.I. (1995) Natural rhizobial populations and nodulation status of woody legumes growing in diverse Kenyan conditions. *Plant and Soil* 173, 211–224.

Odhiambo, H.O., Ong, C.K., Wilson, J., Deans, J.D., Broadhead, J. and Black, C. (1999) Tree–crop interactions for below-ground resources in drylands: root structure and function. *Annals of Arid Zone* 38, 221–237.

Odhiambo, H.O., Ong, C.K., Deans, J.D., Wilson, J., Khan, A.A.H. and Sprent, J.I. (2001) Roots, soil water and crop yield: tree crop interactions in a semi-arid agroforestry system in Kenya. *Plant and Soil* 235, 221–233.

Odum, E.P. (1971) *Fundamentals of Ecology*, 3rd edn. Saunders, Philadelphia, Pennsylvania.

Ohlsson, E.L. (1999) Agroforestry for improved nutrient cycling on small farms in western Kenya. PhD thesis, Swedish University of Agricultural Sciences, Uppsala, Sweden, *Acta Universitatis Agriculturae Sueciae. Agraria* 204. 132 pp.

Oliveira, M.R.D., van Noordwijk, M., Gaze, S.R., Brouwer, G., Bona, S., Mosca, G. and Hairiah, K. (2000) Auger sampling, ingrowth cores and pinboard methods. In: Smit, A.L., Bengough, A.G., Engels, E., van Noordwijk, M., Pellerin, S. and van de Geijn, S.C. (eds) *Root Methods: a Handbook*. Springer-Verlag, Berlin, Germany, pp. 175–210.

Olsen, S.R. and Khasawneh, F.E. (1980) Use and limitations of physical-chemical criteria for assessing the status of phosphorus in soils. In: Khasawneh, F.E., Sample, E.C. and Kamprath, E.J. (eds) *The Role of Phosphorus in Agriculture*. American Society of Agronomy, Madison, Wisconsin, pp. 361–410.

Ong, C.K. (1995) The 'dark side' of intercropping: manipulation of soil resources. In: Sinoquet, H. and Cruz, P. (eds) *Ecophysiology of Tropical Intercropping*. Institute National de la Recherche Agronomique, Paris, pp. 45–65.

Ong, C.K. (1996) A framework for quantifying the various effects of tree-crop interactions. In: Ong, C.K. and Huxley, P. (eds) *Tree–Crop Interactions – a Physiological Approach*. CAB International, Wallingford, UK, pp. 1–23.

Ong, C.K. and Huxley, P. (eds) (1996) *Tree–Crop Interactions – a Physiological Approach*. CAB International, Wallingford, UK, 386 pp.

Ong, C.K. and Leakey, R.R.B. (1999) Why tree crop interactions in agroforestry appear at odds with tree–grass interactions in tropical savannahs. *Agroforestry Systems* 45, 109–129.

Ong, C.K. and Monteith, J.L. (1985) Response of pearl millet to light and temperature. *Field Crops Research* 11, 141–160.

Ong, C.K. and Orego, F. (2003) Links between land management, sedimentation, nutrient flows and smallholder irrigation in the Lake Victoria Basin. In: Blank, H.G., Mutero, C.M. and Murray-Rust, H. (eds) *The Changing Face of Irrigation in Kenya: Opportunities for Anticipating Change in Eastern and Southern Africa*. IWMI, Colombo, Sri Lanka, pp. 135–154.

Ong, C.K., Black, C.R., Marshall, F.M. and Corlett, J.E. (1996) Principles of resource capture and utilization of light and water. In: Ong, C.K. and Huxley, P. (eds) *Tree–crop Interactions – a Physiological Approach*. CAB International, Wallingford, UK, pp. 73–158.

Ong, C.K., Deans, J.D., Wilson, J., Mutua, J., Khan, A.A.H. and Lawson, E.M. (1999) Exploring below-ground complementarity in agroforestry using sap flow and root fractal techniques. *Agroforestry Systems* 44, 87–103.

Onguene, N.A. (2000) Diversity and dynamics of mycorrhizal associations in tropical rain forests with different disturbance regimes in South Cameroon. *Tropenbos-Cameroon Series* 3, 1–167.

Oorts, K., Vanlauwe, B., Cofie, O.O., Sanginga, N. and Merckx, R. (2000) Charge characteristics of soil organic matter fractions in a ferric Lixisol under some multipurpose trees. *Agroforestry Systems* 48, 169–188.

Oorts, K., Merckx, R., Vanlauwe, B., Sanginga, N. and Diels, J. (2002) Dynamics of charge bearing soil organic matter fractions in highly weathered soils. In: Int. Union of Soil Science, The Soil and Fertilizer Society of Thailand, Ministry of Agriculture and Cooperatives of Thailand (eds) *17th World Soil Science Conference 'Soil Science: Confronting New Realities in the 21st Century'*. International Soil Science Society, Bangkok, Thailand, CD – Paper No. 1007, 442 pp.

Oppelt, A.L., Kurth W. and Godbold D.L. (2001) Topology, scaling relations and Leonardo's rule in root systems from African species. *Tree Physiology* 21, 117–128.

Ortiz-Monasterio, J.I., Matson, P.A., Panek, J. and Taylor, R.L. (1996) Nitrogen fertilizer management for $N_2O$ and NO emissions in Mexican irrigated wheat. In: *Transactions 9th Nitrogen Workshop*. Technische Universität, Braunschweig, Germany, pp. 531–534.

Oswald, A., Frost, H., Ransom, J.K., Kroschel, J., Shepherd, K.D. and Sauerborn, J. (1996) Studies on the potential for improved fallows using trees and shrubs to control *Striga* infestation in Kenya. In: Moreno, M.T., Cubero, J.I., Berner, D., Joel, D., Musselman, L.J. and Parker, C. (eds) *Advances in Parasitic Plant Research: Proceedings of the 6th Parasitic Weeds Symposium*. Cordoba, Spain, pp. 835–841.

Otsuka, K., and Place, F. (2001) *Land Tenure and Natural Resource Management: a Comparative Study of Agrarian Communities in Asia and Africa*. John Hopkins University Press, Baltimore, Maryland, 389 pp.

Otter, L.B., Marufu, L. and Scholes, M.C. (2001) Biogenic, biomass and biofuel sources of trace gases in southern Africa. *South African Journal of Science* 97, 131–138.

Otto, W.M. and Kilian, W.H. (2001) Response of soil phosphorus content, growth and yield of wheat to long-term phosphorus fertilization in a conventional cropping system. *Nutrient Cycling in Agroecosystems* 61, 283–292.

Owino, F. (1996) Selection for adaptation in multipurpose trees and shrubs for production and function in agroforestry systems. *Euphytica* 92, 225–234.

Ozier-Lafontaine, H., Lecompte, F. and Sillon, J.F. (1999) Fractal analysis of the root architecture of *Gliricidia sepium* for the spatial prediction of root branching, size and mass: model development and evaluation in agroforestry. *Plant and Soil* 209, 167–180.

Page, S.L.J. and Bridge, J. (1993) Plant nematodes and sustainability in tropical agriculture. *Experimental Agriculture* 29, 139–154.

Pages, L., Jordan, M.O. and Picard, D. (1989) A simultaneous model of three-dimensional architecture of the maize root system. *Plant and Soil* 119, 147–154.

Palm, C.A. (1995) Contribution of agroforestry trees to nutrient requirements of intercropped plants. *Agroforestry Systems* 30, 105–124.

Palm, C.A. and Rowland, A.P. (1997) Chemical characterization of plant quality for decomposition. In: Cadisch, G. and Giller, K.E. (eds) *Driven By Nature: Plant Litter Quality and Decomposition*. CAB International, Wallingford, UK, pp. 379–392.

Palm, C.A. and Sanchez, P.A. (1991) Nitrogen release from leaves of some tropical legumes as affected by their lignin and polyphenolic contents. *Soil Biology and Biochemistry* 23, 83–88.

Palm, C.A., Gachego, C., Delve, R., Cadisch, G. and Giller, K.E. (2001) Organic inputs for soil fertility management in tropical agroecosystems: application of an organic resource database. *Agriculture, Ecosystems and Environment* 83, 27–42.

Palm, C.A., Alegre, J.C., Arevalo, L., Mutuo, P.K., Mosier, A.R. and Coe, R. (2002) Nitrous oxide and methane fluxes in six different land use systems in the Peruvian Amazon. *Global Biogeochemical Cycles* 16(4), 1073.

Palti, J. (1981) *Cultural Practices and Infectious Crop Diseases*, 1st edn. Springer-Verlag, Berlin, Germany, 243 pp.

Pant, H.K., Warman, P.R. and Nowak, J. (1999) Identification of soil organic phosphorus by $^{31}P$ nuclear magnetic resonance spectroscopy. *Communications in Soil Science and Plant Analysis* 30, 757–772.

Paoletti, M.G. (1999) *Invertebrate Biodiversity as Bioindicators of Sustainable Landscapes: Practical Use of Invertebrates to Assess Sustainable Land Use*. Elsevier, Amsterdam, The Netherlands.

Parker, C. and Riches, D.C. (1993) *Parasitic Weeds of the World: Biology and Control*. CAB International, Wallingford, UK, 332 pp.

Parton, W.J., Stewart, J.W.B. and Cole, C.V. (1988) Dynamics of C, N, P and S in grassland soils: a model. *Biogeochemistry* 5, 109–131.

Pashanasi, B., Lavelle, P. and Alegre, J. (1996) Effect of inoculation with the endogeic earthworm *Pontoscolex corethrurus* on soil chemical characteristics and plant growth in low-input agriculture systems of Peruvian Amazonia. *Soil Biology and Biochemistry* 28, 801–810.

Passioura, J.B. (1983) Roots and drought resistance. *Agricultural Water Management* 7, 265–280.

Pate, E. (1998). In: Cadet, P. Gestion écologique des peuplements de nématodes phytoparasites tropicaux: importance des facteurs édaphiques et du ruissellement. *Cahiers Agricultures* 7, 187–194.

Paul, E.A. and Clark, F.E. (1996) *Soil Microbiology and Biochemistry*, 2nd edn. Academic Press, San Diego, 340 pp.

Paustian, K., Ågren, G. and Bosatta, E. (1997) Modelling litter quality effects on decomposition and soil organic matter dynamics. In: Cadisch, G. and Giller, K.E. (eds) *Driven by Nature: Plant Litter Quality and Decomposition*. CAB International, Wallingford, UK, pp. 313–336.

Pearcy, R.W., Ehleringer, J., Mooney, H.A. and Rundel, P.W. (eds) (1991) *Plant Physiological Ecology: Field Methods and Instrumentation*. Chapman and Hall, London, 457 pp.

Peoples, M.B. and Craswell, E.T. (1992) Biological nitrogen fixation – investments, expectations and actual contributions to agriculture. *Plant and Soil* 141, 13–39.

Pfleger, F.L. and Linderman, R.G. (eds) (1994) *Mycorrhizae and Plant Health*. The American Phytopathological Society Press, St Paul, Minnesota, 344 pp.

Pirozynski, K.A. and Malloch, D.W. (1975) The origin of land plants: a matter of mycotrophism. *Biosystems* 6, 153–164.

Plenchette, C., Fortin, J.A. and Furlan, V. (1983) Growth responses of several plant species to mycorrhizae in a soil of low fertility. I. Mycorrhizal dependency under field conditions. *Plant and Soil* 70, 199–209.

Pocknee, S. and Sumner, M.E. (1997) Cation and nitrogen contents of organic matter determine its soil limiting potential. *Soil Science Society of America Journal* 61(1), 86–92.

Poinar, G.O. Jr (1975) *Entomogenous Nematodes. A Manual and Host List of Insect–Nematode Associations*. Brill ed., Leiden, The Netherlands, 317 pp.

Porazinska, D.L. (1998) Nematode communities as indicators of agricultural management practices. PhD thesis, University of Florida, Gainesville, 134 pp.

Prasad, R. and Mishra, C.P. (1984) Studies on root systems of important tree species in dry deciduous teak forest of Sagar (M.P.). *Indian Journal of Forestry* 7, 171–177.

Prasetyo, L.B., Saito, G. and Tsuruta, H. (2000) Estimation of greenhouse gasses emissions using remote sensing and GIS techniques in Sumatra, Indonesia. In: Murdiyarso, D. and Tsuruta, H. (eds) *The Impacts of Land-Use/Cover Change on Greenhouse Gas Emissions in Tropical Asia*. Global Change Impacts Centre, Southeast Asia, Bogor, Indonesia, and National Institute of Agro-Environmental Sciences, Tsukuba, Japan, pp. 53–59.

Preechapanya, P. (1996) Indigenous ecological knowledge about the sustainability of tea gardens in the hill evergreen forest of northern Thailand. PhD thesis, University of Wales, Bangor, UK, 286 pp.

Prematuri, R. (1995) The role of arbuscular mycorrhizal fungi on different species of leguminous trees used in reforestation in South East Asia. Thesis of Post Graduate Diploma in Biotechnology, the University of Kent at Canterbury, UK.

Prematuri, R. and Dodd, J.C. (1997) The effect of arbuscular mycorrhizal fungi on *Albizia saman* and their biotechnical detection in roots. Paper presented at the *International Conference on Mycorrhizas in Sustainable Tropical Agriculture and Forest Ecosystems*, Bogor, Indonesia.

Pritchard, S.G. and Rogers, H.H. (2000) Spatial and temporal deployment of crop roots in $CO_2$-enriched environments. *New Phytologist* 147, 55–71.

Priyanto, J.D. (1996) Mineralisasi nitrogen, biomassa mikrobia dan beberapa sifat biologi tanah akibat masukan berbagai kualitas bahana organik pada ultisol, Lampung Utara. Skripsi S1, Fakultas Pertanian, Universitas Brawijaya, Malang, 75 pp.

Pueppke, S.G. and Broughton, W.J. (1999) *Rhizobium* sp. strain NGR234 and *R. fredii* USDA 257 share exceptionally broad, nested host ranges. *Molecular Plant–Microbe Interactions* 12, 293–318.

Purbopospito, J. and van Rees, K.C.J. (2002) Root distribution at various distances from clove trees growing in Indonesia. *Plant and Soil* 239, 313–320.

Purwanto, E. (1999) Erosion, sediment delivery and soil conservation in upland agricultural catchment in West Java, Indonesia: a hydrological approach in a socio-economic context. PhD Thesis, Vrije Universiteit Amsterdam, The Netherlands.

Quim, F.M. (1996) G × E analyses provide vital insights for IITA crops. *IITA Research Brief* No. 12. IITA, Ibadan, Nigeria, pp. 29–32.

Raab, T.K., Lipson, D.A. and Monson, R.K. (1996) Non-mycorrhizal uptake of amino acids by roots of the alpine sedge *Kobresia myosuroides*: implications for the alpine nitrogen cycle. *Oecologia* 108, 488–494.

Radersma, S. (2002) Tree effects on crop growth on a phosphorus-fixing Ferralsol. PhD thesis, University of Wageningen, Wageningen, The Netherlands, 187 pp.

Randall, P.J., Hayes, J.E., Hocking, P.J. and Richardson, A. (2001) Root exudates in phosphorus acquisition by plants. In: Ae, N., Arihara, J., Okada, K. and Srinivasan, A. (eds) *Plant Nutrient Acquisition: New Perspectives*. Springer-Verlag, Tokyo, Japan, pp. 32–68.

Rao, I.M., Plazas, C. and Ricaurte, J. (2001) Root turnover and nutrient cycling in native and introduced pastures in tropical savannas. In: Horst, W.J., Schenk, M.K., Bürkert, A., Claassen, N., Flessa, H., Frommer, W.B., Goldbach, H.E., Olfs, H.W., Römheld, V., Sattelmacher, B., Schmidhalter, U., Schubert,

S., von Wirén, N. and Wittenmayer, L. (eds) *Plant Nutrition: Food Security and Sustainability of Agro-ecosystems through Basic and Applied Research*. Fourteenth International Plant Nutrition Colloquium, Hanover, Germany. Kluwer Academic Publishers, Dordrecht, The Netherlands, pp. 976–977.

Rao, M.R., Muraya, P. and Huxley, P.A. (1993) Observations of some tree root systems in agroforestry intercrop situations, and their graphical representation. *Experimental Agriculture* 29, 183–194.

Rao, M.R., Nair, P.K.R. and Ong, C.K. (1998) Biophysical interactions in tropical agroforestry systems. *Agroforestry Systems* 38, 3–50.

Rasse, D.P., Smucker, A.J.M. and Santos, D. (2000) Alfalfa root and shoot mulching effects on soil hydraulic properties and aggregation. *Soil Science Society of America Journal* 64, 725–731.

Rasse, D.P., Longdoz, B. and Ceulemans, R. (2001) TRAP: a modelling approach to below-ground carbon allocation in temperate forests. *Plant and Soil* 229, 281–293.

Raussen, T. and Wilson, J. (eds) (2001) *Tree Management for Farming: Pruning Trees Growing in Cropland*. Technical Bulletin 3. Agroforestry Research Network for Africa (AFRENA) – Project Uganda. Kabale, Uganda, 12 pp.

Reeburgh, W.S., Whalen, S.C. and Alperin, M.J. (1993) The role of methyltrophy in the global $CH_4$ budget. In: Murell, J.C. and Kelley, D.P. (eds) *Microbial Growth on C1 Compounds*. Intercept Ltd, Andover, UK, pp. 1–14.

Reiners, W.A., Bouwman, A.F., Parsons, W.F.J. and Keller, M. (1994) Tropical rain forest conversion to pasture: changes in vegetation and soil properties. *Ecological Applications* 4, 363–377.

Rengel, Z. (2002) Genetic control of root exudation. *Plant and Soil* 245, 59–70.

Reynolds, J. (1994) Earthworms of the world. *Global Biodiversity* 4, 11–16.

Richards, A.I. (1939) *Land, Labour and Diet in Northern Rhodesia*. Oxford University Press for the International Institute of African Languages and Culture, Oxford, UK, 424 pp.

Richards, P. (1989) Agriculture as a performance. In: Cambers, R., Pacey, A. and Thrupp, L.A. (eds) *Farmer First*. Intermediate Technology Publications, London, pp. 39–51.

Richards, P. (1994) Local knowledge formation and validation: the case of rice in Sierra Leone. In: Scoones, I. and Thompson, J. (eds) *Beyond Farmer First*. Intermediate Technology Publications, London, pp. 165–170.

Richardson, A.E. (1994) Soil microorganisms and phosphorus availability. In: Pankhurst, C.E., Doube, B.M., Gupta, V.V.S.R. and Grace, E.R. (eds) *Management of the Soil Biota in Sustainable Farming Systems*. CSIRO Publishing, Melbourne, Australia, pp. 50–62.

Riley, D. and Barber, S.A. (1971) Effect of ammonium and nitrate fertilization on phosphorus uptake as related to root-induced pH changes at the soil–root interface. *Soil Science Society of America Proceedings* 35, 301–306.

Rillig, M.C. and Steinberg, P.D. (2002) Glomalin production by an arbuscular mycorrhizal fungus: a mechanism of habitat modification? *Soil Biology and Biochemistry* 34, 1371–1374.

Rillig, M.C., Wright, S.F. and Eviner, V.T. (2002) The role of arbuscular mycorrhizal fungi and glomalin in soil aggregation: comparing effects of five plant species. *Plant and Soil* 238, 325–333.

Ringrose-Voase, A.J. and Cresswell, H.P. (2000) *Measurement and Prediction of Deep Drainage under Current and Alternative Farming Practice*. A final report prepared for the Land and Water Resources Research and Development Corporation, CSIRO Land and Water Consultancy Report.

Risasi, E.L., Kang, B.T. and Opuwaribo, E.E. (1998) Assessment of N availability of roots of selected woody species and maize. *Biological Agriculture and Horticulture* 16, 37–52.

Risch, S.J. (1981) Insect herbivore abundance in tropical monocultures and polycultures: an experimental test of two hypotheses. *Ecology* 62, 1325–1340.

Robertson, M.J., Fukai, S., Ludlow, M.M. and Hammer, G.L. (1993) Water extraction by grain sorghum in a sub-humid environment. I. Analysis of the water extraction pattern. *Field Crops Research* 33, 81–97.

Robertson, M.J., Carberry, P.S. and Lucy, M. (2000) Evaluation of a new cropping option using a participatory approach with on-farm monitoring and simulation: a case study of spring-sown mungbeans. *Australian Journal of Agricultural Research* 51, 1–12.

Robinson, A.C. (1969) Host selection for effective *Rhizobium trifolii* by red clover and subterranean clover in the field. *Australian Journal of Agricultural Research* 20, 1053–1060.

Robinson, R.A. (1976) *Plant Pathosystems*. Advanced Series of Agricultural Science 3. Springer-Verlag, Berlin, 185 pp.

Röckstrom, J., Jansson, P.E. and Barron, J. (1998) Seasonal rainfall partitioning under runon and runoff conditions on sandy soil in Niger. On-farm measurements and water balance modelling. *Journal of Hydrology* 210, 68–92.

Rodrigo, V.H.L., Stirling, C.M., Naranpanawa, R.M.A.K.B. and Herath, P.H.M.U. (2001) Intercropping of immature rubber in Sri Lanka: present status and financial analysis of intercrops planted at three densities of banana. *Agroforestry Systems* 51, 35–48.

Rodrigues, V.G.S., da Costa, R.S.C., Mendes, A.M., Leônidas, F. das C. (2002) Aspectos agronômicos e de sustentabilidade em sistemas agroflorestais com café robusta (*Coffea canephora*) em Rondônia. In: IV Congresso Brasileiro de Sistemas Agroflorestais, 2002, Ilhéus, Anais ... Ilhéus, Brazil: CEPLAC:CEPEC. Extended abstract 1-002, 3 pp. (CD ROM)

Rodríguez-Kábana, R. and Canullo, G.H. (1992) Cropping systems for the management of phytonematodes. *Phytoparasitica* 20, 211–224.

Roessel, B.W.P. (1938) Herbebossching op Java (Reafforestation in Java). Alg. Handelsblad, 1 and 4 March 1938. *Tectona* 31, 595–602.

Roothaert, R.L. and Franzel, S. (2002) Farmers' preferences and use of local fodder trees and shrubs in Kenya. *Agroforestry Systems* 52, 239–252.

Rose, C.W. and Yu, B. (1998) Dynamic process modelling of hydrology and soil erosion. In: Penning de Vries, F.W.T., Agus, F. and Kerr, J. (eds) *Soil Erosion at Multiple Scales: Principles and Methods for Assessing Causes and Impacts*. IBSRAM-CABI, Wallingford, UK, pp. 269–286.

Rossiter, D.G. and Riha, S.J. (1999) Modeling plant competition with the GAPS object-oriented dynamic simulation model. *Agronomy Journal* 91(5), 773–783.

Roughley, R.J., Gault, R.R., Gemell, L.G., Andrews, J.A., Brockwell, J., Dunn, B.W., Griffiths, G.W., Hartley, E.J., Hebb, D.M., Peoples, M.B. and Thompson, J.A. (1995) Autecology of *Bradyrhizobium japonicum* in soybean-rice rotations. *Plant and Soil* 176, 7–14.

Roupsard, O., Ferhi, A., Granier, A., Pallo, F., Depommier, D., Mallet, B., Joly, H.I. and Dreyer, E. (1999) Reverse phenology and dry-season water uptake by *Faidherbia albida* (Del.) A. Chev. in an agroforestry parkland of Sudanese west Africa. *Functional Ecology*, 13, 460–472.

Rouxel, F. (1991) Natural suppressiveness of soils to plant diseases. In: Beemster, A.B.R., Bollen, G.J. and Gerlagh, M. (eds) *Biotic Interactions and Soilborne Diseases. Proceedings of First Conference of European Foundation of Plant Pathology*. Elsevier Science, The Netherlands, pp. 287–296.

Rowe, E.C. (1999) The safety-net role of tree roots in hedgerow intercropping systems. PhD thesis, Wye College, University of London, 288 pp.

Rowe, E., Hairiah, K., Giller, K.E., van Noordwijk, M. and Cadisch, G. (1999) Testing the safety-net role of hedgerow tree roots by $^{15}$N placement at different soil depths. *Agroforestry Systems* 43, 81–93.

Rowe, E.C., van Noordwijk, M., Suprayogo, D., Hairiah, K., Giller, K.E. and Cadisch, G. (2001) Root distributions partially explain $^{15}$N uptake patterns in *Gliricidia* and *Peltophorum* hedgerow intercropping systems. *Plant and Soil* 235, 167–179.

Rowland, V. (1998) Root studies of *Gliricidia sepium* (Jacq.) Steud: BSc Dissertation, University of Wales, Bangor, UK, 23 pp.

Rowse, H.R., Mason, W.K. and Taylor, H.M. (1983) A microcomputer simulation model of soil water extraction by soybeans. *Soil Science* 136, 218–225.

Rubaek, G.H., Guggenberger, G., Zech, W. and Christensen, B.T. (1999) Organic phosphorus in soil size separates characterized by phosphorus-31 nuclear magnetic resonance and resin extraction. *Soil Science Society of America Journal* 63, 1123–1132.

Ruhigwa, B.A., Gichuru, M.P., Mambani, B. and Tariah, N.M. (1992) Root distribution of *Acioa barteri*, *Alchornea cordifolia*, *Cassia siamea* and *Gmelina arborea* in an acid Ultisol. *Agroforestry Systems* 19, 67–78.

Rusmiati, T. (2001) Studi monitoring tingkat populasi dan pola sebaran semut api (*Solenopsis* sp.) pada beberapa umur tanaman nanas (*Ananas comosus* [L.] Merr.). Skripsi S1, Universitas Lampung, Bandar Lampung, 27 pp.

Rusten, E.P. and Gold, M. (1991) Understanding an indigenous knowledge system for tree fodder via a multi-method on-farm research approach. *Agroforestry Systems* 15, 139–165.

Ryan, P.R., Delhaize, E. and Randall, P.J. (1995) Malate efflux from root apices and tolerance to aluminum are highly correlated in wheat. *Australian Journal of Plant Physiology* 122, 531–536.

Ryel, R.J., Caldwell, M.M., Yoder, C.K., Or, D. and Leffler, A.J. (2002) Hydraulic redistribution in a stand of *Artemisia tridentata*: evaluation of benefits to transpiration assessed with a simulation model. *Oecologia* 130, 173–184.

Saggar, S. and Hedley, C.B. (2001) Estimating seasonal and annual carbon inputs and root decomposition rates in a temperate pasture following field $^{14}$C pulse-labelling. *Plant and Soil* 236, 91–103.

Saka, V.W. (1985) *Meloidogyne* spp. research in Region V of the International *Meloidogyne* Project (IMP). In: Sasser, J.N. and Carter, C.C. (eds) *An Advanced Treatise on Meloidogyne*. Volume I. *Biology and Control*. North Carolina State University, pp. 361–368.

Sakuratani, T. (1981) A heat balance method of measuring water flux in the stem of intact plants. *Journal of Agricultural Meteorology* 37, 9–17.

Sample, E.C., Soper, R.J. and Racz, G.J. (1980) Reactions of phosphate fertilizers in soils. In: Khasawneh, F.E., Sample, E.C. and Kamprath, E.J. (eds) *The Role of Phosphorus in Agriculture*. American Society of Agronomy, Madison, Wisconsin, pp. 263–310.

Sanantonio, D. and Grace, J.C. (1987) Estimating fine root production and turnover from biomass and decomposition data: a compartment-flow model. *Canadian Journal of Forest Research* 17, 900–908.

Sanchez, P.A. (1995) Science in agroforestry. *Agroforestry Systems* 30, 5–55.

Sanchez, P.A. (1999a) Improved fallows come of age in the tropics. *Agroforestry Systems* 47, 3–12.

Sanchez, P.A. (1999b) Delivering on the promise of agroforestry. *Environment, Development and Sustainability* 1, 275–284.

Sanchez, P.A. and Logan, T.L. (1992) Myths and science about the chemistry and fertility of soils in the tropics. In: Lal, R. and Sanchez, P.A. (eds) *Myths and Science of Soils of the Tropics*. SSSA Special Publication No. 29. SSSA Inc. and ASA Inc. Madison, Wisconsin, pp. 35–46.

Sanchez, P.A., Shepherd, K.D., Soule, M.J., Place, F.M., Buresh, R.J. and Izac, A.M. (1997) Soil fertility replenishment in Africa: an investment in natural resource capital. In: Buresh, R.J., Sanchez, P.A. and Calhoun, F. (eds) *Replenishing Soil Fertility in Africa*, SSSA Special Publication 51. ASA and SSSA Madison, Wisconsin, pp. 1–46.

Sandor, J.A. and Furbee, L. (1996) Indigenous knowledge and classification of soils in the Andes of Southern Peru. *Soil Science Society of America Journal* 60, 1502–1512.

Sanford, R.L. (1987) Apogeotropic roots in an Amazonian rainforest. *Science* 235, 1062–1064.

Sanford, R.L. and Cuevas, E. (1996) Root growth and rhizosphere interactions in tropical forests. In: Mulkey, S.S., Chazdon, R.L. and Smith, A.P. (eds) *Tropical Forest Plant Ecophysiology*. Chapman and Hall, New York, pp. 268–300.

Sanginga, N., Zapata, F., Danso, S.K.A. and Bowen, G.D. (1990) Effect of successive cuttings on uptake and partitioning of nitrogen 15 among plant parts of *Leucaena leucocephala*. *Biology and Fertility of Soils* 9, 37–42.

Sanginga, N., Bowen, G.D. and Danso, S.K.A. (1991) Intra-specific variation in growth and P accumulation of *Leucaena leucocephala* and *Gliricidia sepium* as influenced by soil phosphate status. *Plant and Soil* 133, 201–208.

Sanginga, N., Danso, S.K.A. and Bowen, C.G. (1992) Variation in growth sources of nitrogen and N-use efficiency by provenances of *Gliricidia sepium*. *Soil Biology and Biochemistry* 24, 1021–1026.

Sanginga, N., Vanlauwe, B. and Danso, S.K.A. (1995) Management of biological $N_2$-fixation in alley cropping systems: estimation and contribution to N balance. *Plant and Soil* 174, 119–141.

Sanhueza, E. (1997) Impact of human activity on NO soil fluxes. *Nutrient Cycling in Agroecosystems* 48, 61–68.

Sass, R.L. (1994) Short summary chapter for methane. In: Minami, K., Mosier, A. and Sass, R. (eds) $CH_4$ and $N_2O$: *Global Emissions and Controls from Rice Fields and Other Agricultural and Industrial Sources*, NIAES. Yokendo Publishers, Tokyo, pp. 1–7.

Saunders, K.E. (2002) Farmers' knowledge of trees used for shade within cocoa farms in Atwima District, Ghana. MSc thesis, University of Wales, Bangor, UK, 85 pp.

Savil, P.S. (1976) The effect of drainage and plowing of surface water on rooting and windthrow of Sitka spruce in Northern Ireland. *Forestry* 49, 133–141.

Savini, I. (1999) The effect of organic and inorganic amendments on phosphorus release and availability from two phosphate rocks and triple superphosphate in phosphorus fixing soils. MSc thesis, University of Nairobi, Nairobi, Kenya.

Saxton, K.E., Rawls, W.J., Romberger, J.S. and Papendick, R.I. (1986) Estimating generalized soil-water characteristics from texture. *Soil Science Society of America Journal* 50, 1031–1036.

Schaefer, M. and Schauermann, J. (1990) The soil fauna of beech forests: comparison between a mull and a moder soil. *Pedobiologia* 34, 299–316.

Schalenbourg, W. (2002) An assessment of farmers' perceptions of soil and watershed functions in Sumberjaya, Sumatra, Indonesia. Internal report, ICRAF SEA, Bogor, Indonesia, 146 pp.

Schaller, M. (2001) Quantification and management of root interactions between fast-growing timber species and coffee in plantations in Central America. PhD thesis, University of Bayreuth, Bayreuth, Germany.

Schaller, M., Schroth, G., Beer, J. and Jiménez, F. (1999) Control de crecimento lateral de las raíces de especies maderables de rápido crecimiento utilizando gramíneas como barreras biológicas. *Agroforestería en las Américas* 6, 36–38.

Schaller, M., Schroth, G., Beer, J. and Jimenez, F. (2003) Species and site factors that permit the association of fast-growing trees with crops: the case of *Eucalyptus deglupta* as coffee shade in Costa Rica. *Forest Ecology and Management* 175, 205–215.

Scheren, P.A.G., Zanting, M.H.A. and Lemmens, A.M.C. (2000) Estimation of water pollution sources in Lake Victoria, East Africa: application and elaboration of the rapid assessment methodology. *Journal of Environmental Management* 58, 235–248.

Schimel, J.P. and Gulledge, J. (1998) Microbial community structure and global trace gases. *Global Change Biology* 4, 745–758.

Schlüßler, A., Schwarzott, D. and Walker, C. (2001) A new fungal phylum, the *Glomeromycota*: phylogeny and evolution. *Mycological Research* 105(12), 1413–1421.

Schreiner, R.P. and Koide, R.T. (1993) Mustards, mustard oils and mycorrhizas. *New Phytologist* 123, 107–113.

Schroth, G. (1995) Tree root characteristics as criteria for species selection and system designs in agroforestry. *Agroforestry Systems* 30, 125–143.

Schroth, G. (1998) A review of belowground interactions in agroforestry, focussing on mechanisms and management options. *Agroforestry Systems* 43, 5–34.

Schroth, G. (2001) Plant–soil interactions in multistrata agroforestry in the humid tropics. *Agroforestry Systems* 53, 85–102.

Schroth, G. (2003) Root systems. In: Schroth, G. and Sinclair, F.L. (eds) *Trees, Crops and Soil Fertility – Concepts and Research Methods*. CAB International, Wallingford, UK, pp. 235–257.

Schroth, G. and Sinclair, F.L. (eds) (2003) *Trees, Crops and Soil Fertility: Concepts and Research Methods*. CAB International, Wallingford, UK.

Schroth, G. and Zech, W. (1995a) Root length dynamics in agroforestry with *Gliricidia sepium* as compared to sole cropping in the semi-deciduous rain forest zone of West Africa. *Plant and Soil* 170, 297–306.

Schroth, G. and Zech, W. (1995b) Above- and below-ground biomass dynamics in a sole cropping and an alley cropping system with *Gliricidia sepium* in the semi deciduous rain forest zone of West Africa. *Agroforestry Systems* 31, 181–198.

Schroth, G., Poidy, N., Morshäuser, T. and Zech, W. (1995) Effects of different methods of soil tillage and biomass application on crop yields and soil properties in agroforestry with high tree competition. *Agriculture, Ecosystems and Environment* 52, 129–140.

Schroth, G., Kolbe, D., Balle, P. and Zech, W. (1996) Root system characteristics with agroforestry relevance of leguminous tree species and a spontaneous fallow in a semi-deciduous rainforest area of West Africa. *Forest Ecology and Management* 84, 199–208.

Schroth, G., da Silva, L.F., Seixas, R., Teixeira, W.G., Macedo, J.L.V. and Zech, W. (1999) Subsoil accumulation of mineral nitrogen under polyculture and monoculture plantations, fallow and primary forest in a ferralitic Amazonian upland soil. *Agriculture, Ecosystems and Environment* 75, 109–120.

Schroth, G., Rodrigues, M.R.L. and D'Angelo, S.A. (2000a) Spatial patterns of nitrogen mineralization, fertilizer distribution and roots explain nitrate leaching from mature Amazonian oil palm plantation. *Soil Use and Management* 16, 222–229.

Schroth, G., Krauss, U., Gasparotto, L., Duarte Aguilar, J.A. and Vohland, K. (2000b) Pests and diseases in agroforestry systems of the humid tropics. *Agroforestry Systems* 50, 199–241.

Schroth, G., Lehmann, J., Rodrigues, M.R.L., Barros, E. and Macêdo, J.L.V. (2001) Plant–soil interactions in multistrata agroforestry in the humid tropics. *Agroforestry Systems* 53, 85–102.

Schulze, E.D. (1983) Root–shoot interactions and plant life forms. *Netherlands Journal of Agricultural Science* 4, 291–303.

Schumann, G.L. (1991) *Plant Diseases: their Biology and Social Impact*. APS Press, St Paul, Minnesota, 397 pp.

Seligman, N.G. and van Keulen, H. (1981) PAPRAN: a simulation model of annual pasture production limited by rainfall and nitrogen. In: Frissel, M.J. and van Veen, J.A. (eds) *Simulation of Nitrogen Behaviour in Soil–Plant Systems*. PUDOC, Wageningen, The Netherlands, pp. 192–221.

Senapati, B.K., Panigrahi, P.K. and Lavelle, P. (1994) Macrofaunal status and restoration strategy in degraded soil under intensive tea cultivation in India. In: *Transactions of the 15th World Congress of Soil Science*, Vol. 4A, ISSS, Acapulco, Mexico, pp. 64–75.

Senapati, B.K., Lavelle, P., Giri, S., Pashanasi, B., Alegre, J., Decaëns, T., Jiménez, J.J., Albrecht, A., Blanchart, E., Mahieux, M., Rousseaux, L., Thomas, R., Panigrahi, P.K. and Venkatachalan, M.

(1999) Soil technologies for tropical ecosystems. In: Lavelle, P., Brussaard, L. and Hendrix, P.F. (eds) *Earthworm Management in Tropical Agroecosystems*, CAB International, Wallingford, UK, pp. 199–237.

Senapati, B.K., Lavelle, P., Panigrahi, P.K., Giri, S. and Brown, G.G. (2002) Restoration of degraded soil and stimulation of tea production with earthworms and organic fertilizers in Indian tea plantations. In: Brown, G.G., Hungria, M.H., Oliveira, L.J., Bunning, S. and Montáñez, A. (eds) *International Technical Workshop on Biological Management of Soil Ecosystems for Sustainable Agriculture: Program, Abstracts and Related Documents*. Embrapa Soja Documentos No. 182, Londrina, pp. 172–190. (Also can be seen at http://www.fao.org/landandwater/agll/soilbiod/caselist.htm)

Senevirathna, A.M.W.K. (2001) The influence of farmer knowledge, shade and planting density on smallholder rubber/banana intercropping in Sri Lanka. PhD thesis, University of Wales, Bangor, UK, 228 pp.

Setiadi, Y. (1996) The practical application of arbuscular mycorrhizal fungi for enhancing tree establishment in degraded nickel mine sites at PT INCO, Soroako. Paper presented at the IUFRO International Symposium on Accelerating Natural Succession of Degraded Tropical Lands, Washington, DC.

Shah, P.B. (1995) Indigenous agricultural and soil classification. In: Schreier, H., Shah, P.B. and Brown, S. (eds) *Challenges to Mountain Resource Management in Nepal: Processes, Trends and Dynamics in Middle Mountain Watersheds. Workshop proceedings, April 22–25, 1995*. ICIMOD, Kathmandu, Nepal, pp. 203–210.

Sharma, R. and Sharma, S.B. (1999) Management of tropical agroecosystems and beneficial soil nematodes. In: Reddy, M.V. (ed.) *Management of Tropical Agroecosystems and the Beneficial Soil Biota*. Science Publishers, Inc., Enfield, New Hampshire, pp. 199–212.

Sharma, S.B., Price, N.S. and Bridge, J. (1997) The past, present and future of plant nematology in International Agricultural Research Centres. *Nematological Abstracts* 66, 119–142.

Sharpley, A.N., Tiessen, H. and Cole, C.V. (1987) Soil phosphorus forms extracted by soil tests as a function of pedogenesis. *Soil Science Society of America Journal* 51, 362–365.

Shepherd, G., Buresh, R.J. and Gregory, P.J. (2000) Land use affects the distribution of soil inorganic nitrogen in smallholder production systems in Kenya. *Biology and Fertility of Soils* 31, 348–355.

Shepherd, G., Buresh, R.J. and Gregory, P.J. (2001) Inorganic soil nitrogen distribution in relation to soil properties in smallholder maize fields in the Kenya highlands. *Geoderma* 101, 87–103.

Shepherd, K. and Walsh, M. (2002) Development of reflectance spectral libraries for characterisation of soil properties. *Soil Science Society of America Journal* 66, 988–998.

Shepherd, K.D., Jefwa, J., Wilson, J., Ndufa, J.F., Ingleby, K. and Mbuthia, K.W. (1996) Infection potential of farm soils as mycorrhizal inocula for *Leucaena leucocephala*. *Biology and Fertility of Soils* 22, 16–21.

Shinozaki, K., Yoda, K., Hozumi, K. and Tira, K. (1964) A quantitative analysis of plant form – the pipe model theory. *Japanese Journal of Ecology* 14, 97–105.

Shipitalo, M.J. and Protz, R. (1988) Factors influencing the dispersibility of clay in worm casts. *Soil Science Society of America Journal* 52, 764–769.

Shipton, P.J. (1977) Monoculture and soilborne pathogens. *Annual Review of Phytopathology* 15, 387–407.

Shrestha, P.K. (2000) Synthesis of the participatory rural appraisal and farmers' knowledge survey at Landruk, Bandipur and Nayatola. LI-BIRD, Pokhara, Nepal, 76 pp.

Sida (Swedish International Development Cooperation Agency) (1999) *Lake Victoria: a Shared Vision*. Sida, Stockholm, Sweden, 11 pp.

Sidle, R.C. (1992) A theoretical model of the effects of timber harvesting on slope stability. *Water Resources Research* 28, 1897–1910.

Sieverding, E. (1991) *Vesicular–arbuscular Mycorrhiza Management in Tropical Agrosystems*. GTZ, Eschborn, Germany, 371 pp.

Sikora, R.A. (1992) Management of the antagonistic potential in agricultural ecosystems for the biological control of plant-parasitic nematodes. *Annual Review of Phytopathology* 30, 245–270.

Sikora, R.A. and Carter, W.W. (1987) Nematode interactions with fungal and bacterial plant pathogens – fact or fantasy? In: Veech, J.A. and Dickson, D.W. (eds) *Vistas on Nematology: a Commemoration of the 25th Anniversary of the Society of Nematologists*. Society of Nematologists, Inc., Hyattsville, Maryland, pp. 307–312.

Silberstein, R.P., Vertessy, R.A. and Hatton, T.J. (2001a) A simple method for hillslope agroforestry design. In: *MODSIM2001, International Congress on Modelling and Simulation*, Canberra, December 2001, Vol. 4, 1883–1888.

Silberstein, R.P., Vertessy, R.A. and McJannet, D.M. (2001b) A parameter space and simulation response surface for agroforestry design. In: *MODSIM2001, International Congress on Modelling and Simulation*, Canberra, December 2001, Vol. 4, 1889–1894.

Sileshi, G. and Mafongoya, P. (2000) Incidence of *Mesoplatys ochroptera* on *Sesbania sesban* in mixed planted fallows in eastern Zambia. In: Ayuk, E., Dery, B., Kwesiga, F. and Makaya, P. (eds) *Proceedings of the 13th Southern Africa Regional Planning and Review Meeting; SADC-ICRAF Zambezi Basin Agroforestry Project, July 1999, Mangochi, Malawi*. ICRAF, Nairobi, Kenya, pp. 64–77.

Sillitoe, P. (1998) The development of indigenous knowledge: a new applied anthropology. *Current Anthropology* 39, 223–252.

Sillon, J.F., Ozier-Lafontaine, H. and Brisson, N. (2000) Modelling daily root interactions for water in a tropical shrub and grass alley cropping system. *Agroforestry Systems* 49, 131–152.

Sinclair, F.L. (2001) *Bridging Knowledge Gaps Between Soils Research and Dissemination in Ghana. Final Technical Report.* DFID Natural Resources Systems Programme, Project R7516, Huntings Technical Services, Hemel Hempstead, UK.

Sinclair, F.L. and Joshi, L. (2000) Taking local knowledge about trees seriously. In: Lawrence, A. (ed.) *Forestry, Forest Users and Research: New Ways of Learning*. European Tropical Forest Research Network (ETFRN) Publication Series 1, pp. 45–61.

Sinclair, F.L. and Walker, D.H. (1998) Acquiring qualitative knowledge about complex agroecosystems. Part 1: representation as natural language. *Agricultural Systems* 56(3), 341–363.

Sinclair, F.L. and Walker, D.H. (1999) A utilitarian approach to the incorporation of local knowledge in agroforestry research and extension. In: Buck, L.E., Lassoie, J.P. and Fernandes, E.C.M. (eds) *Agroforestry in Sustainable Agricultural Systems*. CRC Press LLC, Boca Raton, Florida, pp. 245–275

Sinclair, T.R. and Seligman, N. (2000) Criteria for publishing papers on crop modelling. *Field Crops Research* 68(3), 165–172.

Singh, C.P. and Amberger, A. (1998) Organic acids and phosphorus solubilization in straw composted with rock phosphate. *Bioresource Technology* 63, 13–16.

Singh, J.S., Lauenroth, W.K., Hunt, H.W. and Swift, D.M. (1984) Bias and random errors in estimators of net root production: a simulation approach. *Ecology* 65, 1760–1764.

Singh, R.K., Narain, P., Dhyani, S.K. and Samra, J.S. (2000) The rooting behaviour of four agroforestry species in the western Himalayan valley region. *Journal of Tropical Forest Science* 12, 207–220.

Singh, R.P., Ong, C.K. and Saharan, N. (1989) Above and below ground competitions in alley cropping in semi-arid India. *Agroforestry Systems* 9, 259–274.

Singleton, P.W. and Tavares, J.W. (1986) Inoculation response of legumes in relation to the number and effectiveness of indigenous Rhizobium populations. *Applied and Environmental Microbiology* 51, 1013–1018.

Siriri, D. and Raussen, T. (2003) The agronomic and economic potential of tree fallows on scoured terrace benches in the humid highlands of Southwestern Uganda. *Agriculture, Ecosystems and Environment* 95, 359–369.

Sitompul, S.M. (2002) Analisis Sistem: LovESys Akar Pertanian Sehat. In: Sitompul, S.M. and Utami, S.R. (eds) *Akar Pertanian Sehat*. Seminar Ilmiah, Jurusan Tanah, Fakultas Pertanian, Universitas Brawijaya, Malang, pp. 1–10.

Sitompul, S.M., Hairiah, K., van Noordwijk, M. and Woomer, P.L. (1996) Organic matter dynamics after conversion of forest into food crops or sugarcane: prediction of the Century Model. *Agrivita* 19, 198–206.

Sivapalan, P. (1972) Nematode pests of tea. In: Webster, J.M. (ed.) *Economic Nematology*. Academic Press, London, pp. 285–310.

Six, J., Conant, R.T., Paul, E.A. and Paustian, K. (2002) Stabilization mechanisms of soil organic matter: implications for C-saturation of soils. *Plant and Soil* 241, 155–176.

Skjemstad, J.O., Taylor, J.A. and Smernik, R.J. (1999) Estimation of charcoal (Char) in soils. *Communications in Soil Science and Plant Analysis* 30, 2283–2298.

Slattery, W.J., Edwards, D.G., Bell, L.C., Coventry, D.R. and Helyar, K.R. (1998) Soil acidification and the carbon cycle in a cropping soil of north-eastern Victoria. *Australian Journal of Soil Research* 36, 273–290.

Smaling, E.M.A., Nandwa, S.M. and Janssen, B.H. (1997) Soil fertility in Africa is at stake. In: Buresh, R.J., Sanchez, P.A. and Calhoun, F. (eds) *Replenishing Soil Fertility in Africa*. SSSA Special Publication 51, Soil Science Society of America, Madison, Wisconsin, pp. 47–61.

Smaling, E.M.A., Oenema, O. and Fresco, L.O. (eds) (1999) *Nutrient Disequilibria in Agroecosystems: Concepts and Case Studies*. CAB International, Wallingford, UK, 322 pp.

Smart, G.C., Jr and Perry, V.G. (1968) *Tropical Nematology*. University of Florida Press, Gainesville, Florida, 153 pp.

Smart, G.C. Jr, Nguyen, K.B., Parkman, J.P. and Frank, J.H. (1991) Biological control of mole crickets in the genus *Scapteriscus* with the nematode *Steinernema scapterisci*. In: Pavis, C. and Kermarrec, A. (eds) *Rencontres Caraibes en Lutte Biologique*. Les Colloques No. 58, INRA Paris (Guadeloupe), pp. 151–155.

Smestad, B.T. (2000) Short-duration *Tithonia diversifolia* and *Crotalaria grahamiana* improved fallows for fertility replenishment of an Oxisol in the highlands of Western Kenya. MSc thesis, University of Saskatchewan, Saskatoon, Canada.

Smethurst, P.J. and Comerford, N.B. (1993a) Potassium and phosphorus uptake by competing pine and grass: observations and model verification. *Soil Science Society of America Journal* 57, 1602–1610.

Smethurst, P.J. and Comerford, N.B. (1993b) Simulation of nutrient uptake by single or competing and contrasting root systems. *Soil Science Society of America Journal* 57, 1361–1367.

Smethurst, P.J., Comerford, N.B. and Neary, D.G. (1993) Weed effects on early K and P nutrition and growth of slash pine on a Spodosol. *Forest Ecology and Management* 60(1–2), 15–26.

Smit, A.L., George, E. and Groenwold, J. (2000a) Root observations and measurements at (transparent) interfaces with soil. In: Smit, A.L., Bengough, A.G., Engels, E., van Noordwijk, M., Pellerin, S. and van de Geijn, S.C. (eds) *Root Methods: a Handbook*. Springer-Verlag, Berlin, pp. 235–271.

Smit, A.L., Bengough, A.G., Engels, C., van Noordwijk, M., Pellerin, S. and van de Geijn, S.C. (eds) (2000b) *Root Methods: a Handbook*. Springer-Verlag, Berlin, 587 pp.

Smith, D.M. (2001) Estimation of tree root lengths using fractal branching rules: a comparison with soil coring for *Grevillea robusta*. *Plant and Soil* 229, 295–304.

Smith, D.M. and Allen, S.J. (1996) Measurement of sap flow in plant stems. *Journal of Experimental Botany* 47, 1833–1844.

Smith, D.M. and Roberts, J.M. (2003) Hydraulic conductivities of competing root systems of *Grevillea robusta* and maize in agroforestry. *Plant and Soil* 251, 343–349.

Smith, D.M., Jarvis, P.G. and Odongo, J.C.W. (1997) Sources of water used by trees and millet in Sahelian windbreak systems. *Journal of Hydrology* 198, 140–153.

Smith, D.M., Jarvis, P.G. and Odongo, J.C.W. (1998) Management of windbreaks in the Sahel: the strategic implications of tree water use. *Agroforestry Systems* 40, 83–96.

Smith, D.M., Jackson, N.A., Roberts, J.M. and Ong, C.K. (1999a) Root distribution in a *Grevillea robusta*–maize agroforestry system in semi-arid Kenya. *Plant and Soil* 211, 191–205.

Smith, D.M., Jackson, N.A., Roberts, J.M. and Ong, C.K. (1999b) Reverse flow of sap in tree roots and downward siphoning of water by *Grevillea robusta*. *Functional Ecology* 13, 256–264.

Smith, F.A. and Smith, S.E. (1996) Mutualism and parasitism: biodiversity in form and structure in the 'arbuscular' (VA) mycorrhizal symbiosis. *Advances in Botanical Research* 22, 1–43.

Smith, M.L., Bruhn, J.N. and Anderson, J.B. (1992) The fungus *Armillaria bulbosa* is among the largest and oldest living organisms. *Nature* 356, 428–431.

Smith, P., Smith, J.U., Powlson, D.S., McGill, W.B., Arah, J.R.M., Chertov, O.G., Coleman, K., Franko, U., Frolking, S., Jenkinson, D.S., Jensen, L.S., Kelly, R.H., Klein-Gunnewiek, H., Komarov, A.S., Li, C., Molina, J.A.E., Mueller, T., Parton, W.J., Thornley, J.H.M. and Whitmore, A.P. (1997a) A comparison of the performance of nine organic matter models using datasets from seven long-term experiments. *Geoderma* 81, 153–225.

Smith, S.E. and Read, D.J. (1997) *Mycorrhizal Symbiosis*, 2nd edn. Academic Press, San Diego, California, 605 pp.

Smithson, P.C. (1999) Interactions of organic materials with phosphate rocks and triple superphosphate. *Agroforestry Forum* 9, 37–40.

Smithson, P.C. and Sanchez, P.A. (2001) Plant nutritional problems in marginal soils of developing countries. In: Ae, N., Arihara, J., Okada, K. and Srinivasan, A. (eds) *Plant Nutrient Acquisition: New Perspectives*. Springer-Verlag, Tokyo, Japan, pp. 32–68.

Smithson, P.C., Buresh, R.J. and Sinclair, F.L. (1999) Interactions of organic materials with phosphate rocks and triple superphosphate. *Agroforestry Forum* 9, 37–40.

Smucker, A.J.M. (1993) Soil environmental modifications of root dynamics and measurement. *Annual Review of Phytopathology* 31, 191–216.

Smucker, A.J.M. and Aiken, R.M. (1992) Dynamic root responses to water deficits. *Soil Science* 154, 281–289.

Smucker, A.J.M., Ellis, B.G. and Kang, B.T. (1995) Alley cropping on an Alfisol in the forest-savanna transition zone: root, nutrient and water dynamics. In: Kang, B.T., Osiname, O.A. and Larbi, A. (eds) *Alley Farming Research and Development*. Alley Farming Network for Tropical Africa, Ibadan, Nigeria, pp. 103–121.

Snoeck, D. (1996) Beneficial effects of Leucaena Rhizobium association on coffee in Burundi. *Plantations, Recherche, Développement* 3, 408–417.

Solomon, D. and Lehmann, J. (2000) Loss of phosphorus from soil in semi-arid northern Tanzania as a result of cropping: evidence from sequential extraction and $^{31}$P-NMR spectroscopy. *European Journal of Soil Science* 51, 699–708.

Sousa, G.F., Guimaraes, R.R., Sousa, N.R., Nunes, J.S. and Lourenco, J.N.P. (1999) Multi-strata agroforestry as an alternative for small migrant farmers practicing shifting cultivation in central Amazonian communities in Brazil. In: Jimenez, F. and Beer, J. (eds) *Proceedings of the International Symposium on Multistrata Agroforestry Systems with Perennial Crops, held at CATIE, 22–27 February, 1999*. CATIE, Turrialba, Costa Rica, pp. 243–246.

Southern, A.J. (1994) Acquisition of indigenous ecological knowledge about forest gardens in Kandy district, Sri Lanka. MPhil thesis, University of Wales, Bangor, UK, 152 pp.

Sparovek, G., Ranieri, S.B.L., Gassner, A., de Maria, I.C., Schnug, E., dos Santos, R.F. and Joubert, A. (2001) A conceptual framework for the definition of the optimal width of riparian forests. *Agriculture, Ecosystems and Environment* 90, 169–175.

Spedding, C.R.W. (1990) Agricultural production systems. In: Rabbinge, R., Goudriaan, J., van Keulen, H., Penning de Vries, F.W.T. and van Laar, H.H. (eds) *Theoretical Production Ecology: Reflection and Prospects. Simulation Monographs*. PUDOC, Wageningen, The Netherlands, pp. 239–248.

Sprent, J.I. (1984) Effects of drought and salinity on heterotrophic nitrogen fixing bacteria and on infection of legumes by rhizobia. In: Veeger, C. and Newton, W.E. (eds) *Advances in Nitrogen Fixation Research*. Nijhoff/Dr. W. Junk, The Hague, The Netherlands, pp. 295–302.

Sprent, J.I. and Parsons, R. (1999) Nitrogen fixation by legume and non-legume trees. *Field Crops Research* 65, 183–196.

Squire, G.R. (1990) *The Physiology of Tropical Crop Production*. CAB International, Wallingford, UK, 236 pp.

Steingrobe, B., Schmid, H. and Claassen, N. (2001) Root production and root mortality of winter barley and its implication with regard to phosphate acquisition. *Plant and Soil* 237, 239–248.

Steinmüller, N. (1995) Agronomy of the $N_2$-fixing fodder trees *Sesbania sesban* (L.) Merr. and *Sesbania goetzii* Harms in the Ethiopian highlands. PhD thesis, Universität Hohenheim, Verlag Ulrich E. Grauer, Stuttgart, Germany, 230 pp.

Stephens, W. and Hess, T. (1996) Report on the PARCH evaluation visit to Kenya, Malawi, Zimbabwe and Botswana. Cranfield University, Silsoe, Bedfordshire, UK, 22 pp.

Stephens, W. and Middleton, T. (2002) Why has the uptake of decision-support systems been so poor? In: Matthews, R.B. and Stephens, W. (eds) *Crop–Soil Simulation Models: Applications in Developing Countries*. CAB International, Wallingford, UK, pp. 129–147.

Steudle, E. and Heydt, H. (1997) Water transport across tree roots. In: Rennenberg, H., Eschrich, W. and Ziegler, H. (eds) *Trees – Contributions to Modern Tree Physiology*. Backhuys, Leiden, The Netherlands, pp. 239–255.

Steudler, P.A., Bowden, R.D., Melillo, J.M. and Aber, J.D. (1989) Influence of nitrogen fertilization on methane uptake in temperate forest soils. *Nature* 341, 314–316.

Steudler, P.A., Melillo, J.M., Feigl, B.J., Neill, C., Piccolo, M.C. and Cerri, C. (1996) Consequences of forest-to-pasture conversion on $CH_4$ fluxes in the Brazilian Amazon Basin. *Journal of Geophysical Research*, 101, 547–554.

Stevenson, F.J. (1982) *Humus Chemistry: Genesis, Composition, Reactions*. John Wiley & Sons, New York, 443 pp.

Stirling, G.R. (1990) *Biological Control of Plant Parasitic Nematodes*, 1st edn. CAB International, Wallingford, UK, 282 pp.

Stirzaker, R., Lefroy, T., Keating, B. and Williams, J. (1999) *Emerging Land Use Systems for Managing Dryland Salinity*. CSIRO Land and Water, 24 pp.

Stirzaker, R., Vertessy, R. and Sarre, R. (2002) *Trees, Water and Salt: an Australian guide to using trees for healthy catchments and productive farms*. RIRDC, Canberra, 159 pp.

Stokes, K.L. (2001) Farmers' knowledge about the management and use of trees on livestock farms in the Cañas area of Costa Rica. MSc thesis, University of Wales, Bangor, UK, 75 pp.

Stone, E.L. and Kalisz, P.J. (1991) On the maximum extent of tree roots. *Forest Ecology and Management* 46, 59–102.

Stover, R.H. (1972) *Banana, Plantain and Abaca Diseases*. Commonwealth Mycological Institute, Kew Gardens, UK, 316 pp.

Subedi, K.D. and Gurung, T.B. (1995) Determination of an optimum combination of organic and inorganic fertilizer for barley and wheat grown under irrigated conditions, 1994/95. LARC Working Paper No. 95/48. Lumle Agricultural Research Centre, Kaski, Nepal, 33 pp.

Sudarsono, H., Susilo, F.X., Ginting, C. and Swibawa, I.G. (1995) Comparison of several cultural techniques on early and late soybean surrounded with long-bean. IPM Action Research, UNILA–Clemson University, Soybean IPM Team. Bandar Lampung, 82 pp.

Sukul, N.C. (1992) Plants antagonistic to plant parasitic nematodes. *Indian Review of Life Sciences* 12, 23–52.

Sumann, M., Amelung, W., Haumaier, L. and Zech, W. (1998) Climatic effects on soil organic phosphorus in the North American Great Plains identified by phosphorus-31 nuclear magnetic resonance. *Soil Science Society of America Journal* 62, 1580–1586.

Sumner, D.R. (1994) Cultural management. In: Campbell, C.L. and Benson, D.M. (eds) *Epidemiology and Management of Root Diseases*. Springer-Verlag, Berlin, pp. 309–333.

Suprayogo, D. (2000) The effectiveness of the safety net of hedgerow cropping systems in reducing mineral N-leaching in Ultisols. PhD thesis, Imperial College London, Wye Campus, UK, 316 pp.

Suprayogo, D., van Noordwijk, M., Hairiah, K. and Cadisch, G. (2002) The inherent 'safety-net' of Ultisols: measuring and modelling retarded leaching of mineral nitrogen. *European Journal of Soil Science* 53, 185–194.

Susilo, F.X., Evizal, R., Swibawa, I.G., Murwani, S. and Rustiati, E.L. (1999) Conservation and resource agrobiota: evaluation of current agricultural management practices in Lampung. In: Gafur, A., Susilo, F.X., Utomo, M. and van Noordwijk, M. (eds) *Management of Agrobiodiversity for Sustainable Land Use and Global Environmental Benefits*. ICRAF, Bogor, Indonesia, pp. 1–7.

Swallow, B.M., Garrity, D.P. and van Noordwijk, M. (2001) The effects of scale, flows and filters on property rights and collective action in watershed management. *Water Policy* 3(6), 457–474.

Swift, M.J. (1999) Towards the second paradigm: integrated biological management of soil. In: Siqueira, J.O., Moreira, F.M.S., Lopes, A.S., Guilherme, L.R.G., Faquin, V., Furtani Neto, A.E. and Carvalho, J.G. (eds) *Inter-relação Fertilidade, Biologia do Solo e Nutrição de Plantas*. UFLA, Lavras, Brasil, pp. 11–24.

Swift, M.J. and Anderson, J.M. (1993) Biodiversity and ecosystem function in agricultural systems. In: Schulze, E.D. and Mooney, H.A. (eds) *Biodiversity and Ecosystem Function*. Springer, Berlin, pp. 15–41.

Swift, M.J. and Bignell, D. (2001) Standard methods for assessment of soil biodiversity and land use practice. In: van Noordwijk, M., Williams, S.E. and Verbist, B. (eds) *Toward Integrated Natural Resource Management in Forest Margins of the Humid Tropics: Local Action and Global Concerns*. ASB Lecture Note 06B, ASB-ICRAF/SEA, 34 pp.

Swift, M.J., Vandermeer, J., Ramakrishnan, P.S., Anderson, J.M., Ong, C.K. and Hawkins, B.A. (1996) Biodiversity and agroecosystem function. In: Mooney, H.A., Cushman, J.H., Medina, E., Sala, O.E. and Schulze, E.D. (eds) *Functional Roles of Biodiversity: a Global Perspective*. John Wiley & Sons, London, pp. 261–298.

Swift, M.J., Andren, O., Brussaard, L., Briones, M., Couteaux, M.M., Ekschmitt, K., Kjoller, A., Loiseau, P. and Smith, P. (1998) Global change, soil biodiversity and nitrogen cycling in terrestrial ecosystems: three case studies. *Global Change Biology* 4, 729–744.

Szott, L.T., Palm, C.A. and Sanchez, P.A. (1991) Agroforestry in acid soils of the humid tropics. *Advances in Agronomy* 45, 275–301.

Szott, L.T., Palm, C.A. and Buresh, R.J. (1999) Ecosystem fertility and fallow function in the humid and subhumid tropics. *Agroforestry Systems* 47, 163–196.

Taha, A.H.Y. (1993) Nematode interactions with root-nodule bacteria. In: Khan, M.W. (ed.) *Nematode Interactions*, 1st edn. Chapman and Hall, London, pp. 175–202.

Talawar, S. and Rhoades, R.E. (1998) Scientific and local classification and management of soils. *Agriculture and Human Values* 15, 3–14.

Tamang, D. (1992) *Indigenous Soil Management Practice in the Hills of Nepal: Lessons from East-west Transect*. HMG Ministry of Agriculture – Winrock International, Research Report Series No. 19, 59 pp.

Tarafdar, J.C. and Claassen, N. (1988) Organic phosphorus compounds as a phosphorus source for higher plants through the activity of phosphatases produced by plant roots and microorganisms. *Biology and Fertility of Soils* 5, 308–312.

Taranto, M.T., Adams, M.A. and Polglase, P.J. (2000) Sequential fractionation and characterisation ($^{31}$P-NMR) of phosphorus-amended soils in *Banksia integrifolia* (L.f.) woodland and adjacent pasture. *Soil Biology and Biochemistry* 32, 169–177.

Tate, K.R. (1985) Soil phosphorus. In: Vaughan, D. and Malcolm, R.E. (eds) *Soil Organic Matter and Biological Activity*. Martinus Nijhoff, Dordrecht, The Netherlands.

Tavares, F.C., Beer, F., Jiménez, F., Schroth, G. and Fonseca, C. (1999) Experiencia de agricultores de Costa Rica con la introducción de árboles maderables en plantaciones de café. *Agroforesteria en las Americas* 6, 17–20.

Taylor, J.H. and Gardner, H.R. (1960) Use of wax substrates in root penetration studies. *Soil Science Society of America Journal* 24, 79–81.

Tester, M., Smith, S.E. and Smith, F.A. (1987) The phenomenon of 'non-mycorrhizal' plants. *Canadian Journal of Botany* 65, 419–431.

Thapa, B., Sinclair, F.L. and Walker, D.H. (1995) Incorporation of indigenous knowledge and perspectives in agroforestry development. Part 2: case study on the impact of explicit representation of farmers' knowledge. *Agroforestry Systems* 30, 249–261

Thies, J.E., Singleton, P.W. and Bohlool, B.B. (1990) Influence of size of indigenous rhizobial populations on establishment and symbiotic performance of introduced rhizobia on field-grown legumes. *Applied and Environmental Microbiology* 57, 19–28.

Thompson, J.P. (1996) Correction of dual phosphorus and zinc deficiencies of linseed (*Linum usitatissimum* L.) with cultures of vesicular-arbuscular mycorrhizal fungi. *Soil Biology and Biochemistry* 28, 941–951.

Thorburn, P.J. (1996) Can shallow watertables be controlled by the revegetation of saline lands? *Australian Journal of Soil and Water Conservation* 9, 45–50.

Thorne, P.J., Subba, D.B., Walker, D.H., Thapa, B., Wood, C.D. and Sinclair, F.L. (1999) The basis of indigenous knowledge of tree fodder quality and its implications for improving the use of tree fodder in developing countries. *Animal Feed Science and Technology* 81, 119–131.

Thurston, D.H. (1990) Plant disease management practices of traditional farmers. *Plant Disease* 74, 96–102.

Thurston, D.H. (1992) *Sustainable Practices for Plant Disease Management in Traditional Farming Systems*, 1st edn. Westview Press, Boulder, Colorado, 279 pp.

Thurston, D.H. (1997) *Slash/Mulch Systems: Sustainable Methods for Tropical Agriculture*. Westview Press, Boulder, Colorado, 196 pp.

Tierney, G.L. and Fahey, T.J. (2001) Evaluating minirhizotron estimates of fine root longevity and production in the forest floor of a temperate broadleaf forest. *Plant and Soil* 229, 167–176.

Tiessen, H. and Moir, J.O. (1993) Characterization of available P by sequential extraction. In: Carter, M.R. (ed.) *Soil Sampling and Methods of Analysis*. Canadian Society of Soil Science, Lewis Publishers, Boca Raton, Florida, pp. 75–86.

Tiessen, H., Salcedo, I.H. and Sampaio, E.V.S.B. (1992) Nutrient and soil organic matter dynamics under shifting cultivation in semi-arid north-eastern Brazil. *Agriculture, Ecosystems and Environment* 38, 139–151.

Tiessen, H., Chacon, P., and Cuevas, E. (1993) Phosphorus and nitrogen status in soils and vegetation along a toposequence of dystrophic rainforest on the upper Rio Negro. *Oecologia* 99, 145–150.

Tilman, D. and Lehman, C. (2001) Biodiversity, composition and ecosystem processes: theory and concepts. In: Kinzig, A., Pacala, S. and Tilman, D. (eds) *Functional Consequences of Biodiversity: Empirical Progress and Theoretical Extensions*. Princeton University Press, New Jersey, pp. 9–41.

Tinker, P.B. and Nye, P.H. (2000) *Solute Movement in the Rhizosphere*. Oxford University Press, Oxford, UK, 444 pp.

Tjamos, E.C., Papavizas, G.C. and Cook, R.J. (eds) (1991) Biological control of plant diseases. Progress and challenges for the future. *Proceedings of a NATO Advanced Research Workshop*. NATO ASI Series, Series A: Life Sciences Vol. 230. Plenum Press, New York, 462 pp.

Toan, T.D., Phien, T., Nguyen, L., Phai, D.D. and Ga, N.V. (2001) Soil erosion management at the watershed level for sustainable agriculture and forestry in Vietnam. In: *Soil Erosion Management Research in Asian Catchments: Methodological Approaches and Initial Results*. Proceedings of the 5th Management of Soil Erosion Consortium (MSEC) Assembly, Semarang, Central Java, Indonesia, 7–11 November, 2000. Kasetsart University, Bangkok, pp. 233–253.

Toky, O.P. and Bisht, R.P. (1992) Observations on the rooting patterns of some agroforestry trees in an arid region of Northwestern India. *Agroforestry Systems* 18, 245–263.

Toledo, J.M. (1985) *Pasture Development for Cattle Production in the Major Ecosystems of the Tropical American Lowlands*. Proceedings of the XV International Grassland Congress, Japanese Grassland Society, Kyoto, Japan, pp. 74–81.

Tollefson, L. (1996) Requirements for improved interactive communication between researchers, managers, extensionists and farmers. In: Icid, F. (ed.) *Irrigation Scheduling from Theory to Practice (Water Reports)*. FAO, Rome, pp. 217–226.

Tomich, T.P., van Noordwijk, M., Vosti, S. and Witcover, J. (1998) Agricultural development with rainforest conservation: methods for seeking best bet alternatives to slash-and-burn, with applications to Brazil and Indonesia. *Agricultural Economics* 19, 159–174.

Torquebiau, E.F. and Kwesiga, F. (1997) Root development in *Sesbania sesban* fallow–maize system in eastern Zambia. *Agroforestry Systems* 34, 193–211.

Torsvik, V., Sorheim, R. and Goksoyr, J. (1996) Total bacterial diversity in soil and sediment communities – a review. *Journal of Industrial Microbiology* 17, 170–178.

Touyama, Y., Yamamoto, T. and Nakagoshi, N. (2002) Are ants useful bioindicators? The relationship between ant species richness and soil macrofaunal richness, in Hiroshima prefecture. *Edaphologia* 70, 33–36.

Trenbath, B.R. (1989) The use of mathematical models in the development of shifting cultivation systems. In: Proctor, J. (ed.) *Mineral Nutrients in Tropical Forest and Savanna Ecosystems*. Special publication number 9 of the British Ecological Society. Blackwell Scientific Publications, Oxford, UK, pp. 353–371.

Trimble, S.W. (1999) Decreased rates of alluvial sediment storage in the Coon creek basin, Wisconsin, 1975–93. *Science* 285, 1244–1246.

Tripathi, B.P. (1996) Long-term effect of farm yard manure and mineral fertilizers on rice and wheat yields and nutrient balance in rice–wheat system at Khumaltar condition. LARC Seminar Paper No. 96/35. Lumle Agricultural Research Centre, Kaski, Nepal, 9 pp.

Tromp, J. (1983) Nutrient reserves in roots of fruit trees, in particular carbohydrates and nitrogen. *Plant and Soil* 71, 401–413.

Trumbore, S.E., Davidson, E.A., de Camargo, P.B., Nepstad, D.C. and Martinelli L.A. (1995) Belowground cycling of carbon in forests and pastures of Eastern Amazonia. *Global Biogeochemical Cycles* 9, 515–528.

Tsuji, G. and Balas, S. (eds) (1993) *The IBSNAT Decade: Ten Years of Endeavour at the Frontier of Science and Technology*. Department of Agronomy and Soil Science, University of Hawaii, College of Tropical Agriculture and Human Resources, Honolulu, Hawaii, 178 pp.

Tsuji, G.Y., Uehara, G. and Balas, S. (1994) DSSAT v3, Vol. 2. University of Hawaii, Honolulu, Hawaii, 284 pp.

Tsuruta, H., Ishizuka, S., Ueda, S. and Murdiyarso, D. (2000) Seasonal and spatial variations of $CO_2$, $CH_4$, and $N_2O$ fluxes from the surface soils in different forms of land-use/cover in Jambi, Sumatra. In: Murdiyarso, D. and Tsuruta, H. (eds) *The Impacts of Land-use/cover Change on Greenhouse Gas Emissions in Tropical Asia*, Global Change Impacts Centre for Southeast Asia, Bogor, Indonesia and National Institute of Agro-Environmental Sciences, Tsukuba, Japan, pp. 7–30.

Turco, R.P. (1997) *Earth Under Siege: From Air Pollution to Global Change*. Oxford University Press, New York, pp. 257.

Turk, D., Keyser, H.H. and Singleton, P.W. (1993) Response of tree legumes to rhizobial inoculation in relation to the population of indigenous rhizobia. *Soil Biology and Biochemistry* 25, 75–81.

Turnbull, J. (1995) Organic N uptake by *Eucalyptus*. *Plant Cell and Environment* 18, 1386–1394.

Turton, C. and Sherchan, D.P. (1996) The use of rural peoples' knowledge as a research tool for soil survey in the eastern hills of Nepal. PAC Occasional Paper No. 21. Pakhribas Agricultural Centre, Dhankuta, Nepal, 13 pp.

Tyree, M.T., Patiño, S., Bennink, J. and Alexander, J. (1995) Dynamic measurements of root hydraulic conductance using a high-pressure flowmeter in the laboratory and field. *Journal of Experimental Botany* 46, 83–94.

Umrit, G. and Friesen, D.K. (1994) The effect of the C:P ratio of plant residues added to soils of contrasting phosphate sorption capacities on P uptake by *Panicum maximum* (Jacq.). *Plant and Soil* 158, 275–285.

Unger, P.W. and Kasper, T.C. (1994) Soil compaction and root growth: a review. *Agronomy Journal* 86, 759–766.

Urquiaga, S., Cadisch, G., Alves, B.J.R., Giller, K.E. and Boddey, R.M. (1998) Influence of the decomposition of roots of tropical forage species on the availability of soil nitrogen. *Soil Biology and Biochemistry* 30, 2099–2106.

USDA (1960) Index of Plant Diseases in the United States. *Agriculture Handbook No. 165*, USDA, Washington, DC, 531 pp.

Valentim, J.F., do Amaral, E.F., Cavalcante, M. de J.B., Fazolin, M., Caballero, S.S.C., Boddey, R.M., Sharma, R.D. and de Melo, A.F.W. (2000) Diagnosis and potential socioeconomic and environmental impacts of pasture death in the Western Brazilian Amazon. In: LBA Scientific Conference 1, 2000, Belém, PA, Brazil (Abstracts). MCT/CPTEC/INPE, Belém, PA, Brazil, p. 212.

van Breemen, N. (2002) Natural organic tendency. *Nature* 415, 381–382.

Vance, C.P. (2001) Symbiotic nitrogen fixation and phosphorus acquisition. Plant nutrition in a world of declining renewable resources. *Plant Physiology* 127, 390–397.

Vandenbeldt, R.J. (1991) Rooting systems of western and southern African *Faidherbia albida* (Del.) A. Chev. (syn. *Acacia albida* Del.) – a comparative analysis with biogeographic implications. *Agroforestry Systems* 14, 233–244.

van der Heide, J., Setijono, S., Syekhfani, M.S., Flach, E.N., Hairiah, K., Ismunandar, S., Sitompul, S.M. and van Noordwijk, M. (1992) Can low external input cropping systems on acid upland soils in the humid tropics be sustainable? Backgrounds of the Unibraw/IB nitrogen management project in Bunga Mayang (Sungkai Selatan, Kotabumi, N. Lampung, S. Sumatra, Indonesia). *Agrivita* 15, 1–10.

van der Heijden, M.G.A. and Sanders, I.R. (eds) (2002) *Mycorrhizal Ecology*. Vol. 157, Springer-Verlag, Berlin.

van der Heijden, M.G.A., Klironomos, J.N., Ursic, M., Moutoglis, P., Streitwolf-Engel, R., Boller, T., Wiemken, A. and Sanders, I.R. (1998) Mycorrhizal fungal diversity determines plant diversity, ecosystem variability and productivity. *Nature* 396, 69–72.

Vandermeer, J. (1989) *The Ecology of Intercropping*. Cambridge University Press, Cambridge, UK, 237 pp.

Vandermeer, J. (ed.) (2002) *Tropical Agroecosystems*. CRC Press LLC, Boca Raton, Florida.

Vandermeer J., Van Noordwijk, M., Anderson, J., Ong, C.K. and Perfecto, E. (1998) Global change and multi-species agroecosystems: concepts and issues. *Agriculture, Ecosystems and Environment* 67, 1–22.

van der Molen, W.H. (1983) *Agrohydrology*. Wageningen Agricultural University, Wageningen, The Netherlands.

van der Ploeg, J.D. (1989) Knowledge systems, metaphor and interface: the case of potatoes in the Peruvian highlands. In: Long, N. (ed.) *Encounters at the Interface: a Perspective on Social Discontinuities in Rural Development*. Wageningen Studies in Sociology 27, Agricultural University Wageningen, The Netherlands, pp. 145–163.

van der Putten, W.W., van Dijk, C. and Peters, B.A.M. (1993) Plant-specific soil-borne diseases contribute to succession in foredune vegetation. *Nature* 362, 53–56.

van Duivenbooden, N. (1995) Land use systems analysis as a tool in land use planning, with special reference to North and West African agro-ecosystems. PhD thesis, Wageningen Agricultural University, Wageningen, The Netherlands, 176 pp.

van Genuchten, M.Th. and Dalton, F.N. (1986) Models for simulating salt movement in aggregated field soils. *Geoderma* 38, 165–183.

van Huysstein, L. (1988) Grapevine root growth in response to soil tillage and root pruning practices. In: van Zyl, J.L. (ed.) *The Grapevine Root and its Environment*. Department of Agriculture and Water Supply Technical Communication No. 215, Pretoria, South Africa, pp. 44–56.

van Kessel, C., Farrell, R.E., Roskoski, J.P. and Keane, K.M. (1994) Recycling of the naturally-occurring $^{15}$N in an established stand of *Leucaena leucocephala*. *Soil Biology and Biochemistry* 26, 757–762.

van Keulen, H. (1995) Sustainability and long-term dynamics of soil organic matter and nutrients under alternative management strategies. In: Bouma, J., Kuyvenhoven, A., Bouman, B.A.M., Luten, J.C. and Zandstra H.G. (eds) *Eco-regional Approaches for Sustainable Land Use and Food Production. Systems Approaches for Sustainable Agricultural Development*. Kluwer Academic, Dordrecht, The Netherlands, pp. 353–375.

Vanlauwe, B., Nwoke, O.C., Sanginga, N. and Merckx, R. (1996) Impact of the residue quality on the C and N mineralization of leaf and root residues of three agroforestry species. *Plant and Soil* 183, 221–231.

Vanlauwe, B., Sanginga, N. and Merckx, R. (1998) Soil organic matter dynamics after addition of $^{15}$N labeled *Leucaena* and *Dactyladenia* residues in alley cropping systems. *Soil Science Society of America Journal* 62, 461–466.

Vanlauwe, B., Aman, S., Aihou, K., Tossah, B.K., Adebiyi, V., Sanginga, N., Lyasse, O., Diels, J. and Merckx, R. (1999) Alley cropping in the moist savanna of West-Africa: III. Soil organic matter fractionation and soil productivity. *Agroforestry Systems* 42, 245–264.

Vanlauwe, B., Aihou, K., Aman, S., Tossah, B.K., Diels, J., Lyasse, O., Hauser, S., Sanginga, N. and Merckx, R. (2000) Maize N and P uptake as affected by particulate organic matter and soil characteristics for soils from the West-African moist savanna zone. *Biology and Fertility of Soils* 30, 440–449.

Vanlauwe, B., Akinnifesi, F.K., Tossah, B.K., Lyasse, O., Sanginga, N. and Merckx, R. (2002) Root distribution of *Senna siamea* grown on a series of soils representative for the Togolese derived savanna zone. *Agroforestry Systems* 54, 1–12.

van Noordwijk, M. (1993) Roots: length, biomass, production and mortality. In: Anderson, J.M. and Ingram, J.S.I. (eds) *Tropical Soil Biology and Fertility, a Handbook of Methods*. CAB International, Wallingford, UK, pp. 132–144.

van Noordwijk, M. (1999) Scale effects in crop–fallow rotations. *Agroforestry Systems* 47, 239–251.

van Noordwijk, M. and Brouwer, G. (1991) Review of quantitative root length data in agriculture. In: Persson, H. and McMichael, B.L. (eds) *Plant Roots and Their Environment*. Elsevier, Amsterdam, The Netherlands, pp. 515–525.

van Noordwijk, M. and Brouwer, G. (1997) Roots as sinks and sources of carbon and nutrients in agricultural systems. In: Brussaard, L. and Ferrera-Cerrato, R. (eds) *Soil Ecology in Sustainable Agricultural Systems*. CRC Lewis Publ., Boca Raton, Florida, pp. 71–89.

van Noordwijk, M. and Cadisch, G. (2002) Access and excess problems in plant nutrition. *Plant and Soil* 247, 25–40.

van Noordwijk, M. and de Willigen, P. (1987) Agricultural concepts of roots: from morphogenetic to functional equilibrium. *Netherlands Journal of Agricultural Science* 35, 487–496.

van Noordwijk, M. and Lusiana, B. (1999) WaNuLCAS a model of water, nutrient and light capture in agroforestry systems. *Agroforestry Systems* 43, 217–242.

van Noordwijk, M. and Lusiana, B. (2000) *WaNuLCAS version 2.0, Background on Model of Water, Nutrient and Light Capture in Agroforestry Systems*. ICRAF, Bogor, Indonesia, 186 pp.

van Noordwijk, M. and Mulia, R. (2002) Functional branch analysis as tool for scaling above- and belowground trees for their additive and non-additive properties. *Ecological Modelling* 149, 41–51

van Noordwijk, M. and Ong, C.K. (1999) Can the ecosystem mimic hypothesis be applied to farms in African savannahs? *Agroforestry Systems* 45, 131–158.

van Noordwijk, M. and Purnomosidhi, P. (1995) Root architecture in relation to tree–soil–crop interactions and shoot pruning in agroforestry. *Agroforestry Systems* 30, 161–173.

van Noordwijk, M. and Swift, M.J. (1999) Belowground biodiversity and sustainability of complex agroecosystems. In: Gafur, A., Susilo, F.X., Utomo, M. and van Noordwijk, M. (eds) *Management of Agrobiodiversity for Sustainable Land Use and Global Environmental Benefits*. ICRAF, Bogor, Indonesia, pp. 8–28.

van Noordwijk, M. and van de Geijn, S.C. (1996) Root, shoot and soil parameters required for process-orientated models of crop growth limited by water or nutrients. *Plant and Soil* 183, 1–25.

van Noordwijk, M., Widianto, Heinen, M. and Hairiah, K. (1991a) Old tree root channels in acid soils in the humid tropics: important for crop root penetration, water infiltration and nitrogen management. *Plant and Soil* 134, 37–44.

van Noordwijk, M., Hairiah, K., Syekfani, M.S. and Flach, E.N. (1991b) *Peltophorum pterocarpa* (DC) Back (Caesalpiniaceae), a tree with a root distribution suitable for alley cropping on acid soils in the humid tropics. In: McMichael, B.L. and Persson, H. (eds) *Plant Roots and Their Environment*. Elsevier Science Publishers, Amsterdam, pp. 526–532.

van Noordwijk, M., Brouwer, G. and Harmanny, K. (1993) Concepts and methods for studying interactions of roots and soil structure. *Geoderma* 56, 351–375.

van Noordwijk, M., Spek, L.Y. and De Willigen, P. (1994) Proximal root diameter as a predictor of total root size for fractal branching models I. Theory. *Plant and Soil* 164, 107–117.

van Noordwijk, M., de Jager, A. and Floris, J. (1995) A new dimension to observations in mini-rhizotrons: a stereoscopic view on root photographs. *Plant and Soil* 86, 447–453.

van Noordwijk, M., Lawson, G., Soumare, A., Groot, J.J.R. and Hairiah, K. (1996) Root distribution of trees and crops: competition and/or complementarity. In: Ong, C.K. and Huxley, P. (eds) *Tree–Crop Interactions, a Physiological Approach*. CAB International, Wallingford, UK, pp. 319–364.

van Noordwijk, M., Cerri, C., Woomer, P.L., Nugroho, K. and Bernoux, M. (1997) Soil carbon dynamics in the humid tropical forest zone. *Geoderma* 79, 187–225.

van Noordwijk, M., Hairiah, K., Lusiana, B. and Cadisch, G. (1998a) Tree–soil–crop interactions in sequential and simultaneous agroforestry systems. In: Bergstrom, L. and Kirchman, H. (eds) *Carbon and Nutrient Dynamics in Natural and Agricultural Tropical Ecosystems*. CAB International, Wallingford, UK, pp. 173–190.

van Noordwijk, M., Martikainen, P., Bottner, P., Cuevas, E., Rouland, C. and Dhillion, S.S. (1998b) Global change and root function. *Global Change Biology* 4, 759–772.

van Noordwijk, M., Hairiah, K., Woomer, P.L. and Murdiyarso, D. (1998c) Criteria and indicators of forest soils used for slash-and-burn agriculture and alternative land uses in Indonesia. In: Davidson, E. (ed.) *The Contribution of Soil Science to the Development and Implementation of Criteria and Indicators of Sustainable Forest Management*, ASA Special Publication, Madison, Wisconsin, pp. 137–153.

van Noordwijk, M., Murdiyarso, D., Hairiah, K., Wasrin, U.R., Rachman, A. and Tomich, T.P. (1998d) Forest soils under alternatives to slash-and-burn agriculture in Sumatra, Indonesia. In: Schulte, A. and Ruhiyat, D. (eds) *Soils of Tropical Forest Ecosystems: Characteristics, Ecology and Management*. Springer-Verlag, Berlin, pp. 175–185.

van Noordwijk, M., van Roode, M., McCallie, E.L. and Lusiana, B. (1998e) Erosion and sedimentation as multiscale, fractal processes: implications for models, experiments and the real world. In: Penning de Vries, F.W.T., Agus, F. and Kerr, J. (eds) *Soil Erosion at Multiple Scales*. CAB International, Wallingford, UK, pp. 223–253.

van Noordwijk, M., Tomich, T.P. and Verbist, B. (2001) Negotiation support models for integrated natural resource management in tropical forest margins. *Conservation Ecology* 5(2), 21. Online URL: http://www.consecol.org/vol5/iss2/art21

van Noordwijk, M., Rahayu, S., Hairiah, K., Wulan, Y.C., Farida and Verbist, B.J.P. (2002) Carbon stock assessment for a forest-to-coffee conversion landscape in Sumber Jaya (Lampung, Indonesia): from allometric equations to land use change analysis. *Science in China (Series C)* 45 (Supp.), 75–86.

van Schaik, C.P., Terborgh, J.W. and Wright, S.J. (1993) The phenology of tropical forests: adaptive significance and consequences for primary consumers. *Annual Review of Ecology and Systematics* 24, 353–377.

van Schilfgaarde, J. (ed.) (1974) *Drainage for Agriculture*. Agronomy 17. American Society of Agronomy, Madison, Wisconsin.

van Zyl, J.L. (1988) Response of grapevine roots to soil tillage and root pruning practices. In: van Zyl, J.L. (ed.) *The Grapevine Root and its Environment*. Department of Agriculture and Water Supply. Technical Communication No. 215, Pretoria, South Africa, pp. 30–43.

Varma, A. (ed.) (1998) *Mycorrhiza Manual*. Springer-Verlag, Berlin, 350 pp.

Veldkamp, E. and Keller, M. (1997) Nitrogen oxide emissions from banana plantations in the humid tropics. *Journal of Geophysical Research* 102, 15889–15898.

Verbist, B.J.P., Putra, A.E.D. and Budidarsono, S. (2002) Background paper: Sumberjaya land use change, history and its driving factors. In: Backgrounds for ACIAR Project Planning Meeting, Sumberjaya, 12–16 October, 2002. World Agroforestry Centre (ICRAF), SE Asia, Bogor, Indonesia, pp. 27–42.

Verchot, L.V., Davidson, E.A., Cattânio, J.H., Ackerman, I.L., Erickson, H.E. and Keller, M. (1999) Land use change and biogeochemical controls of nitrogen oxide emissions from soils in eastern Amazonia. *Global Biogeochemical Cycles* 13, 31–46.

Verchot, L.V., Davidson, E.A., Cattânio, J.H. and Ackerman, I.L. (2000) Land-use change and biogeochemical controls of methane fluxes in soils of eastern Amazonia. *Ecosystems* 3, 41–56.

Villenave, C. and Cadet, P. (1997) Interaction of *Helicotylenchus dihystera, Pratylenchus pseudopratensis* and *Tylenchorhynchus gladiolatus* on two plants from the sudano-sahelian zone of West Africa. *Nematropica* 28, 31–39.

Visscher, A. de, Boeckx, P. and van Cleemput, O. (1998) Interaction between nitrous oxide formation and methane oxidation in soils: influence of cation exchange phenomena. *Journal of Environmental Quality* 27, 679–687.

Vogt, K.A. and Persson, H. (1991) Techniques and approaches in forest tree ecophysiology. In: Hinckley, T. and Lassoie, J. (eds) *Root Methods*. CRC Press Inc., Boca Raton, Florida, 151 pp.

Vogt, K.A., Vogt, D.J. and Bloomfield, J. (1998) Analysis of some direct and indirect methods for estimating root biomass and production of forests at an ecosystem level. *Plant and Soil* 200, 71–89.

Vörösmarty, C.J., Fekete, B.M., Meybeck, M. and Lammers, B. (2000) Geomorphometric attributes of the global system of rivers at 30-minute spatial resolution. *Journal of Hydrology* 237, 17–39.

Wahid, P.A. (2001) Radioisotope studies of root activity and root-level interactions in tree-based production systems: a review. *Applied Radiation and Isotopes* 54, 715–736.

Walbridge, M.R. (1990) Phosphorus availability in acid organic soils of the lower North Carolina coastal plain: estimation and control of P availability and plant response. *Ecology* 72, 2083–2100.

Waliszewski, W.S. (2002) Recommendations for the sustainable harvesting and production of *Thaumatococcus daniellii* (Benn.) Benth. in the western region of Ghana. MSc thesis, University of Wales, Bangor, UK.

Waliszewski, W.S. and Sinclair, F.L. (2004) Local knowledge of *Thaumatococcus daniellii*. *Economic Botany* (in press).

Walker, D.H. and Sinclair, F.L. (1998) Acquiring qualitative knowledge about complex agroecosystems. Part 2: formal representation. *Agricultural Systems* 56(3), 365–386.

Walker, D.H., Thapa, B. and Sinclair, F.L. (1995) Incorporation of indigenous knowledge and perspectives in agroforestry development. Part 1: review of methods and their application. *Agroforestry Systems* 30, 235–248.

Walker, D.H., Sinclair, F.L., Joshi, L. and Ambrose, B. (1997) Prospects for the use of corporate knowledge bases in the generation, management and communication of knowledge at a front-line agricultural research centre. *Agricultural Systems* 54(3), 291–312.

Wallace, J.S. (1996) The water balance of mixed tree–crop systems. In: Ong, C.K. and Huxley, P. (eds) *Tree–Crop Interactions: a Physiological Approach*. CAB International, Wallingford, UK, pp. 189–233.

Waller, J.M. (1984) The influence of agricultural development on crop diseases. *Tropical Pest Management* 30, 86–93.

Waller, J.M. and Hillocks, R.J. (1997) Host-plant resistance and integrated control. In: Hillocks, R.J. and Waller, J.M. (eds) *Soilborne Diseases of Tropical Crops*, 1st edn. CAB International, Wallingford, UK, pp. 419–430.

Wang, K.H., Sipes, B.S. and Schmitt, D.P. (2002) *Crotalaria* as cover crop for nematode management: a review. *Nematropica* 32, 35–57.

Waraspati, I.K. (1997) Studi potensi SlNPV dan predator sebagai agensia pengendalian ulat grayak (*Spodoptera litura* F.) pada pertanaman kedelai. Skripsi S1, Universitas Lampung, Bandar Lampung, 31 pp.

Wardle, D.A. and Lavelle, P. (1997) Linkages between soil biota, plant litter quality and decomposition. In: Cadisch, G. and Giller, K.E. (eds) *Driven by Nature: Plant Litter Quality and Decomposition*. CAB International, Wallingford, UK, pp. 107–124.

Warren, G. (1992) *Fertilizer Phosphorus Sorption and Residual Value In Tropical African Soils*. NRI Bulletin 37. Natural Resources Institute, Chatham, UK.

Wassmann, R., Lantin, R.S. and Neue, H.U. (eds) (2000) *Methane Emissions from Major Rice Ecosystems in Asia*. Special Issue of Nutrient Cycling in Agroecosystems 58, Nos 1–3, pp. 1–398.

Wassmann, R., Neue, H.U., Bueno, C., Lantin, R.S., Alberto, M.C.R., Buendia, L.V., Bronson, K., Papen, H. and Rennenberg, H. (1998) Inherent properties of rice soils determining methane production potentials. *Plant and Soil* 203, 227–237.

Watanabe, T., Osaki, M. and Tadano, T. (1998) Effects of nitrogen source and aluminium on growth of tropical tree seedlings adapted to low pH soils. *Soil Science and Plant Nutrition* 44, 655–666.

Wawo, A.L. (2000) Keanekaragaman Jenis Pohon yang Diduga Sebagai Inang Sekunder Cendana di Pulau Timor – Nusa Tenggara Timur. MSc thesis, University of Indonesia, Jakarta, Indonesia, 127 pp.

Weibel, F.P. and Boersma, K. (1995) An improved stem heat balance method using analog heat control. *Agricultural and Forest Meteorology* 75, 191–208.

Weitz, A.M., Linder, E., Frolking, S., Crill, P.M. and Keller, M. (2001) $N_2O$ emissions from humid tropical agricultural soils: effects of soil moisture, texture and nitrogen availability. *Soil Biology and Biochemistry* 33, 1077–1093.

Wellmann, F.L. (1972) *Tropical American Plant Disease (Neotropical Phytopathology Problems)*. The Scarecrow Press, Metuchen, New Jersey.

Wenzel, W.W., Unterfrauner, H., Schulte, A., Ruhiyat, D., Simorangkir, D., Kuraz, V., Brandstetter, A. and Blum, E.H. (eds) (1998) *Hydrology of Acrisols beneath Dipterocarp Forest and Plantations in East Kalimantan, Indonesia*. Springer-Verlag, Berlin.

Whitmore, A.P., Cadisch, G., van Noordwijk, M., Toomsan, B. and Purnomosidhi, P. (2000) An analysis of the economic values of novel cropping systems in N.E. Thailand and S. Sumatra. *Netherlands Journal of Agricultural Science* 48, 105–116.

Whitten, M.G., Wong, M.T.F. and Rate, A.W. (2000) Amelioration of subsurface acidity in the south-west of Western Australia: downward movement and mass balance of surface-incorporated lime after 2–15 years. *Australian Journal of Soil Research* 38, 711–728.

Wibowo, S. (1999) Populasi dan keragaman cacing tanah pada lahan dengan berbagai masukan bahan organik di daerah Lampung Utara. MSc. thesis, Bogor Agriculture University, Bogor, 205 pp.

Wiersum, K.F. (1984) Surface erosion under various tropical agroforestry systems. In: O'Loughlin, C.L. and Pearce, A.J. (eds) *Effects of Forest Land Use on Erosion and Slope Stability*. IUFRO, Vienna, Austria, pp. 231–239.

Wigmosta, M.S., Vail, L. and Lettenmaier, D.P. (1994) A distributed hydrology–vegetation model for complex terrain. *Water Resources Research*, 30, 1665–1679.

Wild, A. (1972) Nitrate leaching under bare fallow at a site in northern Nigeria. *Journal of Soil Science* 23, 315–324.

Williams, C.H. (1980) Soil acidification under clover pasture. *Australian Journal of Experimental Agriculture and Animal Husbandry* 20, 561–567.

Williams, J.R., Jones, C.A. and Dyke, P.T. (1984) A modelling approach to determining the relationship between erosion and soil productivity. *Transactions of the American Society of Engineers* 27, 129–144.

Williams, S.E. (2000) Interactions between components of rubber agroforestry systems in indonesia. PhD thesis, University of Wales, Bangor, UK, 256 pp.

Wilson, K.J. (1944) Over five hundred reasons for abandoning the cross-inoculation groups of the legumes. *Soil Science* 58, 61–69.

Wischmeier, W.H. and Smith, D.D. (1978) Predicting rainfall erosion losses: a guide to conservation planning. USDA Agricultural Handbook 537. USDA, Washington, DC, 57 pp.

Witty, J.F. and Minchin, F.R. (1988) Measurement of nitrogen fixation by the acetylene reduction assay; myths and mysteries. In: Beck, D.P. and Materon, L.A. (eds) *Nitrogen Fixation by Legumes in Mediterranean Agriculture*. Martinus Nijhoff, Dordrecht, The Netherlands, pp. 331–344.

Wong, M.T.F. and Harper, R.J. (1999) Use of on-ground gamma-ray spectrometry to measure plant-available potassium and other topsoil attributes. *Australian Journal of Soil Research* 37, 267–277.

Wong, M.T.F. and Swift, R.S. (2003) Role of organic matter in alleviating soil acidity in farming systems. In: Rangel, Z. (ed.) *Handbook of Soil Acidity*. Marcel Dekker, New York, pp. 337–358.

Wong, M.T.F., Wild, A. and Juo, A.S.R. (1987) Retarded leaching of nitrate measured in monolith lysimeters in south-east Nigeria. *Journal of Soil Science* 38, 511–518.

Wong, M.T.F., Hughes, R. and Rowell, D.L. (1990a) Retarded leaching of nitrate in acid soils from the tropics: measurement of the effective anion exchange capacity. *Journal of Soil Science* 41, 655–663.

Wong, M.T.F., Hughes, R. and Rowell, D.L. (1990b) The retention of nitrate in acid soils from the tropics. *Soil Use and Management* 6, 72–74.

Wong, M.T.F., van der Kruijs, A.C.B.M. and Juo, A.S.R. (1992) Leaching loss of calcium, magnesium, potassium and nitrate derived from soil, lime and fertilizers as influenced by urea applied to undisturbed lysimeters in south-east Nigeria. *Fertilizer Research* 31, 281–289.

Wong, M.T.F., Akyeampong, E., Nortcliff, S., Rao, M.R. and Swift, R.S. (1995) Initial responses of maize and beans to decreased concentrations of monomeric inorganic aluminium with application of manure or tree prunings to an Oxisol in Burundi. *Plant and Soil* 171(2), 275–282.

Wong, M.T.F., Nortcliff, S., Hairiah, K. and van Noordwijk, M. (1997) Role of agroforestry systems in the amelioration of soil acidity, *Proceedings of the International Workshop on Biological Management of Soil Fertility on Acid Upland Soils in the Humid Tropics*. Brawijaya University, Malang, Indonesia, pp. 40–42.

Wong, M.T.F., Nortcliff, S. and Swift, R.S. (1998) Method for determining the acid ameliorating capacity of plant residue compost, urban waste compost, farm yard manure and peat applied to tropical soils. *Communications in Soil Science and Plant Analysis* 29, 2927–2937.

Wong, M.T.F., Gibbs, P., Nortcliff, S. and Swift, R.S. (2000) Measurement of the acid neutralising capacity of agroforestry tree prunings added to tropical soils. *Journal of Agricultural Science (Cambridge)* 134, 269–276.

Wood, T.G. (1988) Termites and soil environment. *Biology and Fertility of Soils* 6, 228–236.

Woods, P.V., Nambiar, E.K.S. and Smethurst, P.J. (1992) Effect of annual weeds on water and nitrogen availability to *Pinus radiata* trees in a young plantation. *Forest Ecology and Management* 48, 145–163.

Woodward, F.I. and Osborne, C.P. (2000) The representation of root processes in models addressing the responses of vegetation to global change. *New Phytologist* 147, 223–232.

Woomer, P. L., Palm, C.A., Alegre, J.C., Castilla, C., Cordeiro, D.G., Hairiah, K., Kotto-Same, J., Moukam, A., Ricse, A., Rodrigues, V. and van Noordwijk, M. (2000) Slash-and-burn effects on carbon stocks in the humid tropics. In: Lal, R., Kimble, J.M. and Stewart, BA. (eds) *Global Climate Change and Tropical Ecosystems*. Advances in Soil Science, CRC Press, Boca Raton, Florida, pp. 99–115.

World Bank (1992) *World Development Report 1992: Development and the Environment*. Oxford University Press, New York, 308 pp.

Wösten, J.H.M., Pachepsky, Y.A. and Rawls, W.J. (2001) Pedotransfer functions: bridging the gap between available basic soil data and missing hydraulic characteristics. *Journal of Hydrology* 251, 123–150.

Wright, S.J. (1996) Phenological responses to seasonality in tropical forest plants. In: Mulkey, S.S., Chazdon, R.L. and Smith, A.P. (eds) *Tropical Forest Plant Ecophysiology*. Chapman and Hall, New York, pp. 440–460.
Wu, K.H., Jentschke, G. and Godbold, D.L. (2001) Contribution of root turnover to nutrient cycling in beech forests. In: Horst, W.J., Schenk, M.K., Bürkert, A., Claassen, N., Flessa, H., Frommer, W.B., Goldbach, H.E., Olfs, H.W., Römheld, V., Sattelmacher, B., Schmidhalter, U., Schubert, S., von Wirén, N. and Wittenmayer, L. (eds) *Plant Nutrition: Food Security and Sustainability of Agro-ecosystems through Basic and Applied Research*. Fourteenth International Plant Nutrition Colloquium, Hanover, Germany. Kluwer Academic, Dordrecht, The Netherlands, pp. 916–917.
Wubet, T., Kottke, I., Teketay, D. and Oberwinkler, F. (2003) Mycorrhizal status of indigenous trees in dry Afromontane forests of Ethiopia. *Forest Ecology and Management* 179, 387–399.
Wunder, S. (2003) *Oil Wealth and the Fate of the Forest: a Comparative Study of Eight Tropical Countries*. Routledge Taylor & Francis Group, UK, 448 pp.
Wyatt-Smith, J. (1982) The agricultural system in the hills of Nepal: the ratio of agricultural to forest land and the problem of animal fodder. APROSC Occasional Paper No. 1, Agricultural Project Service Centre, Kathmandu, Nepal, 17 pp.
Wycherley, P.R. and Chandapillai, M.M. (1969) Effects of cover plants. *Journal of the Rubber Research Institute of Malaya* 21, 140–157.
Xu, Z.H., Saffigna, P.G., Myers, R.J.K. and Chapman, A.L. (1993a) Nitrogen cycling in leucaena (*Leucaena leucocephala*) alley cropping in the semi-arid tropics. I. Mineralization of nitrogen from leucaena residues. *Plant and Soil* 148, 63–72.
Xu, Z.H., Myers, R.J.K., Saffigna, P.G. and Chapman, A.L. (1993b) Nitrogen cycling in leucaena (*Leucaena leucocephala*) alley cropping in the semi-arid tropics. II. Response of maize growth to addition of nitrogen fertilizer and plant residues. *Plant and Soil* 148, 73–82.
Yachi, S. and Loreau, M. (1999) Biodiversity and ecosystem productivity in a fluctuating environment. The insurance hypothesis. *Proceedings of the National Academy of Sciences of the United States of America* 96, 1463–1468.
Yan, F., Schubert, S. and Mengel, K. (1996) Soil pH increase due to biological decarboxylation of organic anions. *Soil Biology and Biochemistry* 28(4/5), 617–624.
Yanai, R.D., Fahey, T.F. and Miller, S.L. (1995) Efficiency of nutrient acquisition by fine roots and mycorrhizae. In: Smith, W.K. and Hinckley, T.M. (eds) *Resource Physiology of Conifers*. Academic Press, New York, pp. 75–103.
Yavitt, J.B. and Wright, S.J (2001) Drought and irrigation effects on fine root dynamics in a tropical moist forest, Panama. *Biotropica* 33, 421–434.
Young, A. (1997) *Agroforestry for Soil Conservation*, 2nd edn. CAB International, Wallingford, UK, 320 pp.
Young, A., Menz, K., Muraya, P. and Smith, C. (1998) *SCUAF Version 4: a Model to Estimate Soil Changes Under Agriculture, Agroforestry and Forestry*. ACIAR Technical Reports Series No. 41. Australian Centre for International Agricultural Research, GPO Box 1571, Canberra, ACT 2601, Australia, 49 pp.
Young, J.P.W. (1992) Phylogenetic classification of nitrogen fixing organisms. In: Stacey, G., Burris, R.H. and Evans, H.J. (eds) *Biological Nitrogen Fixation*. Chapman and Hall, New York, pp. 43–85.
Young, J.P.W. and Haukka, K.E. (1996) Diversity and phylogeny of rhizobia. *New Phytologist* 133, 87–94.
Zhang, F.S., Li, L. and Sun, J.H. (2001) Contribution of above- and belowground interactions to intercropping. In: Horst, W.J., Schenk, M.K. and Buerkert, A. (eds) *Plant Nutrition – Food Security and Sustainability of Agroecosystems*. Kluwer Academic, Dordrecht, The Netherlands, pp. 978–979.
Zhang, T.Q., Mackenzie, A.F. and Sauriol, F. (1999) Nature of soil organic phosphorus as affected by long-term fertilization under continuous corn (*Zea mays* L.): a $^{31}$P NMR study. *Soil Science* 164, 662–670.
Zuckerman, B.M., Dicklow, M.B., Coles, G.C., Garcia-Espinosa, R. and Marban-Mendoza, N. (1989) Suppression of plant parasitic nematodes in the Chinampa agricultural soils. *Journal of Chemical Ecology* 15, 1947–1955.

# Index

Page numbers in **bold** refer to figures in the text; those in *italics* refer to tables or boxes

*Acacia auriculiformis* 75, 237
*Acacia catechu* 64
*Acacia lenticularis* 63
*Acacia mangium* 237
*Acacia nilotica* 63, 316
*Acacia raddiana* 62
*Acacia seyal* 67
acidic soils
    old root channels 85–86
    phosphorus deficiency 154
    *see also* soil acidity
acid neutralizing capacity (ANC) 144
acids, organic 138–139
actinomycetes 229, *289*
'actinorrhizal' symbioses 229
*Afzelia africana* 203, **204**
agricultural intensification
    ecological problems 373
    and mycorrhizal associations 259–260
    and soil biota 290–292, *293*
    and trace gas fluxes 224–225
    western Amazon basin 372–373
Agroecological Knowledge Toolkit software *21*
agroecosystems
    agronomic versus ecological functions 374–375
    as mimic of nature 1–2
Akamba people, Kenya 26, 27–28
*Albizia niopoides* 70
*Albizia procera* 63
*Alchornea cordifolia* 67
alfalfa *see* lucerne
alkalinity
    root exudates 154
    soil sources 144
    transfer from organic matter 146, 149–156
alley cropping
    competition for water 159, **162**
    hedge pruning 320
    impact of organic inputs 204–205
    predicting performance 11–12
    and water balance 334
ALMANAC model *44*, 46
alpine sedge 179
Alternatives to Slash-and-Burn (ASB) Program 221, 291
aluminium phosphates 137

aluminium toxicity 74, 143, 234
    detoxification with organic inputs *151*, *152*, 153
Amazon basin, western 372–374
AMF *see* arbuscular mycorrhizal fungi
ammonification 146, 148, 150
ammonium ions, uptake 145–146
ANC *see* acid neutralizing capacity
Andes, local ecological knowledge 24
*Andira humilis* 62
anecic species *288*, 293
animal manure applications
    local knowledge 28–29
    methane production 219
    soil acidity amelioration 151–153
    and soil pathogens 282
anion exchange capacity (AEC) 176
anions
    adsorption 172–176
    movement in soils 111
    organic 138–139, 150, 152–153, 154
annuals 121–122, 310
ants *290*, *294*, *304–305*
apatites 128–129
aphids 275, *305*
APSIM (Agricultural Production Systems Simulator) *43*, 46, 55
arbuscular mycorrhizal fungi (AMF) 201, 248–251
    inoculation treatments 260
    lifespan 244
    management in agroecosystems 258–260
    mycelial network 252–253
    research methods 245
    specificity 249–251
    taxonomy *251*
armyworm, pestiferous *305*
*Artocarpus hirsutus* 68
ASB *see* Alternatives to Slash-and-Burn (ASB) Program
atmospheric depositions 145
*Azadirachta indica* 163–164, **165**

bacteria, soil 201, *289*
    biodiversity 285
    methanotrophic 212
    pathogenic **265**, *267*, 278

banana aphid *305*
banana production  221, 269, 278, *305*
Banksia *154*
Barber–Cushman type models  256
bare fallows  254
*bari*  27, 28–29
beetles *290*
below-ground biodiversity  285–286
    and above-ground biodiversity  295–296
    and food-web theory  296–302
    functional value  286
    groups of organisms  *288–290*
    indicators  287
    and land use  286, **287**, 290–292, *293*
    optimal levels  375–376
    relevance to farmers  302–303, **304–306**
below-ground plant inputs
    magnitude versus above-ground inputs  193–195
    and soil organic carbon  198–200
bicarbonate, leaching  148
bi-directional flow
    measurement  162–163, **164**
    mechanisms  161–162
    modelling  169
biocontrol agents  **264**, 274
biodiversity
    above : below-ground relationship  295–296
    and disease/pest control  280, 281
    and ecosystem stability  3, 298
    'neutral' theory  301–302
    optimal level  375–376
    and productivity  1
    relevance to farmers  302–303, **303–306**
    *see also* below-ground biodiversity
biodiversity indicators  287
biomass burning  198, 214–215
biomass pyramids  298–299, 299–301
bioturbation  294
boundary trees  *267*, *313*, 318
*Brachiaria decumbens* 194
branched absorbing structures (BAS)  257
branching, roots  71–73
bunchy-top virus *305*
burning, biomass  198, 214–215
bypass flow  176–178, **179**, 331

*Caesalpinioideae*  229
*Cajanus cajan* *64*, 75, 78, *194*
*Calathea allouia see* Guinea arrowroot
calcium  111, 112, 154, 270
calcium phosphates  128, 129
*Calliandra calothyrsus*
    biomass production  *194*
    deep soil nutrient uptake  *114*
    residues  223, 239, **240**
    shoot and root qualities  *195*
capillary fringe  334–335
carbon
    interplant movement in mycorrhizas  252–253
    plant economy  88–90
    soil *see* soil organic carbon (SOC)
carbon cycle  146, 193–195
carbon dioxide, atmospheric  89–90, 210
carbonic acid  145
carbon isotope labelling  96–97
case studies
    aluminium detoxification  *151–152*
    Amazon basin, alternatives to slash-and-burn  372–374
    ants in agroecosystems  *304–305*
    catchment soil erosion management  *341–343*
    earthworms and land-use change  *303*
    earthworms for restoration of soil fertility  *306*
    earthworms in rice production  *304*
    Lake Victoria basin  366–370
    sources of soil acidity  147–149
    Sumberjaya watershed  370–372
    *Tithonia diversifolia* 134
    WaNuLCAS model in maize–tree system  *141–142*
cash crops *267*
cassava  253, 291–292
*Cassia spectabilis* *141–142*
*Casuarina* spp.  229
*Casuarina equisetifolia* **321**
catchments *see* watershed functions; watershed management
cation exchange capacity (CEC)  176, 202–203
cations, soil  111, 112, 172–176
CEC (cation exchange capacity)  176, 202–203
centipedes *290*
CENTURY soil organic matter model  48, **49**
CERES–Maize model  52
charcoal formation  198
chemical pesticide use  275
Chitemene shifting cultivators  26
Cikumutuk catchment, Java *341–342*
citrate, root exudation  106, 138, 153–154
clay–humate complexes  197
clay minerals  197, 199–200, 203
climate  219
*Clitoria fairchildiana* *64*
clubroot fungus  **265**
cluster (proteoid) roots  136, **137**, 154
cocoa
    shade trees  317–318
    soil pests and pathogens  268–269
coffee production
    shade trees  311, 316, 317–318, *353*
    soil pests/pathogens  268–269
    tree–crop interactions  79, 373–374
    watershed management  352–355, 370–372
communication, researchers–farmers  37–38
COMP8 model  *43*, 45
competition  3, 61
    nutrient capture
        environmental modifiers  187–189
        management  186–187
        models  182–186
        plant strategies  181–182
    water use  158–160, 161
complementarity  232, 311
    nutrients  119–123, 171
        spatial  122–123
    water use  158, 159–160, 161
composted materials  151–153, 156
constancy  297
consultants, use of models  52–53
continuous cropping  278
coppicing species  120–121
*Cordia alliodora*  324, **325**
cover crops
    for pest and disease control  280
    pests and diseases affecting *267*, 268
    shoot and root biomass quality  *195*
crop biodiversity  279
cropping systems
    and soil acidification  147–149
    and soilborne pests/disease  266–269
    *see also different cropping systems*
crop rotations
    pest and disease management  276–279

role of legumes 228
  and water balance 334
CROPSYS model 44, 46
*Crotalaria* spp.
  nitrogen fixation 231–232
  rotation and pest control 278
  soil pests/pathogens 268
*Crotalaria grahamiana* 122–123, *194–195*
*Crotalaria paulina* 195
crown gall bacteria **265**
*Cytisus proliferus* 335

*Dactyladenia barteri* 64, 67, 79, 203, **204**
*Dalbergia sissoo* 63
Darwin, Charles 1
da Vinci, Leonardo 71
dead roots 85–86, 176, **177**
decomposers 295
decomposition 195–196
  legume residues 238–239
  and plant secondary metabolites 295–296
  roots 85–86, 97, 176, **177**, 194–195
deep soil nutrients
  improving uptake efficiency 119–123
  mechanisms of accumulation 110–112, 123–124
  nitrate 112, *113*, 120, 121–122, 123
  plant utilization 112–115
  preventing accumulation 123
  quantification of uptake 115–119
deep soil organic matter 111–112, 214
deep soil water uptake 334–335
deforestation *see* forest clearance
denitrification 211, **212**
depletion zones 83, 251
*Desmodium distorum* 326
*Desmodium uncinatum* 95
*Dialium guineense* 63, 70
diameter at breast height (dbh) and rooting depth 73
*Digitaria decumbens* 199
discharge capacity, aquifer 334
disease *see* soil pests and pathogens
disease resistance 269, 279
disease triangle 266
dissolved organic nitrogen (DON) 176, 179
disturbance, ecosystem 258–259, 297
DON (dissolved organic nitrogen) 176, 179
Dong Cao Catchment, Vietnam 355–356
downward siphoning 160
drought stress
  and soil pests/pathogens 70–71
  tree root response 74–75
DSSAT crop models 48, 52, 57
Dupuit's assumption 37

earthworms
  as agricultural pests *304*, 373
  and land-use change 290, *303*
  and soil fertility restoration 306
  and soil structure 176, 199, 201–202, 294
ecosystem stability 3, 297–299
ectomycorrhizal fungi 138, 179, 244
  associations with crop plants 248
  *see also* mycorrhizas
educators, use of models 53
*Eichhornia crassipes* 367
endogeic species *288*
'engineers', soil/ecosystem 201–202, 374
  *see also* earthworms
*Enterolobium cyclocarpum* 64, 70, 76, 80

epigeic species *288*
epiphytes 87
ericoid mycorrhizas 248
ethical issues 37–38
Ethiopia, Nazret 29
*Eucalyptus camaldulensis* **321**
*Eucalyptus deglupta* 79, 311, 318
*Eucalyptus saligna* 194
*Eucalyptus tereticornis* 62–63, *64*
eutrophication, fresh water 367–368
evictions, forest protection 370
extension staff 53–54

*Faidherbia albida* 62, 159, 315, 316
fallows
  bare 254
  natural 281
  *see also* improved (planted) fallows
farmers
  benefits of multi-species agroecosystems 2–3
  biodiversity management 302–303, **303–306**
  chemical fertilizer use 29
  land-use rights 370, 371–372
  livelihoods 58–59, 367
  pest and disease management 275, 283
  reasons for management/non-management of interactions 311
  risk management 189, **190**
  use of simulation models 54–55
  *see also* local ecological knowledge
fast-growing trees 315–316
fertilizers
  local knowledge 27–29
  management of below-ground competition 321–323
  and mycorrhiza 259–260
  nitrogen losses from 178
  organic and inorganic combinated 223–224
  phosphorus 123, **124**, 127, 259–260, 322
  placement 106, 321–322
  and soil acidity 146
  timing of use 106, 322–323
  and trace gas fluxes 213, 217–218
  tree root responses 79–80
  weed uptake 323
  *see also* organic matter inputs
filters, trees as 331, 340–346, **347**, 350
fine roots
  biomass 67
  distribution 68–69
  and mulching 320
  separation of tree–crop 315–316
  and shoot pruning 320
  turnover 88, 98, 194
fire 198, 214–215
fire ant *304, 305*
fish, Lake Victoria 367
flooding
  pest/pathogen control 276
  trace gas emissions 212, 218–221
food-chain concept 296
food-webs
  models 296–297
  relevance to tropical agriculture 299–302
  stability 297–299
  theory 296
forage legumes 230
forest clearance
  and below-ground biodiversity 291–292
  landslide risk 86

forest clearance *continued*
    and mycorrhiza 258–259
    and trace gas fluxes 213–216
    and watershed functions 332, 350–352
forest gardens 23–24, *267*, 269, 280–281
forests
    disturbance, and mycorrhizas 258–259
    hydrological functions 332–336, 340, 344–346, **347**
        perceptions and misunderstandings 350–352
    natural 67, 214
    recovery 105
    riparian 340–341, 343, **344**, 353–354
forest soils 205, **206**
*formae speciales* 278
fractal branching models 71–73
fractional depletion, nutrients 179–180
*Fraxinus* sp. 62
fruit trees 269
functional equilibrium, shoot–root 86–88
functional groups 287, *288*, 293–295, 296
fungal pathogens *268*, 269
    control 276, 277, 278–282
    in different cropping systems *268*, 269
    dispersal 276
    host specificity 278
    hyperparasitism 274
    interactions 271–272, 274
    plant tolerance/resistance 279
    soil and climatic factors affecting 270, 271
    survival 278
fungi
    soil 201, *289*
    *see also* fungal pathogens; mycorrhizas
*Fusarium* spp. 268, 269
*Fusarium* wilt 277, 279, 280
future challenges 378–379

galls, root 263, **264**, **265**
gap replanting 35–36
GAPS model *44*, 46–47
gaseous fluxes
    nitrogen 239–240
    *see also* trace gas fluxes
genetic engineering 279
genetic uniformity, crops 279
genotype-by-environment interactions
    nitrogen fixation 233, 235
    tree root systems 76–77
Geographic Information System (GIS) databases 219–220
Ghana, local ecological knowledge 21–25, 33–34
*Gliricidia sepium* 320
    deep soil nutrient uptake *116*, 117, 120–121
    disease resistance 269
    prunings, nitrous oxide emissions 223
    root architecture 63, *64*, 74, 79–80
    root turnover 99
    safety net function 177, *178*, 188–189
    shoot and root biomass qualities *195*
global issues 59
global warming 210
global warming potential (GWP), rice production 220–221
*Glomales*, taxonomy *251*
glomalin 201, 254
*Glycine max 195*
Gmelina 155
*Gmelina arborea* 67
*Gnetum* 248
grain legumes 227–228

granite, weathering 112
grass barriers 324, **325**
grasslands, transformation to agroforestry *246–247*
greenhouse gases 209–213
    *see also* trace gases
green manures
    nitrogen volatilization 239–240
    and phosphorus availability 133, *134*
    and soil pathogens 282
    trace gas emissions 222–224
    and tree–crop competition 320–321
Green Revolution 365
*Grevillea robusta* 315, 316, **321**
    biomass production *194*
    phosphorus mobilization and uptake 136, **137**, *141–142*
    root morphology 73, 78, 79, 136, **138**
    root sap flow **164**
*Grewia pubescens* 70
groundwater 333–336
    lateral flows 330, 331
    use by plants 334–335
    assessment 163–164, **165**
GUEST (Griffith University Erosion System Template) model 339–340
Guinea arrowroot 280

hairy roots 136
Hanunoo shifting cultivators 19, 26
hardpan layers 71, 75, 270
hedgerow intercropping
    deep soil nutrient uptake 119–121
    pests and pathogens 269
    and soil structure 177–178
    tree–crop competition, management 186–187
    water-uptake partitioning 159, 161, *168*
hedgerows
    pruning 186–187, *314*, 319–320
    root architecture 79–80
    root interactions 78
    'safety-net' functions 116, 177–178, 188–189
    water use 159
Hedley fractionation 131
herbivores, below-ground 294
*Hevea brasiliensis see* rubber
high altitudes 30
high pressure flow meter (HPFM) 166–167
hillsides, water flows 335–336
Himalayas 31
Hindu Kush 31
home gardens 23–24, *267*, 269, 280–281
Hortonian overland flow 338
humic substances 152–153, 154
humid forest soils 205, **206**
humification factor 196
HyCAS model *43*, 45
hydraulic conductance
    roots 160, 166–167
    soils 176–178, **179**
hydraulic lift 160
HyPAR model *43*, 45, 102, 167, 182–184
hyperparasites 273–274


IARCS (International Agricultural Research Centres) 51
IBSNAT project 57
*Imperata cylindrica* 98, *99*, 291, 325
improved (planted) fallows 122
    above and below-ground biomass production *194*
    deep soil nitrate use *114*, 120–121

nitrogen fixation 232
nitrous oxide emissions 223, **224**
soil pests and pathogens 234–235, *267*, 268, 281
species diversity 281
species resource complementarity 122–123
species shoot and root qualities *195*
'index of root shallowness' 69
indigenous knowledge *see* local ecological knowledge
Indonesia
    forest hydrological functions *345*
    jungle rubber systems 22, 26, **27**, 35–36
    watershed management *341–342*, 352–355, 370–372
infiltration theory *345*
'information' 19
information flow, and simulation models 51–52
ingrowth cores 95–96
INRM (integrated natural resource management) 365–366
insect pests **264**, 274, *304–305*
integrated natural resource management (INRM) 365–366
Integrated Soil Acidity Management Strategy 156
intellectual property rights *38*
intensification *see* agricultural intensification
interception
    mobile nutrients 180–181
    rainfall 331
intercropping
    deep soil nutrient uptake 119–121
    fertilizer use 321
    leaching prevention 113–115
    pest and disease control 279–280
    'safety net' function 114–115, *116*
    and soil hydraulic properties 177–178
    tree–crop root interactions 79–80
    use of legumes 227–228
    *see also* hedgerow intercropping
International Agricultural Research Centres (IARCS) 51
interviews *38*
iron concretions 75
iron phosphates 137

Jambi, Indonesia 26, **27**, 33
jarrah 335
jungle rubber systems 221, 311
    local ecological knowledge 24, 26, **27**, 35–36, 37
*Juniperus monosperma* 62

*kamere mato* 23
Kenya, Akamba people 26, 27–28
'knowledge' 18
knowledge-based systems (KBS) approach *38*, 352, 355
knowledge systems 376–378
*Kobresia myosuroides* 179

Lake Victoria basin 366–370
    ecological problems 366–367
    management challenges and opportunities 369–370
Lake Victoria Environmental Management Program 367–368
land equivalence ratio (LER) 4
landslides 86, 370
land use
    and river flows 344–346, **347**
    and root production 105–106
    and soil biota 286, **287**, 290–292, *293*, 303

and soil erosion 338–346, 340, 351–352
and soil organic carbon 197, **198**
and trace gases 213–224
and watershed functions 329, 330, 350–352
land-use rights 370, 371–372
language 23–26, 37, *38*
Lari community, Colca Valley, Peru 23
lateral flows, water 329–330, 331–333
lateral root spread 67–68
LDCs 48, 51, 54, 58–59
leaching 110–111, 112, 124
    of biologically fixed nitrogen 240
    fertilizers 178
    locally derived knowledge 34, 36
    management/prevention 113–115
    nitrogen 111, 112, 146, 148–149, 155, 172–176, 178, 240
    organic residues 178
    root interception of nutrients 180–181
    and soil acidification 146, 148–149, 155
    soil retardation factors 172–176
least developed countries (LDCs) 48, 51, 54, 58–59
legume-based pastures 373
legumes
    management to improve nitrogen fixation 233–235, **236**
    mixed species fallows 232
    organic residues 238–240
    phosphorus mobilization 138
    rhizobial inoculation 235–237, *238*, 241
    rhizobial specificity 230
    role in tropical agriculture 227–228
    trees *see* tree legumes
    uptake of deep soil nitrogen 119
*Leguminosae*, nitrogen fixation 228–229
LER (land equivalence ratio) 4
*Leucaena leucocephala*
    leaf residues *195*, 203, **204**
    root architecture *64*, 68, 74, 75–76, 78, 79
life history strategies, plants 86–87
lignin *195*, 239
liming 143–144, 149, 153, 234
litter
    decomposition 295
    local knowledge 27–29
    and rooting depth 87
litterbag, root incubation 97
livelihoods 67
    enhancement 58–59
livestock 27
local ecological knowledge 17–21, 350
    below-ground interactions 31–34, 37
    cross-cultural regularities 37
    dynamics and evolution of 19–20
    ethical issues 37–38
    forms of 18–19
    limitations of 36–37
    methods for acquiring and evaluating *21*
    nutrient applications 27–29
    participatory design of interventions 35–38
    research 37–38, 352
        ethics *38*
        methods 20–21
        terminology 18, 38
    soil fertility 24–31
    soil types 21–24
    sophistication of 36
    watershed management 352–356
local response 73–76, 90–91
    modelling 102–105

logging
    and mycorrhiza 258–259
    *see also* forest clearance
*Lonchocarpus sericeus* 80
'long-fallow disorder' 254
lopping 77–78
    *see also* shoot pruning
lucerne (alfalfa) 86, 119
'lung branch technique' 106
lupin, white 138

macroaggregates 201
macrofauna 176–177, 199, 201–202, 290, 294
macropores, soil 86, 176–178, **179**
*Macroptilium atropurpureum*, residues 223, **224**, 239, **240**
magnesium 111, 112
malate, root exudation 154
*malilo* 23, 31
malonic acid 138
management of interactions
    and above-ground growth 327
    choice of tree species/provenance 312–316
    fertilizer use/placement 321–323
    reasons for farmers' action/non-action 311
    root barriers 323–324
    scope and options 310, 312, *313–314*
    shoot pruning 77–78, 106, 186–187, 233, *314*, 319–320, **321**
    soil biota 305
    tillage and root pruning 69, *313–314*, 318–319, 327
    tree spacing/planting arrangement *313*, 317–318
    tree species selection 68–69, 77, 80, 106, 312–316
    weed control 324–327
manganese ions 74
*Manihot esculenta* (cassava) 253, 291–292
manures *see* animal manure applications; green manures
Mara river basin 369
*Markhamia lutea* **321**
mealy bug 304
*Melia volkensii* 315
*Meloidogyne* spp. 292
metabolites, plant secondary 195, 222–223, 239, 295–296
methane 211–213
    atmospheric concentrations 211
    emissions from biomass burning 215
    global budget *213*
    global warming potential 220
    soil fluxes 212–213
        agroforestry systems 221–222
        natural vegetation 214
        pasture formation 215–216
        rice production 218–221
        row crops 217
methanogenesis 212
methanotrophy 212, 213
Mexican sunflower *see Tithonia diversifolia*
microclimate 166
micropredators 294
mid-top predators 294
*Millettia thonningii* 70
millipedes *289*
*Mimosoideae* 229
mineral weathering 112, 129, 144
mine sites 260
minirhizotrons 93–95, 98–100
miombo woodlands 248
mites *289*
models/modelling 41–42, 377–378
    ALMANAC *44*, 46
    APSIM *43*, 46
    beneficiaries and target groups 51–55
    CENTURY soil organic matter model 48, **49**
    COMP8 *43*, 45
    CROPSYS *44*, 46
    data requirements 50–52
    DSSAT crop models/modelling 48, 52, 57
    ensuring uptake and impact 55–57
    food webs 296–297
    GAPS *44*, 46–47
    HYCAS *43*, 45
    HYPAR *43*, 45, 102, 182–184
    incorporating below-ground interactions 42–47
    limitations 47, 377
    long-term processes 48–49
    nutrient uptake
        mobile nutrients 173–176, 180, **181**
        and mycorrhizas 255–257
    phosphorus availability and uptake 140–142
    plant competition for nutrients 182–186
    reflection of farming conditions 47–48
    relevance to larger systems 57–59
    root bi-directional flow 169
    root turnover 101–105
    SCUAF 42, *43*
    soil erosion 339–340
    and spatial variability of agroecosystems 49–50
    TRAP 102
    WANULCAS 42–45, 167–169, 255–256
    water uptake 167–169
    WIMISA *43*, 45–46
mole cricket nematode 274
molybdenum deficiency 233
monocrops, risks of 189, **190**
morphospecies 292
*Mucuna* 57, 98, *99*
mulching
    soil pest/disease control 281–282
    tree–crop interactions *314*, 320–321
multispecies agroecosystems
    drawbacks 3
    framework for predictive understanding 10–14
    as mimic of nature 1–2
    potential benefits to farmers 2–3
    risk management 189, **190**
multistrata complex systems 280–281
mushrooms, edible 248
mycorrhizal responsiveness 249, *250*
mycorrhizas 243
    and agricultural practices 246
    arbuscular 244, *245*, 248–251
    determining abundance *245*
    edible mushroom production 248
    effects of decline 254
    function and benefits in agroecosystems 251–254
    identification *246*
    inoculation treatments 246–247, 260
    introduced species 248
    and land-use change **291**
    lifespan 244
    management 258–260
    mycelial network 252–253, 257
    and nematodes 272
    plant disease protection 274
    and plant nutrient uptake 136, 179, 180, 255–257, 260
    research methods 244–246
    and root longevity 96
    sheathing (ectomycorrhizas) 138, 179, 244, 248
    species diversity 246, 249
    specificity 249–251
    superstrains/superspecies 246
    types of associations 248

natural ecosystems
    agriculture as mimic of 1–2
    soil carbon 197
natural forests 67, 214
*Nauclea latifolia* 64, *70*, 80
Nazret, Ethiopia 29
nematodes 289, **291**, 292, 294
    dispersal 276
    entomoparasitic **264**, 274
    free-living 272
    fungivorous 272
    interactions with other soil organisms 271–272
    pathogenic
        and cropping systems 267–269
        soil and climatic factors affecting 270–271
    potato cyst 276
    root-knot 263, **264–265**, 268, 275
    tree legumes 234–235
Nepal, local ecological knowledge
    below-ground interactions 31–34
    soil fertility 25–26, 27–30
    soil types 21–23
'neutral' theory of biodiversity/biogeography 301–302
NGOs *see* non-governmental organizations
nickel-mine sites 260
Nile perch 367
nitrate
    deep soil 112, *113*, *114*, 120–122, 123
    diffusion and adsorption coefficients *174*
    leaching 112, *113*, 146, 148–149, 155, 173–176
    preventing deep soil accumulation 123, **124**
nitric oxide 211, 216, 217
nitrification 211, **212**
nitrogen (soil) 227
    from nitrogen fixation 231–232
    dissolved organic (DON) 176, 179
    effects on nitrogen fixation 234
    gaseous losses 239–240
    immobilization 195
    leaching 112, *113*, 146, 148–149, 155, 173–176, 178
    from organic inputs 203–205, 238–239
nitrogen cycle 145–146
nitrogen fertilization
    and mycorrhiza 260
    and nitrate leaching 111
    placement 322
    and soil acidity 146
    and soil trace gas fluxes 213, 223–224
    and tree root architecture 79–80
nitrogen fixation 227–228
    and available soil nitrogen 234
    contribution to nitrogen balance 231–232
    fate of fixed nitrogen 238–240
    in *Leguminosae* 228–229
    management to improve 233–235, **236**
    measurement 230–231
    and nematodes 272
    non-legume trees 229
    rhizobial inoculation 235–237, *238*, 241
    and soil acidification 146
    and soil phosphorus 234
    trees 230, 231–232
    and uptake of deep soil nitrogen 119
nitrogen isotope tracer 117
nitrogen : lignin ratio 195
nitrous oxide 210–211
nitrous oxide fluxes 216
    agroforestry systems 221–223
    and biomass burning 215
    effects of organic and inorganic applications 223–224

    from natural vegetation 214
    pasture formation 216
    rice production 220–221
    row crops 217, 218
nodulation
    effects of pruning 230, 233
    *see also* nitrogen fixation
non-governmental organizations (NGOs) 53–54
no-till treatments 218, 318
nuclear magnetic resonance, $^{31}$P 130–131
nursery beds, disease control 275
nutrient cycles
    nitrogen 145–146
    phosphorus **129**, 132–134, **135**
    and soil acidity 145–146
    sulphur 146
    and *T. diversifolia* 134
nutrient mobility 171, 172–181
    heterogeneous soil conditions 176–178, **179**
    subsurface flows 336–338
    in uniform soil conditions 172
    *see also* leaching
'nutrient pumping' 113
nutrient transfers, livestock 27
nutrient uptake 179–181
    competition 3, 61, 181–189
    crop, and large tree roots 69
    fractional depletion 179–180
    modelling 47, 182–186, 255–257
    and mycorrhiza 136, 179, 180, 255–257, 260
    and organic matter inputs 203–205
    perennial species 109–110, 113–115
    plant strategies 86–87
    and root turnover 100
    tree roots 68
    uptake potential 255

Ohm's Law analogue 160, **161**
oil palm 121–122
orchid mycorrhizas 248
organic anions 138–139, 150, 152–153, 154
organic matter inputs
    above- versus below-ground 193–195
    alkalinity transfer 149–156
    aluminium detoxification *151*, *152*, 153
    below-ground 193–195, 198–200
    and below-ground biodiversity **305**, *306*
    composted materials 151–153, 156
    land-cover types *300*
    nutrient release 238–239
    and nutrient uptake 203–205
    and trace gas fluxes 219, 222–224
    *see also* prunings; soil organic matter (SOM)
oxygen isotope tracing 163–164, **165**
ozone, tropospheric 211

*Papilionoideae* 228
PAPRAN pasture model 48
*Paraserianthes falcataria* 311
parasitic trees 257
parasitic weeds 325–326
*Parkinsonia* 229
parkland trees 316
Participatory Landscape Analysis (PaLA) 355
participatory research 35–38
particulate organic mater (POM) 203–205, **206**
pastures
    formation from forest 215–216, 373
    root turnover 106
    soil carbon 214

pastures *continued*
    species biomass production *194*
    trace gas fluxes 215–216, *217*
pathovars 278
peach palm agroforestry 147–149
*Peltophorum dasyrrachis*
    nutrient uptake *116*, 117, 181
    residues 223, **224**
    root distribution *64*
    root turnover *99*
    safety net function 177, *178*, 188–189
perennials
    nutrient uptake 109–110, 113–115
    shade trees 317–318
    *see also* trees
pesticides, soilborne pests 275
pests *see* soil pests and pathogens
phosphatase enzymes 132, 139–140
phosphate rock 127, 128–129, 154
phosphorus 127
    conceptual versus operationally defined fractions 131–132
    cycling **129**, 132–134, **135**
    estimating soil availablility 128
    inorganic forms 128–130
    organic forms 130–131
    plant uptake 100, 128, 135–140
        deep soil 117
        modelling 140–142
        and mycorrhiza 252, 260
    soil retention 129–130
phosphorus deficiency 127, 154, 233, 254
phosphorus fertilizers 127, 322
    effects on soil nitrate 123, **124**
    and mycorrhiza 259–260
    rock phosphates 127, 128–129, 154
pH, soil, *see* soil acidity
pinho cuiabano 374
*Pinus edulis* 62
*Pinus ponderosa* 62
'pipe stem' theory 71
piscidic acid 138
*Pithecellobium dulce* 63
plant biomass
    above- versus below-ground production 193–195
    alkalinity transfer 149–150, *151–152*, 155
    legumes 238–239
    nitrous oxide emissions 222–224
    nutrient release 238–239
    and soil pathogens 282
    weed control 326–327
plant metabolites, secondary 195, 222–223, 239, 295–296
plant pathogens, soilborne 263–266
ploughing
    root pruning 318–319
    *see also* tillage
policy makers 53
pollarding 320
polyphenols *195*, 222–223, 239
*Pontoscolex corethrurus* 304, 373
potassium 12, 111
potato, disease resistance 279
potato cyst nematode 276
predators, below-ground 294, 300
productivity
    and biodiversity 1
    potential of multispecies systems 2–3
    and soil organic carbon 203–205, **206–207**
*Prosopis cineraria* 62–63, *64*
*Prosopis juliflora* 62, 316
protection forest 370

proteoid (cluster) roots 136, **137**, 154
protozoa *289*, 294
provenance, choice of 312–316
proximal root direction 69
pruning
    combined root and top 319
    root 69, *313–314*, 318–319, 327
    shoot 186–187, *314*, 319–320, 327
        legume trees 233
        root responses 77–78, 106
        and tree–crop competition 186–187, 320, **321**
prunings
    aluminium detoxification *151, 152*
    nitrous oxide emissions 222–224
    release of nutrients 238–239
    and soil pH 149–151, 155
*Pterocarpus mildbraedii* 64, 70
*Pterocarpus santalinoides* 203, **204**
*Pterocarpus* spp. 63, 228
pulse-chase experiments 164
pulse labelling 96–97

radioisotope tracers 117
*Radopholus similis* 269, 278, 280
rainfall
    shoot:root ratios 87
    and soilborne pests/pathogens 270–271
    and tree–crop nutrient competition 187–189
    tree filter effects 331, 340–346, **347**, 350
recycled wastes 151–153, 156
reflectance spectroscopy 368–369
reforestation *345–346*
Relative Agronomic Function (RAF) 374, **375**
Relative Ecological Function (REF) 374, **375**
reporter genes 89
researchers
    communication with farmers 37–38
    use of simulation models 52
resource balance concept 8–9
resource capture
    concepts and rules 6–14
    *see also* nutrient uptake; water uptake
respiration, roots 90
Rhizobia
    classification 229–230
    nodule appearance 272
    acid-tolerance 234
    indigenous populations 237
    inoculation treatments 235–237, *238*, 241
    and nematodes 272
    plant disease protection 274
    specificity 230
rhizosphere biota 294
    *see also* mycorrhizas
rhizosphere modifications
    pH 138, 145–146, 154
    phosphorus mobilization and uptake 128, 132, 135–140
rhizotrons 93–95, 98–100
rhizovory 83
*Rhus viminalis* 62
rice production
    earthworms as pests *304*
    global warming potential 220–221
    trace gas emissions 218–221
riparian vegetation 340–341, *343*, **344**, 353–355
risk management 189, **190**
rivers
    bank vegetation 340–341, *343*, **344**, 353–355
    flow and land use 344–346, **347**
    sediments 340–343, 353–355

rock phosphate  127, 128–129, 154
root architecture, trees *see* tree root architecture
root barriers  323–324, **325**
root : crown spread ratio  67
root exudates
    organic acids  106, 138–139, 153–154
    phosphatase enzymes  132, 139–140
root-knot nematodes  263, **264**, **265**, 268, 275
    control  278, 281
    legume trees  234–235
root litterbag incubation  97
root 'mats'  87
root–rhizosphere boundary  89
root rots  269, 270
root : shoot ratio  68, 73–74, 86–88, *194*
root turnover
    after land-use change  105–106
    costs and benefits  90
    definitions  84–85
    empirical data  97–100
    fine roots  88, 98, 194
    importance in agroecosystems  83–84
    management  106
    measurement  84, 91–97
    and nutrient uptake  100
    research issues  106–107
    simulation models  101–105
roots
    activity  68, 316
    biomass  67, 71–73, 84, 194
    carbohydrate allocation  88–90
    contribution to soil carbon  194–195, 198–200
    decomposition  85–86, 176, **177**, 194–195
        measurement  97
    depth  62, 63–67, 87, 113
    distribution  62–69
        indicators  69–73
        lateral spread  67–68
    fractal branching models  71–73
    genotype-by-environment (G × E) interactions  76–77
    hydraulic properties  166–167
    local responses  73–76, 90–91, 102–105
    longevity  90, 93–95, **96**, 98
    organic C and N release  294
    proximal  69
    pruning  69, *313–314*, 318–319, 327
    respiration  90
    safety net function  114–115, 171, 177, *178*, 180–181, 188–189
    and soil structure  176–178, **179**
    and soil temperature  76
    and stem pruning  77–78
    turnover *see* root turnover
    water flow, bi-directional  161–162, 163, **164**
    water uptake, principles  160–164
row crops, trace gas fluxes  216–218
rubber agroforestry
    fertilizer use  322
    local knowledge  24, 26, **27**, 33, 35–36, 37
    root turnover  98–100
    trace gas emissions  221
    weed control  98–100, 311, 324–325
Rubber Research Institute, Sri Lanka  37
*rukho*  23, 29, 31
*rukhopan-malilopan* concepts  31, **32**
runoff  330
    soil erosion  338–343
run-on  330


safety net
    soils  172–176

tree roots  114–115, 171, 177, *178*, 180–181, 188–189, 338
saline water  335
*Salix viminalis*  96–97
sandalwood  257
sanitation, soil  274–276
*Santalaceae* (sandalwood)  257
sap flow measurement  169
    roots  162–163, **164**, 169
    stems  162
SARP (System Analysis for Rice Production) project  47–48, 57
saturation overland flow  338
savannah soils  205, **206**
*Sclerocarya birrea*  67
*Scleroderma sinnamariense*  248
scorpions  *290*
SCUAF (Soil Changes Under Agroforestry) model  42, *43*
seasonality  87, 106
security of tenure  370, 371–372
sedimentation  198, 340–343, 353–355, 367–370
seedling establishment  254
seeds, disease control  275
*Senna fistula*  63
*Senna siamea*
    biomass quality  *195*
    root architecture  *64–65*, 67, 71, 73, 74, 75, 79, 80
*Senna spectabilis*
    biomass production  *194*
    deep nutrient uptake  118
    water use  158
*Sesbania grandiflora*  63
*Sesbania sesban*
    biomass production  *194*
    deep soil nutrient uptake  *114*, 117–118, 120
    nitrogen fixation  231–232
    organic residues  223, 224, 239, **240**, 282, 326–327
    root architecture  63
    soil pests and pathogens  **264**, 268, 270–271, 278–279
shade trees  311
    coffee plantations  311, 316, 317–318, *353*
    and soil pests/pathogens  268–269
    spacing/arrangement  317–318
shelterbelts  323
shifting cultivation
    local knowledge  19, 26
    soil acidification  144, 147–149
    soil organic carbon  198
    *see also* slash-and-burn systems
shoot pruning  *314*, 319–320, 327
    legume trees  233
    root responses  77–78, 106
    and tree–crop competition  186–187, 320, **321**
shoot : root ratio  68, 73–74, 86–88, *194*
short rotations  334
silt fractions  197, 203
simulation models *see* models/modelling
SIRATAC dial-up crop management system  54–55
siratro  12
*sisipan*  35–36
slash-and-burn systems
    and mycorrhiza  259
    trace gas emissions  214–215, 221
    *see also* shifting cultivation
soil acidity  77, 143
    and aluminium toxicity  153–154
    amelioration  155–156
        plant residues  149–150, *151–152*, 155
        recycled wastes  150–153, 156
        root exudates  154
    and cropping system  147–149
    Integrated Management Strategy  156

soil acidity *continued*
    Integrated Management Strategy 156
    and nitrate leaching 155
    (and) phosphorus availability 130, 137–138, 154
    rates of 147, 148
    soil buffering capacity 144, 147
    sources 144–149
    tree root responses 74
soil aggregation 200–201
soil amendments *see* organic matter inputs
soil biota
    effects of land-use changes 290–292, *293*
    functional groups 287, *288*, 293–295, 296
    interactions 271–274
    management **305**
    pathogenic and beneficial groups **265**
    relationship to above-ground biota 295–296
    and soil structure 199, 201–202, 294
    *see also* below-ground biodiversity; soil pests and pathogens *and named organisms*
Soil Changes Under Agroforestry (SCUAF) model 42
soil charges 202–203
soil classification *see* soil types
soil compaction 75–76, 213, 270
soil cover, and soil erosion 340
soil erosion 338–343
    Lake Victoria basin 367–370
    local knowledge 352–356
    models 339–340
    role of land use and vegetation 338–346, 351–352, 353, 355–356
    and soil organic carbon 198, 202
soil fertility
    decline in tropics 143–144
    locally derived concepts and knowledge 24–31
    restoration in tea plantations 306
    and tree root architecture 79–80
    *see also* fertilizers
soil hydraulic conductivity 176–178, **179**
soil macropores 86, 176–178, **179**
soil nutrients
    deep *see* deep soil nutrients
    uptake *see* nutrient uptake
soil organic carbon (SOC) 195–198
    below-ground inputs and activity 193–195, 198–200
    deep soil 214
    effects on soil properties 200–203, **204**
    global stocks 196
    and land use change 213–214
    and plant productivity 203–205, **206–207**
soil organic matter (SOM)
    alkalinity 146, 149–153
    ammonification 146
    decomposition 195–196
    deep soil 111–112
    modelling 48, **49**
    nitrogen release and losses 178, 238–239
    particulate (POM) 203–205, **206**
    and soil pathogens 281–282
    *see also* organic matter inputs
soil pests and pathogens
    and cropping systems 266–269
    disease triangle 266
    dispersal 276
    interactions 271–274
    knowledge of 283
    management
        avoidance strategies 276–279
        control strategies 279–282
        general sanitation 274–276
    plant tolerance/resistance 279
    soil antagonists 273–274
    soil and climatic factors 269–271
soil structure
    compaction 75–76, 213, 270
    and soil biota 199, 201–202, 294
    and soil carbon 200–202
    and water/nutrient mobility 176–178, **179**
soil temperatures
    local knowledge 29–30
    and mycorrhiza 259
    and pests/pathogens 270, 275
    root responses 76
soil texture 196–200, 203
    and soil carbon 196–197
    and soilborne pests/pathogens 270
soil types
    and lateral water flow 338
    local knowledge 21–24
    root responses 63, 73–74
solarization, soil 275
soybean 228
species richness, soil biota 286–287, *303*
species selection, trees 68–69, 77, 80, 106, 312–316
specific root length 76
spiders *290*
'sponge' model of forests 344–346, **347**
springtails *289*
Sri Lanka, local ecological knowledge 23
stability, ecosystem 3, 297–299
statistical tests, root turnover estimates **2**, 91
stress sensitivity, legumes 233
*Striga* spp. 254, 278–279, 325–327
subsoil nutrients *see* deep soil nutrients
subsurface flow 336–338
sulphur cycle 146
Sumatra
    local knowledge 24
    Sumberjaya valley 352–355, 370–372
'supernatural' knowledge 19
sustainability 3, 365
sustainable livelihoods 58–59
System Analysis for Rice Production (SARP) project 47–48, 57
*Syzygium cumini* 63

tagasaste 335
'Tanggamus' operation 370
taproot systems 69–71, 160
*taungya* system 77
tea plantations *306*
temperatures *see* soil temperatures
*Tephrosia candida 194–195*, 295
*Tephrosia vogelii*
    biomass production and quality *194–195*
    pests and pathogens 268
terminology
    local knowledge 23–26, 38
    local knowledge research 18
termites *290*, 294
terraces, soil erosion *341–342*
*Tetrapleura tetraptera* 70
Thailand, local knowledge 23–24
*Thaumatococcus daniellii* 29, **30**
throughflow 336–338
tillage
    nitrous oxide emissions 218
    root pruning *314*, 318–319
*Tithonia diversifolia* 134
    biomass residues *134*, 154, 282
    soil pests/pathogens 268

top predator biomass 300, **301**
trace gases 209–213
　　and agricultural intensification 224–225
　　and fertilizer application 217–218
　　and land use 213–224
tracers, soil nutrients 117
traditional knowledge *see* local ecological knowledge
trainers, use of models 53
translation *38*
transpiration 165
　　in mixed-species systems **162**, 165–166
transport equation 172
trap crops 276, *277*, 280
TRAP (Tree Root Allocation of Photosynthates) model 102
*Treculia africana*, residues 203, **204**
tree belts 334, 335–336
tree–crop interactions
　　local knowledge and concepts 31–34, 352, *353*
　　nutrient capture 3, 61, 181–189
　　phosphorus mobilization and uptake 133, *134*, 140, *141–142*
　　root growth and distribution 79–80
　　and tree shoot pruning 320, **321**
　　water use 158–160, 161
tree legumes 229
　　estimates of nitrogen fixation 231–232
　　pruning and nitrogen fixation 233
　　root-knot nematodes 234–235
　　uptake of deep soil nitrogen 117–119
tree root architecture 62–63
　　depth 62–63
　　fractal branching models 71–73
　　genotype–environment interactions 76–77
　　lateral spread 67–68
　　and phosphorus uptake 136
　　responses
　　　　to drought 74–75
　　　　to fertilization 79–80
　　　　intercropping and fallows 79
　　　　mulching 320–321
　　　　to pruning and lopping 77–78
　　　　to soil compaction 75–76
　　　　and root barriers 323–324
　　root distribution 62–69
　　　　indicators 69–73
　　root size class distribution 68–69
　　　　soil and site conditions 73–74
　　and soil temperatures 76
　　and spacing/density 78–79, *313*, 317–318
　　species/provenance selection 68–69, 77, 315–316
　　taproots 69–71
　　tree spacing 78–79
trees
　　belts 334, 335–336
　　choice of species/provenance 312–316
　　hydrological functions 332–336, 340, 344–346, **347**, 350–352
　　legumes *see* tree legumes
　　litter *see* litter
　　nitrogen fixation 229, 230, 231–232
　　parasitic 257
　　prunings *see* prunings
　　root architecture *see* tree root architecture
　　spacing and arrangement 78–79, *313*, 317–318, 341
　　species selection 68–69, 77, 80, 312–316
　　*see also named tree species*
trenches *313*
　　for root competition control 319, 324
　　soil amelioration treatments *306*
*Triplochiton scleroxylon* 70

trophic pyramids 296
tropical rainforest ecosystems 62
*t*-test 91, **92**

'understanding' 18
Universal Soil Loss Equation (USLE) 339–340
upland soils 212
UPTAKE 140
uptake *see* nutrient uptake; water uptake
uptake potential 255
urea 146
urea fertilizer 147–148
ureides 146
USLE (Universal Soil Loss Equation) 339–340

*Verticillium* spp. 269
vesicular-arbuscular mycorrhiza *see* arbuscular mycorrhizal fungi (AMF)
viruses, soilborne 271
volatilization, nitrogen 239–240

WaNuLCAS (Water, Nutrient, Light Capture in Agroforestry Systems) 42–45
　　mobile nutrient capture 173–176, 180, **181**
　　mycorrhizas 255–256
　　phosphorus mobilization and uptake *141–142*
　　root dynamics 102–105
　　water uptake 167–169
waste materials, soil amelioration 151–153, 156
water availability
　　and shoot : root ratio 87
　　tree root responses 74–75
water balance 331–336
　　catchment level 333–336
　　plot level **330**, 331–332
water budget 157
water hyacinth 367
water movements
　　hillsides 335–336
　　lateral 329–330, 331–333
　　roots 159–160, 161–162, 163, **164**
　　soils 176–178, **179**
　　subsurface 336–338
water potential 160
　　measurement 165
water stress *see* drought stress
water uptake
　　competition and complementarity 158–160, 161
　　groundwater 163–164, **165**, 334–335
　　partitioning 159, 161, *168*
　　　　control 164–167
　　　　modelling 167–169
　　principles of plant 160–164
　　root hydraulic properties 166–167
watershed functions 329
　　and land use 329, 330, 350–352
　　riparian vegetation 340–341, 343, **344**, 353–354
　　role of forests 350–352
　　role of trees 332–336, 340, 344–346, **347**, 350–352
watershed management
　　information and knowledge 369
　　Lake Victoria basin 369–370
　　local ecological knowledge 355–356
　　Sumberjaya valley 352–355, 370–372
water-stable aggregates 199–200
water-use efficiency 158–159, 332
wax, root penetration 76
Way Besai catchment 370

weathering, minerals  112, 129, 144
weed control  280, *314*
    biological  278–279
    plant residues  326–327
    rubber  98–100, 311, 324–325
weeds
    below-ground interactions  324–327
    fertilizer uptake  323
    parasitic  325–326
    soilborne pathogens  276
WEPP erosion model  *343*
wetlands  211, 212, 368, 369
whitefly  275

wild jack tree  68
willow, basket  96–97
WIMISA model  *43*, 45–46
windbreak trees  311, 312, 323
witchweed  254, 278–279, 325–327
wood lice  *289*
'wood wide web'  252–253

zero-tillage  218, 318
zinc deficiency  254
zinc uptake  251
Zuni Indians, New Mexico  23